Ascomycete Systematics

Problems and Perspectives in the Nineties

NATO ASI Series

Advanced Science Institutes Series

A series presenting the results of activities sponsored by the NATO Science Committee, which aims at the dissemination of advanced scientific and technological knowledge, with a view to strengthening links between scientific communities.

The series is published by an international board of publishers in conjunction with the NATO Scientific Affairs Division

A	**Life Sciences**	Plenum Publishing Corporation
B	**Physics**	New York and London
C	**Mathematical and Physical Sciences**	Kluwer Academic Publishers
D	**Behavioral and Social Sciences**	Dordrecht, Boston, and London
E	**Applied Sciences**	
F	**Computer and Systems Sciences**	Springer-Verlag
G	**Ecological Sciences**	Berlin, Heidelberg, New York, London,
H	**Cell Biology**	Paris, Tokyo, Hong Kong, and Barcelona
I	**Global Environmental Change**	

Recent Volumes in this Series

Volume 263 — Angiogenesis: Molecular Biology, Clinical Aspects
edited by Michael E. Maragoudakis, Pietro M. Gullino, and Peter I. Lelkes

Volume 264 — Magnetic Resonance Scanning and Epilepsy
edited by S.D. Shorvon, D.R. Fish, F. Andermann. G.M. Bydder, and H. Stefan

Volume 265 — Basic Mechanisms of Physiological and Aberrant Lymphoproliferation in the Skin
edited by W.C. Lambert and W.A. van Vloten

Volume 266 — Esterases, Lipases, and Phospholipases: From Structure to Clinical Significance
edited by M.I. Mackness and M. Clerc

Volume 267 — Bioelectrochemistry IV: Nerve Muscle Function—Bioelectrochemistry, Mechanisms, Bioenergetics, and Control
edited by Bruno Andrea Melandri, Giulio Milazzo, and Martin Blank

Volume 268 — Advances in Molecular Plant Nematology
edited by F. Lamberti, C. De Giorgi, and D. McK Bird

Volume 269 — Ascomycete Systematics: Problems and Perspectives in the Nineties
edited by David L. Hawksworth

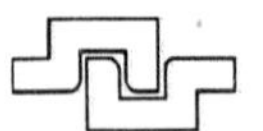

Series A: Life Sciences

Ascomycete Systematics

Problems and Perspectives in the Nineties

Edited by

David L. Hawksworth

International Mycological Institute
Egham, Surrey, United Kingdom

Plenum Press
New York and London
Published in cooperation with NATO Scientific Affairs Division

Proceedings of a NATO Advanced Research Workshop on
Ascomycete Systematics,
held May 11–14, 1993,
in Paris, France

NATO-PCO-DATA BASE

The electronic index to the NATO ASI Series provides full bibliographical references (with keywords and/or abstracts) to more than 30,000 contributions from international scientists published in all sections of the NATO ASI Series. Access to the NATO-PCO-DATA BASE is possible in two ways:

—via online FILE 128 (NATO-PCO-DATA BASE) hosted by ESRIN, Via Galileo Galilei, I-00044 Frascati, Italy

—via CD-ROM "NATO Science and Technology Disk" with user-friendly retrieval software in English, French, and German (©WTV GmbH and DATAWARE Technologies, Inc. 1989). The CD-ROM also contains the AGARD Aerospace Database.

The CD-ROM can be ordered through any member of the Board of Publishers or through NATO-PCO, Overijse, Belgium.

Library of Congress Cataloging in Publication Information

On file

ISBN 0-306-44882-3

A Division of Plenum Publishing Corporation
233 Spring Street, New York, N.Y. 10013

Printed in the United States of America

CONTENTS

ULTRASTRUCTURE

SECONDARY CHEMISTRY

PALEOMYCOLOGY, CLADISTICS, AND BIOGEOGRAPHY

MOLECULAR SYSTEMATICS

BIOLOGY AND SPECIES CONCEPTS

PROBLEMS AND PERSPECTIVES

OUTLINE OF THE ASCOMYCETES

ANNEX I

FOREWORD

In the important world of the Fungi, the ascomycete group is not only the largest in the number of species, but also the one with the most medical, pharmaceutical, and nutritional applications.

Systematics of this group is difficult because criteria are rather few and show slight variations. Moreover, about half of the taxa have been separately studied as "lichens", and general systematic books are dated and often imperfect. However, during the last decade, an important revival has taken place in the study of ascomycete systematics. This advance stems from several different factors. First, Eriksson, and later Eriksson and Hawksworth, have tentatively elaborated a new systematic synthesis and published, initially every year, a documented "Outline of the ascomycetes". Second, new technical possibilities, especially chemical, electron microscopical and molecular have allowed more specific criteria to be used in the definition of taxa, and led to the introduction of new systematic concepts. Third, a methodical revision of some important groups, especially the lichen-forming ascomycetes, resulted in marked systematic modifications. Fourth, contacts between lichenologists and other mycologists have become more frequent, partly due to a renewal of interest in lichenicolous fungi and those of varied biologies. Finally, new discoveries are being made in countries or environments previously insufficiently explored, resulting in the recognition both of numerous new taxa, and of new systematic problems.

In this revival, researchers have encountered increasing difficulty both in being correctly informed of new systematic data, and in obtaining a coherent overview of the whole of the ascomycetes. An authoritative review had moreover become essential as the intensity and pace of the application of molecular techniques to ascomycete systematics increased. As a response to this need, the three of us, prompted by J.M., proposed to organize a workshop exclusively devoted to ascomycete systematics in Paris. In doing so, we also aimed to honour the memory of some eminent French masters particularly the Tulasne's brothers, Boudier, and especially Chadefaud (1900-1984), whose research contributed so much to the revisiting of basic concepts in ascomycete systematics.

Coincidently, the editors of the "Outline of the ascomycetes" had proposed at the Fourth International Mycological Congress (IMC4) in Regensburg in 1990 that a meeting of specialists should be held before the next Congress in order to obtain as many inputs as possible into this detailed systematic project, which could then be considered further at the Fifth International Mycological Congress (IMC5) in Vancouver in 1994.

The project was supported by the French Lichenological Society (AFL), and discussed with D.L. Hawksworth who suggested these two elements were brought together in the workshop we planned and accepted our invitation to serve as Workshop Director. We had anticipated a modest meeting in the best traditions of the NATO Advanced Research Workshop (ARW) series, and they agreed to incorporate it into their programme. In the event, the First International Workshop on Ascomycete Systematics in Paris on 11-14 May 1993 exceeded all our expectations in attracting 140 researchers from 24 countries. After considerable discussion a structure was adopted which both preserved the dialogue so important in the NATO ARW's and at the same time did not exclude the chance of inputs from others. We were most anxious not to be open to a charge of limiting debates to a chosen few!

The Workshop thus fell into two distinct parts, each with a different objective. The first two days of the meeting consisted of authoritative and up to date reviews of either the main criteria used in ascomycete systematics or the general characteristics and problems in the major taxa of this group. We were fortunate in that these communications were presented by the most competent world specialists from many different countries. Presentations were followed by broad and often lively discussions. Special provision was reserved for recent research relating to the implications of molecular biology for ascomycete systematics. The last two days were more specialized and more technical, consisting essentially in structured discussions relative to changes proposed in the classification of each of the orders constituting the *Ascomycota*. It particularly aimed at obtaining as wide as possible a view on the "Outline of the ascomycetes", to be considered further at IMC5 in Vancouver. The detailed changes arising from this part of the Workshop will be enumerated in *Systema Ascomycetum* 13(2) in December 1994, but the Discussions and main "Outline" are included here.

All the communications and debates of the Workshop are thus incorporated into the present volume, edited by D.L. Hawksworth from the authors' papers and taped record. This book is not therefore only a juxtaposition of reports made by different authors concerning their current personal researches, but due to the original concept of the Workshop and briefing notes provided this book gives, on the contrary, a comprehensive and coherent view of the present state of ascomycete systematics.

Moreover, and this is another original feature of the present book, it includes non-lichenized and lichenized taxa in an integrated way. Consequently, as a result of this collaboration between lichenologists and other mycologists, and of their determination to realize here the best information possible, this work will be a fundamental value as a basis for up to date documentation and views for all biological researchers whose work is concerned with ascomycetes to a greater or lesser extent.

This First International Workshop on Ascomycete Systematics could not have been held without the funds generously made available by NATO, and we are very pleased to strongly acknowledge this support. The assistance provided by some official French organizations has been especially appreciated notably: CNRS, INRA, Université Pierre et Marie Curie (Paris), Laboratoire de Cryptogamie du Museum National d'Histoire Naturelle, École Normale Supérieure Fontenay de St Cloud. We also thank Dior-Parfums and Rhone-Poulenc-Fongicides for active support. We will not forget to recall the initial support given by the Association Française de Lichénologie, nor that subsequently provided by the International Mycological Association (IMA), the International Association for Lichenology (IAL), and the International Commission for the Taxonomy and Nomenclature of Fungi (ICTF), and the International Mycological Institute (IMI). We, and other members of the Organizing Committee, are also pleased to thank all those who gave us their trust to convene this Workshop, and all who spontaneously and willingly gave us their help.

There is abundant evidence that the Workshop was successful. We are confident that readers of its proceedings will not only appreciate its scientific value but will also discover that researchers devoting their work to ascomycetes are driven by faith and enthusiasm, and that they are desirous of working together to advance our knowledge in these important fungi.

André Bellemère (President Association Française de Lichénologie)
Marie Agnès Letrouit-Galinou (CNRS, Université Pierre et Marie Curie, Paris)
Jean Mouchacca (Muséum National d'Histoire Naturelle, Paris)

ORGANIZING COMMITTEE

D.L. Hawksworth (International Mycological Institute, Egham) *Director*
A. Bellemère (Association Française de Lichénologie)
H. Hertel (Botanische Staatssammlung, München)
M.-A. Letrouit-Galinou (Université Pierre & Marie Curie, Paris)
J. Mouchacca (Museum National d'Histoire Naturelle, Paris)
J.W. Taylor (University of California, Berkeley)

Note: K.A. Pirozynski (Canadian Museum of Nature) was originally a member of the Organizing Committee, but had to relinquish this role in early 1993.

Introduction

ASCOMYCETE SYSTEMATICS IN THE NINETIES

D.L. Hawksworth[1] and J. Mouchacca[2]

[1]International Mycological Institute
Bakeham Lane, Egham
Surrey TW20 9TY, UK

[2]Muséum National d'Histoire Naturelle
Laboratoire de Cryptogamie
12, rue Buffon, 75005 Paris
France

SUMMARY

An overview of key issues that emerged at the First International Workshop on Ascomycete Systematics is presented. The value of various characters used in, and approaches to, ascomycetes in various groups are highlighted. The main challenges identified for the future are the holomorph concept, the use of supraordinal ranks, alternative terminologies, improving working practices and data capture, the meeting of user needs, and confronting the mismatch between the available resources and the knowledge gap.

INTRODUCTION

The First International Workshop on Ascomycete Systematics was an unprecedented gathering. Almost all the leading world specialists on these fungi (including the lichen-forming groups and yeasts) participated in intensive focussed discussions for four days. The proceedings are in effect a major "state of the art" review. In introducing this volume, we reflect on its assessment of recent achievements, problems yet to be resolved, and the challenges for the future.

THE SYSTEMATIC VALUE OF DIFFERENT CATEGORIES OF CHARACTERS

Ascoma, Conidioma, and Thallus Structure and Ontogeny

Immense systematic significance has been accorded to the ontogeny of the ascomata for over six decades. The late M. Chadefaud (Parguey-Leduc *et al.*, this volume) and his students and co-workers in Paris have undertaken detailed and protracted studies on ascomycetes of diverse groups during the latter half of this period. They now conclude (Letrouit-Galinou *et al.*, this volume) that ascoma ontogeny can be viewed as occurring in three steps involving sterile elements. Variation occurs in the ébauche (primary corpus) which can mature before all steps have taken place, and also in the site of the primordium, but the stroma type has no influence on the rest of the development. The variation can be considerable, to the extent that caution is needed where few species have

Ascomycete Systematics: Problems and Perspectives in the Nineties
Edited by D.L. Hawksworth, Plenum Press, New York, 1994

been studied ontogenetically. The danger of emphasizing particular ontogenetic interpretations on the basis of inadequate data sets is apparent in the presentation of Henssen and Thor (this volume). These authors not only thoroughly document the "Zwischengruppe" between Nannfeldt's (1932) *Ascoloculares* and *Ascohymeniales*, but show that the classical interpretation of the *Pleospora*-type centrum may be incorrect.

In the case of vegetative characters in lichens, it is the developmental features rather than the mature shape which must be emphasized (Jahns and Ott, this volume). The principles of "multifunctionality" and "plasticity" need to be recognized, and for this reason vegetative characters should only be used with caution at the generic level and above - and never alone.

Increased precision in distinguishing morphological features may increase their value as systematic tools. In the case of conidial fungi, Hennebert and Sutton (this volume) were able to recognize 20 unitary parameters. When all such fungi have been coded by this new scheme, the soundness of reported correlations at higher ranks can be tested more critically (Sutton and Hennebert, this volume). Kendrick and Murase (this volume) highlight the problem of monophyl *vs* convergence in the elucidation of hyphomycete relationships, and so the caution needed in the use of these states, as either supportive characters or tests of systems based on other data sets.

Ascus and Ascospore Structure

Variations in ascus structure are currently accorded paramount importance in the classification of ascomycetes, especially at the level of family and above. Regrettably, most interpretations are based on light microscopy rather than the electron microscopy needed for a critical evaluation and interpretation (Bellemère, this volume). The emphasis placed on iodine reactions was of particular concern to many participants where the homology of the actual tissues giving those reactions at the ultrastructural level was unknown (see below). In any case, asci need to be examined in as fresh a condition as possible (Baral, 1992), through their developmental stages, with care taken to compare the same stages. It must not be forgotten that considerable variation in ascus structure does occur in some well-circumscribed orders (e.g. *Pezizales*, *Rhytismatales*).

The mixing of structural and functional terms when describing asci is frequent; both should be employed. For example, an ascus could be described as "bitunicate and fissitunicate", the terms referring to the structure and function respectively - neither word is satisfactory alone.

Ascospores are particularly subject to environmental selection, and are consequently of value at lower ranks than ascus structure. Ultrastructural comparisons are again important as perispore features can, for example, arise in parallel ways in different groups. Superficial studies of ascospore walls not backed by ultrastructural and developmental studies could lead to erroneous conclusions. Pigmentation is a metabolic process occurring rather late in ascosporogenesis and, as with ascospore septation, it is therefore of most value at lower taxonomic levels.

Septal Structure

The value of differences in the ultrastructure of the septum at the base of the ascus and in ascogenous hyphae has become apparent through the elegant studies of Kimbrough (this volume) and his co-workers. In general, the correlation between septal type and family is good in discomycete groups, although heterogeneity does occur in the *Otideaceae*. Definite septal types also occur in other ascomycete groups, for example the *Neurospora*-type, and there are indications that some pyrenocarpous and loculoascomycete fungi may have similar pore types.

The septae at the base of the asci provide a wealth of information that still requires investigation and analysis in most ascomycete orders. Septal structure has also proved of immense value in elucidating relationships in basidiomycetes (Moore, 1994), and this feature clearly merits more general attention in ascomycetes.

Secondary Metabolites

Mycologists have been rather slow to employ secondary metabolites in developing higher level classifications. Mantle (this volume) points out that production of these compounds requires a great deal of energy and many different enzymatic steps. While this means that they are potentially very important characters in reflecting the actions of many genes, it also follows that their expression may be thwarted by small genetic changes. Growth conditions can also affect what compounds are formed.

Progress has been more rapid in the case of lichenized ascomycetes, partly as a consequence of the development and widespread adoption of standardized thin-layer chromatographic procedures, a practice which most mycologists have been slow to emulate although compilations of microchemical data on fungal products are at last becoming available (Paterson and Bridge, 1994). Provided that the methodology is used with an awareness of its limitations, it is a powerful tool, especially at the species level (Frisvad, 1994). The potential for the approach to shed new light on the nature of species, barriers between species, and within-species variation is illustrated here by two exquisite case studies based on intensive sampling and culturing programmes (Culberson and Culberson, this volume). While the taxonomic interpretations of the results from such investigations might be open to debate, their contribution to a clearer understanding of the biology of complex situations is undeniable.

In the future, it will also be important for lichenologists increasingly to consider a wider range of secondary metabolites. Triterpenoids in particular can contribute to the detection of biogeographic patterns and evolutionary research (Galloway, this volume).

Palaeomycology and Biogeography

While the fossil record for ascomycetes is poor, it does provide some clues that can be used as a check for molecular clock datings; some interesting discrepancies arise (T. Taylor, this volume). The fossil evidence indicates that ascomycetes were well-established and diverse by the Late Silurian and associated with the earliest terrestrial vegetation. The molecular evidence suggests later dates (Berbee and J. Taylor, this volume) but this could be largely attributable to the limited number of orders and families so far sequenced.

The application of biogeographic principles, especially the relationship between continental movements and current distributions, to ascomycete systematics is severely constrained by the lack of data sets in which we can have confidence. Notable exceptions occur in certain lichenized orders and families where they have proved of value in the detection of ancient groups (Galloway, this volume). Biogeographic analyses are becoming standard components of monographs of lichenized families, and these should increasingly feature in revisions of non-lichenized groups.

Ecological Strategies, Species Concepts, and Population Biology

Mycologists have been slow to integrate population biology theory into systematic studies, although there are some important exceptions (e.g. Culberson and Culberson, this volume). Rayner (this volume) points out that in the ascomycetes only limited attention has been accorded even to the elucidation of reproductive mechanisms, colony characteristics, or their relationships in community development. It is unclear whether outcrossing is routinely involved in ascospore formation, whether hyphal fusions have a role in genetic exchange, mycelial systems are rarely considered, and population studies are almost unknown outside plant pathogens. Many of the approaches employed in basidiomycete ecology and population biology during the last 15 years merit application also in ascomycete fungi, as evidenced by preliminary work in *Xylariaceae*.

Species concepts pose a particular problem in lichenized groups, especially with respect to the rank to be used for asexually reproducing "secondary" species. Poelt (this volume) argues for the retention of species rank in the latter case on practical grounds. We see no reason to depart from that view, especially in a period when other mycologists give separate binary names to stages of pleomorphic fungi (see also below). At the same time, the numerous infraspecific taxa proposed for certain lichenized species in the past often represent only environmental modifications or developmental stages and uses are to be discouraged.

Poelt also draws attention to the boundaries which arise between individuals in mosaics of sexual species such as *Fuscidea cyathoides*, and further to the variation in other characters seen between such individuals. At the same time, significant amounts of variation can be found within largely asexual species, for example *Parmelia omphalodes*. Incompatibility studies and molecular investigations would be of particular interest in such situations.

Cladistics

The rigour of the cladistic methodology, the formation of groups on the basis of nested sets of derived characters, is only gradually being taken up by non-molecular fungal systematists. In employing this, as any technique, it is essential to be fully conversant with the limitations and pitfalls as well as the potential when it comes to interpretation. In cladistics, the transparency as to how conclusions are reached is in stark contrast to the practice of traditional "intuitive" approaches. The output can be repeated and tested in the best scientific traditions, something possible for phenetics but not traditional methods. For these reasons cladistics merits wide application and experimentation amongst taxonomists. However, we would find it scientifically more honest to present all equally parsimonious trees obtained, rather than a computed consensus tree or one selected by the investigator.

Tehler (this volume) demonstrates the tremendous potential of this tool, and compares it with phenetic and evolutionary approaches. The patterns of relationship revealed by cladistic studies on morphological data sets produce hypotheses that can then be tested by both molecular methods and any other independent additional characters. Cladistic programmes have come to the fore in the analysis of sequence data and are used almost without exception in the generation of molecular phylogenies (see below).

Molecular Biology

Molecular methods, especially the analysis of sequence data, have immense power which can shed light on issues almost unresolvable by other approaches, for example the position of the different groups historically viewed as "fungi" (J. Taylor *et al.*, this volume). Of particular interest is the clear separation obtained between two groups for which the names "*Pyrenomycetes*" and "*Plectomycetes*" were employed (see below), and the hypothesis that *Pneumocystis, Schizosaccharomyces*, and *Taphrina* all belong to a first radiation of the ascomycetes. The sequence selected for comparison needs to be one appropriate to the question addressed. In the case of *Pneumocystis*, analyses of different sequences suggest different affinities; this type of situation may become more frequent in future and independent data sets are then needed to resolve the situation.

The strength of molecular sequence data is in the recognition of monophyletic groups, not in determining how and at what ranks those groups should be named. This problem, addressed by Berbee and J. Taylor (this volume), is acute at two ends of a spectrum: with extant species which appear to be of great geological age, and deciding at what level of sequence similarity individual strains should be regarded as conspecific. In order to provide additional hierarchical ranks, Berbee and Taylor add the prefix "super-" to the ranks recognized in the Code, although that is currently not provided for in the Botanical Code. For example, while there can be little doubt as to the monophyly of the two groups for which the authors use the class names *Plectomycetes* and *Pyrenomycetes*, the questions as to whether the rank of "class" is appropriate and if the range of taxa sampled is sufficient to merit taking up those names are issues that remain a matter of personal judgement.

The ability of sequence data to shed light on taxonomic uncertainties through careful taxon sampling is demonstrated in this volume by Landvik and Eriksson for a wide range of discomycetes, Spatafora and Blackwell in the unitunicate pyrenomycetes, Blackwell and Spatafora for arthropod-dispersed taxa, and Kurtzman and Robnett in the ascomycete yeasts.

The extent to which the yeasts have been studied in comparison with most other ascomycetes enabled Kurtzman and Robnett (this volume) to provide an indication of the variation to be expected at different taxonomic ranks in 25S rRNA sequences: 1% within a species, 1-5% between closely related species, and 3-20% between genera. The method has been especially useful in clarifying generic synonymies. They found ascospore shape to be the least reliable character for separating families and genera (see above); differences in septal types and coenyzme Q provided better correlations with the molecular data.

However, sequence data could pose problems for the phylogenetic species concept as ever finer cuts are possible with DNA sequences (Discussion 1). There is also the question as to how molecular data should be weighted in comparison to morphological criteria (Korf, Discussion 8).

PROBLEMS IN PARTICULAR GROUPS

Currah (this volume) considers that passive ascospore release has led to the production of similar peridial types in the "prototunicate" ascomycetes. The "tumbleweed" ascomata of the *Gymnoascaceae* are interpreted as polyphyletic. The anamorphs and substrates (particularly the ability to degrade keratin) suggest that two different lineages are involved, one related to the discomycetes and the other to *Hypocreales*. Ascospore type, especially as viewed in the SEM, had been emphasized in the taxonomy of the group, but is not congruent with other characters (except in *Eurotiaceae*; see above); amongst the anamorphs, the production of arthroconidia may have been repeatedly evolved. The realignments proposed provide a new hypothesis for testing by molecular testing.

In the *Leotiales*, Huhtinen (this volume) argues for a fresh approach to traditional techniques in order to maximize data gathering and its utility (see below). Particular care is needed in the interpretation of microscopic characters as some are profoundly affected by the procedures employed (Baral, 1992).

In *Pezizales*, van Brummelen (this volume) emphasizes the need to employ multiple data sets in order to achieve a stable classification. There had been a shift in importance from morphological to microscopical characters. The correspondence between septal types, ascus apices, and also ascospore nuclear number, now provides the basis of family separations in the order. The structure of the ascus wall varies in importance but can be of value at a wide range of levels, the early stages being the most important at higher ranks (see above).

Particular concern was expressed over the circumscription of *Lecanorales* by Hafellner (this volume) who preferred to treat *Peltigera-*, *Pertusaria-* and *Teloschistes-*type asci as variants of the lecanoralean type. A subordinal system was proposed as an alternative following discussions between several specialists in the order, but it was considered premature by other participants without either further ultrastructural or molecular evidence to support it (Discussion 4). *Pertusariales* and *Peltigerales* had strong advocates (Discussion 5). Particular concern was expressed again as to the emphasis given to iodine reactions as compared to ultrastructural studies when interpreting ascus types (Discussion 5; see above).

While generic concepts in crustose lichens are increasingly based on microscopic features of the ascomata and asci, these characters are only exceptionally examined critically in foliose and fruticose groups. As a result, generic concepts in the *Parmeliaceae* in particular are currently far from clear (Discussion 4). The host ranges of lichenicolous fungi can provide one independent data set by which to test some of the macrolichen systems proposed.

In the non-lichenized pyrenomycetes, Rogers (this volume) considered that many of the problems in classification arise because of the age of the group and convergent evolution. As more taxa have been studied, intermediates and other variants in what were thought to be clear-cut characters have come to light, for example in the case of centrum types. It is also unclear to what extent differences in ascogenous systems can be used as a family criterion. Most fungi have only been examined once and superficially so that we have to face an increasing "ignorance load" (*cfr*. Huhtinen, this volume). He stresses that restraint should be used in erecting new systems based on too few taxa, especially as when new methods are used characters may be found to be more complex than generally assumed; for example, asci of *Diaporthales* as seen in fluorescence microscopy. Particular caution is needed in these fungi as a result of parallel and convergent evolution to adapt to particular substrata. Molecular data may be required to resolve many cases, for example the relationship between *Diatrypales* and *Xylariales*.

In the case of the *Clavicipitales* and *Hypocreales*, however, in view of the molecular data (Spatafora and Blackwell, this volume) there was general agreement that those long-discussed orders should be united even thought the asci were somewhat different (Discussion 3).

Interestingly, evidence is growing that some orders that have long been viewed as monophyletic may not be after all, notably *Caliciales* (Tibell, Discussion 6) and *Halosphaeriales* (Discussion 3).

The extent to which groups in which convergence has been a major factor can be resolved with the armory of techniques now available is illustrated in a case-study of the ophiostomatoid fungi (Wingfield *et al.*, this volume). The evidence presented that two quite different and distantly related groups of taxa exist in what was regarded as a single genus until relatively recently is overwhelming and generally accepted (Discussion 3). Parallel investigations in equal depth on other ascomycete groups currently united by putatively convergent characters can be expected to be similarly illuminating.

The scale of the problem of grappling with the fissitunicate ascomycetes is emphasized by Eriksson (this volume). The group includes around 7200 species referred to 818 genera and 72 families. Just how many orders to recognize continues to be a cause of controversy, current opinions ranging from 6 to 11. Particular difficulties arise because of apparent intermediates between major groups and the large number of taxa never studied in detail. Different interpretations can be placed on characters depending on whether fresh material has been studied and the techniques used. In a number of cases series can be recognized, but molecular data is needed to determine polarity.

The question as to whether more orders should be accepted for dothidealean fungi in the next *Outline* prompted considerable debate (Discussion 9). While certain groups can be expected to be accepted at higher ranks in due course, there are dangers in emphasizing characters of uncertain phylogenetic significance. We suspect that no entirely satisfactory resolution of these fungi will be achieved without molecular data from a much larger proportion of the accepted families than is currently available. In the interim, it is prudent to be cautious.

CHALLENGES FOR THE FUTURE

The Holomorph Concept

The most challenging long-term task in ascomycete systematics is the complete integration of mitosporic fungi in which no sexual stage is known into the ascomycete system. The logic of this is inescapable, especially with the molecular tools now at our disposal (Reynolds and Taylor, 1993). Nevertheless, there is as yet a tremendous reluctance to formally combining meiosporic and mitosporic fungi under a single generic name, although this can be done with increasing confidence even when no sexual stage is known (J. Taylor, Discussion 2). The main barrier is not a scientific issue but a nomenclatural one. Mycologists have developed a complex system of regulations to enable different phases of pleomorphic fungi to bear separate names; many of these names are familiar and relate to organisms of economic importance.

Reynolds (this volume, Discussion 2) eloquently argues that merger should be an objective for the 1990s, and that the relevant Art. 59 of the Code should be deleted. We concur with these objectives, but also strive to be pragmatic. Science advances much more rapidly than nomenclature and we do not see a solution to the nomenclatural problems being effected without considerable pain. It is an anathema that nomenclatural rules should constrain the incorporation of scientific advances into taxonomic systems, but that is the reality at this time.

Supraordinal Ranks

A recurring theme at the Workshop was whether, and if so what, supraordinal categories might be recognized (see above). While considerable progress was being made, many taxa remain unstudied so any systems produced will inevitably be unstable. Eriksson (Discussion 2) felt that the formalization of higher taxon names should be left for 10 years. Berbee (Discussion 9), in contrast, considered that "best guess" systems should be produced so that they could be challenged by new methods. However, these two opinions are not totally incompatible. In this debate, it is important to distinguish between classifications produced as speculative hypotheses for testing, and generalist reasonably robust systems which can be commended for non-specialist use (Hawksworth, 1985, Discussion 9).

Huhtinen (Discussion 8) was concerned that at our present state of knowledge, or rather ignorance, energy was being wasted looking for a phylogenetically more accurate system, rather than being expended in improving the knowledge base. Such views are pertinent in view of the current knowledge gap (see below).

Terminology

A particularly controversial topic at the Workshop was that of the terminology of "imperfect", "asexual", "conidial", or "anamorphic" fungi. The proposal to use "mitosporic", with "meiosporic" for sexual (teleomophic) fungi, which had been well received by most participants at a 1992 workshop (Reynolds and Taylor, 1993), had vociferous opponents on this occasion (see also Korf and Hennebert, 1993), as well as strong statements in its favour (Reynolds, this volume). The issue requires consensus to limit students being inflicted with an increasing array of conflicting unfamiliar terms. Our personal preference is for mitosporic *vs* meiosporic (with the latter term not used pedantically but referring to the state in which meioitic cell division is to be found) as it immediately conveys information to the general biologist, something that could never be claimed for "anamorph" and "teleomorph". The issue merits much more debate.

Working Practices and Data Capture

A recurrent theme in the Workshop, whether considering supraordinal ranks or genera, was the need to treat classifications as hypotheses to be subjected to tests aimed at falsifying them, in the best traditions of the scientific method. Hypotheses prepared by any one method can be tested by all others. Molecular methods provide an immensely powerful tool for the testing of classifications that have been proposed intuitively or by the rigourous procedures of either phenetics or cladistics. Taxonomy has often been regarded as an art rather than a science. While there will always be some subjectivity in translating the complexities of natural relationships into a formal hierarchical system, there is no reason why the elucidation of those same relationships should not proceed step-wise by a succession of hypotheses and tests. If systematists wish the classifications, or changes in classifications, they propose to be widely adopted, it is essential that they explain the bases of their hypotheses and present the results of independent data sets used as tests of the robustness of these hypotheses. A more rigourous scientific approach will not only be of benefit to the advancement of our science and minimize the publication of speculative changes, but contribute to a raising of the way systematics is perceived by other biologists.

Changes in working practices are also required at the data collection level. Huhtinen (this volume) argued for new ways of working in order to maximize data capture and exchange. He identified nine guidelines for the future which should be reflected on by all systematic mycologists before they embark on a collecting trip or start to make a microscopic preparation. The question of data exchange is particularly important, and new technologies and approaches are essential adjuncts, or even alternatives to, traditional publication if we are ever really to come to grips with the diversity of tropical fungi.

There is a need for systematists to collaborate to resolve particular key issues in a timely manner. This approach has been advocated by the International Commission on the Taxonomy of Fungi (ICTF) which currently has subcommissions working on *Aspergillus* and *Penicillium*, *Colletotrichum*, *Fusarium*, and *Trichoderma*. The principle merits more general adoption.

Meeting the Needs of Users

A major user need is the provision of a reference framework. The challenge is how this can be achieved at a time when "a diversity of opinion is the essence of ascomycete systematics in the 1990s" (Reynolds, Discussion following Hawksworth and Eriksson, this volume). The periodically revised *Outline of the Ascomycetes* (Eriksson and Hawksworth, 1993; Appendix 1), developed through *Systema Ascomycetum*, aims to provide an eclectic generalist system for the ascomycetes which gradually evolves step-wise. This approach was a key issue for debate during the Workshop and many important points were made in the discussion of Hawksworth and Eriksson's (this volume) presentation which are being evaluated by its compilers.

Name changes are a particular cause of irritation amongst users. It is important to distinguish between name changes made for scientific reasons (taxonomic reasons), and those due to applications of the International Code of Botanical Nomenclature (nomenclatural reasons). When to make changes for taxonomic reasons is a matter of scientific responsibility; it is better to be cautious while data accumulates until the case for a realignment is overwhelming, rather than to speculate or introduce changes without any presentation of the underlying reasons. The international attitude towards the changing of names for nomenclatural reasons has altered dramatically in the last 4-5 years. The International Code of Botanical Nomenclature adopted by the Tokyo Congress in September 1993 includes a whole raft of provisions which collectively mean that it is not now necessary to change names in most ranks for non-scientific reasons (Hawksworth, 1993a). It is the responsibility of each taxonomist to use the new nomenclatural, as well as the scientific, armory now available.

As stressed by Tibell (Discussion 4), if systematists continue to argue and not to produce reasonably stable systems the subject will increasingly be brought into disrepute.

The Mismatch Between Resources and the Knowledge Gap

It is now generally accepted that the estimate of 1.5 million species for the fungi on Earth (Hawksworth, 1991) is not unreasonable (Hawksworth, 1993b; Rossman, 1994). Assuming that ascomycetes will constitute about the same proportion amongst undescribed as they do amongst the 71 000 described fungi, and the genus:species ratio also holds, that implies that the number of undescribed ascomycetes may be as high as 62 000 genera and 669 000 species. The World's entire systematic mycological workforce currently describes around 1700 species each year. Even if that effort was focussed only on ascomycetes, it would still take over 390 years to document even those fungi using current practice.

The mismatch between human resources and the knowledge gap can consequently only be described as immense. It also leads to the questioning as to how the existing systematic resource is most effectively deployed, and how that cadre can be extended. This is not an issue peculiar to ascomycetes, or even fungi as whole, but rather to the entire field of systematics (Janzen, 1993). Some attempt to prioritize fungal groups for study has already been made (Hawksworth and Ritchie, 1993).

There is an urgent need to assert the relevance of our subject, to upgrade its image in the eyes of funding bodies, to make the subject scientifically challenging and attractive, and to organize our work and resources so that we can squarely address the issues of the 1990s. These are roles for the systematist and cannot be left to others.

REFERENCES

Baral, H.O., 1992, Vital versus herbarium taxonomy: morphological differences between living and dead cells of ascomycetes and taxonomic implications, *Mycotaxon* 44: 333-390.

Eriksson, O.E. and D.L. Hawksworth, 1993, Outline of the ascomycetes - 1993, *Systema Ascomycetum* 12: 51-257.

Frisvad, J.C., 1994, Classification of organisms by secondary metabolites, *In*: *The Identification and Characterization of Pest Organisms* (D.L. Hawksworth, ed.): 303-321, CAB INTERNATIONAL, Wallingford.

Hawksworth, D.L., 1985, Problems and prospects in the systematics of the *Ascomycotina, Proceedings of the Indian National Academy of Science, Plant Sciences* 94: 319-339.

Hawksworth, D.L., 1991, The fungal dimension of biodiversity: magnitude, significance, and conservation, *Mycological Research* 95: 641-655.

Hawksworth, D.L., 1993a, Name changes for purely nomenclatural reasons are now avoidable, *Systema Ascomycetum* 12: 1-6.

Hawksworth, D.L., 1993b, The tropical fungal biota: census, pertinence, prophylaxis, and prognosis, *In: Tropical Mycology* (S. Isaac, J.C. Frankland, A.J. Whalley and R. Watling, eds): 265-293, Cambridge University Press, Cambridge.

Hawksworth, D.L.. and J.M. Ritchie, 1993, *Biodiversity and Biosystematic Priorities: Microorganisms and Invertebrates*, CAB International, Wallingford.

Janzen, D.H., 1993, Taxonomy: universal and essential infrastructure for development and management of tropical wildland biodiversity, *In*: *Proceedings of the Norway/UNEP Experts Conference on Biodiversity* (O.T. Sandlund and P.J. Schei, eds): 100-113. Oslo, NINA.

Korf, R.P., and G.L. Hennebert, 1993, A disastrous decision to suppress the terms anamorph and teleomorph, *Mycotaxon* 48: 539-542.

Moore, R.T., 1994, Third order morphology: TEM in the service of taxonomy, *In*: *The Identification and Characterization of Pest Organisms* (D.L. Hawksworth, ed.): 249-259. CAB INTERNATIONAL, Wallingford.

Nannfeldt, J.A., 1932, Studien über die Morphologie und Systematik der nicht-lichenisierten inoperculaten Discomyceten, *Nova Acta Regiae Societatis Scientarum Upsaliensis, ser.* IV, 8(2): 1-368.

Paterson, R.R.M. and Bridge, P.D., 1994, *Biochemical Techniques for Filamentous Fungi* [IMI Technical Handbooks No. 1.] CAB INTERNATIONAL, Wallingford.

Reynolds, D.R. and J.W. Taylor (eds), 1993, *The Fungal Holomorph: Mitotic, Meiotic and Pleomorphic Speciation in Fungal Systematics*, CAB INTERNATIONAL, Wallingford.

Rossman, A.Y., 1994, The need for identification services in agriculture, *In*: *The Identification and Characterization of Pest Organisms* (D.L. Hawksworth, ed.): 35-46. CAB INTERNATIONAL, Wallingford.

IMPLICATIONS OF THE HOLOMORPH CONCEPT FOR ASCOMYCETE SYSTEMATICS

D.R. Reynolds

Natural History Museum of Los Angeles County
900 Exposition Boulevard
Los Angeles, California 90007, USA

SUMMARY

The expectation for ascomycetous species expressed by de Bary in 1887 is still prevalent. A life cycle with morphological expression of the ascal stage was the norm. The mitotic fungus was considered as a sometimes more frequently expressed intermittant life cycle stage. The mitotic and meiotic species were acknowledged as rare and derived from the loss of a pleomorphous life cycle stage.

In a contemporary sense, the ascomycete species with a pleomorphic life cycle are fewer in number than those species that morphologically express only meiotic or only mitotic reproduction. Yet, the anamorph and the teleomorph stages of the pleomorphic species have a predictive role as indicators of phylogenetic trends that can provide a basis for a single system of classification.

The recognition of meiotic, mitotic and pleomorphic holomorphs in fungal systematics has both taxonomic and nomenclatural implications which are discussed with examples from the *Capnodiaceae* and *Mycosphaerella*.

The merger of mitotic and meiotic ascomycetes into a single system of classification is recommended as a specific goal for ascomycete systematics in the 1990s.

INTRODUCTION

A major problem for the systematics of the 1990s is a better recognition of the relationships between sexual and asexual ascomycetes. The resolution lies in the consideration of all lineages of ascomycete fungi as a matter of evolutionary coherence. The perspective is of fungi with an expressed or an inferred ascus.

The holomorph is a core concept underlying a coherent system of classification which unifies sexual and asexual ascomycetes.

The holomorph concept is historically one of morphological derivation (Kendrick, 1979). Progress in ascomycete systematics for the 1990s can be tied to the increasing ability to understand the holomorph on an evolutionary basis. Phylogenetic assessment of the holomorph in all its kinship manifestations is now possible. Assessment of morphology-based preceps with independent molecular data is a critical activity.

The consideration of meiotic, mitotic, and pleomorphic speciation was undertaken in a "Holomorph Conference" held in Newport, Oregon, on 4-7 August 1992. The papers presented at that conference, the participant's discussions, and the conference recommendations have been published as *The Fungal Holomorph: a consideration of meiotic, mitotic and pleomorphic speciation* (Reynolds and Taylor, 1993). This essay derives from these sources.

HOLOMORPH TERMINOLOGY

The special terminology utilized in *The Fungal Holomorph* (Reynolds and Taylor, 1993) deliberations is at once familiar, but with some modification.

(1) *Holomorph.* This term is still used in the sense that was proposed in the Second Kananasksis Conference (Kendrick, 1979). It refers to the whole fungus. It is the basis of the fungal species.
(2) *Mitosporic fungi.* Sutton (1993) proposed this term for fungi which sporulate via the mitotic production of nuclei. The mitosporic fungus is expressed as a mitotic species or as the anamorph of a pleomorphic species. This term is scheduled to be utilized in the next edition of the *Dictionary of the Fungi.*
(3) *Meiosporic fungi.* This is a counterpart term to the mitosporic fungus for fungi which sporulate via the meiotic production of nuclei. The meiosporic fungus is expressed as a meiotic species or as the teleomorph of a pleomorphic species.
(4) *Meiotic holomorph.* A holomorph in which only sexual reproduction is morphologically expressed.
(5) *Mitotic holomorphic.* A holomorph in which only asexual reproduction is morphologically expressed. This is the "sterile species" or "secondary species" of lichenized ascomycetes.
(6) *Pleomorphic holomorph.* A holomorph in which both sexual (the teleomorph) and asexual (the anamorph) reproduction is morphologically expressed.
(7) *Anamorph.* Hennebert (1993) provided clarification of the pleomorphic life cycle reproductive states. The anamorph is the mitosporic expression of a pleomorphic species.
(8) *Teleomorph.* The teleomorph is the meiosporic expresssion of a pleomorphic species.

THE PLEOMORPHIC HOLOMORPH

The perception of the pleomorphic holomorph is an important issue to be resolved. The predictive value of the pleomorphic species in defining taxa is a basic concern in morphology-based ascomycete systematics. Assessment of the pleomorphic potential of a species is critical in bringing a phylogenetic dimension to a morphological concept of the holomorph. The phylogenetic value of the pleomorphic species in the context of ascomycete systematics is a problem for the present decade.

Specific points can be drawn from *The Holomorph Conference* deliberations that have implications for the pleomorphic holomorph perspective: (1) The mitotic species is a reality; and (2) the holomorph concept incorporates a recognition of phylogenetic relationships of all holomorphic expression.

TWO HOLOMORPH CONCEPT IMPLICATIONS FOR ASCOMYCETE SYSTEMATICS

Two implicatations of the the holomorph concept can be posed as questions.

Question 1: Will Every Mitosporic Ascomycete Have a Sexual Connection?

Since de Bary (1887), pleomorphic expression of both sexual and asexual reproduction in a life cycle has been considered a widespread but optional criterion of speciation. This is the morphology-based view of the ascomycete species. De Bary admitted that the strictly mitotic fungus was likely, but considered it exceptional. The existence of the mitotic species was attributed to the rare nonexpression of sexual reproduction in the pleomorphic species.

The idea can be stated as the "de Bary Hypothesis". The mitotic species is derived from a loss of sexual reproduction expression (the teleomorphic state) in pleomorphic species.

The species-pair concept developed by Poelt (1970, 1972) reflects this hypothesis. The "sterile species" is the lichen equivalent of the mitotic species; sexual reproduction is morphologically unexpressed. The sterile, "secondary," species was said to be paired with chemically identical sexual, "primary," species. Culberson (1986) reviewed this concept as well as the counter arguments of Robinson (1975) and Tehler (1982).

The loss of sexual reproduction is supported by the observations of Turgeon et al. (1993). Molecular evidence from non-lichenized ascomycetes suggested that mating-type genes may indicate loss of sexual reproduction with certain levels of mutation.

A phylogenetic aspect is also implicit in de Bary's explanation. Mitotic species are derived from species having the capability to morphologically express sexual reproduction. Berbee and Taylor (1993) found with molecular evidence that meiosporic fungi were older than mitotic species. LoBuglio and Taylor (1993) suggested that *Aspergillus* and *Penicillium* species are of more recent origin than *Talaromyces flavus* and *Eurotium rubrum*. Significantly, these molecular data also reveal mitotic species that are derived from other mitotic species. Is this asexual from asexual derivation the phylogenetic equivalent of the pleoanamorph as morphologically defined by Hennebert (1991)?

Blackwell (pers. comm.) suggested an appealing phylogenetic extrapolation of the morphological term "anamorph". The phylogenetic derivation of a mitotic species is from a meiosporic antecedent. The implication is of a pleomorphic association in an evolutionary sense. The temporal relationship is obviously a linear progression rather than a repetative cycle.

The evolutionary aspect of the de Bary Hypothesis suggests that pleomorphic species might yield both meiotic and mitotic species at a predictable rate. The ratio of pleomorphic to meiotic and mitotic species in a taxon might serve as a morphological criterion of phylogenetic soundness.

The frequency for the occurrence of pleomorphic holomorphs was found to be 5% for 10 596 ascomycetes surveyed by Reynolds (1993). Pleomorphic fungi occurred in a ratio of 1:9 to all other meiosporic and mitosporic ascomycetes in the 9247 ascomycetes included in an assessment of fungi on USA plant and plant products (Farr et al., 1989; Rossman, 1993). Sterile species were estimated to comprise 5% of lichenized fungi by S.C. Tucker (pers. comm.). Thus, a benchmark range for ascomycete pleomorphic frequency of 5-11% is suggested from these estimates.

Data from Sivanesan (1984) and Hawksworth et al. (1983) were utilized to estimate the pleomorphic frequency in ascomstromatic taxa at the genus and family levels. The percentage of pleomorphic species were calculated for the generic and familial taxa. The number of genera and families with a particular percentage of pleomorphic species can be examined as a reflection of the pleomorphic frequency. Pleomorphic species comprised 13% of the ascomycetes in Sivanesan's book. The range of pleomorphic content in the genera was from 1 to 100%; 12% fell within the 5-13% range. The pleomorphic content of only one family out of 20 fell within the benchmark range; the sample range was from 1-30%.

We conclude from this survey that the compartmentalization of pleomorphic species as genera and families lessens a reflection of the putative frequency benchmark.

Question 2. Is the pleomorphic fungus indicative of related asexual and sexual ascomycetes?

Rossman et al. (1993) proposed the "genus-for-genus" hypothesis as a model of ascomycete taxon definition. According to this approach, "a single teleomorph genus corresponds to a comparable anamorph genus." A meiosporic/mitosporic taxon pair is tailored from a pleomorphic holomorph pattern. The mitosporic fungus expression as an anamorph anticipates that all morphologically similar mitosporic fungi will eventually be found as anamorphs in the predicted association.

This ascomycete systematic practice is illustrated by the work of Alcorn (1983). *Dreschlera-Pyrenophora*, *Bipolaris-Cochliobolus*, and *Exoserohilum-Setophaeria* were all taxon pairs. The predictive aspects of mitosporic fungus classification were said to be established. A correlated set of characters for the mitosporic and the meiosporic genera were found as represented in their pleomorphic connection.

A recent application of this hypothesis is Rossman *et al.* (1993) introduction of the generic name *Leuconectria*. A major basis of the taxon was a newly discovered pleomorphic holomorph in which the meiosporic component was a *Nectria*-like species in the genus *Pseudonectria* and the mitosporic component was a *Gliocephalotrichum*. The mitotic ascomycete as anamorph was the preferred character in the introduction of the new genus. The presumption was that all five of the *Gleiocephalotrichium* species would eventually be proven anamorphs rather than recognized as mitotic species. The authors offered no explaination as to why the pleomorphic species are so important in revisionary work. The relatedness of meiosporic and mitosporic fungi was not at issue, merely the recognition of a special taxon for a genus-to-genus relationship.

This approach is short-sighted on a morphological level in that there is a presumption of a universal holomorph (Reynolds, 1993). The mitosporic taxon is presumed to comprise both proven and anticipated anamorphs. The segregation of pleomorphic species taxonomically obscures the relationships, both potential and real, of related meiotic and mitotic holomorphic species. The relationship of the meiotic genus to mitotic species is not indicated. The relationship of the mitotic genus to meiotic holomorphic species is based on a the bias toward the sexual system; phylogenetic masking of broader relationships may well be created. At a phylogenetic level, a taxon of the obvious relationships is created. The pleomorphic holomorphs point the way to the mitosporic genus. On the other hand, the mitotic taxon could well comprise closely related species, but of a mitotic holomorphic nature.

The pleomorphic holomorph is a dynamic association of asexual and sexual phases. Anamorphic diversity is high. The range of number of anamorph taxa associated at the generic level ranges from 1 to 23 (Reynolds, 1993).

PHYLOGENETIC MASKING

Rossman's morphology-limited approach to ascomycete systematics raises an interesting issue relating to the extrapolation of kinship demonstrated in pleomorphic species: "Although some problems are due to the undocumented nature of the connections, the generic names of both teleomorph and anamorph may mask the relationships between holomorph species."

Phylogenetic masking was first suggested by Simmons (1952). In this case, a morphologically identical mitosporic species was said to be associated with several diverse meiosporic fungi. From this example, phylogenetic masking occurs if a morphologically identical stage of the pleomorphic species is taxonomically split among taxa of a diverse classification. Although Simmon's data have been since reinterpreted, the principle can be seen with both mitosporic and meiosporic examples.

Luttrell's consideration of the deuteromycetes and their relationships (Luttrell, 1979) is an illustration of the mitosporic problem. A classification was contrived with the anamorph taxon as the basis of the classification. The result was incongruous with accepted ascomycete classification. Meiosporic taxa were reclustered from divergent genera and families. For example, genera of the *Fusicladiaceae* were clustered with a genus from the *Eurotiaceae*, 5 genera in 5 families of the *Pyrenomycetes*, 8 genera in 5 families of *Loculoascomycetes*, and with a genus in a discomycete family.

Phylogenetic masking of a meiosporic taxon occurs with the implication that *Mycosphaerella* species are derived from the *Pleosporales* as well as the *Dothideales* (sensu Barr, 1987). Of the 1239 species recognized in *Mycosphaerella* by Corlett (1991), only 107 are pleomorphic. Of the 107 pleomorphic species, 56 would fall into the genus-to-genus group. The problem is how to place the remaining 1219 meiotic holomorphic species. Which of these will have a kinship with the "*Cercospora* complex" pleomorphic species? Which of these would have a relationship of some of the 51 pleomorphic species that are largely associated with meiosporic fungi that are classified in the *Pleosporales*?

The masking indicated by the *Mycosphaerella* species with the taxonomically divergent anamorphs brings a focus to the characters of the telemorph. On a morphological level the characters of the teleomorph are similar. On a phylogenetic level, the critical lack of a hamathecium remains unevaluated.

The implication is that *Mycosphaerella* species are derived from the *Pleosporales* as well as the *Dothideales* (sensu Barr, 1987). Creation of a separately named taxon for the genus-to-genus species of the *Mycosphaerella*-"*Cercospora* complex" does not resolve the problem.

MERGER OF ASEXUAL AND SEXUAL ASCOMYCETES INTO ONE SYSTEM

The merger of asexual and sexual ascomycetes into one system of classification has been attempted in the past without resolve. The more common example of Rossman et al. (1993) has been discussed above. Other examples can be found in the literature. Further efforts are needed under reconsidered guidelines. Recognition of the mitosporic ascomycetes in the same classification as meiosporic ascomycetes has routinely been done on a morphological basis in several ways:

Previous Examples

(1) *Acanthostigmella brevispina* anamorph (Barr and Rogerson, 1983). This format leaves the mitotic ascomycete without a name. This is an accepted practice, allowed under the Art. 59 of the *International Code of Botanical Nomenclature* (Greuter *et al.*, 1988). The disadvantage of this practice is the lack of a reference to any mitosporic kinship.
(2) *Drechslera* state of *Cochliobolus spicifer* (Ellis, 1971); *Pleospora* state of *Stemphyllium globuliferum* (Sivanesan, 1984). These are examples of an attempt to find an alternative to providing a binominal for mitosporic fungi with anamorph status. The mention of the deuteromycete genus name as a "state" was to denote morphological similarity to other asexual species.
(3) *Leptosphaerulina*, anamorph *Pithomyces* (Hanlin, 1990). This example of kinship indicated between mitosporic and meiosporic fungi is a common one at the generic level. The unfortunate implication is that the entire mitosporic taxon comprises no mitotic species. The alternative implication is the historical expectation that all mitotic fungi will eventually be found to be associated with a meiosporic species as an anamorph.

Proposed Guidelines

A guideline for a reasonable effort to merge meiotic, mitotic, and pleomorphic species into a single ascomycete classification is proposed:

(1) Recognize the mitotic species.
(2) Make a clear distinction between the mitosporic ascomycete as a mitotic species and the mitosporic ascomycete as an anamorph.
(3) Merge mitotic and meiotic species both as recognized taxa and as teleomorph/anamorph of pleomorphic species.

Merger is most likely at the genus level and requires that a mixture of meiotic, mitotic, and holomorphic species be considered under a single name. Art. 59 of the Code applies only to the pleomorphic species.

An example is the placement of a new species in the *Capnodiaceae* (Reynolds and Taylor, 1994). A genus, *Leptoxyphium*, is comprised of mitotic species. A new pleomorphic species can be placed in the genus as *Leptoxyphium littoralis*. The genus would then comprise both mitotic and pleomorphic holomorphs. The type species of the genus is *Leptoxyphium graminum*. An alternative is to create a new genus, *Capnoseta*, for the pleomorphic species, *Capnoseta littoralis*. This establishes a genus-to-genus relationship between *Leptoxyphium* and *Capnoseta*. To fully demonstrate the relationship, both *Capnoseta* and *Leptoxyphium* are recognized in the family *Capnodiaceae*.

Mycosphaerella is a genus that has great potential for the merger of meiotic, mitotic, and pleomorphic species. Following the genus-to-genus hypothesis, the *Mycosphaerella* and "*Cercospora* complex" fungi would be recognized as a monophyletic group. The taxonomic disposition should initially include the 54 pleomorphic species indicated by the morphological correlations. Also, the 1944 nonpleomorphic species in 14 genera of the "*Cercospora* complex" should be placed in the same family, the *Mycosphaerellaceae*. The inclusion of the nonpleomorphic *Mycosphaeraella* species would await other, likely molecular, evidence.

A FINAL POINT

I propose a specific goal toward advancing ascomycete systematics in the 1990s:

Integrate meiotic, mitotic, and pleomorphic holomophs as fully as possible into the next classification of orders, families and genera of the ascomycetes that is periodically prolmugated as the "Outline of the Ascomycetes" (Eriksson and Hawksworth, 1991) for *Systema Ascomycetum.*

REFERENCES

Alcorn, J.L., 1983, Generic concepts in *Drechslera, Bipolaris* and *Exoserohilum, Mycotaxon* 27: 1-86.

Barr, M.E., 1987, *Prodromus to Class Loculoascomycetes,* Privately printed, M E Barr, Amherst, Mass.

Barr, M.E. and C.T. Rogerson, 1983, Two new species of loculoascomycetes, *Mycotaxon* 27: 247-252.

Bary, A. de, 1887, *Comparative Morphology and Biology of the Fungi, Mycetozoa and Bacteria,* [English transl. H.E.F. Garnsey, revised I.B. Balfour], Clarendon Press, Oxford.

Berbee, M. and J.W. Taylor, 1993, Ascomycete relationships: dating the origin of asexual lineages with 18s ribosomal RNA gene sequence data, *In: The Fungal Holomorph* (D.R. Reynolds and J.W. Taylor, eds): 67-78. CAB International, Wallingford.

Corlett, M., 1991. An annotated list of the published names in *Mycosphaerella* and *Sphaerella.* [Mycological Monograph No. 18.] J. Cramer, Stuttgart.

Culberson, W.L., 1986, Chemistry and sibling speciation in the lichen-forming fungi: ecological and biological considerations, *The Bryologist,* 89: 123-131.

Ellis, M.B., 1971, *Dematiaceous Hyphomycetes,* Commonwealth Mycological Institute, Kew.

Eriksson, O.E. and D.L. Hawksworth, 1991, Outline of the ascomycetes - 1990, *Systema Ascomycetum* 9: 39-271.

Farr, D.F., G.F. Bills, G.P. Chamuris, and A.Y. Rossman, 1989, *Fungi on plants and plant products in the United States,* American Phytopathological Society Press, St. Paul, Minnesota.

Greuter, W., H.M. Burdet, W.G. Chaloner, V. Demoulin, R. Grolle, D.L. Hawksworth, D.H. Nicholson, P.C. Silva, F.A. Stafleu, E.G. Voss, and J. McNeill, eds., 1988., *International Code of Botanical Nomenclature adopted by the Fourteenth International Botanical Congress, Berlin, July-August, 1987.* [Regnum Vegetabile No. 118.] Koeltz Scientific Books, Königstein.

Hanlin, R.T., 1990. *Ilustrated Genera of Ascomycetes.* American Phytopathological Society Press, St. Paul, Minnesota.

Hawksworth, D.L., B.C. Sutton, and G.C. Ainsworth, 1983, *Ainsworth & Bisby's Dictionary of the Fungi,* 7th edn, Commonwealth Mycological Institute, Kew.

Hennebert, G.L., 1991. Article 59 and the problem with pleoanamorphic fungi, *Mycotaxon* 40: 479-496.

Hennebert, G.L., 1993, Toward a natural classification of the fungi, *In: The Fungal Holomorph* (D.R. Reynolds and J.W. Taylor, eds): 283-294. CAB International, Wallingford.

Kendrick, [W.] B., ed., 1979, *The Whole Fungus,* National Museums of Canada, Ottawa.

LoBuglio, K. and J.W. Taylor 1993, *Penicillium* species, subgenus *Biverticillium, In: The Fungal Holomorph* (D.R. Reynolds and J.W. Taylor, eds): 115-119. CAB International, Wallingford.

Luttrell, E.S., 1979, *Deuteromycetes* and their relationships, *In: The Whole Fungus* ([W.] B. Kendrick, ed.) 1: 241-264. National Museums of Canada, Ottawa.

Poelt, J., 1970, Das Konzept der Artenpaare bei den Flechten, *Vörtrage aus dem Gesamtgebeit der Botanik, N.F.* 4: 187-198.

Poelt, J. 1972. Die taxonomische Behandlung von Artenparren bei den Flechten. *Botaniska Notiser* 125: 77-81.

Reynolds, D.R., 1993, The fungal holomorph: an overview, *In: The Fungal Holomorph* (D.R. Reynolds and J.W. Taylor, eds): 15-25. CAB International, Wallingford.

Reynolds, D.R., and J.W. Taylor 1993, *The Fungal Holomorph: A Consideration of Meiotic, Mitotic and Pleomorphic speciation,* CAB International, Wallingford.

Reynolds, D.R., and J.W. Taylor, 1994, Higher fungi genera: their holomorphic content, *Mycologia Helvetica* 6: 123-140.

Robinson, H., 1975, Considerations on the evolution of lichens, *Phytologia* 32: 407-413.

Rossman, A.Y., G.J. Samuels, and R. Lowen. *Leuconectria clusiae* gen. nov. and its anamorph *Gliocephalotrichum bulbilium* with notes on *Pseudonectria, Mycologia* 85: 685-704.

Simmons, E.G., 1952, Culture studies in the genus *Pleospora, Clathrospora,* and *Leptosphaeria, Mycologia* 44: 330-365.

Sivanesan, A. 1984, *The Bitunicate Ascomycetes and their Anamorphs,* J. Cramer, Vaduz.

Sutton, B. 1993, Mitosporic fungi (deuteromycetes) in the *Dictionary of the Fungi*, *In*: *The Fungal Holomorph* (D.R. Reynolds and J.W. Taylor, eds): 27-55. CAB International, Wallingford.

Tehler, A. 1982, The species pair concept in lichenology, *Taxon* 31: 708-717.

Turgeon, G., S.K. Christiansen, and O.C. Yoder, 1993, Mating type genes in ascomycetes and their imperfect relatives, *In*: *The Fungal Holomorph* (D.R. Reynolds and J.W. Taylor, eds): 199-215. CAB International, Wallingford.

DISCUSSION

Gams: The dichotomy between mitosporic/ meiosporic and that between anamorphic/ teleomorphic do not coincide. The former is based on genetic information not often available, and the latter on observed morphological differences. In the case of certain basidiomycetes, and surely in some ascomycetes, mitosporic teleomorphs are known where meiosis has not been proved. A switch in terminology could be confusing, especially as the current nomenclature is morphologically based.

Reynolds: There is certainly a widespread use of anamorph/ teleomorph, but the use of mitosporic/ meiosporic in the next edition of the *Dictionary of the Fungi* is likely to contribute to a much wider understanding of these concepts.

Blackwell: Are there so few pleomorphic ascomycetes because connexions have not been made or is this a real phenomenon ?

Reynolds: Differences can emerge from how results are boxed into genera based on pleomorphic, mitotic or sexual concepts. What is clear from the molecular data is that there are distinct mitotic species. I wonder if there is a constant rate of pleomorphy giving rise to mitotic species but that cannot be analysed from how the data is compartmentalized.

Hennebert: I wish to confirm Gams' statement that there is not a complete synonymy between anamorphic/ teleomorphic and mitotic/ meiotic. Soon after Weresub and myself had established the former terms, we stressed that these were not cytological. Further, parasexuality is known in some "mitotic" fungi. I like very much the expression of Hawksworth (this volume) that every name is a hypothesis; terms such as mitotic/meiotic also include a hypothesis. We need to work with those terms cautiously and experimentally. Further, using a mitotic/ meiotic terminology reduces the content of the holomorph to what we know or suppose we know of it - a holomorph name includes also that which is unknown (e.g. anamorphs not discovered or correlated).

Reynolds: I agree, but this points out again the clarification needed in the use of the term anamorph.

Malloch: The life histories of ascomycetes are extremely poorly known, as those of us who undertake fieldwork appreciate. Only one group of ascomycetes, major plant pathogens, can be considered to have their life histories well-worked out. In these, the number which are pleomorphic is certainly higher than the estimates given. I believe this could be so for the majority of ascomycetes when their life-cycles have been worked out. Much of the ascomycete literature is based on fruit-bodies being collected and brought back and not on cultural studies outside plant pathology and medical mycology.

Reynolds: That may well be the case, but is not that reflected in the literature at this time. The main point being made is a difficult one, especially with those who have been working with the anamorph/ teleomorph concept for some time.

Kendrick: A further problem is that many of those who work with ascomycetous or basidiomycetous teleomorphs regard themselves as specialists on those stages and are not concerned to search for other stages of the life-cycle.

Rossman: The question of training is certainly a real one, and nowadays there is a need to integrate independent data sets, including that from molecular sources, in developing synthetic hypotheses.

Ascoma and Thallus Structure and Ontogeny

ASCOMA STRUCTURE AND ONTOGENESIS IN ASCOMYCETE SYSTEMATICS

M.-A. Letrouit-Galinou, A. Parguey-Leduc, and M.-C. Janex-Favre

Université Pierre et Marie Curie
Laboratoire de Cryptogamie, Case 33
7 quai Saint-Bernard, F-75252
Paris cedex 05, France

SUMMARY

General features of ascoma ontogeny are reviewed. Ascomata comprise both fertile and sterile elements. Three steps involving the sterile elements complete ascoma development: primordium formation, production of an "ébauche" (primary corpus), and formation of a parathecial crown. Variations in the sterile elements are presented, with an emphasis on the ébauche. In some cases ascoma may mature before all developmental processes have taken place. New light can be thrown on classifications by ontogenetic data, for example in providing a clear distinction between *Nephromataceae* and *Peltigeraceae*. Caution is needed because of the variability that may occur, and as few species have been studied in the required detail. Phylogenetic hypotheses relating to ascomatal ontogenetic types are discussed, but more explanation is needed to ascertain which is most likely to be correct.

INTRODUCTION

This contribution concentrates on the ontogeny of ascomata which, in a way, covers other features used for a long time as criteria in systematics, such as characters of the adult apothecia (shape, colour, opening of the mature ascoma). Likewise, the distinction between *Ascohymeniales* and *Ascoloculares*, a feature still now considered as essential, will be approached in this presentation.

No comprehensive exploration of the question will be attempted, and we will refer mainly to the concepts developed in Paris by Chadefaud (1960, 1982, 1984; see also Parguey-Leduc *et al.*, this volume) and his students: Bellemère and the three present authors.

ASCOMA ONTOGENESIS: SOME GENERALITIES

An ascoma is no more than a female fruitbody developed on a haploid vegetative thallus, and is reminiscent of the general name "gynocarp" proposed by Chadefaud (1982).

Ascomata are formed of two types of elements: sterile elements enveloping fertile elements; the last finally produce asci and ascospores.

The Fertile Elements (Fig. 1A)

At the origin of the fertile elements pluricellular female elements, the "ascogonia", are usually found; their morphology and number in one ascoma are variable. They lead to the formation of "dikaryotic ascosporophytic" filaments, generally provided with hooks and generating asci, within which haploid ascospores are borne.

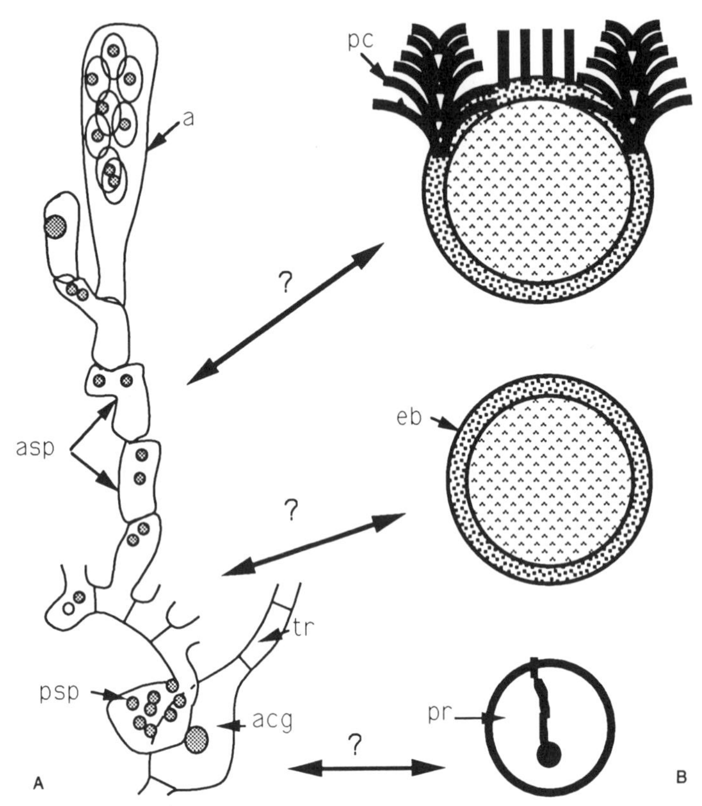

Figure 1. General features of ascoma development. **A**, Development of the fertile elements. **B**, Development of the sterile elements. *a*, ascus; *acg*, ascogonium; *asp*, ascosporophytic apparatus at the top with hooks and dikaryons; *eb*, ébauche; *pc*, parathecial crown; *pr*, primordium; *psp*, prosporophytic elements with haploid not paired nuclei; *tr*, trichogyne; arrows, possible links between developing fertile and sterile elements.

Rather well-documented, although more discussed, is the existence of "prosporophytic" elements between the ascogonial cells and the dikaryotic filaments. The prosporophytic elements have been compared to the carposporophytic ones described in some *Rhodophyceae*. In this connection, the existence, in rare cases, of an inner sporulation of some prosporophytic cells should be pointed out.

A detailed study of the fertile elements led Chadefaud (1960) to consider that the ascomycetes have fundamentally a trigenetic cycle, whose phase II (prosporophytic) and III (ascosporophytic, this sometimes subdivided into two subphases) develop as parasites one upon the other and inside the enveloping sterile elements of phase I (gametophytic) which evolve in a parallel manner.

The development of the sporophytic elements remains moderate, and is opposite to that in algae and other cryptogams; no vegetative thallus develops from these elements.

For further details see Chadefaud (1960, 1982), Parguey-Leduc (1966, 1967a, 1967b, 1970, 1972, 1973), Letrouit-Galinou (1966, 1974), Letrouit-Galinou and Bellemère (1989). Differences in the fertile elements, especially the ascogonial ones, have sometimes been used to segregate families, for instance, Moreau (1928) isolated *Nephromataceae* from *Peltigeraceae* on this basis.

The Sterile Elements (Fig. 1B)

In the last decades, several works dealing with ascoma ontogeny have all recognized three major steps to complete ascoma development.

(1) The formation of a "primordium", arising from the haploid vegetative mycelium, and essentially composed of sterile elements, which are all alike and surround the ascogonial apparatus.

(2) Later, probably when prosporophytic elements arise from the ascogonial cells, differentiation and growth lead the primordium to the "ébauche" (primary corpus) stage whose development will be examined further in detail.

(3) The formation of a "parathecial crown" (or "brush") made of divaricate hyphae.

This general outline puts aside the notion of a "stroma" which played an important part in the original distinction between *Ascoloculares* and *Ascohymeniales* (Nannfeldt, 1932). In fact, Chadefaud (1982) suggested the fusion of the concepts of stroma, lichen thallus and primordium into that of gynocarp, an hypothesis which merits examination.

A question then arises: is the existence of three stages in the development of the sterile elements of the ascoma, as well as in the fertile ones, only coincidence or is there an ontogenetical link between those facts?

VARIATION IN THE STERILE ELEMENTS OF THE ASCOMATA

Although it seems possible to describe the ascoma development of any ascomycete by reference to the above outline, a great number of variations is observed. Only those dealing with the sterile elements will be examined here. They concern mainly the primordium, the ébauche, the parathecial apparatus, the ontogenetical stage at which the ascoma achieves its definitive structure and some other points.

Variation in the Primordium (Fig. 2)

Numerous variations are observed in the primordium. Some of these relate to the place in which the primordium forms. In many lichenized fungi, it forms in the thallus, while in non-lichenized ascomycetes it appears either on the mycelium or in a stroma, the last being a differentiated vegetative formation. Great importance was attached to this in the definition of *Ascohymeniales* and *Ascoloculares*. However, it has been demonstrated (Bellemère, 1967) that in the discomycetes this difference has no influence on the rest of the development, and in particular on the nature of the interascal filaments.

Great importance has also been attached to whether the primordium forms from filaments issued from the ascogonial foot (as demonstrated in *Pyronema*, *Sordariales*, other non-lichenized fungi, and also *Collema*), or from surrounding mycelial elements. Letrouit-Galinou (1966) and Bellemère (1967), amongst others, have shown that in the discomycetes, contrary to what is the rule in pyrenomycetes*, there is no link between this observation and the presence or absence of interascal filaments in the mature ascoma, another character used to delimit *Ascoloculares* from *Ascohymeniales*.

The structure of the primordium is variable. It may be arbuscular, a type considered primitive by Parguey-Leduc (1967), palisadic, with parallely raised filaments, or plectenchymatous. This last structure, which is the rule in lichenized groups (pyrenolichens as well as discolichens; Henssen, 1963; Letrouit-Galinou, 1966; Janex-

* "Pyrenomycetes" is used here to refer to all ascomycetes with flask-shaped ascomata, including both *Ascoloculares* and *Ascohymeniales*.

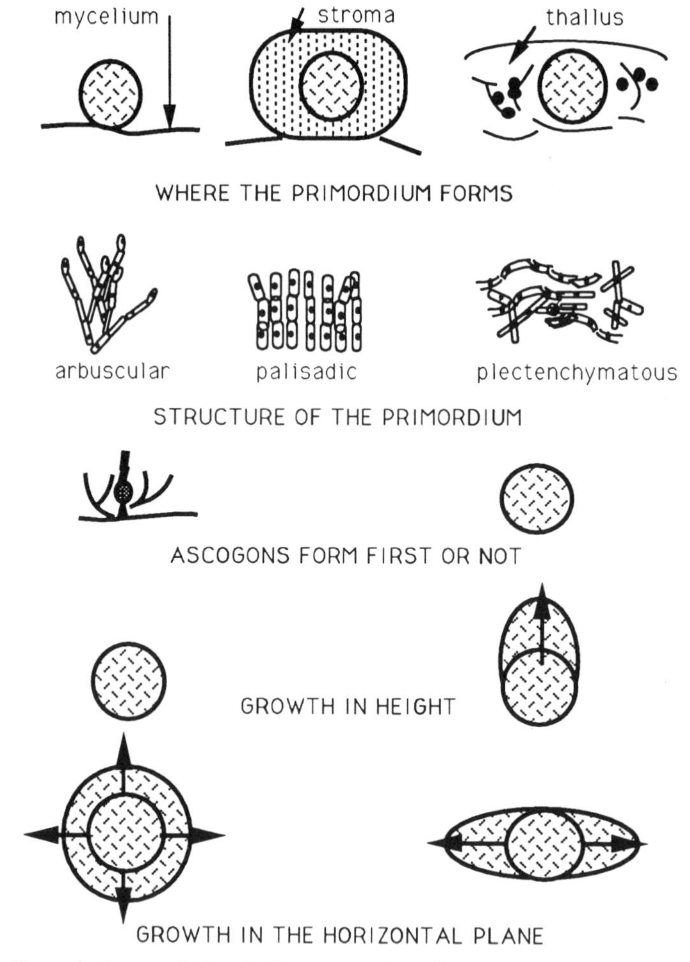

Figure 2. Some variations in the ascoma primordium. See text for explanation.

Favre, 1970; Henssen and Jahns, 1973) and in inoperculate discomycetes (Bellemère, 1967), is in contrast extremely rare in non-lichenized pyrenomycetes (ascolocular and as cohymenial) where other types of primordia have been described (Parguey-Leduc, 1966, 1967a, 1967b; Parguey-Leduc and Janex-Favre, 1981).

The primordium usually enlarges. Commonly, multiplication of the hyphae occurs only at the periphery. Rarely, primordia develop equally in all directions, so that they remain more or less spherical. More often, growth takes place preferentially either in height or in the horizontal plane. In this case, it occurs either all around the primordium, which then becomes circular, or it may be limited to some points, often two opposite ones: in this case the primordium elongates (i.e. *Hysteriaceae*, *Graphidaceae*, *Opegraphaceae*).

Variation in the Ébauche

The most variable structure of the ascoma is the ébauche. To examine its variations efficiently, it is best first to consider the general features of differentiation and growth at this stage.

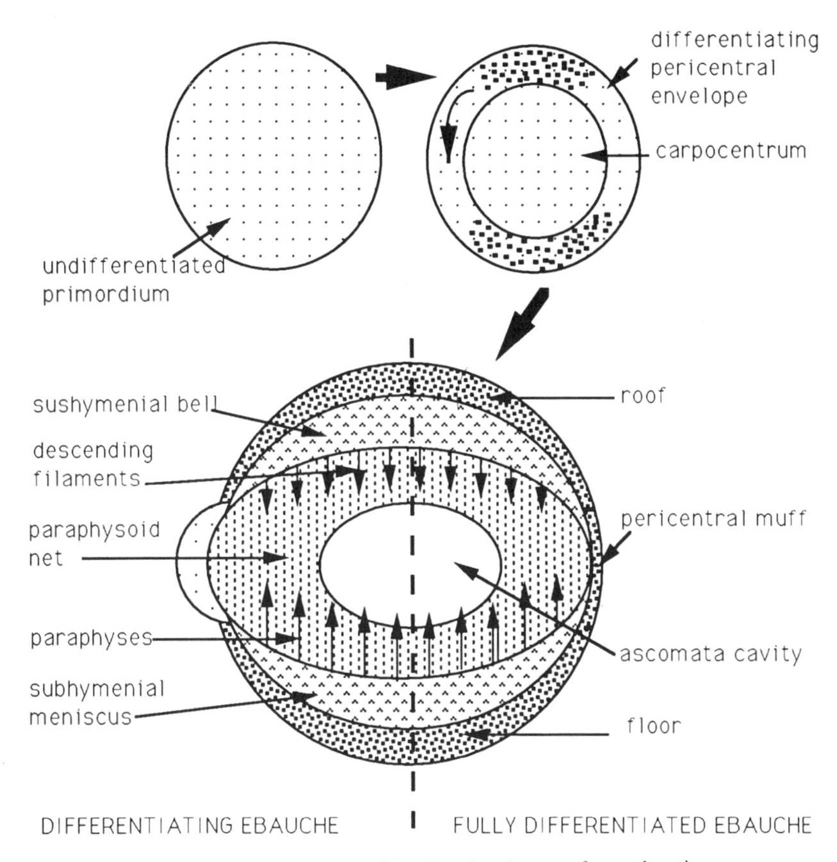

Figure 3. Basic development of the ébauche. See text for explanation.

Basic Developmental Scheme (Fig. 3). In the basic developmental scheme of the ébauche, differentiation occurs in two ways: concentrically (from the inner- to the outer-side), and in a superposed way, from bottom to top. The primordium divides into a peripheral "pericentral" (primary) "envelope" devoid of fertile elements, and an inner "carpocentrum" that contains them. The outer part of the carpocentrum may form a secondary envelope lining the pericentral envelope (Janex-Favre, 1970). Then, both rapidly differentiate into three superposed parts.

The "carpocentrum" divides into: (1) at the base, a "subhymenial meniscus" which eventually generates paraphyses (filaments with free tips growing upwards) and where the sporophytic elements remain confined (it will become the subhymenium after the asci formed); (2) at the top, a "sushymenial bell", which eventually gives rise to descending filaments which may remain short ("periphyses") or be longer and become another type of interascal filaments ("pseudoparaphyses"); and (3) joining them, a "paraphysoid system" whose structure is that of a net in most lichenized and non-lichenized discomycetes, but varies in pyrenomycetes. The paraphysoidal system may be either transitory or persistent, and then plays the role of interascal filaments ("paraphysoids").

Similarly, the "pericentral envelope" divides, from bottom to top, into a floor, a lateral muff, and a roof.

Widening of the ébauche is an important feature in many ascomycetes. It generally results from the activity of a peripheral growth zone, usually ring-shaped. In some cases, as shown in Fig. 3 (left), its structure recalls that of the primordium. New undifferentiated hyphae are born outwards, the differentiating processes occurring inwards. As observed, for example, in the *Arthoniales*, *Opegraphales* and *Graphidales* (Janex-Favre and Letrouit-Galinou, 1969), this type of growth goes on until the pericentral muff is fully developed.

Some Examples of Variation in the Ébauche. Variations in the ébauche are considerable. Of particular interest are those concerning the degree of differentiation reached, the structure of the different parts, the mode of growth (in width and height), and the elements affected by growth (Fig. 4).

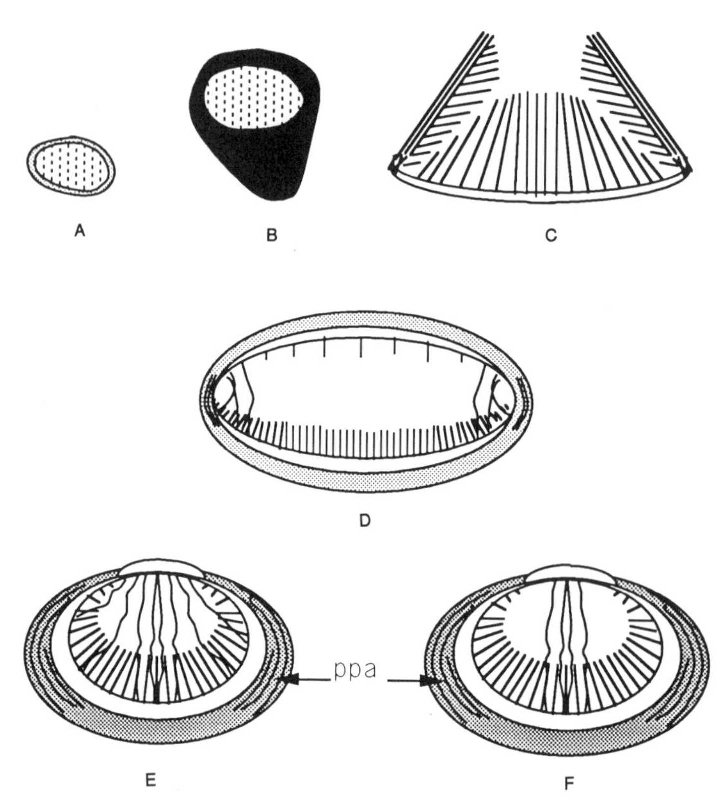

Figure 4. Examples of variation at the ébauche stage. **A**, Poorly differentiated ébauche made of a carpocentrum and a pericentral envelope. **B**, Ébauche with a thickened pericentral floor (podium; e.g. *Opegrapha calcicola*); **C**, Ébauche reduced to a highly differentiated carpocentrum (e.g. *Thelotremataceae*). **D**, All the parts of the ébauche grown equally in width (e.g. *Peltigeraceae*). **E-F**, The elements of the ébauche developed differently; in **E** (e.g. *Lobaria*), the roof does not develop while the floor, the subhymenium, and the hymenium grow, and new paraphysoids are formed; **F** is similar to **E**, except that the paraphysoid net does not enlarge (e.g. *Diploicia canescens*, *Calloria fusiparum*). *ppa*, newly formed part of the floor (=proparathecium).

Not rarely, differentiation stops at the stage with only a carpocentrum and a pericentral envelope. This is observed not only in species whose differentiation will remain feeble even in the adult stage (i.e. *Arthonia* and probably some *Dothideales*; Fig. 4A) but also in species characterized at the mature stage by the presence of parathecial elements as shown in Fig. 6C, *Cyathicula coronata* and in Fig. 7D, *Mitrula pusilla*, *Bisporella citrina*.

In other cases, all the elements described above are formed, for instance in the *Peltigeraceae* (Fig. 8A) or in some inoperculate discomycetes (e.g. *Godronia, Durandiella*; Bellemère, 1967), but later certain parts may vanish. This concerns especially the paraphysoid net which usually disappears making room for the ascoma cavity (Fig. 4D, Fig. 7Aa, 7Ad). Also, the roof can disappear very soon to open the ascoma.

Most often, only some of these elements develop. Consequently, paraphyses and descending filaments are produced together only rarely in discomycetes, and never in pyrenomycetes. In *Thelotrema lepadinum* (Fig. 4C), the pericentral envelope is absent while the carpocentrum is highly differentiated with both paraphyses and periphyses, even an ostiolar apparatus is produced.

Other variations in the structure of the elements include the carpocentrum being paraplectenchymatous (as in *Dothideales*) or plectenchymatous (as in many discomycetes, including numerous lichens).

Variations in the texture and colour of the different formations have to be mentioned, particularly those of the pericentral envelope which is often dark coloured. In many cases, the subhymenium may also be dark, and in this event it has been confused with the envelope. However, the absence in the envelope of sporophytic elements permits their separation.

Growth in height may occur in the ébauche, as in the primordium. For instance, in species where the ébauche is hardly differentiated, it leads to the formation of pedunculate ascomata in which either the carpocentrum elongates (*Geoglossum*, Fig. 7Db) or only the pericentral floor thickens in a "podium" (e.g. *Opegrapha calcicola*, Fig. 4B; *Baeomyces*, Fig. 7Da). In many perithecial and lireliform ascomycetes with more evolved ébauches, the growth in height of the envelope, generally modest, results from the lengthening of the muff. Sometimes it is the subhymenium that thickens, its lower part becoming a "hypothecium" (e.g. *Pyrenopeziza escharodes*, Fig. 6C', top line).

Growth in width is affected by three major causes of variation:

(1) The form of the growing zone may differ: as in the primordium (Fig. 2), it may be present either all around the ébauche or only at some points (two or more), then giving rise to elongate or stellate ébauches.

(2) The structure of the growing zone varies: only rarely does it keep the undifferentiated primordial appearance described above, more often it is made of parallely arranged hyphae growing at one or both ends.

(3) Widening rarely affects all the elements of the ébauche equally and this is at the origin of a great number of morphological and anatomical variants (Fig. 4E-F). The part which enlarges along the base often produces filaments outwards and inwards, the last increasing the subhymenium and the hymenium, so that it recalls one of the parathecial formations and has been called a "proparathecium".

In addition, other variations can be noted. For instance, certain parts of the ébauche may give rise to special formations such as the ostiolar apparatus in perithecia whose development has been studied by Parguey-Leduc (1966, 1967a, 1967b, 1970, 1972, 1973), Janex-Favre (1970), and Parguey-Leduc and Janex-Favre (1981). Also, the possible production by the carpocentrum of "epicentral filaments" has been noted (*Baeomyces, Geoglossum,* several *Pyrenulales*); these do not intervene in the hymenium widening, but may have to do with the parathecial formations.

Variation in the Secondary Parathecial Formations (Fig. 5)

The parathecial formations are made up of a system of divaricate hyphae which ramify intensely at the top while they lengthen at the bases. This ensures that all parathecial formations have a rather similar appearance. However, some differences can be observed.

First, the parathecial elements can be totally lacking (Fig. 6A, 6A'). When present, they may appear at any degree of development of the ébauche. They usually arise at the top of the pericentral muff. When the roof is narrow, they form an apical brush (Fig. 5A) or, more often, when it is wider, an annular crown (Fig. 6B-C, upper lines).

Sometimes, the parathecial formations do not develop significantly (Fig. 6B-B'). More often, the parathecial initial grows either in height or in width, then taking an important part in the enlargement of the hymenium. The most regular growth is in width (Fig. 5B). New hyphae develop at the top of the crown. Those developing outwards (amphithecial hyphae) form an "amphithecium", which adds to the envelope, and no fertile elements develop amongst these. Their bases, more or less parallel, form a cup-shaped "parathecium", also devoid of fertile cells. The branches arising inwards add to the hymenium at their top, and to the subhymenium at their bases, in which asci and sporophytic elements respectively are observed.

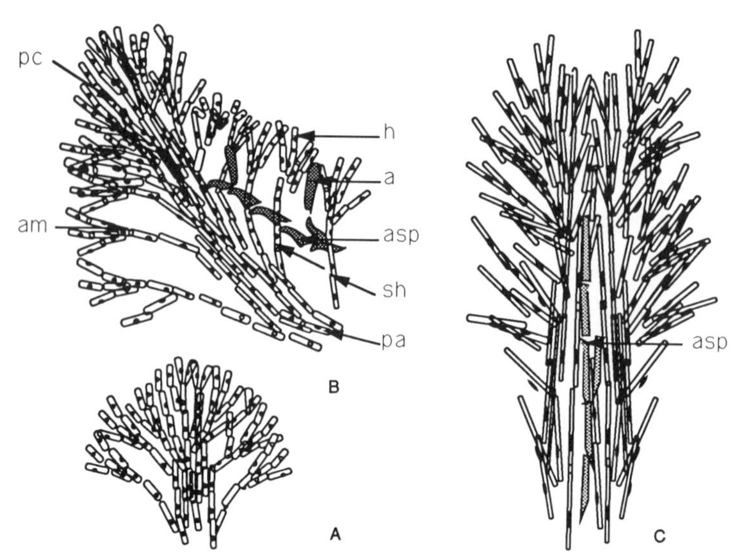

Figure 5. Variation in the secondary formations of ascomata. **A**, Parathecial initial (apical brush or marginal crown). **B**, Enlargement of the crown which takes part or not in the widening of the hymenium. **C**, Extension of the apical brush which forms a discopodium. *a*, ascus; *am*, amphithecium; *asp*, ascosporophytic hyphae; *h*, hymenium; *pa*, parathecium; *pc*, parathecial crown; *sh*, subhymenium.

When the initial is an apical brush, it may at first develop in height forming a stalk ("eudiscopodium") inside which sporophytic elements lengthen (Fig. 5C) as it has been shown in numerous discopodian non-lichenized discomycetes (Bellemère, 1967). Only when the dikaryotic hyphae are fully mature and ready to produce asci, does the apical brush turn into a parathecial crown, which develops as described above, and whose internal branches take part in the formation of the hymenium and the subhymenium.

Besides these major variations, some other may be mentioned: the colour and structure of the amphithecium, and the presence of algae in the amphithecium which characterizes several families and genera of lichenized ascomycetes with lecanorine apothecia.

The divaricate structure of the parathecial formations are by no way exceptional in fungi. Some arbuscular primordia have a similar appearance, as do various thalline appendages in lichenized fungi, for instance isidia (Letrouit-Galinou, 1969) or fibrils. In some cases, this makes interpretation difficult: so that in the genus *Cladonia,* the nature, ascomatal or thalline stroma of the podetium is always under discussion.

Variation Related to the Stage of Development Reached when the Ascoma is Fully Developed: Ascomatal Ontogenetic Types (Fig. 6)

A major cause of variation in ascomycetes is the stage of development reached when the ascoma becomes mature; that is, when its sterile components have attained their complete structure. From this point of view, it is possible to distinguish, with Bellemère (1968, 1978), three major degrees of differentiation (discostromian, preparathecian and parathecian, respectively Fig. 6A-C), each including sessile (upper line) and stipitate ascomata (lower line).

The "discostromian-type" (Fig. 6A) is characterized by the absence of secondary parathecial formation, the mature ascocarp keeping the structure of an ébauche more or less evolved. Belonging to this type, for instance, are the lichenized orders *Pertusariales*, *Opegraphales*, *Graphidales*, (incl. *Thelotrematales*) and the non-lichenized *Rhytismatales*, *Ostropales*. Most pyrenomycetes (ascolocular and ascohymenial) can be referred to this type, in which an important variation is related to the type of sterile filaments intermingled with the asci (interascal filaments): there may be no filaments at

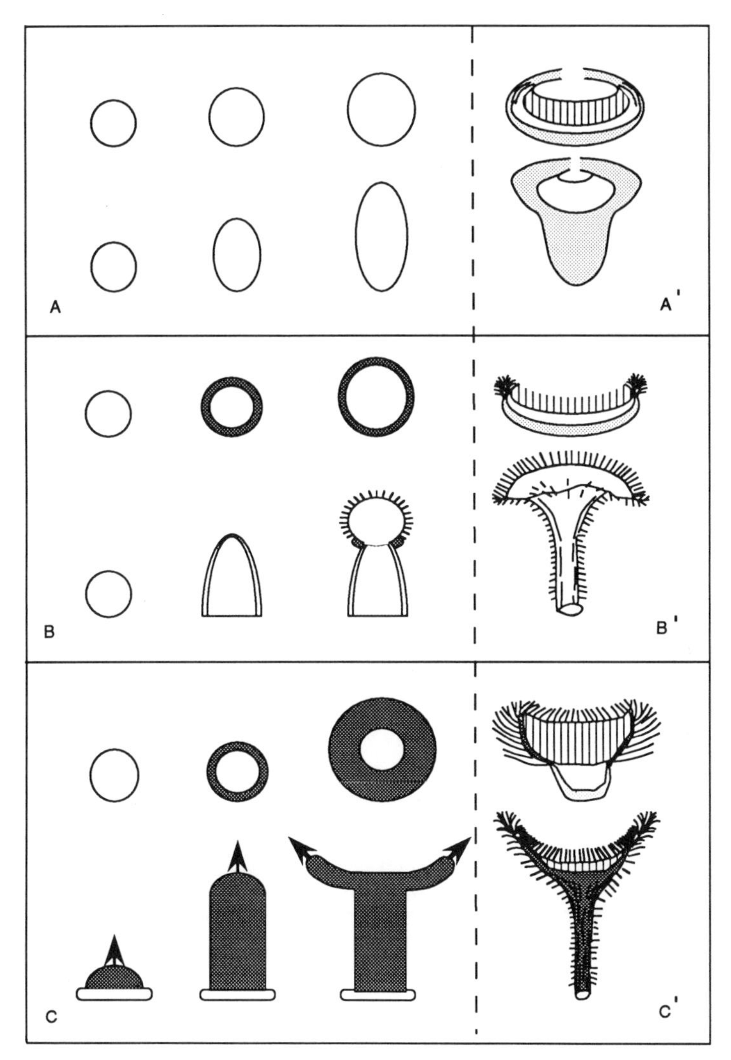

Figure 6. Ontogenetic types according to the developmental stage reached when the ascoma is fully developed. **A**, Discostromian types whose ascomata entirely form from the ébauche; **A'** examples, upper line sessile ascomata (e.g. *Phacidiostroma multivalve*, *Graphidales*); lower line stipitate pseudodiscopodian ascomata (e.g. *Bulgaria inquinans*, *Opegrapha calcarea*). **B**, Preparathecian types with a small parathecial crown which does not take part in the widening of the hymenium; **B'** examples, upper line sessile ascomata (e.g. *Calycella* sp., *Gyalectaceae*), lower line stipitate prediscopodian ascomata (e.g. *Mitrula pusilla*, *Cladonia* sp.). **C**. Parathecian types whose parathecial elements take an important part to the widening of the hymenium and to the stipe formation (eudiscopodian type); **C'** examples, upper line sessile ascomata (e.g. *Pyrenopeziza escharodes*); lower line discopodian types (e.g. *Cyathicula coronata*). carpocentrum, pericentral envelope, parathecial elements, parathecial hypothecium, parathecial crown. Adapted from Chadefaud (1960), Bellemère (1968), Letrouit-Galinou and Bellemère (1989), and Clauzade and Roux (1985).

all, the asci being lodged in a locule, while in other cases persistent paraphysoids, paraphyses, or pseudoparaphyses occur (Fig. 7). The matching stipitate forms ("pseudodiscopodian-type") are also devoid of parathecial elements, the stipe being formed by certain elements of the ébauche, generally the base as in *Opegrapha calcicola*, *Bulgaria inquinans* (Fig. 6A', lower line) and *Baeomyces* (Fig. 7Da), or sometimes the carpocentrum as in *Geoglossum difforme* (Duff, 1922; Fig. 7Db). However, the presence

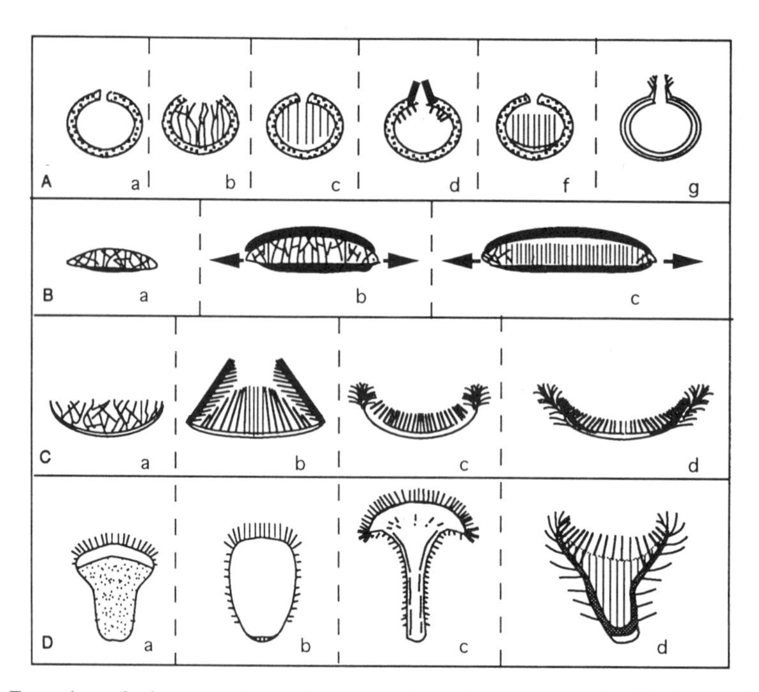

Figure 7. Examples of shapes and structural variations in ascomata. **A**, Flask-shaped ascomata (pyrenocarpous); *a*, only a stromatic or pericentral envelope exists, the evanescent paraplectenchymatous carpocentrum giving rise to a locule, (e.g. *Dothideales*); b, a pericentral envelope and a persistent paraphysoid net are present (e.g. *Trypetheliaceae*); c, the interascal filaments are pseudoparaphyses (e.g. *Pleosporales*, *Hypocreales*); d, ascomata have short descending filaments and no interascal filaments (e.g. *Verrucariaceae*); e, with a pericentral envelope and true paraphyses (e.g. *Pyrenula*, *Diatrypa*); f, as in a, there are no interascal filaments, but the envelope comes from an arbuscular primordium arising from the ascogonial foot and perhaps equivalent to a parathecial initial (e.g. *Clavicipitales*); note that except for f, all the flask-shaped ascomata lack parathecial formations, and also that all the pyrenocarpous lichens have developed from a plectenchymatous primordium and have bitunicate asci. **B**, lirelliform ascomata (the structure of the adult ascoma is also that of an ébauche); a, the envelope is weakly developed, and the hymenium is derived from the paraphysoid net (e.g. *Arthonia*); b, the structure is similar except that the pericentral envelope is well-developed (e.g. *Opegrapha*); c, the structure is still that of an ébauche but here are true paraphyses (e.g. *Graphis*, *Rhytisma*). **C**, sessile discoid ascomata; a, a weakly differentiated ébauche with only a base and a persistent paraphysoid net (e.g. *Pertusaria*, *Lichina*); b, the structure of the mature apothecia is that of an evolved carpocentrum (e.g. *Thelotrema*, *Stictis*); c, a parathecial crown which does not develop is present (e.g. *Gyalecta*); d, true parathecian apothecium with a developed parathecial apparatus (e.g. *Lecidella elaeochroma*, most *Lecanorales*, also numerous non-lichenized ascomycetes). **D**, Stipitate ascomata; a, the stipe comes from the base of the ébauche and there is no parathecial crown (e.g. *Baeomyces rufus*); b, the stipe is derived from the carpocentrum which is covered by epicentral filaments (e.g. *Geoglossum difforme*); c, the development is somewhat similar, but a small parathecial crown surrounds the hymenium (e.g. *Mitrula pusilla, Cladonia*); d, the stipe formed by elongation of a parathecial brush (eudiscopodium) (e.g. *Bisporella citrina* or *Dasyscyphus niveus;* this type is unknown amongst lichenized fungi). Adapted from Chadefaud (1960), Bellemère (1968), Letrouit-Galinou (1966, 1969), Parguey-Leduc (1966-73), and Janex-Favre (1970, 1974).

in the two last species of short epicentral filaments covering the carpocentrum may represent a step towards the following type.

In the "preparathecian" ascomata (Fig. 6B), a parathecial apparatus is present, but it does not develop and does not take part in the growth of the hymenium. Here also there are on one hand, sessile forms (Fig. 6B, upper line), as in *Gyalectales* and various non-lichenized discomycetes (e.g. *Bisporella*), and on the other, stipitate ascomata (Fig. 6B, lower line) as *Mitrula pusilla* and *Cladonia* species where the stalk, usually covered with epicentral filaments, is formed by a lengthening of the carpocentrum while the hymenium is surrounded by a weakly developed parathecial crown (the "prediscopodian-type").

In the "parathecian-type" (Fig. 6C), the hymenium arises mainly or totally from the development of a parathecial apparatus, according to the process described above. Here again the ascomata are either sessile (Fig. 6C, upper line) possibly with a thickened base derived from the ébauche (e.g. *Bisporella citrina*), or stipitate, with the stalk and the hymenium arising entirely from the elongation of a parathecial brush (Fig. 6C, lower line; e.g. *Dasyscyphys niveus*, *Cyathicula coronata*, Fig. 7Dd).

It is generally not too difficult to assign a development pattern to one of these types, although there are some exceptions, for instance when the primordium and the ébauche are reduced and when the only sterile elements observed are arbuscular, as in *Cladonia*; it is then not always easy to decide if it is an arbuscular primordium, a precocious parathecial brush, or even a thalline appendage

Others Variations

Other causes of variation are observed. For example, the hymenium can be largely exposed from the beginning ("gymnocarpic development"), or remain always enclosed ("cleiostocarpic"), or becomes exposed through a narrow aperture ("angiocarpic"), or is at first enclosed and later largely exposed ("hemiangiocarpic"). These terms have been used to differentiate ontogenical types, but those appear rather coarse compared with the general outline presented above which permits segregations to be made.

However, various ascomata develop in peculiar ways whose links with the general scheme are not always clearly perceptible. Truffles are an example amongst the non-lichenized fungi (Parguey-Leduc et al., 1987, 1989), and amongst the lichenized groups the *Parmeliaceae* include species which form a special "husk" in the course of their ascomatal development (Letrouit-Galinou, 1970; Henssen and Jahns, 1973).

It is also pertinent to note the ability of the ascomata in certain species to partially or totally regenerate. This has been observed in ascomata of very different types, above all in the lichenized *Graphis elegans*, and *Pertusaria pertusa*; it is the rule in *Umbilicaria* (Henssen, 1970; Janex-Favre, 1974).

LINKS WITH CLASSIFICATION

The link between ascomatal ontogeny and classification arises from the great number of variations and their possible combinations, to which the general outline classification can be tested. It is possible to sort ascomycetes according to their ascomatal ontogeny, which leads to more fundamental differences than purely morphological observations. Figure 7, for instance, shows that different types of development may end in similar shapes.

Therefore, if to classify is to group taxa according to their similarities, ascomatal development can obviously be used as an efficient criterion with which to do it. This may lead to new delimitations of families or genera, as, for example, in the *Peltigeraceae*, *Parmeliaceae* and *Cladoniaceae*, from which *Nephroma*, *Ramalina*, and *Baeomyces* respectively have been excluded on this basis.

In *Peltigeraceae* (Fig. 8), whose apothecia lie at the upper face of the thallus, there is a long angiocarpic stage during which all the parts of the very complete ébauche grow; later, a parathecial crown arises and its development ensures the ultimate widening of the ascoma. Conversely, in *Nephroma*, the ébauche is established in the growing point of the thallus lobe and is hardly differentiated, while at its border a parathecial crown soon appears; then only the upper side grows giving rise to the whole mature ascoma. Differences are also observed in the ascogonium and ascus structures, so that the separation of the family *Nephromataceae* seems to be fully justified.

Jahns (1970) excluded *Baeomyces* from the *Cladoniaceae*, a family once characterized by the ascomata born on podetia (a special lichenized stipe), because of differences in ascoma ontogeny; differences in the asci can also be observed. Interestingly, in the non-lichenized fungi, all species with a stipitate apothecioid ascoma are placed in the same family, *Leotiaceae*, whether they are of a pseudo-, pre- or eudiscopodian-type, and whatever the ascal type.

Henssen and Jahns (1973) united into the family *Parmeliaceae* all taxa whose ascomata had a paraplectenchymatous "husk", namely most genera of the two previously accepted families, *Parmeliaceae* and *Usneaceae*, excluding *Ramalina* which is devoid of such a husk and which they placed in a separate family, *Ramalinaceae*; in this last case too, differences in ascus structure also exist.

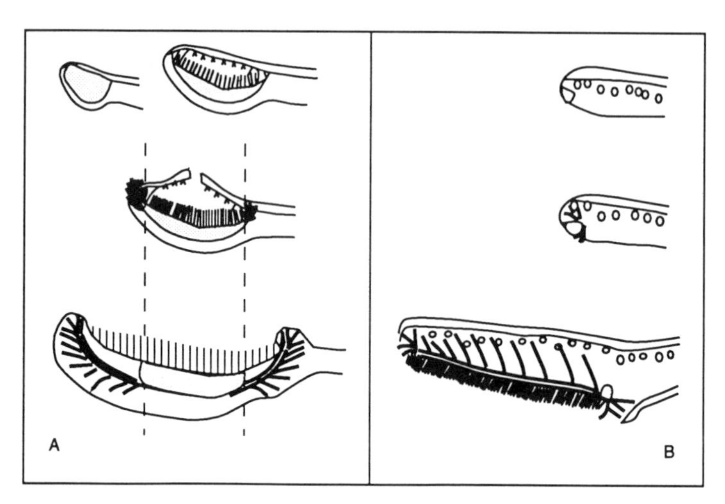

Figure 8. Comparison of ascomatal ontogeny in *Peltigeraceae* (with a well-developed ébauche) and *Nephromataceae* (with a reduced one). Adapted from Letrouit-Galinou and Lallemant (1970, 1971).

However, prudence is recommended when using ascoma ontogeny as a systematic criterion, especially because the degree of variability inside accepted taxonomic units is very poorly known. As the study of ascoma development is time-consuming and arduous, generally only a few taxa of a family or a genus are ever studied. Amongst the features open to infrageneric variability, can be mentioned the more or less important development of the ébauche floor, as in *Opegrapha* where the floor can be pale and thin in dimidiate species (e.g. *O. saxatilis*), or dark and well-developed (e.g. *O. herbarum*), while in others, it thickens into a kind of podium (e.g. *O. calcicola*).

LINKS WITH PHYLOGENY

Systematics is not simply classification; it intends also to reflect phylogeny and that is quite another problem. Can the ontogeny of ascomata be used as an Ariadne's thread to reconstitute the history of the ascomycetes? Because of the well-ordered development described above, and of all the possible variations, it seems sensible to say "yes", or at least "probably", but in fact one must note that no study on ascoma ontogeny has been done with the aim of answering this question first. In consequence, too many points remain either unsolved or nearly so. For instance, the possible homology between stromas, lichen thalli, pericentral envelopes, discopodia, and apothecia is a recurrent problem. Too rarely has the development of stromas been considered, although sometimes there is striking morphological similarities between them and discopodia (e.g. *Xylariaceae*). Also, what must be considered primitive; should the palisade-like disposition of the hyphae, which recalls some algal thalli, be considered as primitive, or should the arbuscular arrangement? Because these questions, and others, are unanswered, the problem of using developmental features as phylogenetical tracers remains.

However, some hypotheses may be proposed. One theory, the most generally developed, puts at the base of the ascomycetes the less differentiated forms with a discostromian ascomata (at first the less differentiated, and then the more complex), a little above are placed the preparathecian and prediscopodian forms, while the parathecian and discopodian ones are arranged at the top. Such an evolution (which remains a possibility), proceeding from the simplest to the more complex forms, does not seem to be in perfect accordance with the fact that parasitism usually evolves in the opposite way.

Another theory places a kind of synthetic archetype, uniting discostromian and parathecian features at the base. From that point, regressive evolution would lead either to purely discostromian types, or to parathecian types with a weakly developed ébauche, while progressive evolution would give rise on the one hand to complex discostromian types, and on the other to discopodian species. Most discomycetes have such synthetic types of ascomata, and this led some authors to suggest that the discomycetes were the most primitive ascomycetes. The *Peltigeraceae* (Fig. 8) whose ascomata combine an evolved ébauche and important parathecial elements are as an example of such a type; moreover, in *Peltigera* the asci also have composite characters having together an annelasceous apex and a fissitunicate dehiscence.

Chadefaud (1982), putting together all available data, proposed another type of evolution resting on two concepts, that of the gynocarp (which subsumes the notions of stroma, lichen thallus, and primordium), and that of the carpocentrum. This led him to consider an early divergence between a phylum mostly including species with a largely exposed hymenium (i.e. more or less the discomycetes), and a second divergence with all flask-shaped ascomata (i.e. pyrenocarps). For further explanations see Parguey-Leduc et al. (this volume).

It is to be noted that none of these theories really takes into account the fact that three stages are recognizable in ascomatal development.

CONCLUSIONS

The development of ascomata is in no way simple. However, the ability to distinguish three stages within it (primordium, ébauche, parathecial apparatus), and to define precisely some fundamental elements, appears to provide a powerful tool for ontogenetical analysis and clearly facilitates comparisons between taxa.

However various points remain difficult to interpret. We have drawn attention to the stromas and other peculiar types which do not fit perfectly into the above scheme. Also, the possibility that the successive elements of the ascomata have similar structures, as shown by Delattre-Durand and Parguey-Leduc (1979) in *Anthracobia nitida*, must be mentioned.

The great number of variations which may affect the general developmental outline are very appealing to researchers, and it has been shown that they can usefully be used to improve classifications, even if care is necessary.

It is also clear that this great variability hides something which has to do with phylogeny; unhappily this point still remains largely to be explored, and it will be valuable for young and less young mycologists to undertake such work.

REFERENCES

Bellemère, A., 1968 ["1967"], Contribution à l'étude du développement de l'apothécie chez les discomycètes inoperculés, *Bulletin de la Société Mycologique de France* 83: 395-931.

Bellemère, A., 1978, Les fructifications à asques (ascocarpes) et leur développement. *In*: *Mycologie et Pathologie Forestière*. Vol. 1. *Mycologiè Forestière* (L. Lanier, P. Bondou and A. Bellemère, eds): 154-166. Masson, Paris

Chadefaud, M., 1960, *Les Végétaux non Vasculaires (Cryptogamie)*. [Traité de Botanique, Vol. 1.] Masson, Paris.

Chadefaud, M., 1982, Les principaux types d'ascocarpes : leur organisation et leur évolution. I-III, *Cryptogamie, Mycologie* 3: 1-9, 103-144, 199-235.

Chadefaud, M., 1984, Le gyno-carpophore gamétophytique des asco- et basidiomycètes et son évolution, *Cryptogamie, Mycologie* 5: 1-11.

Clauzade, G., and C. Roux, 1985, Likenoj de okcidenta Europo. *Bulletin de la Société Botanique du Centre-Ouest, n.s. Numero special* 7: 1-893.

Duff, G.H., 1922, Development of the *Geoglossaceae*, *Botanical Gazette* 74: 264-290.

Delattre-Durand, F., and A. Parguey-Leduc, 1979, Developpement et structure de l'apothecie d'*Anthracobia nitida* (discomycète operculé), *Bulletin de la Société Mycologique de France* 95: 355-375.

Henssen, A., 1963, Eine Revision der Flechtenfamilien *Lichinaceae* und *Ephebaceae*, *Symbolae Botanicae Upsalienses* 18 (1): 1-120.
Henssen, A., 1970, Die Apothecienenwicklung bei *Umbilicaria* Hoffm. emend Frey, *Vorträge der Botanik, Deutsch Botanische Gesellschaft, n.f.* 4: 103-126.
Henssen, A., and H.M. Jahns, 1973 ["1974"], *Lichenes. Eine Einfürhung in der Flechtenkunde.* Georg Thieme, Stuttgart.
Jahns, H., 1970, Untersuchungen zur Entwicklungsgeschichte der Cladoniaceen unter besonderer Berücksichtigung des Podetion-problems, *Nova Hedwigia* 20: 1-177.
Janex-Favre, M.C., 1970, Recherches sur l'ontogénie, l'organisation et les asques de quelques pyrénolichens. *Revue Bryologique et Lichénologie* 37: 421-650.
Janex-Favre, M.C., 1974, L'ontogénie et la structure des apothécies de l'*Umbilicaria cylindrica, Revue Bryologique et Lichénologie* 40: 59-86.
Janex-Favre, M.-C., and M.-A. Letrouit-Galinou, 1969, Sur la morphogenèse des apothécies lirelliformes des Graphidacées et ses mécanismes, *Mémoires de la Société Botanique de France* 115: 156-167.
Letrouit-Galinou, M.-A., 1966, Recherches sur l'ontogénie et l'anatomie comparée des apothécies de quelques discolichens, *Revue Bryologique et Lichénologique* 34: 3-4, 423-588.
Letrouit-Galinou, M.-A., 1969, Remarques sur le thalle, les isidies et les rhizines du *Parmelia conspersa* Ach. (discolichen, Parmeliacée), *Bulletin de la Société Botanique de France* 116: 1-14.
Letrouit-Galinou, M.-A., 1970, Les apothécies et les asques du *Parmelia conspersa* (discolichen, Parmeliacée), *Bryologist* 73: 39-58.
Letrouit-Galinou, M.-A., 1974 ["1973"], Sexual reproduction. *In*: *The Lichens* (V. Ahmadjian and M.E. Hale, eds): 59-90. Academic Press, New-York.
Letrouit-Galinou, M.-A., and A. Bellemère, 1989, Ascomatal development in lichens: a review, *Cryptogamie, Bryologie et Lichénologie* 10: 189-233.
Letrouit-Galinou, M.-A., and R. Lallemant, 1970, Le développement des apothécies du *Nephroma resupinatum* (L.) Ach., lichen, Néphromacée, *Revue Génerale de Botanique* 77: 331-351.
Letrouit-Galinou, M.-A., and R. Lallemant, 1971, Le thalle, les apothécies et les asques du *Peltigera rufescens* (Weis) Humb. (discolichen, *Peltigeraceae*), *Lichenologist* 5: 59-68.
Moreau, F., 1928, Les phénomènes cytologiques de la reproduction chez les champignons des lichens, *Le Botaniste* 20: 1-67.
Nannfeldt, J.A., 1932, Studien über die Morphologie und Systematik der nicht-lichenisierten inoperculaten Discomyceten, *Novae Acta Regiae Societas Scientiarum Upsaliense*, sér. IV, 8 (2): 1-368.
Parguey-Leduc, A., 1966, Recherches sur l'ontogénie et l'anatomie comparée des ascocarpes des pyrénomycètes ascoloculaires. I, *Annales de Sciences Naturelles, Botanique*, sér. 12, 7: 505-690; 8: 1-110.
Parguey-Leduc, A., 1967a, Recherches sur l'ontogénie et l'anatomie comparée des ascocarpes des Pyrénomycètes ascoloculaires. II, *Annales de Sciences Naturelles, Botanique*, sér. 12, 8: 1-110.
Parguey-Leduc, A., 1967b, Recherches préliminaire sur l'ontogénie et l'organisation des ascocarpes des Pyrénomycètes ascohyméniaux. I-III, *Revue de Mycologie* 32: 57-68, 259-277, 369-407.
Parguey-Leduc, A., 1970, Recherches préliminaire sur l'ontogénie et l'organisation des ascocarpes des pyrénomycètes ascohyméniaux. IV. Les asques des diatrypacées et leurs ascothécies du type "*Eutypa*", *Revue de Mycologie* 35: 90-129.
Parguey-Leduc, A., 1972, Recherches préliminaire sur l'ontogénie et l'organisation des ascocarpes des pyrénomycètes ascohyméniaux. V. Les asques des *Xylariales* et leurs ascothécies du type "*Xylaria*", *Revue de Mycologie* 36: 194-237.
Parguey-Leduc, A., 1973, Recherches préliminaires sur l'ontogénie et l'anatomie comparée des ascocarpes des pyrénomycètes ascohyméniaux. VI. Conclusions générales, *Revue de Mycologie* 37: 60-82.
Parguey-Leduc, A., and M.-C. Janex-Favre, 1981, The ascocarps of ascohymenial pyrenomycetes, *In*: *Ascomycete Systematics: The Luttrellian Concept* (D.R. Reynolds, ed.): 102-123. Springer-Verlag, New York.
Parguey-Leduc, A., C. Montant, and M. Kulifaj, 1987, Morphologie et structure de l'ascocarpe adulte du *Tuber melanosporum* Vitt. (Truffe noire du Périgord, discomycètes), *Cryptogamie, Mycologie* 8: 173-202.
Parguey-Leduc, A., M.-C. Janex-Favre, C. Montant, and M. Kulifaj, 1989, Ontogénie et structure de l'ascocarpe du *Tuber melanosporum* Vitt. (Truffe noire du Périgord, Discomycètes), *Bulletin de la Société Mycologique de France* 105: 227-246.

M. CHADEFAUD AND ASCOMYCETE SYSTEMATICS

A. Parguey-Leduc, M.-C. Janex-Favre, M.-A. Letrouit-Galinou
and A. Bellemère

Université Pierre & Marie Curie
Laboratoire de Cryptogamie
7 quai St. Bernard
75252 Paris Cedex 05
France

SUMMARY

An illustrated synopsis of the concepts introduced by M. Chadefaud into ascomycete systematics and phylogeny is presented. The different types of ascus apical structures and ascocarp ontogeny he recognized are described, and his views on the development of the sporophytic apparatus explained.

INTRODUCTION

New concepts in ascomycete systematics and phylogeny arose from the great volume of observations made by M. Chadefaud (1900-1984) and his students on discomycetes and pyrenomycetes. They were based on ascal and ascocarp[1] features: the apical apparatus of the asci, ascocarp ontogenesis, and development of the elements of the sporophytic apparatus.

APICAL APPARATUS OF THE ASCI

Fig. 1 (Chadefaud, 1942, 1973; Chadefaud et al., 1969)

The most simple and primitive ascus type from which all asci were considered to derive is the "préarchaeascé". In the "préarchaeascé I" type the ascus wall is composed of two layers of regular thickness (exoascus and endoascus). In the "préarchaeascé II" type, which was viewed as slightly more evolved, the ascus wall also has a "doublure interne" (i.e. an internal sheath) which is thickened at its apex.

Further evolution was considered to have led to the "archaeascé" type and one of its variants, the "nassascé" type. From the latter was derived the "annellascé" type. These three fundamental types of ascal apical apparatus led Chadefaud to propose the division of the ascomycetes into three groups:

(1) "*Archaeascés*": asci belonging to the synthetic ancestral type characterized by both a "nasse" and an "anneau" (= ring) as the apical apparatus; the nasse caps

[1]"Ascocarp" is used in this contribution in preference to "ascoma" in accordance with the practice of Chaudefaud and discussed in Chadefaud (1982) [Ed.].

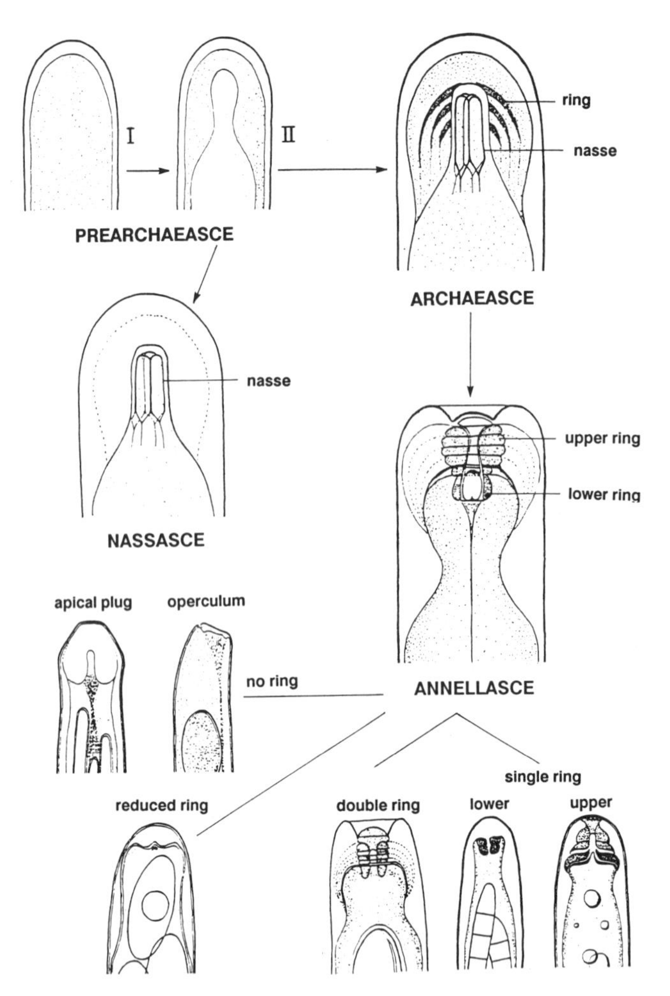

Figure 1. Types of apical apparatus in asci distinguished by Chadefaud and co-workers. See text for explanation.

the epiplasmic apex while the ring differentiates within the internal sheath ring and is formed of superimposed parts.

(2) "*Nassascés*": as a rule, with the ascal apical apparatus only a "nasse".

(3) "*Annellascés*": the ascal apical apparatus is a ring, showing a great deal of variation. In the most complex cases it is double (a lower and an upper ring, with one or more superimposed parts each). In other cases, there is a single ring (equivalent either to the lower or the upper one). Finally, an operculum or an apical plug may differentiate in the place of a typical ring.

STRUCTURAL ONTOGENESIS OF THE ASCOCARPS

Fig. 2 (Chadefaud, 1960, 1982)

In order to describe the development of the ascocarp in ascomycetes, Chadefaud proposed the two concepts of "gynocarpe" (i.e. gynocarp) and "carpocentre" (i.e. carpocentrum). They led him to consider any ascocarp as a variant of a single fundamental type.

The gynocarp is the young female fruitbody at the origin of the ascocarp. It is formed of two parts: the carpocentrum (grey-coloured in Fig. 2) containing the fertile el-

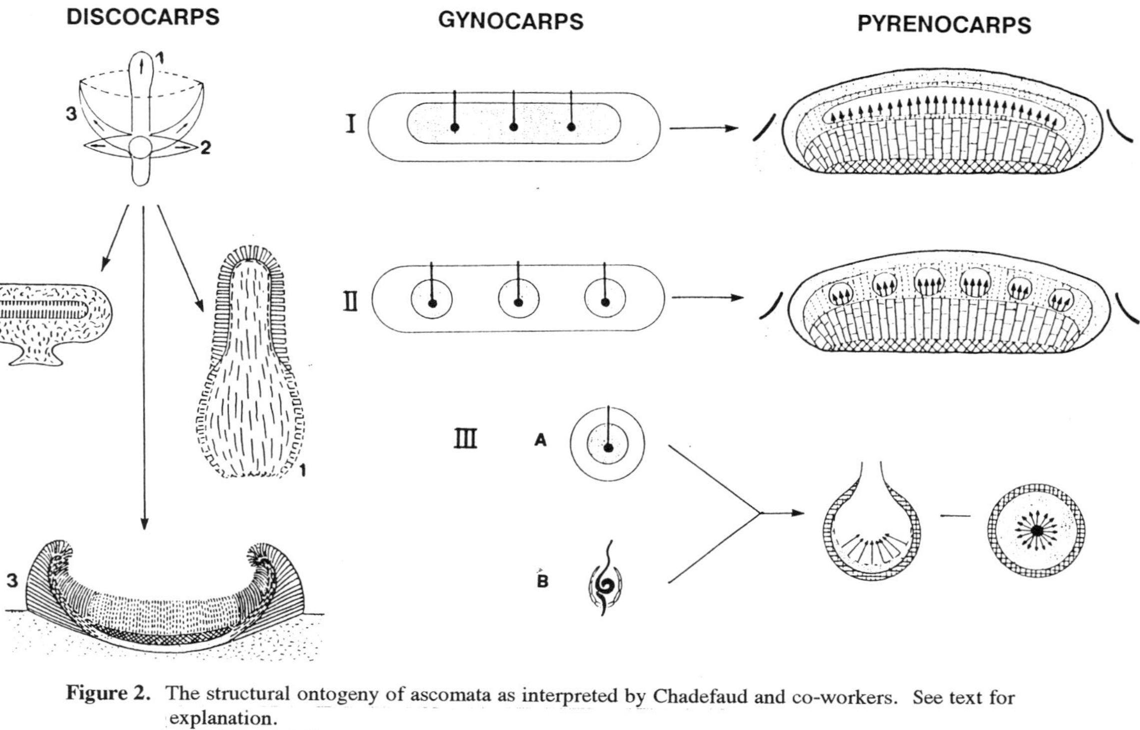

Figure 2. The structural ontogeny of ascomata as interpreted by Chadefaud and co-workers. See text for explanation.

ements of the ascogonial apparatus (dark coloured) and the surrounding envelope (light coloured).

Several types of gynocarps are found in ascomycetes: discoidal gynocarps, provided with either one (Fig. 2I) or more (Fig. 2II) carpocentra, and globular gynocarps in which (Fig. 2III) the single carpocentrum may be surrounded by an envelope (A) or not (B). In the latter case, the ascogonium appears first and the carpocentrum second, as well as the envelope which arises from the base of the ascogonium.

Chadefaud named the different ascocarps types as "discocarpes" (i.e. discocarps) in discomycetes and "pyrénocarpes" (i.e. pyrenocarps) in pyrenomycetes.

In discomycetes three major types of discocarps develop from the original lenticular or globular gynocarp:

(1) *"Discopodien" type*: the discocarp increases in height and forms a "discopode" with a cupular palisadic hymenium.
(2) *"Discostromien" type*: lenticular or not, the discocarp enlarges in width from a peripheral growth zone; the hymenium is palisadic.
(3) *"Parathécien" type*: the growth zone differentiates into a well-developed "parathecioid muff" which is completed, or not, by parathecial formations enlarging both the palisadic hymenium and the envelope.

In pyrenomycetes, each of the three basic types of gynocarps is related to a type of pyrenocarp. Discoidal gynocarps produce disc-shaped stromas hollowed with either a large single locule (type I) or several small locules (type II). In both cases, the base of the locule bears a palisadic hymenium. Globular gynocarps (Fig. 2III) produce "pyrénosphères" (i.e. perithecia), generally provided with a neck and an ostiole (flask-shaped perithecia) and a palisade of asci; a few perithecia are "périsporiés" (i.e. devoid of an ostiole); others are, in addition, "plectascés" (the locule containing disordered asci). The ontogenetic data led Chadefaud to propose phylogenetical concepts in the pyrenomycetes, which he divided into three related groups: "Ascoloculaires", "Ascohyméniaux" and "Périsporiés". The latter are considered to be the result of a regressive evolution from the two previous groups.

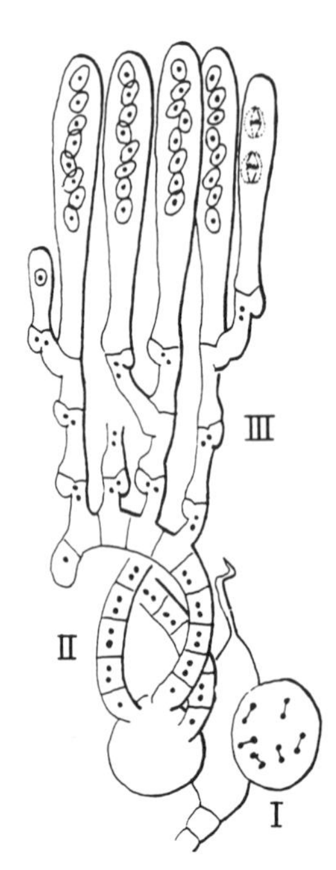

Figure 3. Development of the sporophytic apparatus in ascomycetes as interpreted by Chadefaud. See text for explanation.

DEVELOPMENT OF THE SPOROPHYTIC APPARATUS

Fig. 3 (Chadefaud, 1953, 1960)

Contrary to previous views relating to the sexual cycle of ascomycetes, Chadefaud distinguished successive developmental phases between the ascogonium and the asci:

(1) *The prosporophyte* (I): The fertilized ascogonial nucleus divides several times. The sister nuclei migrate into vesicular ascogonial buddings. This prosporophyte is "micto-haploide" (i.e. comprises mixed scattered haploid nuclei).

(2) *The ascosporophyte* comprises two successive phases (II and III). Phase II derives from phase I and consists of dikaryotic filaments. Phase III arises from the distal cells of these filaments: it is formed of superimposed dikaryotic "dangeardies" (i.e. cells provided with croziers) generating asci.

The fundamental sexual cycle of ascomycetes is trigenetic (gamétophyte, prosporophyte, ascosporophyte); it may be compared with the trigenetic sexual cycle described in floridean *Rhodophyceae*.

ACKNOWLEDGEMENTS

We thank F. Guilloux and M. Avnaim for their friendly technical cooperation.

REFERENCES

Chadefaud, M., 1942, Étude d'asques. Structure et anatomie comparée de l'appareil apical des asques chez divers Disco-et Pyrénomycètes, *Revue de Mycologie* 7: 57-88.

Chadefaud, M., 1953, Le cycle et les sporophytes des ascomycètes, *Bulletin de la Société Mycologique de France* 69: 201-219.

Chadefaud, M., 1960, *Les Végétaux non Vasculaires (Cryptogamie).* [Traité de Botanique Vol. 1.] Masson, Paris.

Chadefaud, M., 1973, Les asques et la systématique des ascomycètes, *Bulletin de la Société Mycologique de France* 89: 127-170.

Chadefaud, M., 1982, Les principaux types d'ascocarpes: leur organisation et leur évolution. I-II. *Cryptogamie, Mycologie* 3: 1-9, 103-144.

Chadefaud, M., M.-A. Letrouit-Galinou, and M.-C. Janex-Favre, 1969, Sur l'origine phylogénétique et l'évolution des ascomycètes des lichens, *Mémoires de la Société Mycologique de France, Colloque sur les Lichens*: 79-111.

DEVELOPMENTAL MORPHOLOGY OF THE "ZWISCHENGRUPPE" BETWEEN *ASCOHYMENIALES* AND *ASCOLOCULARES*

A. Henssen[1] and G. Thor[2]

[1]Fachbereich Biologie der Philipps-Universität
D-35032 Marburg
Germany

[2]Sveriges Lantbruksuniversitet
Institutionen för ekologi och miljövård
Box 7072
S-750 07 Uppsala
Sweden

SUMMARY

Characteristic features of the "Zwischengruppe", the *Arthoniales* s.l., are the ascohymenial ascocarp ontogeny and fissitunicate asci. The asci have a thick exotunica and a thin, expanding, hemiamyloid endotunica with an apical amyloid structure. The ascocarps are apothecioid, lirelliform, or in some genera stromatic. They develop gradually from a generative tissue that includes ± bent ascogonia ending in a trichogyne. The hamathecium consists of paraphysoids, and the exciple is annular and rudimentary to cupular and carbonized. The ascocarp ontogeny is briefly discussed in the type species of *Arthonia, Arthothelium, Opegrapha*, and *Roccellina,* and more comprehensively in species of *Melampylidium*, *Dichosporidium*, and *Erythrodecton*. The developmental morphology of the "Zwischengruppe" is compared with that of *Pleospora herbarum* and *Myriangium duriaei*, members of the *Ascoloculares* s.str. with ± pseudoparenchymatous ascostromata. Diagnoses are provided for the new species *Dichosporidium latisporum* Thor & Henssen and *Melampylidium redonii* Henssen. The new combination *Melampylidium macrosporum* (R.C. Harris) Henssen is made.

INTRODUCTION

In his basic treatment of inoperculate discomycetes, Nannfeldt (1932) distinguished two large groups, the *Ascohymeniales* and *Ascoloculares*, the latter being subsequently named *Bitunicatae* (Luttrell, 1951) or *Loculoascomycetes* (Luttrell, 1973). The term "Zwischengruppe" was introduced by Henssen and Jahns (1973) for the *Arthoniales*, an order of mainly lichenized ascomycetes, to emphasize their intermediate systematic position, and to separate them from the remaining *Ascoloculares* s.str. that have a ± pseudoparenchymatous ascostroma. The families of the *Arthoniales* combine the presence of fissitunicate (bitunicate) asci, a feature of ascolocular fungi, with a paraphysoid hamathecium that develops from a generative tissue, a character of ascohymenial fungi. In the *Outline*, Eriksson and Hawksworth (1988) placed the families of the "Zwischengruppe" in two orders, *Arthoniales* and *Opegraphales*. Tehler (1983, 1990, this volume), on the other hand, included the families of the *Opegraphales* in the *Arthoniales*. Tehler (1990) provides a detailed historical survey of the systematic treatment of the *Arthoniales* by different authors.

The families of the *Ascoloculares* in the restricted sense are kept by Eriksson and Hawksworth (1988) in *Dothideales* s.lat. The developmental morphology of the ascostromata in members of the *Dothideales* was discussed particularly by Luttrell (1951, 1973, 1981) and Parguey-Leduc (1966). In regard to the differentiation of the ascostromata, Henssen and Jahns (1973) recognized a disintegration type (Zerfalltyp) and an aggregation type (Aggregatyp). The first is represented, for example, by *Pleospora herbarum*, and the second in *Myriangium duriaei*. The developmental morphology in those fungi is briefly described here to elucidate the deviating ascocarp ontogeny in the "Zwischengruppe", *Arthoniales* s.lat.

MATERIAL AND METHODS

This list of specimens is restricted to those photographed: *Arthonia radiata*: Finland, Nylandia, Helsingfors, on *Sorbus aucuparia*, 1949, *Fagerström* (hb Henssen 3119).- *Arthothelium spectabile*: Switzerland, Basel. *Hepp* (MB).- *Dichosporidium latisporum*: The Seychelles, Praslin, Vallée de Mai, Coco de Mer forest, on *Lodoicea sechellarum*, 150-200 m, 1981, *Henssen* 26840a (hb Henssen).- *D. microsporum*: Fiji, Viti Levu, Rewa Distr., Wailoku *c.* 7. km N of Suva, 1985, *Thor* 6288 (S).- *Erythrodecton granulatum*, Panama, Barbour-Lathroup trail, Barro Colorado island, 1977, *Hale* 47659 (US).- *E. kurzii*: Australia, Queensland, Atherton Tableland, E of Tinaroo Falls Dam, 1985, *Thor* 5226 (S).- *E. malacum*: Australia, Queensland, Atherton Tableland, *c.* 2.5 km N of Tinaroo Falls Dam, 1985, Thor 5194 (S).- *E.* sp.: The Seychelles, Praslin, Vallée de Mai, Coco de Mer forest, on *Lodoicea sechellarum*, 150-200 m, 1981, *Henssen* 26840b (hb Henssen).- *Melampylidium cerei*: Chile, Prov. Coquimbo, La Serena, Guanaqueros, dwarf-shrub succulent desert, on candelabrum cacti, 1973, *Henssen* (24225a), *Vobis & Redon* (MB).- *M. macrosporum*: USA, North Carolina, Carteret Co., Morehead City, Bogue Banks, on *Quercus laurifolia*, 1977, *Henssen* 25884h (hb Henssen).- *M. metabolum*: New Zealand, North Island, Rangitoto Island, Kowhai Grove, *Metrosideros excelsa* forest, on *Melicytus ramiflorus*, 1985, *Henssen* (30011a) & *Lumbsch* (MB).- *M. redonii:* Chile, Prov. Coquimbo, Parque Nacional Fray Jorge, Cordillera de Talinay, fog oasis Fray Forge, fog forest between 300 to 400 m, on bark, 1973, *Henssen* (24204b), *Vobis & Redon* (MB).- *Myriangium duriaei*: British Isles, South Devon, Wembury near Plymouth, on *Fraxinus*, 1964, *Henssen* 17590a (hb Henssen).- *Opegrapha vulgata*: Austria, Steiermark, Grundlsee, Sattelhöhe, on *Abies*, 1952 *Henssen* 3125 (hb Henssen). - *Pleospora herbarum*: Sweden, Öland, Borgholm, on *Melilotus officialis*, 1928, *Eliasson* (UPS).- *Roccellina cerebriformis*: Chile, Prov. Coquimbo, La Serena, Guanaqueros, W exposed slope of peninsula, on S exposed cliff, 1973, *Henssen* (24229a), *Vobis & Redon* (MB).- Undescribed fungus: Switzerland, Gandria near Lugano, S slope of M. Brè, in seepage lines of *c.* 10 m tall, calcareous rock face at roadside, 1989, *Henssen* 33054e (hb Henssen).

Freezing microtome sections, 18-22 μm thick, were mounted in lactic/glycerine with cotton/blue (LB). For the iodine reaction Lugol's solution was added *without* pretreatment with potassium hydroxide. A Wild M7 dissecting microscope was used for habit photographs and a Wild M20 compound microscope for micrographs of sections. For both a Kodak professional film 5-TMX 120, 6 x 6 cm format was used.

TAXONOMIC NOTES

Diagnoses for two new species included in the study are provided to validate the names employed, a new combination is made, and the exclusion of *Melampylidium cerei* from the genus is suggested.

Dichosporidium latisporum Thor & Henssen sp.nov.: Thallus byssoideus, excorticatus. Apothecia pseudoangiocarpia, in structura stromati aggregati sine strato nigro. Sporae obovatae, *c.* 4-septatae, (33-)35-41(-44) x 4-)4-5(-5) μm. Conidia bacilliformia, (4-)5-6(-6) x *c.* 1 μm. Acida salazinicum, norstictum et protecetraricum vel acida salazinicum et norsticticum continens. - Typus: The Seychelles, Praslin, Vallée de Mai, Coco de Mer forest, on *Lodoicea sechellarum*, 150-200 m, 1981, *Henssen* 26840a (hb Henssen - holotypus).

Melampylidium redonii Henssen sp.nov.: Thallus crustaceus, albescens. Apothecia nigra, rotundata, stipitata, usque ad 1.2 mm diam. Hymenium 190 μm, subhymenium 215 μm altum, excipulum cupulare et nigrum. Sporae muriformes, 30-37 x 11-14 μm. - Typus: Chile, Prov. Coquimbo, Parque Nacional Fray Jorge, Cordillera de Talinay, fog oasis Fray Jorge, fog forest between 300 and 400 m, on bark, 1973, *Henssen* (24204b), *Vobis & Redon* (hb Henssen - holotypus).

Melampylidium macrosporum (R.C. Harris) Henssen comb.nov.: Basionym: *Bactrospora macrospora* R.C. Harris, *Some Florida Lichens*: 40 (1990).- Type: USA, Florida, Nassau Co. E of Lofton Creek, on *Acer*, 1987, *Harris* (NY - holotype, not seen). - The collection from North Carolina, *Henssen* 25884h, differs in the somewhat longer and narrower spores, 75-126 x 6-8.5 μm (80-90 x 7-10 μm *fide* Harris, 1990; 60-98 x 6-10 μm *fide* Egea and Torrente, 1993).

The genus *Bactrospora* was recently treated by Egea and Torrente (1993). The correspondence on shape and structure of the apothecia between *Melampylidium* and *Bactrospora* is quite remarkable. However, Harris (1990) pointed out the deviating spore type of *B. macrospora* compared with other species assigned to the genus (*Homalotropa*-type in Egea and Torrente, 1993). In *M. macrosporum*, not only the ascospores but also the asci differ by the thick-walled exotunica from *Bactrospora* s.str.; on the other hand, the asci and ascospores are of the same type as in *Melampylidium*. In our opinion, *M. macrosporum* is most closely related to *M. metabolum*, the type species of the genus. *M. redonii* differs by the prominent cupular exciple, the hymenium interspersed with oil drops, and the short, multi-celled spores.

Melampylidium cerei (Schim.-Czeika) Redon & Follm. (*Philippia* **1**: 189, 1972; as "*Melampydium*") is certainly not related to *Melampylidium metabolum* and should be excluded from the genus. The species has quite different ascocarps (see below). The systematic position of the lichen is at present unsolved.

TERMINOLOGY

Ascocarp is used as a general, comprehensive term for fruit-bodies of ascomycetes; *ascostroma* for the ± pseudoparenchymatous fruit-bodies of the *Ascoloculares* in the restricted sense excluding the "Zwischengruppe" (Henssen and Jahns, 1973; Henssen, 1976). Ascostromata might be uniloculate or multiloculate. We use the word paraphyses for any type of interascal filaments, and hamathecium after Eriksson (1981) to cover all kinds of interascal hyphae or tissue, and hyphae projecting into the locule or ostiole. *True paraphyses* are the characteristic hymenial filaments of the *Ascohymeniales*; they arise from the subhymenial layer and have free upper ends from the beginning (Henssen and Jahns, 1973). *Pseudoparaphyses* are the interascal filaments of the uniloculate *Dothideales* that are attached by both ends in the cavity of the fruit-body. They arise from rows of short cells of reticulately branched hyphae (Luttrell, 1965; Henssen and Jahns, 1973; Henssen 1976, 1987). *Paraphysoids* develop from the generative tissue, and later mainly at the margin of the hymeniun. They are, more or less densely, reticulately branched. The ends become secondarily free or remain united to form a more or less distinctly limited epithecium (Henssen and Jahns, 1973). *Apical paraphyses* arise at the roof on an ascocarp and grow downward into the preformed cavity of the fruit-body (Henssen, 1994).

DEVELOPMENTAL MORPHOLOGY IN ASCOLOCULARES S. STR.

Disintegration type of ascostroma formation. In the uniloculate ascostroma of *Pleospora herbarum*, the development starts with the formation of a globose primordium of isodiametric cells. In the centre, the walls of the cells disintegrate (Fig. 1) and ascogonial cells are differentiated. The vegetative cells in the centre start to stretch and form pseudoparaphyses around and between the ascogonial cells and ascus initials (Fig. 2). The process continues, the pseudoparaphyses remain connected to the base as well as to the meristematic plug of the ascostroma (Figs 3-4). They elongate by intercalarly cell division and the elongation of the attached meristematic cells. In the young ascostroma the pseudoparaphyses are relatively thick. Due to further stretching of the cells, they become thinner in the large ascostromata (Fig. 5). A corresponding development of pseudoparaphyses was observed in the ascostromata of *Lichenothelia* species (Henssen, 1987).

Our interpretation of the development of pseudoparaphyses in *Pleospora herbarum* does not conform with observations made by Parguey-Leduc (1966) and Corlett (1973) who both reported a downward growth of the pseudoparaphyses from the roof to the centrum (see also Luttrell, 1965, 1973, 1981; and the terminology in Eriksson, 1981). This type of paraphyses, for which we use the term apical paraphyses, was observed, for example, in the ascostromata of *Dacampia* (Henssen, 1994) and an undescribed fungus (*Henssen* 33054e). In both cases, a cavity arose in the centre before the apical paraphyses were differentiated. In *Dacampia*, they subsequently grew downward to the bottom

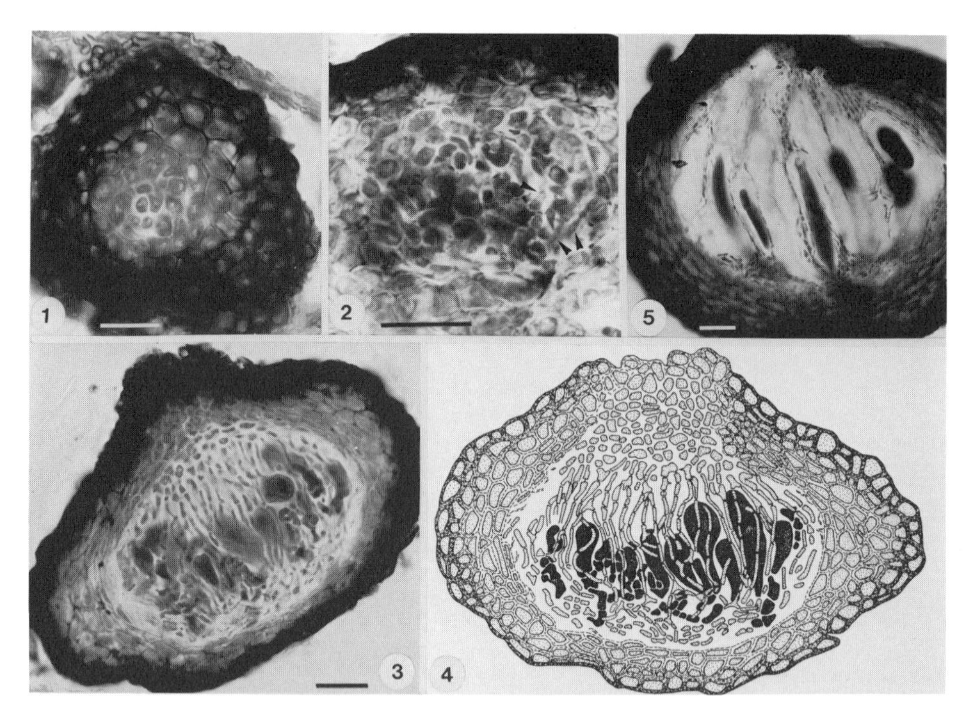

Figures 1-5. Disintegration type of ascostroma development in *Pleospora herbarum*. **Fig. 1**. Initial stage with dissolving centre. **Fig. 2**. Subsequent stage with developing pseudoparaphyses (arrows), ascogenous cells, and ascus initials. **Figs 3-4**. Micrograph and drawing of young ascostroma with developing asci and pseudoparaphyses that are connected to the base and the meristematic plug of the ascostroma. **Fig. 5**. Mature, uniloculate ascostroma with thin pseudoparaphyes. Scale 1-3 and 5 = 20 μm. Fig. 4 after Henssen and Jahns (1973).

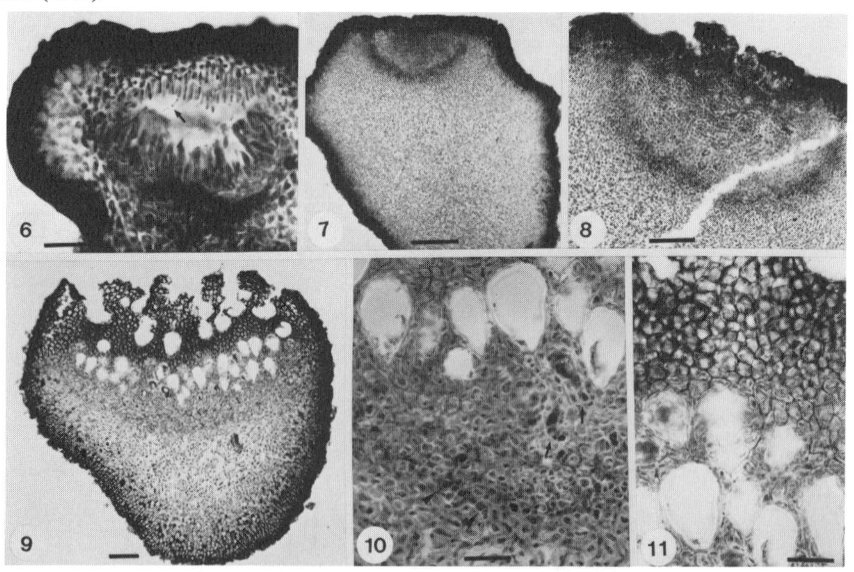

Figures 6-11. Ascocarp development in *Ascoloculares* s.str. **Fig. 6**. Undescribed fungus (*Henssen* 33054e); young ascostroma with apical paraphyses growing into a preformed cavity; remnant of pseudoparaphyses indicated by an arrow. **Figs 7-11**. Aggregation type of ascostroma development in *Myriangium duraei*. **Fig. 7**. Young, cup-shaped ascostroma arising in the upper part of the vegetative stroma. **Fig. 8**. Young ascostroma in higher magnification. **Figs 9-11**. View and details of mature, multiloculate ascostromata; young asci on foot-cells indicated by arrows, invading ascogenous hyphae by an arrowhead. Scale 6 and 10-11 = 20 μm; 7 = 100 μm; 8-9 = 50 μm.

of the centrum; in the undescribed fungus they remained short (Fig. 6) some pseudoparaphyses developing prior to the formation of the cavity and gradually disintegrating.

Aggregation type of ascostroma formation. The multiloculate ascostroma of *Myriangium duriaei* arises in the upper part of the vegetative stroma (Fig. 7). The vegetative stroma is composed of reticulately branched hyphae with short cells embedded in a gelatinous matrix. In the cup-shaped ascostroma the cells are densely aggregated, and a basal zone of plasma-rich, meristematic cells limits the cup at the base (Fig. 8). The asci are produced individually in a basipetetal direction (Fig. 9). The cells between the developing asci transform gradually into a pseudoparenchyma with isodiametric cells (Figs 10-11). This tissue weathers away and the asci become exposed. Ascogenous hyphae may be observed invading from the vegetative tissue into the ascostroma (Fig. 10).

ASCOCARP DEVELOPMENT IN THE "ZWISCHENGRUPPE"

Ascocarp spot-like or lirelliform. The first stage observed in *Arthothelium spectabile* is the formation of the hyphal web of generative tissue. The hyphae are plasma-rich and stain darker in LB (Fig. 12). In the mature ascocarp the paraphysoids are densely branched and the asci arise singly from groups of aggregated ascogenous cells (Fig. 13). In *Arthonia radiata* the ascogenous hyphae grow horizontally (Fig. 14). As in *Arthothelium*, the sites of the emptied asci remain visible as cavities in the hymenium,

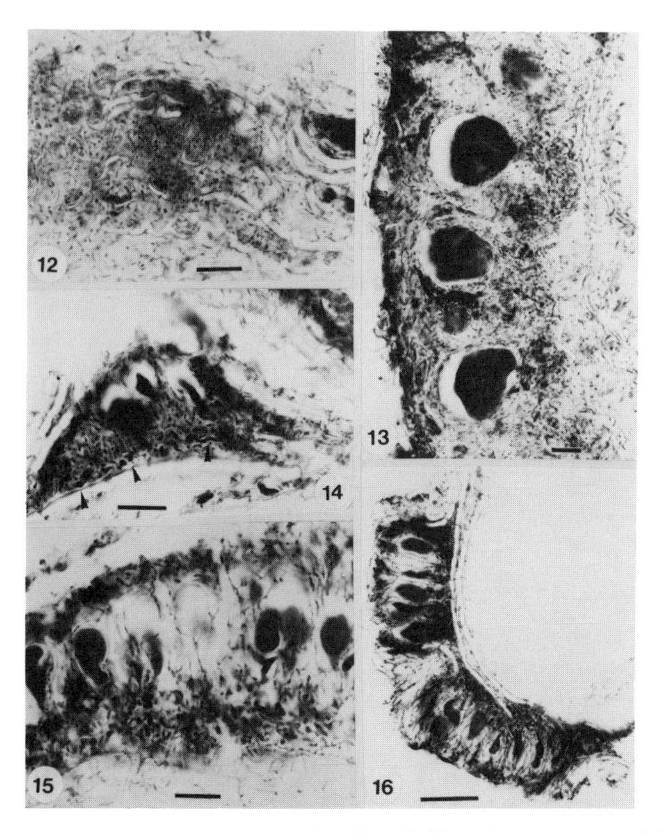

Figures 12-16. Ascocarp development in *Arthoniales*. **Figs 12-13.** *Arthothelium spectabile*. **Fig. 12.** Generative tissue with plasma-rich hyphae arising in the vegetative thallus. **Fig. 13.** Mature ascocarp; asci arising singly from groups of ascogenous hyphae. **Figs 14-16.** *Arthonia radiata*. **Fig. 14.** Edge of ascocarp; horizontal growth of ascogenous hyphae (arrows). **Fig. 15.** Hymenium with empty ascus cavities. **Fig. 16.** Section through branched ascocarp with a sterile centre. Scale 12-15 = 20 μm; 16 = 50 μm.

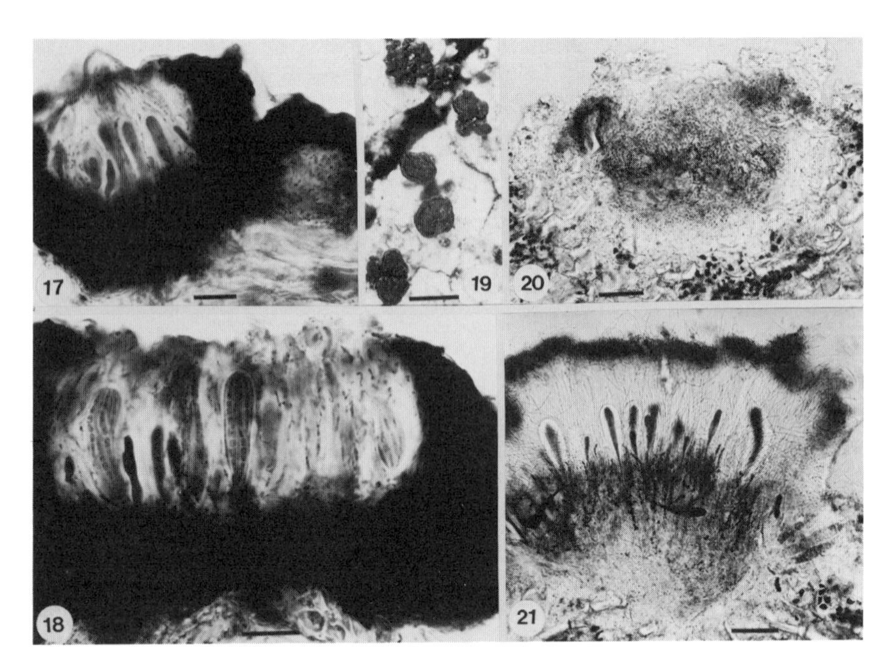

Figures 17-21. Ascocarp development in *Arthoniales*. **Figs 17-18**. *Opegrapha vulgata*, transverse sections of lirellae. **Fig. 17**. Initial stage with generative tissue adjacent to young fruit-body; in the latter the hymenium surface is covered by mucilage. **Fig. 18**. Mature ascocarp with thick, black, cupular exciple. **Figs 18-21**. *Melampylidium metabolum*. **Fig. 19**. Habit photograph of rounded to irregular apothecia. **Figs 20-21**. Young and mature apothecia. Scale 17 and 20-21 = 50 μm; 18 = 20 μm; 19 = 1 mm.

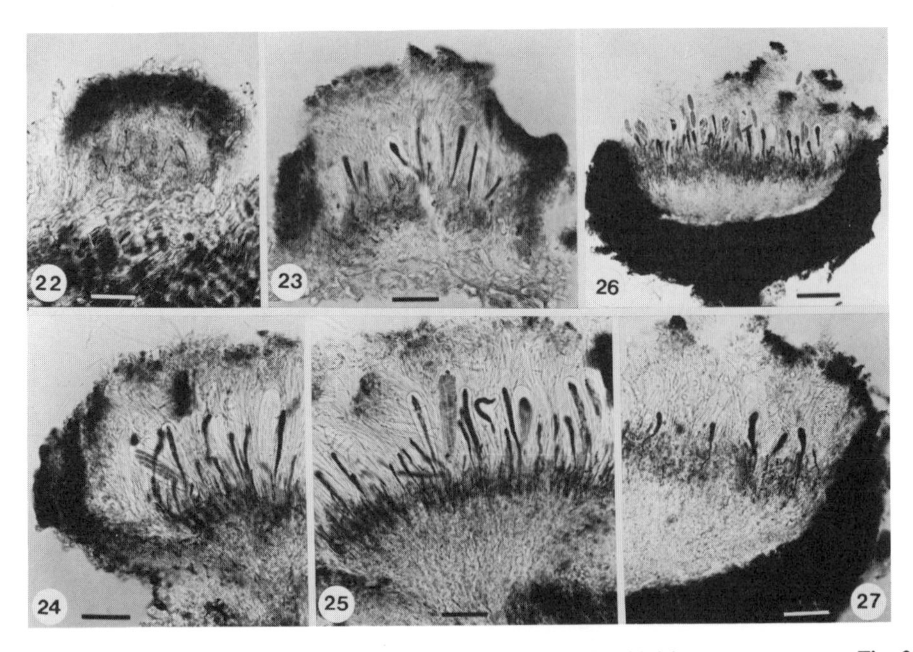

Figures 22-27. Apothecial development in *Melampylidium* species. **Figs 22-24**. *M. macrosporum*. **Fig. 22**. Primordium with young ascus. **Fig. 23**. Young apothecium with annular exciple. **Fig. 24**. Part of mature apothecium. **Figs 25-27**. *M. redonii*. **Fig. 26**. Mature apothecium with thick, black, cupular exciple. **Figs 25-27**. Central and marginal part of mature apothecia; hymenium with oil drops. Scale 22-25 and 27 = 50 μm; 26 = 100 μm.

and the tips of the branched paraphysoids form an epithecium (Fig. 15). The exciple is rudimentary, comprising only a few pigmented hyphae (Fig. 16). In *Opegrapha vulgata* a thick, black, cupular exciple surrounds the hymenium (Fig. 18). In the primordium, the dark pigmentation is restricted to a layer covering the generative tissue. The covering layer bursts when the fruit-body enlarges, and the hymenium surface is slimy (Fig. 17).

Development and structure of the apothecia in Melampylidium. In *Melampylidium* the lecideine apothecia are rounded or irregular in outline, and have an uneven to gyrate disc (Fig. 19). The primordia arise in the centre of the thallus, the pigmentation of the margin and upper surface starting when the ascogenous hyphae differentiate (Figs 20, 22). The hymenium is very delicate, strongly gelatinous, and ruptures easily when sectioned (Figs 21, 23-27). The thin paraphysoids are sparsely branched, except at the ± strongly branched tips in the pigmented epithecium. The exciple is poorly to well-developed and annular in *M. metabolum* and *M. macrosporum* (Figs 21, 23), and thick and cupular in *M. redonii* (Figs 26, 27). The asci are long-stalked, and the exotunica is thick-walled (Figs 28-33). In iodine, the tip of the endotunica is provided with a broad, reddish ring-like structure, which is also evident after dehiscence (Figs 28-29, 31-32); a rose-colour limits the whole length of the endotunica. The ascospores are long and muriform in the type species, *M. metabulum*; long and pluriseptate with occasionally one longitudinal septum in *M. macrosporum*; and short and muriform in *M. redonii*; in *M. metabolum* and *M. macrosporum*, they are constricted at several septa, fragmenting into segments (Figs 28-33).

Figures 28-33. Asci and spores in *Melampylidium* species (in iodine). **Figs 28-30**. *M. metabolum*. **Fig. 28**. Tips of asci with amyloid, ring-like structures. **Figs 29-30**. Asci with a thick exotunica and expanding endotunica; note remnant of the ejaculated, amyloid ring-like structure (arrow). **Fig. 31**. Long-stalked ascus with multi-celled, muriform ascospores in *M. redonii*. **Figs 32-33**. Asci and long, pluriseptate ascospores in *M. macrosporum*; ascus in Fig. 33 undergoing dehiscence. Scale 28 and 30-33 = 20 μm; 29 = 50 μm.

Melampylidium cerei has quite different ascocarps and should be excluded from the genus (see above). The apothecia are lecanorine (Fig. 37). The thalline exciple has a peculiar structure; *Trentepohlia* filaments are restricted to the inner and basal parts, while the upper part is filled with deposits of secondary metabolites (Figs 34-36). The submuriform ascospores are also of a deviating type; fusiform, and provided with a gelatinous sheath. The paraphysoids have pigmented tips (Figs 35-36) as in many other genera of *Arthoniales*, for instance *Roccellina* (Fig. 66). The grouping of the ascogenous hyphae in young ascocarps (Fig. 34) corresponds to the general development pattern in the "Zwischengruppe".

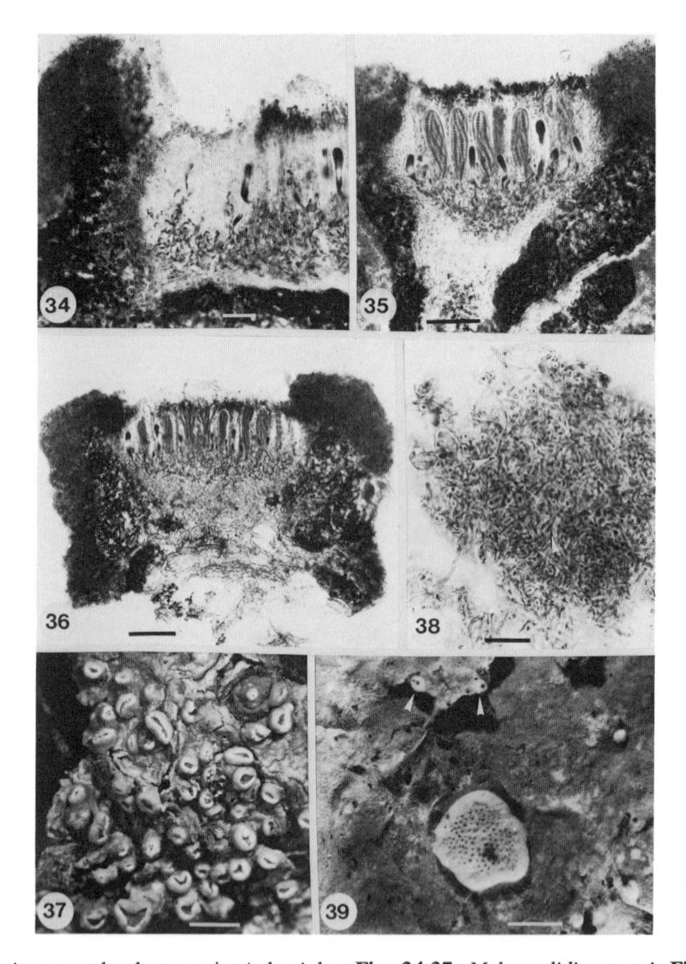

Figures 34-39. Ascocarp development in *Arthoniales*. **Figs 34-37**. *Melampylidium cerei*. **Fig. 34**. Young apothecium; ascogenous hyphae and asci aggregated in groups. **Figs 35-36**. Mature apothecia; tips of paraphysoids dark pigmented. **Fig. 37**. Habit photograph; note the thick, apothecial thalline exciple. **Figs 38-39**. *Dichosporidium latisporum*. **Fig. 38**. Globose generative tissue with differentiating ascogonia (arrows). **Fig. 39**. Thallus with a large stromatic ascocarp and pycnidia (arrows). Scale 34-35 = 50 μm; 36 = 100 μm; 37 and 39 = 1 mm; 38 = 20 μm.

The developmental morphology of stromatic ascocarps has been studied in *Dichosporidium* and *Erythrodecton* genera recently described by Thor (1990). In *Dichosporidium* the thallus is byssoid, and the ascospores are hooked or biclavate. The pycnidia and ascocarps are both stromatic. The pycnidia are enclosed in small, wart-like outgrowths of the thallus (Fig. 39). The stromatic ascocarps are large, flat, slightly or distinctly elevated structures, and the surface is covered by dark dots, the small discs of the pseudoangiocarpic apothecia (Fig. 39). The generative tissue forms a web of thin, plasma-rich hyphae, in which ascogonia and trichogynes are not easily discerned (Fig. 38). The apothecia arise individually, and diverge, or are paired in the stroma (Figs 40-42). In the young primordia, the paraphysoids are richly branched; ascogenous hyphae accumulate at the base, but also run through the whole globose structure (Figs 40-41). The mature apothecia are surrounded by a 10-20 μm thick, black exciple (Figs 42-43). In the stromata of *D. microsporum*, the development of individual, pseudoangiocarpic apothecia corresponds to that of *D. latisporum*. In addition, however, a black layer is formed underneath the apothecia that extends to the base of the thallus (Figs 44-45).

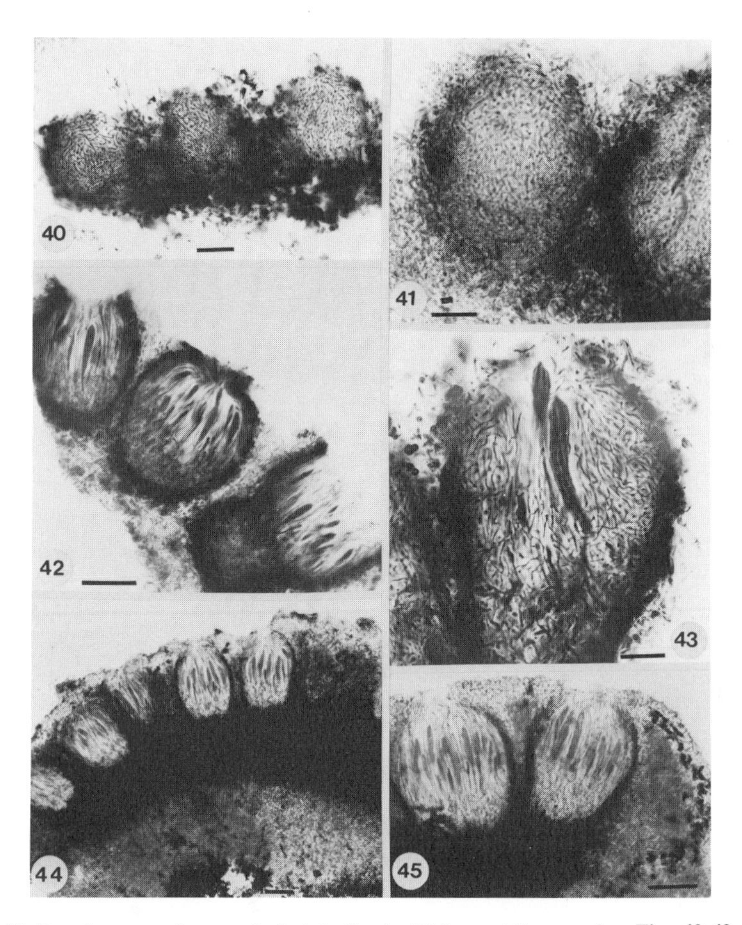

Figures 40-45. Development of stromatic fruit-bodies in *Dichosporidium* species. **Figs 40-43**. *D. latisporum*. **Figs 40-41**. Primordia developing individually in the byssoid thallus; paraphysoids richly branched. **Figs 42-43**. Mature, pseudoangiocarpic apothecia. **Figs 44-45**. *D. microsporum*, pseudoangiocarpic apothecia united by a thick, black, basal layer. Scale 40, 42 and 44-45 = 50 μm; 41 and 43 = 20 μm.

In *Erythrodecton* two different developmental types of stromatic ascocarps were recognized. In Type I, represented by *E. malacum* and *E. granulatum*, the ontogeny corresponds to that in *Dichosporidium*. Numerous pseudoangiocarpic apothecia develop individually within a stroma that is slightly elevated around each of the small discs (Figs 46-47). The generative tissue differentiates in the thallus centre (Figs 48-49). The ascogonia are thin and straight to slightly bent (Fig. 48). In *E. malacum*, a thick, black, basal layer is observed underneath the apothecia (Fig. 50), as in *Dichosporidium microsporum*, while in *E. granulatum* such a layer is lacking, as in *D. latisporum* (Fig. 51).

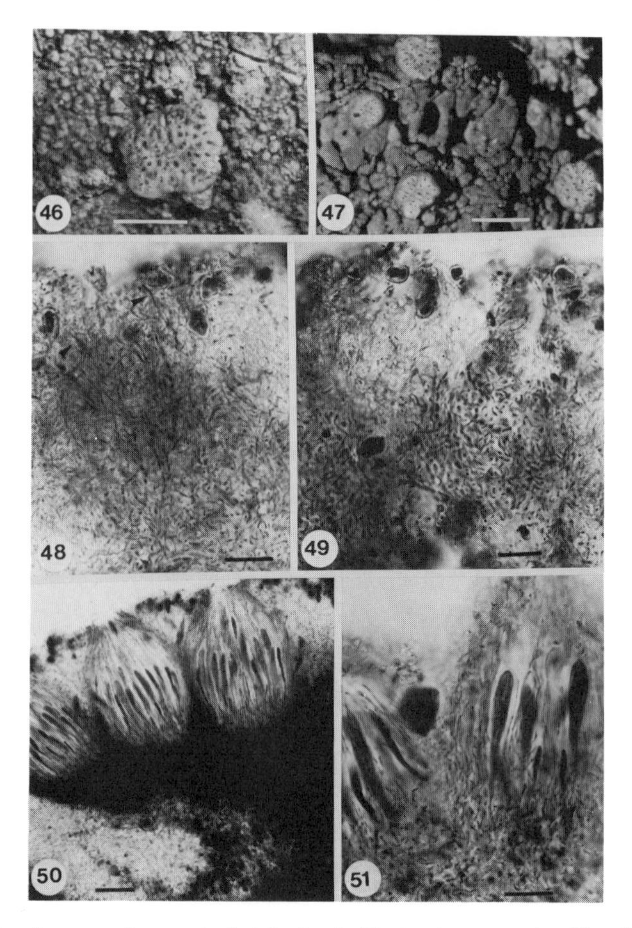

Figures 46-51. Development of stromatic fruit-bodies in *Erythrodecton* species. **Fig. 46**. *E. granulatum*, thallus with large stroma and soralia. **Figs 47-50**. *E. malacum*. **Fig. 47**. Thallus with stromata. **Fig. 48**. Generative tissue with ascogonia and trichogynes (arrows). **Fig. 49**. Generative tissue with developing ascogenous hyphae. **Fig. 50**. Part of the stroma, pseudoangiocarpic apothecia on a black basal layer. **Fig. 51**. *E. granulatum*, mature pseudoangiocarpic apothecia. Scale 46-47 = 1 mm; 48-49 and 51 = 20 μm; 50 = 50 μm.

In Type II, represented by *E. kurzii* and an undescribed *Erythrodecton* species (*Henssen* 26840b), a single large apothecium only develops in the stroma. In *E. kurzii*, the generative tissue is situated in the top of a small thallus outgrowth (Fig. 52) that enlarges considerably around the developing apothecium (Fig. 53). A great quantity of secondary metabolites is deposited underneath and between the *Trentepohlia* filaments. The young apothecium is surrounded by a black exciple of unequal thickness (Fig. 53). During the subsequent vertical growth of the stroma, a black subhymenial stalk is gradually formed (Figs 54-55). In mature apothecia, the tips of the paraphysoids are densely branched, forming an epithecium, in which secondary metabolites are also deposited (Figs 55-56). *E.* sp. differs from *E. kurzii* in the superimposed, stromatic apothecia (Figs 57-58). In the hymenium, secondary metabolites are deposited predominantly in the lower part (Fig. 62). The superimposed apothecia develop either by a partial, continuous vertical growth of the preformed apothecium (Fig. 60), or from generative tissue newly differentiated above those parts of old apothecia in which growth has ceased (Figs 59-61). In section the development of the stromatic fruit-bodies resembles that in lecanorine apothecia with an extensive vertical growth. The stromatic appearance in surface view is caused by the surface pattern (Fig. 58): the whitish spots are secondary metabolites, and

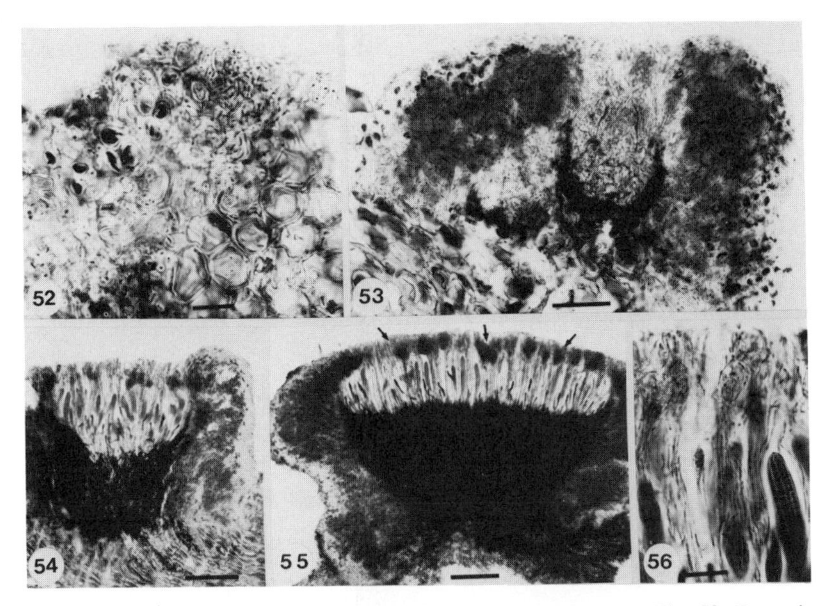

Figures 52-56. Development of stromatic apothecium in *Erythrodecton kurzii*. **Fig. 52**. Formation of generative tissue in a thallus outgrowth. **Fig. 53**. Young apothecium in stroma that is filled with deposited secondary metabolites. **Figs 54-55**. Small and large, adnate apothecia; in the hymenium, deposited secondary metabolites are indicated by arrows. **Fig. 56**. Upper part of hymenium. Scale 52 and 56 = 20 μm; 54-55 = 100 μm.

the dark stripes the parts of the hymenium lacking such compounds, and through which the underlying, black, subhymenial layer is apparent.

E. kurzii was included in *Erythrodecton* on the basis of the corresponding thallus structure and the presence of rhodocladonic acid (Thor, 1992). The ascospores are, however, fusiform (Fig. 56) and not of clavate as in the other species. The correlating fundamental deviation in the ascocarp ontogeny may necessitate the future exclusion of both *E. kurzii* and the new species from the genus *Erythrodecton*.

Ascocarp development in Roccellina cerebriformis. According to Tehler (1983), the ascocarps in *R. cerebriformis* are stromatoid or apothecioid. In sections of stromata with lirelliform discs (Fig. 63) we observed a great number of initial stages, including generative tissue with ascogonia and trichogynes (Fig. 64) as well as young primordia (Fig. 65). In the mature apothecium, the hymenium is divided by dark-pigmented, sterile strands (Fig. 66).

In his terminology of ascomatal types, Tehler (1990: 2466) described a fruit-body as it occurs in *R. cerebriformis* as "Multiascal locules, pluricarpocentral, discothecium, solitary". Apothecia with a divided hymenium are, however, rather common in lichenized ascomycetes with ascohymenial development. Cavities in the hymenium left after the dehiscence of asci are not restricted to the *Arthoniales* but occur also, for instance, in *Pertusariales*, in which the paraphysoids are richly branched and intertwined. We prefer to use the term "locules" only for *Ascoloculares* s.str.

CONCLUSIONS

The studies on the developmental morphology that are presented here, confirm the previous results obtained for the "Zwischengruppe" (i.e. *Arthoniales* s.lat.; Henssen and Jahns, 1973). We consider the *Arthoniales* s.lat., inclusive of *Arthothelium*, to be a well defined systematic group that is not related to the *Dothideales* s.lat. or *Myriangiales*, orders of the *Ascoloculares* s.str., contrary of the view of Tehler (1990). The "Zwischengruppe" differs from the *Ascoloculares* s.str. in: (1) the type of the

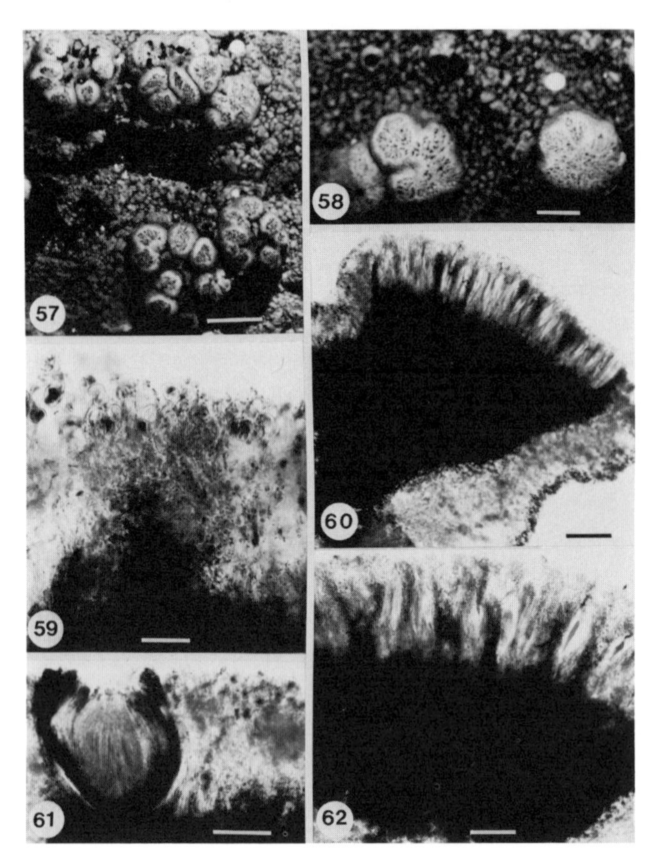

Figures 57-62. Development of superimposed, stromatic apothecia in *Erythrodecton* sp. **Figs 57-58**. Habit photographs of superimposed, stromatic apothecia. **Fig. 59**. Initial stage of generative tissue developing above a black basal layer. **Fig. 60**. Superimposed, mature apothecium. **Fig. 61**. Small apothecium and two initial stages in a thallus above the black basal layer. **Fig. 62**. Hymenium with a sterile hyphal strand and deposited secondary metabolites. Scale 57 = 1 mm; 58 = 0.5 mm; 59 = 20 μm; 60 = 100 μm; 61-62 = 50 μm.

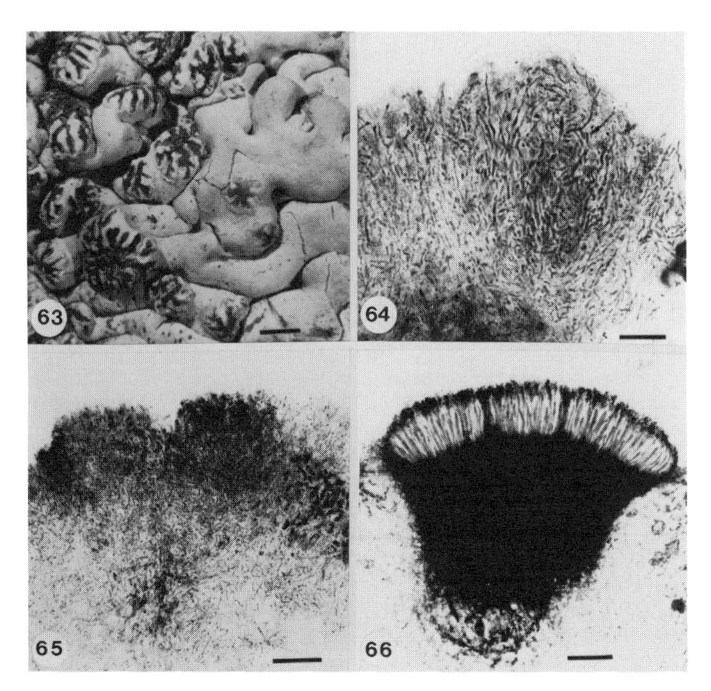

Figures 63-66. Ascocarp development in *Roccellina cerebriformis*. **Fig. 63**. Stromata with star-like discs. **Fig. 64**. Generative tissue with ascogonia and trichogynes. **Fig. 65**. Longitudinal section of star-like disc and two adjacent primordia; algal filaments (arrowed). **Fig. 66**. Mature apothecium; hymenium divided by sterile hyphal strands; tips of paraphysoids dark-pigmented. Scale 63 = 2 mm, 64 = 20 μm; 65 = 50 μm; 66 = 100 μm.

fissitunicate (bitunicate) asci; and (2) the ascohymenial development from a generative tissue, in contrast to a differentiation of an ascostromatic centre according to the disintegration or aggregation pattern.

ACKNOWLEDGEMENTS

Financial support was received from the Deutsche Forschungsgemeinschaft by A.H. We thank the curators of collections cited for the loan of material in their care, Dr G. Vobis for Fig. 64, Dr H. Czeika-Shiman and Dr K. Kalb for help with literature, and Mrs H. Brandtner for assistance in the photographic work.

REFERENCES

Corlett, M., 1973, Observations and comments on the *Pleospora* centrum type, *Nova Hedwigia* 24: 347-366.

Egea, J.M. and P. Torrente, 1993, The lichen genus *Bactrospora*, *Lichenologist* 25: 211-255.

Eriksson, O. [E.], 1981, The families of bitunicate ascomycetes. *Opera Botanica* 60: 1-220.

Eriksson, O.E. and D.L. Hawksworth, 1988, Outline of the ascomycetes - 1988, *Systema Ascomycetum* 7: 119-315.

Harris, R.C., 1990, *Some Florida Lichens*. R.C. Harris, New York.

Henssen, A., 1976, Studies in the developmental morphology of lichenized ascomycetes, *In*: *Lichenology: Progress and Problems* (D.H. Brown, D.L. Hawksworth and R.H. Bailey, eds): 107-138. Academic Press, London.

Henssen, A., 1987, *Lichenothelia*, a genus of microfungi on rocks, *Bibliotheca Lichenologica* 25: 257-293.

Henssen, A., 1994, Studies on the biology and structure of *Dacampia* (*Dothideales*), a genus with lichenized and lichenicolous species, *Cryptogamic Botany*: in press.

Henssen, A. and H.M. Jahns, 1973 ["1974"] *Lichenes, eine Einführung in die Flechtenkunde*, Thieme Verlag, Stuttgart.

Henssen, A. (in cooperation with G. Keuck, B. Renner and G. Vobis), 1981, The lecanoralean centrum, *In*: *Ascomycetes Systematics: The Luttrellian Concept* (D.R. Reynolds, ed.): 138-234, Springer Verlag, New York.

Luttrell, E.S., 1951, Taxonomy of the pyrenomycetes, *University of Missouri Studies* 24: 1-120.

Luttrell, E.S., 1955, The ascostromatic ascomycetes, *Mycologia* 47: 511-531.

Luttrell, E.S., 1965, Paraphysoids, pseudoparaphyses and apical paraphyses, *Transactions of the British Mycological Society* 48: 135-144.

Luttrell, E.S., 1973, Loculoascomycetes, *In*: *The Fungi*, Vol. IV A (G.C. Ainsworth, G.C. Sparrow and A.S. Sussman, eds): 135-219. Academic Press, New York, San Francisco, London.

Luttrell, E.S., 1981, The pyrenomycete centrum-*Loculoascomycetes*, *In*: *Ascomycete Systematics: The Luttrellian Concept* (D.R. Reynolds, ed.): 124-137. Springer Verlag, New York.

Nannfeldt, J.A., 1932, Studien über die Morphologie und Systematik der nicht-lichenisierten inoperculaten Discomyceten, *Nova Acta Regiae Societas Scientarum Upsaliensis*, ser. 4, 8(2): 1-368.

Parguey-Leduc, A., 1966, Recherches sur l'ontogénie et l'anatomie comparée des ascocarpes des pyrénomycètes ascoloculaires, *Annales des Sciences Naturelle, Botanique*, sér. 12, 7: 505-690.

Redon, J. and G. Follmann, 1972, Beobachtungen zur Verbreitung chilenischer Flechten VI, Revision einiger Arten der Krustenflechtenfamilie *Lecanactidaceae*, *Philippia* 1/4: 186-193.

Tehler, A., 1990, A new approach to the phylogeny of *Euascomycetes* with a cladistic outline of *Arthoniales* focussing on *Roccellaceae*, *Canadian Journal of Botany* 68: 2458-2492.

Thor, G., 1990, The lichen genus *Chiodecton* and five allied genera, *Opera Botanica* 103: 1-92.

Thor, G., 1992, *Erythrodecton kurzii*, comb. nov., *Nordic Journal of Botany* 12: 733-735.

THALLIC MYCELIAL AND CYTOLOGICAL CHARACTERS IN ASCOMYCETE SYSTEMATICS

H.M. Jahns and S. Ott

Botanisches Institut
Universitätsstrasse 1
40225 Düsseldorf, Germany

SUMMARY

The lichen system is primarily based on characteristics of the ascomata and especially of the ascus, while structures of the vegetative thallus are of minor importance. In comparison to other cryptogams, the degree of differentiation in lichen thalli is high while the number of specialized cells and tissues is low. Differentiation is often achieved by multifunctionality and low determination of structures. This variability of characteristics reduces their value for systematic purposes. Successful use of vegetative characteristics depends primarily on their development and not on their mature shape.

INTRODUCTION

In this paper we discuss some aspects of the lichen thallus which can be of importance for the evaluation of vegetative characteristics in ascomycete systematics. We will not present many new results or observations, but try to explain some principles which may perhaps contribute to the understanding of the unique nature of the lichen thallus. After all, the real problem of systematic work is not the observation and registration of characteristics, but rather their evaluation and interpretation. Especially in lichens, the insight into the systematic significance of structures is made difficult by contradictory aspects of their life and development which we will try to explain. So perhaps a preliminary discussion of the facts, which in themselves are well-known, or even trivial, may be necessary.

RELIABILITY OF CHARACTERS

Systematics, if understood as a mirror of evolution, is faced with a principal problem: we were not there to watch it happening. Therefore, we must try to reconstruct the course of evolution from characteristics of the present day organisms, and every system depends on the comparative evaluation of characteristics. This means that a certain subjective component is inevitable. This is also true for the cladistic approach and for systematics based on molecular data. Both methods are attractive for natural scientists, probably because the final results are reached by mathematical methods. Numbers are always impressive and there is a tendency to forget that the results in these fields also are based on a subjective evaluation of the importance of characteristics. The subjective element is better hidden but, nevertheless, present. Cladistic and molecular approaches are an important step forward. The cladistic method has forced on us a more logical way of thinking, and molecular data have presented us with a completely new set of exact comparative characteristics. But subjective evaluation is still inevitable.

For this necessary evaluation, biologists always try to find characteristics with a high degree of constancy and low variability. Especially for the definition of higher taxa, conservative characteristics are needed which show a low mutation rate. In vascular plants the classic example are the structures of the flower. As functional sexual reproduction is a fundamental prerequisite for survival, it is logical that many of these essential structures are coded by several genes. They do not mutate easily and are good systematic characters. In the ascomycetes the same principle has been accepted and the definition of higher taxa is mostly based on the structures of the ascomata, and especially on the ascus. Here conservative characteristics can be expected to exist, as any change in the apical structures which prevents the distribution of ascospores would eliminate itself. But in this context also we should be careful. In vascular plants it has recently been shown by molecular genetic experiments that some aspects of the perianth are conservative while others change easily (Coen and Meyerowitz, 1991). It can be concluded that the latter characteristics are not of vital importance to survival and, therefore, among the reproductive structures we can expect to find a mixture of conservative and variable characteristics. This could mean that in the asci of the ascomycetes, changes in the apical structure are insignificant in so far as they do not prevent the distribution of ascospores. It is not impossible that some of the observed apical structures are conservative while others are variable and of small consequence for the definition of higher taxa. This does not question the primary importance of ascus structure for the systematics of ascomycetes, but the principal importance of a set of characteristics does not imply that all these characteristics are of equal importance and validity. Interpretation and evaluation - the subjective components - are again inevitable.

These problems are well-known, and as a result the vegetative characteristics, which can be expected to show a high degree of variability, are considered to be of lesser value than the generative structures. Nevertheless, in the vascular plants vegetative characteristics such as the anatomy of the cormus are generally used for systematics. The importance of these vegetative characteristics in vascular plants is based on the fact that, in connection with the adaptation to terrestrial life, a significant number of different cell types and tissues with special functions have been developed. The systematic importance of these structures is high as in the ontogeny of vascular plants many types of cells and tissues are finally determined and, therefore, show a low degree of inconvenient variation.

In the ascomycetes a very different picture emerges when vegetative characteristics are considered. The cytological characteristics such as the type and number of pores between cells of the hyphae will not be discussed here. These are certain to be of importance (see Kimbrough, this volume) and are more reliable than the characteristics which will be treated below. It is difficult to estimate the variability of these cytological characteristics and further investigations of this aspect will be necessary.

Many ascomycetes consist only of a diffuse mycelium and ascomata and, therefore, have hardly any vegetative characteristics. A slightly different case is taxa with sclerotia. Here the shape of the sclerotium, the development of pseudoparenchymatous or pseudosclerenchymatous tissues, and pigmentation, are interesting features and the occurrence of a sclerotium in itself is an important characteristic. But the number of these features is relatively small, probably because the functions of the structures are limited. Sclerotia mainly serve for survival under unfavourable conditions and for carrying the fruitbodies. The functional requirements are limited to a certain durability and there is no developmental pressure for higher differentiation of cells and tissues. Variability of the structures can be accepted as no unfavourable consequences for survival are probable. These are unsatisfactory conditions for the development of valid taxonomic criteria.

VEGETATIVE STRUCTURES IN LICHENS

In the large group of lichenized ascomycetes the circumstances are different. Here we have vegetative structures of a complexity comparable only to vascular plants. A comparison between the structures of vascular plants and lichens is very interesting as both had to adapt to the same habitat requirements and many similar solutions can be found. For example, lichens as photosynthetically active units developed leaf-like, foliose, structures. In both types of "leaves" the chlorophyll-containing cells are arranged in a layer at the point of optimal light intensity. Cortical structures of lichens are analogous with epidermal tissues of angiosperms, serving for protection and in connection with water relations. The gaseous exchange of lichens is regulated by different types of pores and by impregnation of the cell walls inside the medulla. Recent investigations have shown that the complexity of these internal cavities is of great importance. The capillary cavities are an exogenous space for the cells of the bionts and are used for excretion, but at the same time these spaces are often isolated from the surrounding atmosphere by the cortex, especially if this tissue is saturated with water. This is often forgotten in the interpretation of physiological activities of the thallus.

As the lichen belongs in the category of "primitive" poikilohydrous organisms, the thallus is expected to react as a uniform tissue. This is certainly not true. For example, while we know the water content of the thallus, the location of the water in the tissues and cells is mostly unknown. Further, as the investigations by Lange *et al.* (1990) have shown, the question of water distribution inside the thallus is crucial for the photosynthesis of lichens, either with green algae or cyanobacteria as photobiont. Not only the distribution of water inside the thallus, but also the diffusion of gases inside the lichen is unknown. We know hardly anything about the internal concentration of carbon dioxide, a problem which is not only influenced by structural aspects but also by the respiration of two bionts. The internal concentration of carbon dioxide is not only important for photosynthesis, but also for other processes such as ethylene production. Here the lichen exhibits a high degree of complex interactions of structures and functions which have only recently been suspected and are still uninvestigated. All these problems of adaptation have connections with differentiation and, inevitably, with systematic characteristics.

The complexity of vegetative structures can also be found in supporting tissues of shrubby, fruticose, thalli. Adaptations to upright and to pendulous growth, and supporting tissues of separate strands of hyphae, have been described. All these structures are adaptations to requirements of the habitat and are mostly analogous developments to organs of vascular plants.

Many more examples could be mentioned in relation to appendaged organs. Of special interest are the typical lichen structures serving as vegetative propagules. Poelt (1993) has recently pointed out the heterogeneity in structure and development of these organs. We will, therefore, only mention a few examples. *Lobaria pulmonaria* has corticate soredia which are pushed out from the soralium by growth processes. At the same time, elongated isidia, and lobules developing from isidia, can be observed. Another type of lobule is formed by limited growth of the thallus margin. The terms "isidium", "soredium", and "lobule" are used here in a broad sense and are purely descriptive; we do not want do discuss definitions in this context. In other lichens secondary thalli develop from isidia still attached to a lichen, or small secondary thalli with rhizines are detached from the thallus and function in distribution. The list of specialized structures can be lengthened at will, and we should perhaps show some restraint with the naming of these regenerative organs otherwise we will become lost in a flood of already swelling special terms.

Complexity of structures can also be described in connection with longevity, with adaptation to extreme habitats and with interactions between two or more bionts. But a most important point has become clear: lichens as photosynthetic units have adapted to ecological requirements with a number of vegetative structures. The complexity of these structures is surpassed only by cormophytes. This means that lichen systematics has a large number of vegetative characteristics at its disposal and only the systematic value of these structures must be decided.

MULTIFUNCTIONALITY AND PLASTICITY

While specialized structures are common in lichens, another aspect of specialization is completely lacking. Zimmermann (1969), in a now somewhat outdated developmental tree of the plant kingdom, shows that the number of specialized cells and tissues has increased with the evolution of higher taxa. This specialization is the base for the complexity of vascular plants. The specialization of tissues and cells in fungi, including the lichens, is low by comparison. This does not mean that the shape of cells and tissues is uniform. The opposite is true. But the tissues of lichens lack the final determination which is characteristic of specialized structures of vascular plants. Lichens succeed in developing a large number of distinct vegetative structures, but have an unusually small number of specialized and determined cells and tissues. It could be viewed as a complicated building constructed from very few types of stones.

This contradiction between simple means and complicated results is solved by two methods which are very successful, but which have unpleasant consequences for the taxonomist. The first is the principle of multifunctionality, which can be explained with a few simple examples. In *Parmelia saxatilis* the marginal parts of the thallus develop elongated pseudocyphellae from which lateral isidia can emerge. The places from which isidia have become detached serve as openings for gaseous exchange. Other isidia which remain attached to the thallus lose their circular organization and become dorsiventral lobules. Finally, they can develop into lobes of a secondary thallus. Rhizines attach the thallus to the substratum and at the same time they may form a capillary space between the lower surface of the lichen and the substratum which may serve for water storage. The principle of multifunctionality is not restricted to the vegetative structures but also occurs in generative development. One well known example are pycnoascocarps, old pycnidia used for the development of ascomata.

Multifunctionality can be seen as a pragmatic, not an optimal, method of construction. Probably, the different requirements are only just met, but that is a common biological principle: in the plant kingdom, structures which serve sufficiently are usually not optimized. The search for the optimal solution seems to be typical for human kind. Multifunctionality mostly does not occur simultaneously but the structures serve different purposes during ontogeny, as isidia growing into new thalli.

A second aspect of lichen differentiation, the lack of a final determination, is still more important than the principle of multifunctionality. It is often forgotten that all lichen tissues, in the course of ontogeny, can resume growth and develop into completely new structures. As most of the hyphae are connected only by gelatinous substances they can move and rearrange themselves. For example, in the podetium of *Cladonia* the tip grows by cell division forming a dense tissue of short-celled and thin-walled hyphae. In the lower parts of the podetium, growth is by elongation and the hyphae arrange themselves in a circular layer. The cells become long and thick-walled. If this part of the podetium is damaged, these hyphae grow out and can develop into rhizines. In other lichens compact rhizines or rhizinomorphs may acquire algae and develop into layered thalli. Even a highly differentiated cortex is not finally determined, as demonstrated by *Peltigera aphthosa* where outgrowing cortical cells start the development of cephalodia. This example demonstrates the danger of describing non-vascular plants with terms originally used for vascular ones. As biologists generally start their scientific education with vascular plants, they tend to transfer not only the names of structures but also their properties to other groups. A cortex is subconsciously understood as a determined protective layer, and it is easily forgotten that in the lichens every structure and every tissue can be changed, transformed and developed during ontogeny. Anatomical descriptions which only include static observations of one, usually adult, stage of development can, therefore, lead to oversimplified or even wrong conclusions.

The principle of plasticity seems to apply to all vegetative structures and also to pycnidia. Old pycnidia of *Cetraria islandica* can develop into regenerative lobules. In *Cladonia rangiferina*, nearly all the tips of the branched thallus end in pycnidia without terminating their growth. Old pycnidia can sometimes be observed in a subapical position inside the branches before they finally disappear completely as a result of the continued growth processes. At the tips of the branches new pycnidia are formed, and in this way submatial conidia are always present for fertilization in case short-lived trichogynes are developed. Usually, only in mature ascomata does development of the tissues seems to be terminated, but even here outgrowing apothecia can be found.

As with the principle of multifunctionality, the principle of plasticity is also connected with ontogeny and development. The change in the differentiation of cells and tissues occurs during the life of the organism.

SYSTEMATIC IMPLICATIONS

What significance do these principles have for systematic research? In lichens there are an unusually large number of vegetative characteristics which are nearly absent in non-lichenized ascomycetes. While it would be folly not to use them, plasticity can be the cause of grave errors if a static observation and interpretation is used. The small number of basic elements used for the construction of numerous organs and tissues inevitably results in a large number of analogous developments. Also, the multifunctionality of structures may in the course of ontogeny result in a pronounced change in the phenotype. Both aspects complicate the use of vegetative characteristics in systematics. Moreover, as the plasticity is the answer of the organism to the requirements of the habitat it can be expected that the situation is further complicated by the occurrence of numerous environmental modifications.

The solution to these problems has already been presented by Poelt (1993); in connection with vegetative characteristics, the ontogenetic aspect must always be included. It is not sufficient to describe adult organs. Their development must be investigated to check the reliability of the characteristics. If this is done, vegetative characteristics can be of importance to lichen systematics, especially for the definition of species where ascomata may show only small or no variability. For the definition of orders, families and genera vegetative characteristics should only be used with great care. Vegetative characteristics alone should never be sufficient for the delimination of suprageneric taxa, and even on the level of genera great care is necessary.

These objections to the use of vegetative characteristics of lichens, which can be seen as adaptations to life as photosynthetically active "land-plants", does not apply to other groups of characteristics. Cytology, and lichen chemistry in the broadest sense, are probably more reliable. We have the paradoxical situation that those characteristics which are most typical for the lichens as a group are the least reliable in macrosystematics inside the group. This does not mean that vegetative characteristics should be ignored, but no general statement on the value of a single characteristic is possible. Instead, in every case, the value of a characteristic must be examined for the specific systematic problem being investigated.

CONCLUSIONS

In conclusion, the following major points have been identified in regard to the use of vegetative characters in the systematics of lichenized fungi:

(1) In addition to the ascomata, cytological and chemical characteristics are of especial importance.
(2) Vegetative characteristics of the thallus are difficult to use as:
  (a) The high complexity of the thallus is achieved by a minimum of basal structures.
  (b) The adaptation to numerous requirements of the habitat is facilitated by the multifunctionality of structures.
  (c) Most structures are not finally determined but show a high degree of plasticity.
  (d) The vegetative structures are changed by environmental modifications.
(3) Structures of the lichen thallus should only be used as systematic characteristics in connection with ontogenetic observations.
(4) Vegetative characteristics of lichens should primarily be used for the delimination of lower taxa.

REFERENCES

Coen, E.S. and E.M. Meyerowitz, 1991, The war of the whorls: genetic interactions controlling flower development, *Nature* 353: 31-37.

Lange, O.L., H. Pfanz, E. Kilian, and A. Meyer, 1990, Effect of low water potential on photosynthesis in intact lichens and their liberated algal components, *Planta* 182: 467-472.

Poelt, J., 1993, On lichenized asexual diaspores in foliose lichens - a contribution towards a more differentiated nomenclature (lichens, *Lecanorales*), *Cryptogamic Botany*: in press.

Zimmermann, W., 1969, *Geschichte der Pflanzen*. 2nd edn. Georg Thieme, Stuttgart.

DISCUSSION

Letrouit-Galinou: Why do lichens have a thallus?

Jahns: I believe that they developed that way by chance and it proved to be functional.

Letrouit-Galinou: My opinion is rather different. I believe ascomycetes are derived from algae and that the lichen thallus is a relic of the ability of algal ancestors to form a vegetative thallus.

Poelt: While I agree with many statements in this contribution, I disagree on one. The basic structure of many fruticose and especially foliose lichens is the cortex. For the most part the cortex is a dead structure which is not able to change. When different types of cells are found in the cortex, they can thus be used as taxonomic characters not only for species as you suggest but also for genera and in some cases for families. In some flowering plants cells may stop development at different stages to give different tissues, so this phenomenon is not an absolute one.

Jahns: I only wished to point out that for the higher taxa it would be advisable not to use vegetative characteristics alone, but in combination with others. With regard to cortex determination, I agree with your view that this is somewhat linked to the ascomata, so that we can expect some types of cortical cells to be as well-determined as the ascomata. It is always necessary to look for changes, however, as some can occur rather easily.

Blackwell: In relation to the question of algal ancestry for ascomycetes, there is no modern evidence for this to my knowledge but there is evidence against it. Sachs proposed that each group of fungi arose from a different one of algae. In the case of the *Oomycota* (now placed in the kingdom *Chromista*) he was a prophet, but not for other groups. However, the proposals that ascomycetes originate from red algae have been refuted by numerous molecular and cladistic studies as well as older ultrastructural investigations.

Tehler: I agree that the red algae are not related to the ascomycetes. To my knowledge all phylogenetic analyses place the red algae in a very basal position, usually as sister group to the remaining eukaryotes. However, there is evidence that the chitinous fungi (i.e. *Eumycota*) should be included in a clade together with heterokont brown algae (Tehler, *Cladistics* 4: 227, 1988).

Conidiomata and Mitosporic States

UNITARY PARAMETERS IN CONIDIOGENESIS

G.L. Hennebert[1] and B.C. Sutton[2]

[1]Mycothèque de l'Université Catholique de Louvain
Laboratoire de Mycologie Systématique et Appliquée
Université Catholique de Louvain
Faculté des Sciences Agronomiques
Place Croix du Sud 3, B-1348 Louvain-la-Neuve, Belgium

[2]International Mycological Institute
Bakeham Lane, Egham, Surrey TW20 9TY, UK

SUMMARY

The reasons for the absence of consistent correlations between systematic systems for ascomycetes and deuteromycetes are discussed, especially in relation to the inadequate descriptive terminology for conidiogenous events. Twenty unitary parameters in conidiogenesis are defined and diagrammatically illustrated. Their application to anamorphic genera is viewed as a necessary precursor to the future reassessments of correlations with ascomycetes.

INTRODUCTION

From published records compiled by Kendrick amd DiCosmo (1979) and by Hennebert and Bellemère (1979), it is known that at generic level there is no consistent correlation between the taxonomies of the ascomycetes and of the deuteromycetes. It is *a fortiori* true that at the level of families of ascomycetes, the scenario is even more heterogeneous (see Sutton and Hennebert, this volume).

This situation arises from a variety of causes. The first that might be proposed is that the families and genera of ascomycetes are polyphyletic as a result of convergent evolution. This is possible, but there are no convincing arguments that are citable at the moment. On the other hand, appearing polyphyletic may also mean that they are artificial and heterogeneous, grouping unrelated organisms.

Another relevant cause is that not all records of teleomorph-anamorph connections compiled to date have the same degree of reliability, not necessarily because they may be erroneous but because the description, classification and identification of anamorphs has been mainly by the use of descriptive terminologies, diagnostic criteria, taxonomic classifications and nomenclatures that are now considered confused or obsolete. Even if such published connections could be erased from the record, or better, revised and updated according to present day criteria, classifications, and nomenclatures, an heterogeneity of teleomorph-anamorph connections would probably still be observed in many ascomycete genera and families.

There are three reasons for this in regard to the deuteromycetes.

One is that the present descriptions of anamorphs are far from being accurate, either because not all possible characters are looked for and the descriptions of the anamorphs are therefore incomplete, or because the characters themselves are not analysed by their

Ascomycete Systematics: Problems and Perspectives in the Nineties
Edited by D.L. Hawksworth, Plenum Press, New York, 1994

elementary parameters and the terminology used is therefore multivocal and eventually equivocal.

A second reason is that the descriptions and diagnoses do not include characters of any other morphs of the fungus than that of the anamorph being described. For instance, in the description of conidial *Botrytis*, characters of the *Sclerotium* or *Myrioconium* anamorphs or of any other morphs such as appressoria are ignored. This is in full agreement with the anatomical taxonomy commended by Art. 59 of the Code. As a result, some consider it a heresy to accept the connection with a particular synanamorph and *a fortiori* with a teleomorph as a valid taxonomic argument in systematics of deuteromycetes. But Art. 59 should itself rather be viewed as the heretical element in fungal taxonomy.

A third reason, a consequence of the previous ones, is that, because of incomplete and unsatisfactory descriptions of the anamorphs, gross similarity between anamorphs is regarded as an expression of relationship, and leads to the lumping together of unrelated organisms within enlarged generic concepts that are found represented in unrelated ascomycetes.

Similar situations are to be found in ascomycete taxonomy. According to Art. 59, taxonomy of the *Ascomycotina* and *Basidiomycotina* is to be based on the teleomorph, the teleomorph providing the name for the holomorph. Consequently there is a current opinion that including anamorphic characters in the diagnoses of teleomorphs is an error. The result is that the taxonomy of the ascomycetes has sometimes been and perhaps still is rudimentary. The diagnoses of *Ascomycotina* and *Basidiomycotina* should include all characters of the fungus, characters of all distinguishable conidial anamorphs, of spermatial anamorphs, of microconidial anamorphs if distinct from spermatia, and also of any vegetative states (vegetative anamorphs).

Although generally there is no consistent correlation between the taxonomies of the ascomycetes and of the deuteromycetes, there are exceptions. In the *Sclerotiniaceae* (Sutton and Hennebert, this volume). There is parallel and interactive refinement of the taxonomies of the conidial anamorphs and the corresponding ascomycetes which has allowed some homogeneity to emerge in teleomorph-anamorph connections, presumably as an expression of monophylogeny.

Refinement in deuteromycete taxonomy will only result from a full analysis of every character by its basic unitary parameters and from the complete and consistent description of large numbers of species. Unitary parameters are the one to several basic elements which when present or acting together result in the observed character.

In describing ascomycetes and anamorphic deuteromycetes, conidiogenesis is now considered to be one of the most important sets of characters to observe after the ascoma when present. Other sets of characters, such as conidiophore and conidial morphology, genesis and morphology of the conidiomata, features of various vegetative structures, and any other associated morphs (synanamorphs) are also of value, but their analysis is beyond the scope of this discussion.

In this paper an attempt is made to redefine the unitary parameters of conidiogenesis. The term "conidium" is understood in the broad sense adopted at the First Kananaskis Conference (Kendrick, 1971) and in general use today. Although conidia are generally immediately resulting from mitosis and may be referred to as mitospores, as being one case of mitospores (and conidiogenesis one case of mitosporogenesis; Sutton, 1993), the terminology currently in use will be employed.

PARAMETERS

Parameter 1. Conidium origin and ontogeny (Fig. 1)

Formation of a conidium is in four stages; initiation, elongation, swelling, and delimitation. Stages occur in various sequences. Initiation occurs from a point of the conidiogenous cell or from an area as wide as the conidiogenous cell. Initiation and elongation include apical wall building and/or (when intercalary) ring wall building. Swelling is by diffuse wall building. Concepts of different types of wall building are from Minter et al. (1983a).

Three expressions of this parameter occur depending on the sequence of the stages. The conidium is:

(a) *blastic*, when initiation occurs from a point on the conidiogenous cell wall, followed by elongation and swelling and finally delimitation by basal septation.
(b) *thalloblastic*, when initiation and elongation occurs from an area as wide as the conidiogenous cell followed by swelling and later delimitation by basal septation.
(c) *thallic*, when initiation and elongation are from an area as wide as the conidiogenous cell, followed by delimitation by basal septation. Swelling occurs eventually (not always) after delimitation or after secession.

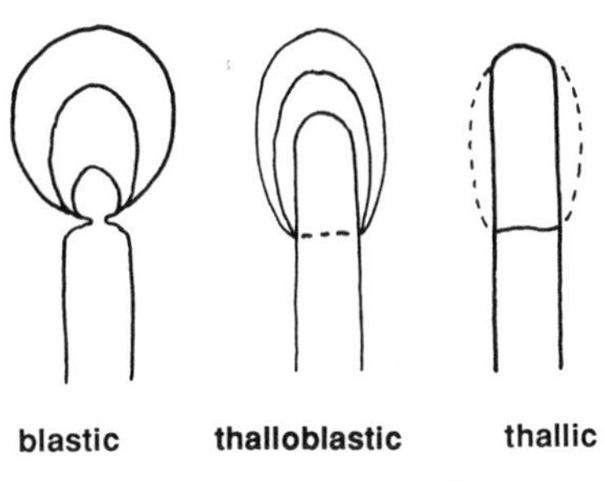

Figure 1. Conidium origin and ontogeny.

The term *thallic* is derived from "thallospore", used by Vuillemin (1910a) to designate a spore that is a portion of the thallus. It is different from the term *arthric*. The latter is derived from "arthrospore", another term from Vuillemin (1910b) designated for conidia that are individualised by disarticulation (fragmentation) of a conidiogenous hypha, as in *Geotrichum*. Arthric conidia are therefore one case of thallic conidiogenesis.

Parameter 2. Origin of conidium wall in relation to the immediate conidiogenous cell wall (Fig. 2)

Two terms are proposed here in order to distinguish the two parameters involved in the terms *holoblastic* or *holothallic* and *enteroblastic* or *enterothallic* coined by the First Kananaskis Conference (Kendrick, 1971). The conidium is:

(a) *hologenous*, when outer and inner wall layers of conidiogenous cell and conidium are in continuity.
(b) *enterogenous*, when the inner wall layer of the conidiogenous cell forms the outer wall of the conidium, the outer wall layer(s) of the conidiogenous cell having been ruptured during formation of the first or previous conidium. A new inner wall layer is deposited in the conidium and inside the septal wall layer at the apex of the conidiogenous cell.

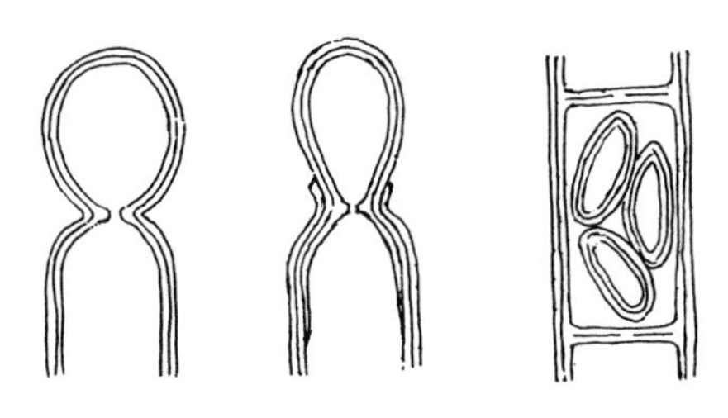

Figure 2. Origin of mitospore wall.

Another expression of this parameter is *endogenous*: when no wall layers are in continuity between the conidium and the conidiogenous cell. The distinction between endogenesis and enterogenesis is evident, and was the reason why the prefix *entero-* was proposed at the First Kananaskis Conference (Kendrick, 1971).

This parameter expression is apparently unknown in ascomycete conidiogenesis. It might be the situation for the "endospores"in *Aureobasidium* (Hermanides-Nijhof, 1984) but this needs verification, since internal conidia known in ascomycetes are formed inside dead hyphal cells from the septal pore of adjacent living cells, e.g. internal phialidic spermatia in *Botrytis*. Ascospores, including mitotic ascospores in eight or more spored asci, are typically endogenous spores. Endosporogenesis occurs in the sporangia of the *Mastigomycotina* and *Zygomycotina*.

Parameter 3. Number of loci on the conidiogenous cell (Fig. 3)

The conidiogenous locus is the point or area on the conidiogenous cell from which formation of the conidium originates. The term results from papers and discussions at the First Kananaskis Conference (Kendrick, 1971). It is not clear whether the locus is the site itself or the concentration of cytoplasmic vesicles responsible for wall building. We opt for the latter interpretation. It is therefore an active site rather than the site itself. The conidiogenous cell is:

(a) *unilocal*, bearing one locus.
(b) *multilocal*, bearing several loci.

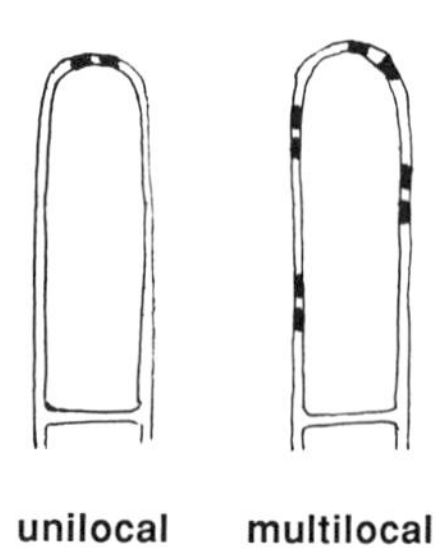

Figure 3. Number of loci.

Parameter 4. Locus position on conidiogenous cell (Fig. 4)

The locus is:

(a) *apical*, when at the apex of the conidiogenous cell. A locus that was originally apical may become subapical or lateral when a new locus or proliferation occurs at the apex of the conidiogenous cell.
(b) *lateral*, when on the side of the conidiogenous cell.
(c) *circumspersed*, when all around the conidiogenous cell, including apical, subapical and lateral positions.

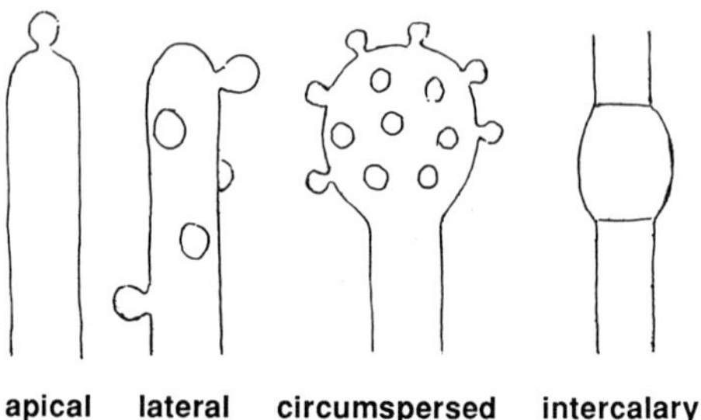

Figure 4. Locus position.

(d) *intercalary*, when in a hypha (intercalary "chlamydospores"; the term chlamydospore is not an ontogenetic term but a morphological one).

These adjectives may be applied to conidia when they are single on a locus position.

Parameter 5. Temporal order of appearance of loci on multilocal conidiogenous cell (Fig. 5)

Loci are:

(a) *simultaneous*, or synchronous, when appearing more or less at the same time.
(b) *successive*, when appearing one after another. When loci are successive, the spatial sequence of their positions must be considered.

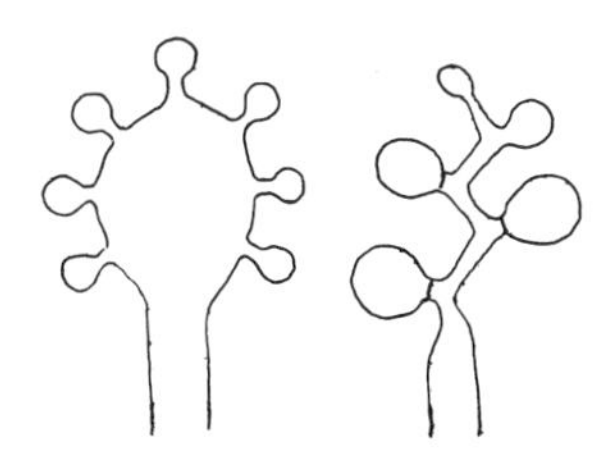

Figure 5. Temporal order of loci.

Parameter 6. Spatial sequence of successive locus positions on conidiogenous cell (Fig. 6)

Successive loci may be produced at increasingly higher or lower levels, or at the same level of the conidiogenous cell, successive locus each producing one conidium.

Loci are:

(a) *progressive*, arising successively in an apical position along the cell axis. The sequence implies obligate proliferation of the conidiogenous cell.
(b) *sympodial*, arising subapically and becoming apical. The sequence involves either a proliferation of the conidiogenous cell or just an accumulation of conidium attachments (and of scars after conidium secession) near each other.
(c) *stationary*, stable or oscillating, arising at the same level.

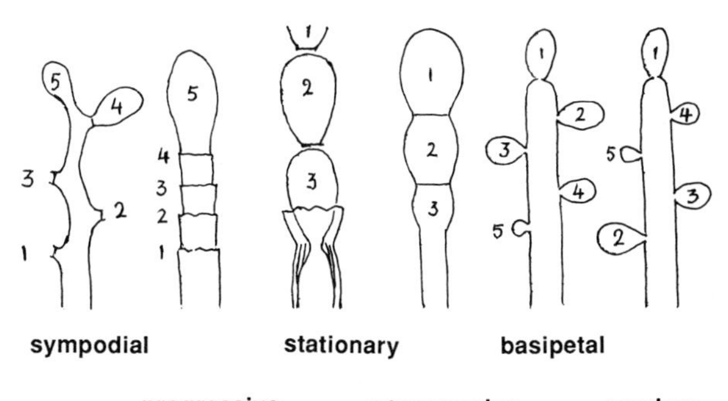

Figure 6. Spatial sequence of successive loci.

(d) *retrogressive*, arising axially towards the base. The sequence implies the shortening of the conidiogenous cell (*Trichothecium*).
(e) *basipetal*, arising laterally towards the base (*Arthrinium*).
(f) *random*, arising at any level.

The stationary successive conidiogenous loci each producing one conidium have been considered as one single locus remaining active and producing several conidia successively. These two divergent interpretations depend on the concept of the locus. The locus is the site where conidium formation commences. It might be visualised as the point or area of concentration of the microvesicles described by Bracker (1977) and presumably responsible for wall building of the new conidium. While the conidium is expanding and becoming delimited, the microvesicles are dispersed in the conidium. Once it is delimited, a new concentration of microvesicles at a new locus will appear under the delimiting septum to intitiate the next conidium. Electron microscopy of phialides has shown that the sites where conidia are initiated are not always at the same level but oscillate around it (Sutton and Sandhu, 1969).

Parameter 7. Arrangement of conidia on conidiogenous cell locus (Fig. 7)

At the position of a locus or of successive stationary loci on a conidiogenous cell or on its proliferation, one or several conidia may develop. Conidia are:

(a) *solitary*, a single one to each locus of the conidiogenous cell.
(b) *catenate*, or in true chains, i.e. chains in which the walls of all conidia are in continuity (hologenous ontogeny) with those of the previous and the following one. In this case each conidium, before or at maturity, functions as a conidiogenous cell with one apical locus or several successive sympodial loci. Catenate conidia may arise from a stationary locus forming either a basipetal hologenous chain (*Oidium*), or a basipetal enterogenous chain (*Aspergillus*), from simultaneous loci or from successive sympodial loci forming an acropetal chain (*Nematogonium* or *Cladosporium*) or from retrogressive loci forming a basipetal chain (*Basipetospora*).
(c) *seriate*, or serial (Kendrick, 1971), in false chains or spore heads in which conidia have no wall layers in common but eventually are maintained in series by the medial lamella of the delimiting septum as disjunctor. Seriate conidia are formed by successive loci moving up (*Scopulariopsis*) or down (*Basipetospora*) in successive proliferation or regression of the conidiogenous cell or by loci oscillating around a stationary position (*Acremonium*).

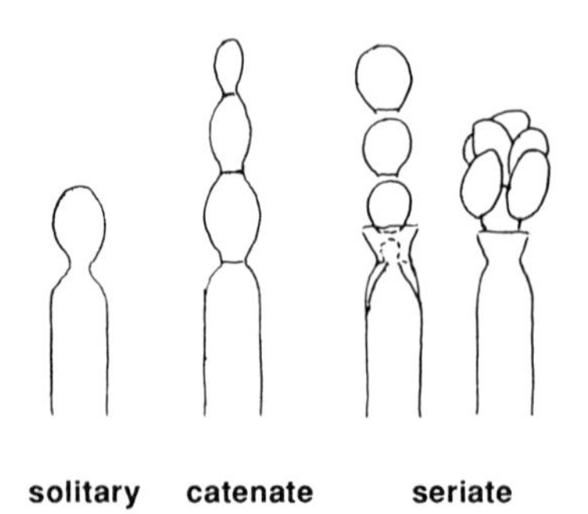

Figure 7. Arrangement of conidia.

Parameter 8. Wall origin of true chain (Fig. 8)

True chains can originate hologenously or enterogenously. They are:

(a) *hologenous*, when the first conidium wall layers are in continuity with the wall layers of the conidiogenous cell or of its proliferation. The chain can develop acropetally or basipetally (see parameter 9). Examples are *Cladosporium* and *Oidium*.

(b) *enterogenous*, when the inner wall layer only of the first conidium is in continuity with the inner wall of the conidiogenous cell or of its proliferation. Examples are *Penicillium* and *Aspergillus*.

Parameter 9. Mode of catenation, i.e. formation of true chain (Fig. 8)

True chains are:

(a) *monopodial*, acropetal unipolar (*Bispora*).
(b) *sympodial*, acropetal multipolar, branched (*Cladosporium*, *Candida*).
(c) *basipetal*, developing from the base, either hologenous (*Oidium*, *Basipetospora*), or enterogenous (*Aspergillus*).

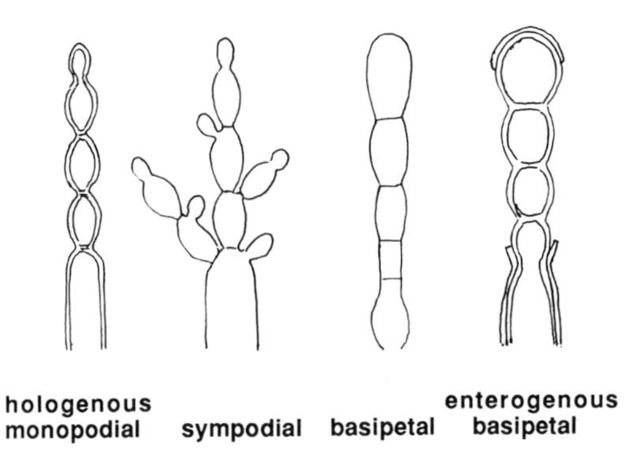

Figures 8. Wall origin of catenate conidia and mode of catenation.

Parameter 10. Growth of conidiogenous cell (Fig. 9)

Conidiogenous cells are:

(a) *determinate*, when no growth occurs during conidiogenesis.
(b) *indeterminate*, when proliferation occurs during conidiogenesis, without delimiting a new conidiogenous cell by a septum. Such conidiogenous cells proliferate either regularly between formation of each conidium or spasmodically with alternation of conidial production and vegetative growth.
(c) *resumptive*, when the proliferation becomes septate, so delimiting a new conidiogenous cell. This might be considered an indeterminate conidiogenous cell.

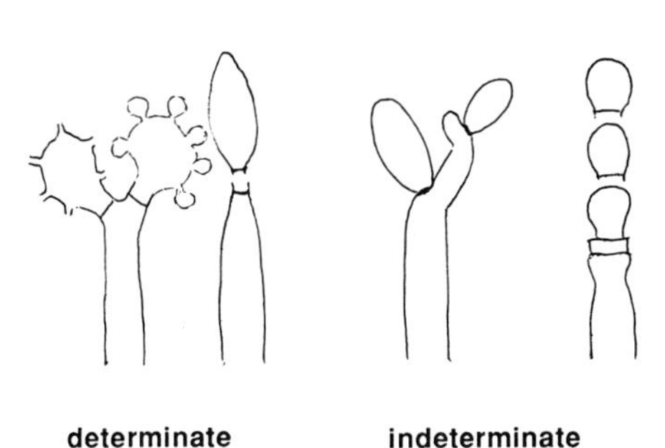

Figure 9. Growth of conidiogenous cell.

Parameter 11. Wall origin of proliferation of indeterminate or resumptive conidiogenous cell (Fig. 10)

Wall origin is interpreted in relation to the conidiogenous cell wall. Conidiogenous cell proliferation is:

(a) *hologenous*.
(b) *enterogenous*.

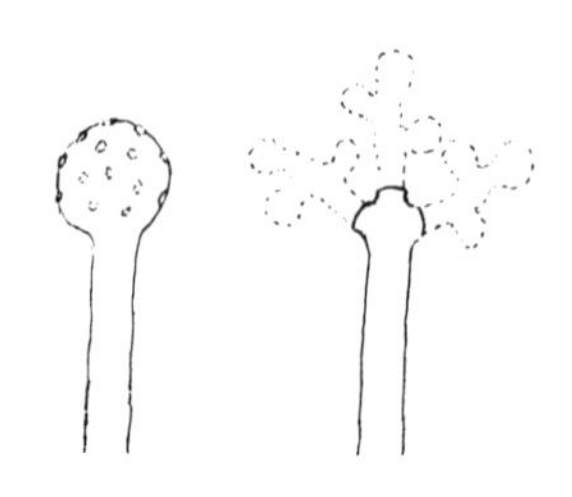

Figure 10. Evolution of determinate conidiogenous cell.

Parameter 12. Type of wall fracture giving rise to enterogenous proliferations (Fig. 11)

Proliferations are:

(a) *percurrent*, proliferating through a conidial scar which becomes open.
(b) *erumpent*, proliferating through hyphal wall fracture, either monopodially or sympodially.

Parameter 13. Mode of proliferation of indeterminate or resumptive conidiogenous cell (Fig. 11)

Conidiogenous cell proliferation is:

(a) *monopodial*, axially proliferating.
(b) *sympodial*, subterminally or laterally proliferating.
(c) *basauxic*, proliferating enterogenously from the base.

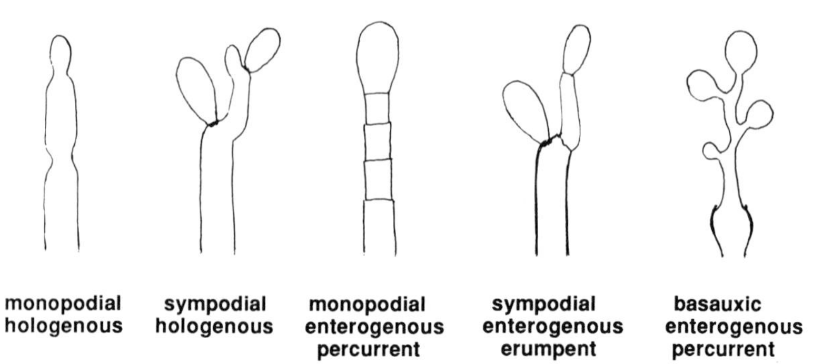

Figure 11. Mode of proliferation of indeterminate mitosporogenous cell, wall origin and type of wall fracture.

Parameter 14. Evolution of a determinate conidiogenous cell (Fig. 11)

A determinate conidiogenous cell is:

(a) *persistent*, although inactive.
(b) *collapsing*, perishing and disappearing after function.

Parameter 15. Number of basal septa delimiting conidium (Fig. 12)

(a) *one*, the septum being susceptible to cleavage at conidial secession or not.
(b) *two*, the two septa delimiting a separating cell. The separating cell generally lacks a nucleus and is therefore not viable as soon as the delimiting septa are fully formed.

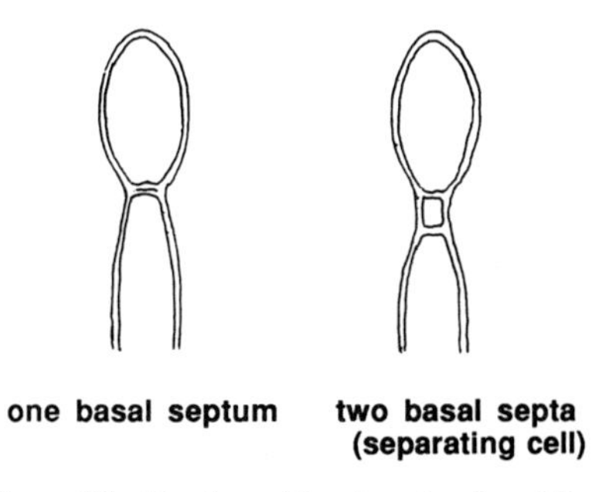

Figure 12. Number of basal septa of conidium.

Parameter 16. Secession of conidium (Fig. 13)

A conidium is:

(a) *schizolytic*, when the two layers of the basal septum disjunct.
(b) *rhexolytic*, when the denticle of the conidiogenous cell breaks beneath the the basal septum of the conidium, or the separating cell collapses or breaks apart.

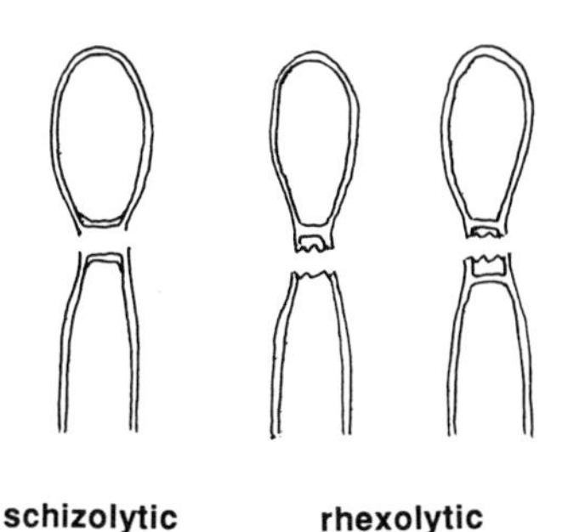

Figure 13. Conidium secession.

Parameter 17. Type of scar on the conidiogenous cell (Fig. 14)

The point or area at which a conidium was attached to a conidiogenous cell is the abscission or secession scar. It is a:

(a) *closed denticle without a frill*, sealed by the remaining layer of the delimiting septum.
(b) *closed denticle with a frill*, derived from the fractured separating cell.
(c) *open denticle with a frill*, derived from the broken wall beneath the delimiting septum.

(d) *pore*, with circumferential thickening.
(e) *collarette*, or frill derived from the outer wall layer of the conidiogenous cell breaking and detaching from the first conidium at a higher level than that of the delimiting basal septum. Length is dependent on where the break occurs. It may be associated with periclinal thickening at the apex of the conidiogenous cell or not.
(f) *annellation*, or a succession of collarettes.

Scars on conidia correspond in a, b, c and d, but not in e or f.

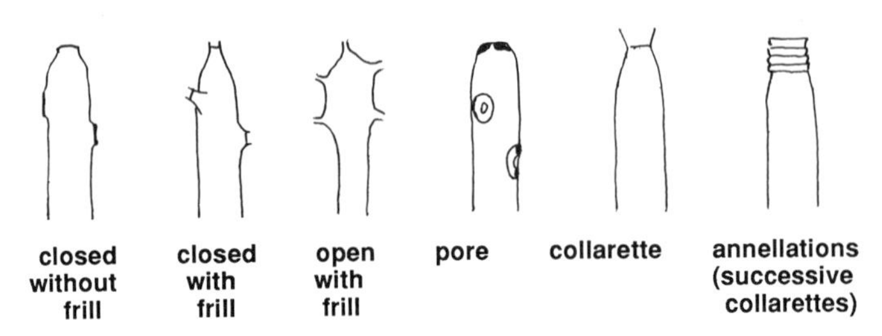

Figure 14. Scar types.

Parameter 18. Scar relief on the conidiogenous cell (Fig. 15)

The scar is:

(a) *applanate*, flat, flush with the wall of the conidiogenous cell.
(b) *protuberant*, projecting above the wall of the conidiogenous cell.

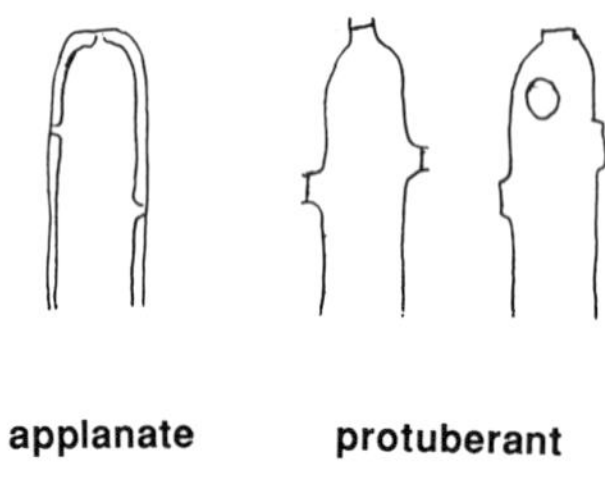

Figure 15. Scar relief.

Parameter 19. Wall thickening at scar on conidiogenous cell (Fig. 16)

The scar is:

(a) *unthickened*.
(b) *thickened, melanized*, material deposited on and adjacent to the delimiting septal layer.
(c) *periclinally thickened*, non-melanized, deposited on the inner periclinal wall layers.

Scars on conidia correspond in a and b, but not in c.

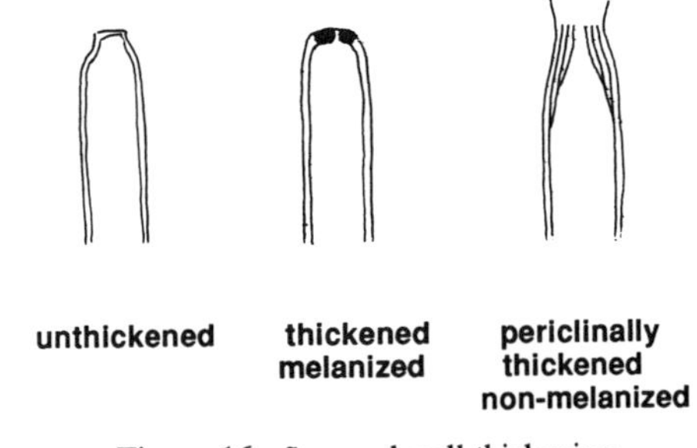

Figure 16. Scar and wall thickening.

Parameter 20. Formation of new conidiogenous cells in relation to an exhausted conidiogenous cell (Fig. 17)

New conidiogenous cells are:

(a) *resumptive*, when formed by proliferation of the previous conidiogenous cell and delimited by a septum.
(b) *adventitious*, when proliferating from the cells supporting a collapsed conidiogenous cell.

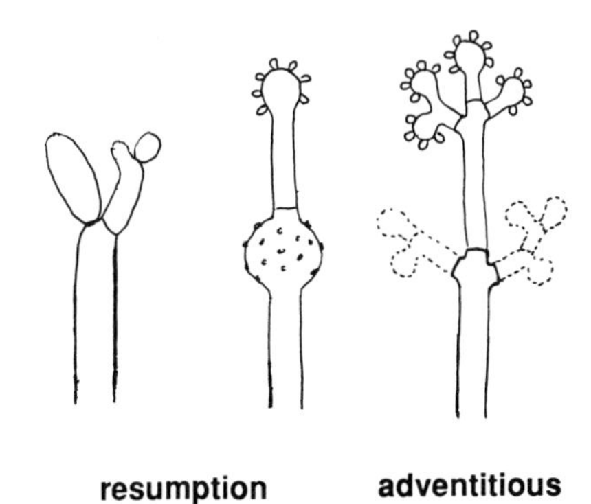

Figure 17. Development of new conidiogenous cell by proliferation.

This parameter is close to those that govern development of the conidiogenous cell on the supporting hyphal structure (conidiophore). Characterization of wall origin and mode of development of repetitive or adventitious new conidiogenous cells is by parameters 11 and 12. The event that this parameter governs is identical with that of conidiogenous cell proliferation.

CONCLUSIONS

Parameters governing the development of conidiophores are beyond the scope of this treatment and are not considered here. The parameters for conidiogenesis actually described govern the situation in ascomycetes, i.e. three of the five stages of development defined by Minter et al. (1982, 1983a, 1983b) are dealt with, initiation, growth, and secession. Other parameters may be added in order to describe the differentiation of the conidium itself during the stages of maturation, germination, and the morphogenesis of conidiophores and conidiomata.

The scheme is an attempt based on a small number of cases of conidiogenesis. We encourage constructive criticism and suggestions for improvement.

To achieve definition of the unitary parameters for a dynamic reference point for anamorphic fungi, an accurate re-examination of the deuteromycetes in the light of the already proposed parameters will be necesary more than ever. Then complete descriptive records of the type species of genera and of any synanamorphs connected with teleomorphs will be possible and will allow computer analysis of similarity or eventually phylogenetic relationships. On the other hand such a re-examination of congeneric species will facilitate testing homogeneity versus heterogeneity of anamorphic genera which should lead to taxonomic revisions of deuteromycetes as a precursor to a better correlation with the taxonomy of the ascomycetes.

REFERENCES

Bracker, C.E., 1977, Structure and transformation of chitosomes during chitin microfibril synthesis, *In*: *2nd International Mycological Congress Abstracts* (H.E. Bigelow and E.G. Simmons, eds) 1: 64. Second International Mycological Congress, Tampa.

Hennebert, G.L., and A. Bellemère, 1979, Les formes conidiennes des Discomycètes. Essai taxonomique, *Revue de Mycologie* 43: 259-315.

Hermanides-Nijhof, E.J., 1984, *Aureobasidium* and allied genera, *Studies in Mycology, Baarn* 15: 141-177.

Kendrick, W.B. (ed.), 1971, *Taxonomy of Fungi Imperfecti*, University of Toronto Press, Toronto.

Kendrick, W.B., and F. DiCosmo, 1979, Teleomorph-Anamorph connections in Ascomycetes, *In*: *The Whole Fungus* (W.B. Kendrick, ed.) 1: 283-410. National Museum of Natural Sciences, Canada.

Minter, D.W., P.M. Kirk, and B.C. Sutton, 1982, Holoblastic phialides, *Transactions of the British Mycological Society* 79: 75-93.

Minter, D.W., P.M. Kirk, and B.C. Sutton, 1983a, Thallic phialides, *Transactions of the British Mycological Society* 80: 39-66.

Minter, D.W., B.C. Sutton, and B.L. Brady, 1983b, What are phialides anyway?, *Transactions of the British Mycological Society* 81: 109-120.

Sutton, B.C., 1993, Mitosporic fungi (deuteromycetes) in the *Dictionary of the Fungi*, *In*: *The Fungal Holomorph: mitotic, meiotic and pleomorphic speciation in fungal systematics* (D.R. Reynolds and J.W. Taylor, eds): 27-55. CAB International, Wallingford.

Sutton, B.C., and D.K. Sandhu, 1969, Electron microscopy of conidial development and secession in *Cryptosporiopsis* sp., *Phoma fumosa*, *Melanconium bicolor*, and *M. apiocarpum*, *Canadian Journal of Botany* 47: 745-749.

Vuillemin, P., 1910a, Les conidiosporés, *Bulletin de la Société des Sciences de Nancy* 3: 151-175.

Vuillemin, P., 1910b, Materiaux pour une classification rationelle des Fungi Imperfecti, *Compte Rendu Hebdomadaire des Séances de l'Academie des Sciences Paris* 150: 882-884.

INTERCONNECTIONS AMONGST ANAMORPHS AND THEIR POSSIBLE CONTRIBUTION TO ASCOMYCETE SYSTEMATICS

B.C. Sutton[1] and G.L. Hennebert[2]

[1]International Mycological Institute
Bakeham Lane, Egham, Surrey TW20 9TY, UK

[2]Mycothèque de L'Université Catholique de Louvain
Laboratoire de Mycologie Systématique et Appliquée
Université Catholique de Louvain, Faculté des Sciences Agronomiques
Place Croix du Sud 3, B-1348 Louvain-la-Neuve, Belgium

SUMMARY

Previous attempts at syntheses between ascomycete classifications and systems proposed for deuteromycetes (mainly based on conidiogenesis) are reviewed. Problems surrounding previous and current attempts at correlative work are emphasized. The relationships between non-lichenized ascomycetes and the latest concepts and terminology on conidiogenesis in anamorphs are reassessed. The anamorph-teleomorph connections in the *Sclerotiniaceae*, *Pezizaceae* and *Mycosphaerella* are treated in more detail.

BACKGROUND

An interpretation of ascomycete systematics based solely on teleomorphs runs the risk of only representing a partial analysis of a very complex problem. Consideration of ascomycete teleomorphs together with their various types of anamorphs and synanamorphs is a logical position from which to advance hypotheses and test them. It is the genome in all its expression that must be brought to bear on systematic rearrangements, whether one is dealing with an anamorph alone (that is a taxon for which no teleomorphic correlation has yet been reliably made), a teleomorph associated with anamorphs or *vice versa* (where correlations have been substantiated), or a solely teleomorphic fungus (for which there is no known anamorph).

Various general/colloquial terms have been used in the past for fungi which are anamorphic, some of which have been correlated with ascomycetes. These include fungi imperfecti, asexual fungi, mitosporic fungi (presumptively mitotic, Sutton, 1993), and deuteromycetes and a number of its derivatives. The preferred term for the eighth edition of the *Dictionary of the Fungi* is "mitosporic fungi" (Sutton, 1993).

In trying to assess the contribution that anamorphs have made in the past and can make to ascomycete systematics in the future it is important to establish not only reliable foundations but also the framework which will allow extrapolation and synthesis. Kendrick (1979) edited two volumes entitled *The Whole Fungus*. These focussed attention on "the fungus" in all its reproductive states but excluded information about the mycelium. In this same work Kendrick and DiCosmo (1979) listed the known correla-

Ascomycete Systematics: Problems and Perspectives in the Nineties
Edited by D.L. Hawksworth, Plenum Press, New York, 1994

tions reported between many ascomycetes and their anamorphs. They envisaged the compilations for ascomycetes as the basic raw material from which future syntheses may be forged. This provides one starting point. The other concerns the systematics of anamorphs themselves and this must to a large extent revolve around conidiogenesis and associated events, though not exclusively so, for there are many other taxonomic criteria that can be brought to bear on this problem. The most recent published work in this area is by Sutton (1993) who has extended the concepts advanced by Minter et al. (1982, 1983a, 1983b). A further development is the unitary parameters for conidiogenous events which are outlined by Hennebert and Sutton (this volume).

Until comparatively recently there has been increasing dichotomy in the classificatory schemes that have evolved for ascomycetes and their anamorphs. These originated with Fuckel (1873) who formalized the differences when he separated them into two groups, the *Fungi Perfecti* and the *Fungi Imperfecti*. The impetus for rectifying this dichotomy provided by *The Whole Fungus* was continued in *The Holomorph* (Reynolds and Taylor, 1993). Here, a rapprochement was also sought between morphologically based taxonomic systems and the results which are increasingly emerging from the use of molecular techniques in systematics at all levels. There is now an increasing momentum toward rationalizing the systematics of ascomycetes with those of their anamorphs.

The lists of ascomycete connections by Kendrick and DiCosmo (1979) were arranged in a number of ways. A separation was made between unitunicate and bitunicate groups and although this is an oversimplification of the actual situation that obtains in ascus structure, it still serves as the only comprehensive reference point that we have. Within these two groups the teleomorph-anamorph connections were listed alphabetically by genus and each entry was coded for the reliability of the correlation. These were followed by teleomorph-anamorph connections arranged alphabetically by ascomycete order, family and genus, with an indication of the number of connections in each anamorph genus. There was discussion and some generalization about the bitunicate relationships but little on the unitunicate correlations.

There is a series of well-known landmarks in the ontogenetic approach to systematics of conidial fungi. These will not be dealt with extensively since they have been fully documented and reviewed elsewhere by Sutton (1973), Nag Raj (1981), Subramanian (1983), and Kendrick (1981). They include primarily the work of Costantin (1888), Vuillemin (1910a, 1910b, 1911), Höhnel (1923), Mason (1937), Hughes (1953), Kendrick (1971), Cole and Samson (1979), Sutton (1980), and latterly the work of Minter et al. (1982, 1983a, 1983b), and Sutton (1993). Until the work of Hughes (1953), the classification of these fungi was firmly based on concepts used by Saccardo (1884, 1886). They concerned various aspects of structure and behaviour, such as spore morphology, conidiomatal type and host-fungus relationships. Although still used extensively in identification these characters are not now considered to be of primary importance in determining relationships either within the group or in working out relationships with ascomycetes and basidiomycetes. The end products of all pre-Hughesian and most post-Hughesian systematic arrangements were hierarchical classifications where the fungi were assigned to classes, orders, families, sections and groups etc. with clear, or what was hoped were clear, hiatuses between taxa at the same levels. That all conidial fungi known did not conveniently fit into such systems has been a continuing problem and inevitably this was the spur to further refinement of such taxonomic systems. A quite different end-product was envisioned by Minter et al. (1982, 1983a, 1983b) who emphasized the fluidity/plasticity/malleability of conidiogenous events, i.e. from several perspectives the events surrounding the production of spores represented continua to varying degrees. This work defined a number of basic events:- conidium initiation, maturation, delimitation and secession, apical, diffuse and ring wall building and replacement wall growth, collarette production, conidiogenous cell proliferation and regeneration. Minter et al. (1982, 1983a, 1983b) concluded that accurate description of these events would relieve deuteromycete taxonomy of some of the difficulties that then encumbered it, especially in relation to terminology. For the purposes of coding these events in the *Dictionary of Fungi*, Sutton (1993) used mitospore terminology and applied it to a range of described genera. For the *Holomorph Conference* held in August 1992 (Reynolds and Taylor, 1993), thirty four different combinations of these events were identified. Since this date several more have been recognized and the total now stands at least at 43. There are almost certainly more to be recognized.

PREVIOUS ATTEMPTS AT SYNTHESIS

There have been relatively few modern attempts to equate ascomycete genera, families and orders with classificatory schemes for conidial fungi. Tulasne and Tulasne (1861-65) and Klebahn (1918) stand out as having made significant earlier contributions to this type of correlative work although they did not specifically deal with equating ascomycete classifications with those then known for conidial fungi.

Tubaki (1958) was concerned only with hyphomycete anamorph correlations. Based on the premise that if classification of the perfect states is truly phylogenetic, he suggested that some parallel phylogenetic relationships must be found among their imperfect states. He took the type of conidiophore and conidial development and structure to be the most stable systematic criteria in the hyphomycetes. Various modifications of Hughes' groups at sectional level were combined with conidial type and overlaid on 11 orders of ascomycetes in which correlations had been reported. His later publication (Tubaki, 1981), which also summarized hyphomycetes and their perfect-imperfect correlations, was essentially based on the same parameters but used a greater number of examples.

Subramanian (1971) observed a number of relationships between hyphomycetes and their perfect states but refrained from "weaving" a natural classification with reference to their possible perfect states because "the relationships of various ascomycetous genera themselves are not in all cases clear and the taxonomy of this group itself is in a state of flux and continual change".

Müller (1971) was of the same opinion and acknowledged that despite advances in both groups the systematics of ascomycetes and conidial fungi had not then become stable. He arranged hyphomycete and coelomycete anamorphs according to a modified system of Hughes' (1953) scheme. His conclusion was that the information on connections between ascomycetes and their anamorphs was still in a poor state.

In *The Whole Fungus*, Luttrell (1979) concluded that an Adansonian excercise in defining relationships by the number of characteristics organisms have in common had not been conspicuously successful when conidiogenesis was overlaid on ascomycete orders and families. The reciprocal testing of anamorph and teleomorph classifications and their degrees of fitness in his opinion had left both open to question. He then went on to discuss the various anomalies that this approach had thrown up. He saw no reason to assume, in advance of testing, that characters of conidiogenesis are somehow more fundamental than, for example, characters of conidium septation.

De Hoog (1979), in addition to a limited number of criteria involving conidiophores, conidiogenous cells and conidiogenesis, used a few other features such as conidiomata, sterile elements (setae) and spore morphology. These were combined with a much wider selection of ascomycete characters to produce a deductive classification for ascomycetes based on 340 genera from 48 families.

Hennebert and Bellemère (1979) restricted their synthesis to non-lichenized and lichenized discomycetes. Data on anamorphs consisted of four conidiomatal types and six conidiogenous types. The degree of homogeneity and heterogeneity varied within orders.

Vobis (1980) and Vobis and Hawksworth (1981) brought together the then known information on morphological and developmental aspects of anamorphs of lichenized ascomycetes. They concluded that too few have been studied in detail to enable generalizations to be made about the types found in the various ascomycete groups involved.

It is likely that the reasons for what might be interpreted as the failures of these attempts to divine an overall scheme incorporating anamorphs in the ascomycetes is due to a combination of factors. The most obvious of these was the use of concepts and ideas that from present day knowledge no longer remain entirely tenable. A lack of appreciation of the extent to which events surrounding conidiogenesis are plastic is central to this problem, such as the maintenance of firm distinctions between "phialides" and "annellides" when a continuum is operating, an unawareness of the heterogeneity of what have been called phialides, lack of appreciation of the different origins of chains of conidia etc. Another reason might have been the acceptance of orders, families and other suprageneric categories for the conidial fungi, partly based on the continued unrealistic use of conidiomatal structure, for instance to distinguish between hyphomycetes and coelomycetes. The polyphyletic origin of certain patterns of conidiogenous events and the ability of conidial fungi to evolve independently of meiotic states must also have an influence on the extent to which any syntheses can be made.

CAVEATS AND PROBLEMS WITH CORRELATIVE INVESTIGATIONS

There are still serious difficulties in attempting syntheses of information on ascomycete systematics, the reported correlations with anamorphs, and relationships within the conidial fungi. Such problems are diverse and concern the correlations that have been reported and the state of ascomycete and conidial fungal systematics.

Kendrick and DiCosmo (1979) took care to point out the widely differing reliability of the information they compiled. This was by instituting a coding system to assess evidence for correlation or affiliation between anamorph and teleomorph. Any synthesis must take this into account.

The Kendrick and DiCosmo (1979) lists show a dearth of information on correlations. For the 1018 genera listed, in 699 (68%) cases only a single anamorph correlation of any sort per ascomycete genus has been reported. Ascomycete genera in which there are 1-5 recorded correlations account for 951 out of the total, i.e. 93%. It is only in cases like *Diaporthe-Phomopsis* where 139 correlations have been reported, *Eupenicillium-Penicillium* 42, *Valsa-Cytospora* 32, *Nectria-Acremonium* 21, *Melanconis-Melanconium* 19, and *Talaromyces-Penicillium* 17 in the unitunicates, and *Mycosphaerella-Septoria* 25, *Leptosphaeria-Phoma* 19, *Elsinoe-Sphaceloma* 17, *Mycosphaerella-Cercospora* 16 and *Cucurbitaria-Camarosporium* 15 in the bitunicates, that there is any weight of evidence for consistently repeatable unambiguous correlation within genera. This brings into question the feasibility of developing any synthesis.

There is wide variation in the numbers of correlations reported for individual ascomycete genera. Of the 362 unitunicate genera, 250 have only a single anamorph genus correlated, while 350 have 5 anamorph genera linked or less. In some ascomycetes however, the situation is much more complex. In *Nectria* 19 different anamorph genera have been correlated, for *Hypomyces* 15, for *Ceratocystis* also 15, and for *Physalospora* 14. The pattern is similar in the bitunicate genera, only there is a proportionately larger number with more than a single correlated anamorph genus. Out of 143 bitunicate ascomycete genera 80 have only a single anamorph genus correlated and a further 55 with 2-4 linked anamorph genera. Again there are some genera with much greater numbers, *Mycosphaerella* has 27 different correlated anamorph genera, *Leptosphaeria* 19, *Guignardia* 14, *Pleospora* 11, and *Cucurbitaria* 10. The heterogeneity of such multiple correlations raises questions about reliability of identification of both anamorphs and teleomorphs, and by extrapolation the status of systematics in the respective fungal groups, especially at generic level.

There is similarly wide variation in the extent to which individual anamorph genera have been reported as correlated with ascomycetes. A sizeable proportion (304, 64%) of the 473 anamorph genera listed as correlated with ascomycetes have been linked only once. A total of 441 (98%) have been listed as correlated with 5 ascomycete genera or less. At the other extreme a small proportion of genera have been linked with several different ascomycete genera, such as *Acremonium* which is reported for 31 ascomycete genera, *Phoma* for 21, *Chrysosporium* for 16, *Phialophora* for 13, *Aspergillus* for 13, and *Cytospora* for 12. The possibility of polyphylogeny and independent convergent and divergent evolution of similar anamorphic states in different groups of not only ascomycetes but other groups, such as the basidiomycetes, must be taken into account.

The number of anamorph genera which have been reported as correlated with ascomycetes but for which we do not have reliable information on conidiogenesis is depressingly high. In Kendrick and DiCosmo (1979) these number 39 in the unitunicates and 33 in the bitunicates. Reasons for such inadequacy include nomenclatural problems in the application of names, absence of typification or lectotypification, the fact that some names used are already formally rejected and no reliable substitutes are known, and absence of data because original accounts were either poor or inadequate for modern interpretation and material has not been re-examined since description. The status of ascomycete genera in this respect, that is, accurate information about morphology and development, may be similar.

A NEW ATTEMPT AT SYNTHESIS

In addition to the report by the bitunicate committee on the correlations listed in Kendrick and DiCosmo (1979) some general observations follow.

Amongst some of the more common conidial genera some have only been reported as correlated with unitunicate teleomorphs. However, within the unitunicates there is wide variation in occurrence. *Aspergillus* has only been reported for the *Trichocomaceae* in the *Eurotiales*. *Penicillium*, although reliably only known from the *Eurotiales*, has also been reported as correlated in the *Hypocreales* in the *Hypomycetaceae* and *Hypocreaceae*. *Oidium* and *Oidiopsis* on the other hand have only been reported in the *Erysiphaceae* of the *Erysiphales*. Genera such as *Paecilomyces, Fusarium, Chalara, Phialophora* and *Graphium* have been reported as anamorphs of a wide range of orders, families and genera in the unitunicates. *Geotrichum* and *Chrysosporium* also have been reported as anamorphs of different orders and families in the unitunicates but not as wide a range as the aforementioned group of genera.

Some genera have been correlated not only with unitunicates but also bitunicates. *Acremonium* has been reported as an anamorph in many orders and families of both groups of ascomycetes. It is more abundantly reported in the unitunicates where correlations in several families of the *Eurotiales* and "*Sphaeriales*" are known. *Phoma* is similarly correlated in both groups, but less frequently than *Acremonium*. The links are more with bitunicate orders, especially of the *Pleosporales* where several families have been reported with correlations. *Cytospora* is also a genus which has teleomorphs in several families in both the unitunicates and bitunicates.

Anamorph genera which have been exclusively correlated with the bitunicates include the range of unusual fungi linked to the *Capnodiaceae* by Hughes (1976). Others are *Curvularia, Drechslera, Exserohilum, Sphaceloma,* all of which have a one to one relationship with ascomycete genera. *Cercospora* is also linked only with the bitunicate group but restricted to genera in the *Dothideales*, *Dothideaceae*.

There have been major advances theoretically and practically in the conidial fungi fungi since Cole and Samson (1979) provided much experimental evidence supporting ontogenetic initiatives in systematics of the group. It coincided with the last flush of attempts to synthesize ascomycete and anamorph correlations.

Taxonomic ranks above the level of genus are not now used in the conidial fungi, despite there being a long tradition for doing so. Indeed for the last 10 years or so there have been no further attemps to propose supra-generic classificatory systems in the group. The unitary parameters for conidiogenesis and the 43 combinations of conidiogenous events at present recognized do not comprise classifications, neither are they substitutes for classifications. Furthermore, no existing taxonomic categories are being adapted to cope with this way of presenting information and no new ones are to be proposed by these authors. The schemes are simply descriptive means of bringing together groups of genera which share similarities in the combinations of events surrounding conidiogenesis. It would be surprising if this did not give some indications as to the similarity and differences between groups of genera.

The unitary parameters for conidiogenesis (Hennebert and Sutton, this volume; Table 1) provide a synopsis of present ideas on conidiogenesis and allow some combinations of events which share certain parameters to be brought together into recognisable groups. Table 1 shows the 43 combinations of conidiogenous events with a typical genus for each. They are organised into twelve groups. These consist of **core** groups which contain the vast majority of conidial fungi, combinations numbered:

1, 2
6-9
10-14, 30, 31
15-21, 32, 33

and **outlier** groups which contain far fewer genera, combinations numbered:

38-41
24-29
37
3-5
34-36
23
22
42, 43.

Table 1. Unitary parameters for conidiogenesis.

Conidiogenous Events	1	2	3	4	5	6	7	8	9	10
1,2	blastic thalloblastic thallic	hologenous	unilocar	apical lateral	NA	NA	solitary	NA	NA	determinate
6-9	blastic	hologenous	multilocar	circumspersed	simultaneous	NA	solitary catenate	hologenous	monopodial	determinate resumptive
10-14,30,31	blastic thalloblastic	hologenous enterogenous	multilocar	apical lateral	successive	progressive sympodial random	solitary	hologenous	NA	determinate indeterminate resumptive
15-21,32,33	blastic	enterogenous	unilocar multilocar	apical lateral	simultaneous successive	progressive sympodial stationary random	catenate seriate	enterogenous	basipetal	determinate indeterminate
38-41	thallic	hologenous enterogenous	NA multilocar	apical lateral intercalary	simultaneous successive	NA sympodial	solitary catenate	hologenous enterogenous	monopodial sympodial	determinate indeterminate
24-29	blastic	enterogenous	unilocar multilocar	apical laterial intercalary	simultaneous successive	* progressive sympodial stationary	solitary catenate	enterogenous	monopodial	determinate indeterminate resumptive
37	blastic	hologenous	unilocar	apical laterial	successive	retrogressive	solitary	NA	NA	determinate
3-5	blastic	hologenous	unilocar multilocar	apical laterial	successive	progressive sympodial random retrogressive	catenate	hologenous	monopodial sympodial	determinate
34-36	blastic holothallic	hologenous enterogenous	unilocar	apical	* successive	* retrogressive	solitary catenate	hologenous	basipetal	NA
23	holothallic	hologenous	unilocar	apical	NA	NA	catenate	hologenous	basipetal	determinate
22	holothallic	enterogenous	unilocar	apical	NA	NA	catenate	enterogenous	basipetal	determinate
42,43	blastic thallic	hologenous	NA	NA	NA	NA	solitary catenate	hologenous	monopodial	NA

Table 1. Cont.

Conidiogenous Events	11	12	13	14	15	16	17	18	19	20
1,2	NA	NA	NA	persistent	one two	schizolytic rhexolytic	closed-frill closed+frill open + frill	applanate protuberant	unthickened thickened + melanin	resumptive
6-9	hologenous	NA	monopodial	persistent	one two	schizolytic rhexolytic	closed-frill closed+frill open + frill	applanate protuberant	unthickened thickened + melanin	resumptive
10-14,30,31	hologenous enterogenous	percurrent erumpent	sympodial	persistent	one two	schizolytic rhexolytic	closed-frill closed+frill	applanate protuberant	unthickened thickened + melanin	resumptive
15-27,32,33	enterogenous	percurrent	monopodial	persistent	one	schizolytic	collarette annellation	applanate	thickened periclinally	resumptive
38-41	hologenous	NA	NA sympodial	NA	one two	schizolytic rhexolytic	closed-frill closed+frill	applanate	unthickened	*
24-29	hologenous	percurrent	monopodial sympodial	persistent	one	schizolytic	pore	applanate	thickened + melanin	NA
37	hologenous	NA	basauxic	collapsed	one	schizolytic	closed-frill	applanate	unthickened	NA
3-5	NA	NA	NA	persistent	one	schizolytic	closed-frill	protuberant	unthickened thickened + melanin	NA
34-36	NA	NA	NA	NA	one	schizolytic	closed-frill	applanate	unthickened	NA
23	NA	NA	NA	persistent	one	schizolytic	closed-frill	applanate	unthickened	*
22	NA	NA	NA	persistent	one	schizolytic	closed-frill	applanate	nthickened	resumptive adventitious
42,43	NA	NA	NA	NA	multiple	schizolytic rhexolytic	NA	NA	NA	NA

Examples of conodiogenous events.
1, *Clasterosporium*; 2, *Pithomyces*; 3, *Cladosporium*; 4, *Bispora*; 5, *Oidiodendron*; 6, *Gonatobotrys*; 7, *Botrytis*; 8, *Balanium*; 9, *Gonatobotryum*; 10, *Tritirachium*; 11, *Dematophora*; 12, *Leptostroma*; 13, *Wardomyces*; 14, *Pyricularia*; 15, *Acremonium*; 16, *Cylindrotrichum*; 17, *Catenularia*; 18, *Chloridium*; 19, *Scopulariopsis*; 20, *Conioscypha*; 21, *Graphium*; 22, *Chalara*; 23, *Oidium*; 24, *Piricaada*; 25, *Corynespora*; 26, *Drechslera*; 27, *Spadicoides*; 28, *Diplococcium*; 29, *Alternaria*; 30, *Pseudospiropes*; 31, *Cercospora*; 32, *Aspergillus*; 32, *Sagenomella*; 34, *Trichothecium*; 35, *Cladobotryum*; 36, *Phragmotrichum*; 37, *Arthrinium*; 38, *Ampulliferina*, 30, *Geotrichum*; 40, *Rhexoampullifera*; 41, *Ptychogaster*; 42, *Dactuliophora*; 43, *Barnettella*.

Whereas there are quite rigid hiatuses between outlier groups based on perhaps one difference, such as multiple secession of conidia (42, 43), the retrogressive sequence of loci (34-36), or the pore-like scar (24-29), there are fewer differences between the core groups, especially 6-9 compared with 10-14, 30, 31. Here the distinctions are in the circumspersed loci which produce conidia simultaneously rather than the apical or lateral loci which form conidia successively. There are also many genera in which development of the first-formed conidium in a sequence has most, if not all the combinations of features shown by those genera in which only a single conidium is formed on each conidiogenous cell (1, 2).

CORRELATIONS ANALYSED BY ORDER/FAMILY COMBINED WITH CONIDIOGENOUS EVENTS

An analysis of correlations has been made by listing data on conidiogenous events (rather than anamorph generic names) for various orders and families of non-lichenized ascomycetes for which information is available. This draws on data compiled by Kendrick and DiCosmo (1979), updated as appropriate, and is assessed by reference to the latest edition of the *Outline of the ascomycetes* (Eriksson and Hawksworth (1993; this volume, Annex 1). In assessing these data the various relationships within the groups of conidiogenous events were taken into account. This allowed a subjective appraisal of the **homogeneous** or **heterogeneous** status of the correlations reported. Some correlations were mostly homogeneous except for isolated cases where the conidiogenous event is almost certainly erroneous. In such instances the decision is listed as **?homogeneous**.

DIAPORTHALES

Diaporthaceae: Except for the report of the outlier *Sporoschisma* with *Melanochaeta* there is a preponderance of "phialidic" anamorphs with coelomycetous conidiomata in this family. Many of the ascomycete genera are correlated with several anamorph genera and a substantial proportion of these are incorrect due to modern synonymy, or dubious owing to use of antequated names. **Conclusion: ? homogeneous.**

Valsaceae: A single correlation genus to genus indicates a relationship in conidiogenous terms of this family to the *Xylariaceae*.

DIATRYPALES

Diatrypaceae: Anamorphs with sympodially or irregularly proliferating conidiogenous cells predominate but it is difficult to reconcile this with the reports of correlations of *Cytospora* with *Eutypella*. **Conclusion:**
? homogeneous.

ERYSIPHALES

Erysiphaceae: There are consistently proved correlations in this single family, although *Oidium* is reported as the anamorph of several different genera. Conidiogenesis in *Oidium* and *Ovulariopsis* is closely related. **Conclusion: homogeneous.**

EUROTIALES

Monascaceae: There is only one reported correlation, *Basipetospora* with *Monascus*.

Trichocomaceae: With some exceptions (see below), all conidiogenous events are closely linked, and this, combined with the repeated correlations between *Aspergillus* and *Penicillium* and their teleomorphs gives a stable system. What is contentious is the linking of *Aspergillus* with a large number and *Penicillium* with just two different teleomorphic genera. The contribution that anamorphs might make to ascomycete systematics in this situation requires further study. In other genera now included in the family, anamorphs show very much a mixture in terms of conidiogenous events incorporating both what used to be termed phialidic and sympodial hologenous types. The renaming of the anamorph of *Sagenoma* from *Acre-*

monium to *Sagenomella* illustrates the plasticity of events in *Acremonium* sensu lato. **Conclusion: ? homogeneous.**

HALOSPHAERIALES

Halosphaeriaceae: A combination of unnamed or very diverse anamorphs and few correlations makes conclusions premature.

HYPOCREALES

Clavicipitaceae: There are some reliable correlations genus to genus such as *Balansia-Ephelis, Hypocrella-Aschersonia, Claviceps-Sphacelia*, but for others, such as *Cordyceps* the large number of anamorph genera reported indicate either erroneous correlations or a situation where the systematics of the genus might be helped by using anamorph criteria. - **Conclusion: heterogeneous**.

Hypocreaceae: The spectrum of conidiogenous events is dominated wholly by "phialidic" anamorphs, even though the conidiomata are quite varied. There are a few reports which clearly are incorrect, that is ones involving *Sphaeropsis, Graphium, Penicillium* and *Sporothrix*. Also some genera, such as *Nectria* have an enormous number of reported anamorphs. Nevertheless, this family is one in which more experimental correlative studies have been made than most and the anamorphs have played a major role in resolving the systematics of their ascomycete teleomorphs. *Peckiella* is linked with *Acremonium* and *Sepedonium* and the latter is also correlated with *Apiocrea*. Conidiogenous events are very mixed in taxa formerly segregated into *Hypomycetaceae*, and especially in *Hypomyces*. **Conclusion: ? homogeneous**.

LEOTIALES

Ascocorticiaceae: With only two correlations and one of those linked to a questionable ascomycete genus the status is unclear. **Conclusion: ? homogeneous**.

Dermateaceae: There is a thoroughly diverse range of conidiogenous events but with "phialidic" types the most frequently reported. Conidiomata are invariably, though not exclusively, coelomycetous. Several ascomycete genera have 4 or more different anamorphs reported and these are very confused. Spurious relationships obscure some very clear correlations, e.g. *Pezicula*, reported with 8 different anamorphic genera, is now known to have *Cryptosporiopsis* anamorphs. **Conclusion: heterogeneous.**

Hemiphacidiaceae: A link between *Kabatiella* (now a synonym of *Aureobasidium*) with *Sarcotrochila* might seem at variance with *Rhabdogloeum* correlated with *Rhabdocline*. However, the conidiogenous events are closely related. **Conclusion: homogeneous.**

Hyaloscyphaceae: The correlations are few and dubious, with the reports of *Cytospora* linked with *Dasyscypha* and *Trichoscyphella* most unlikely. **Conclusion: heterogeneous**.

Leotiaceae: The correlations are thoroughly mixed, with no combination of conidiogenous events dominant. Although anamorphs for *Godronia* are particularly varied, there are a number of one to one relationships, such as *Ascocalyx* and *Bothrodiscus, Crumenulopsis* and *Digitosporium, Heterosphaeria* and *Heteropatella, Pragmopora* and *Pragmopycnis, Scleroderris* and *Brunchorstia, Tympanis* and *Sirodothis* which are sufficiently well-established for problematical correlations reported in the literature to be eliminated. Even so there are far too many anamorphs in the outlier groups for reported relationships in the family to be accepted. **Conclusion: heterogeneous.**

Orbiliaceae: There are only two correlations and these involve widely disparate anamorphs in terms of conidiogenous events. **Conclusion: heterogeneous.**

Phacidiaceae: The reliable correlations are all of a "phialidic" nature, except for the link of *Polymorphum* with *Ascodichaena* which has a quite different set of conidiogenous events. *Ascodichaena* was placed in this family pro tem in Kendrick and DiCosmo (1979). **Conclusion: ? homogeneous**.

Sclerotiniaceae: For a fuller discussion of relationships see below. **Conclusion: homogeneous.**

MICROASCALES

Microasceae: Anamorph relationships here are problematic because rhexolytic secession in *Dematophora* which is linked to *Microascus* cannot be aligned closely with development in the other anamorphic genera reported for the genus. The heterogeneity of *Graphium* casts some doubt on the reported links between this and *Kernia, Petriella* and *Petriellidium*, bearing in mind that it has also been linked with some genera of the *Ophiostomataceae*. **Conclusion: ? homogeneous.**

ONYGENALES

Gymnoascaceae: There is almost overwhelming consistency in correlations between a variety of thallic and holothallic anamorphs and ascomycete genera in this family. The exceptions are the links to *Arachniotus* which are suspected to be erroneous, and the reports of *Gliocladium* and *Phialocephala* for *Lilliputia*. These are not consistent with other correlations in the family and may be incorrect, or, the placement of *Lilliputia* should be reassessed. **Conclusion: ? homogeneous.**

Onygenaceae: Most of the correlations are so overwhelmingly typical of the family that the "phialidic" correlations of *Paecilomyces* with *Dactylomyces* and *Aphanoascus* seem dubious. **Conclusion: ? homogeneous.**

OPHIOSTOMATALES

Ophiostomataceae: The biggest problem here is the 15 different anamorphic genera listed for *Ceratocystis*. Although they follow a number of recognized patterns in conidiogenous events and conidiomatal structure their value in contributing to the rationalisation of *Ceratocystis* and *Ophiostoma* systematics is yet to be fully realized. **Conclusion: ? homogeneous.**

OSTROPALES

Stictidaceae: There are too few correlations to form conclusions, but the linking of *Cylindrocarpon* with *Biostictis* appears manifestly incorrect.

PATELLARIALES

Patellariaceae: There are few correlations and those that have been made involve *Diplodia*. The listing of "*Phoma* vel aff." may refer to a microconidial state. **Conclusion: homogeneous.**

PEZIZALES

Ascobolaceae: There is a range of anamorphic genera correlated in this family but apart from the obvious incorrect linking of *Stemphylium* with *Ascobolus* there is a pattern of conidiogenous activity which unites them. **Conclusion: ? homogeneous.**

Morchellaceae: The twice reported link between *Costantinella* and *Morchella* is the only correlation for this family.

Otideaceae: Only one correlation is known, involving *Geniculodendron* and *Caloscypha*, but the anamorph is most closely related to those of the *Xylariaceae* (*Sphaeriales*).

Pezizaceae: The correlations mainly involve the genus *Peziza* but this includes a number of very different anamorphic genera for which there is no uniting pattern of conidiogenous events. Correlations are complicated by micro- and macro-conidial states. **Conclusion: heterogeneous.**

Pyronemataceae: There are a number of genus to genus correlations that involve differing conidiogenous events which cannot be even tenuously linked to genera like *Oedocephalum, Botrytis* and *Dichobotrys*. **Conclusion: ? homogeneous.**

Sarcoscyphaceae: Although a number of different anamorph genera are correlated on a one to one basis in this family the conidiogenous events they represent are closely related. **Conclusion: homogeneous.**

PHYLLACHORALES

Phyllachoraceae: This family is correlated with several anamorphs, the names of which are of dubious application. In *Phyllachora*, probably due to the fact that its species have not been cultured and conidiomata are often colonized by mycoparasitic fungi, the situation is very confused. **Conclusion: heterogeneous.**

RHYTISMATALES

Rhytismataceae: Several of the correlations reported in this family are missing data on conidiogenesis. Even within ascomycete genera where a number of correlations have been made and there is supporting evidence on conidiogenesis the situation is confused because different anamorph generic names have been used. **Conclusion: heterogeneous.**

SORDARIALES

Ceratostomataceae: Most correlations involve anamorphs with the simplest of conidiogenous developments but with some reported as "phialidic" and others with sympodial proliferation of conidiogenous cells the situation is confused and no traits are discernible. *Chaetomium* and *Melanospora* are reported with a number of diverse anamorphs which need checking. **Conclusion: heterogeneous.**

Lasiosphaeriaceae: The majority of correlations show variations on the "phialidic" theme. So within *Chaetosphaeria* there is a recognizable pattern with respect to conidiogenous events. However, there are a number of one to one correlations involving quite unrelated anamorphic genera such as *Sporidesmium, Cytospora, Oedemium, Chaetosticta* and *Diplodia* and these invite further work, especially on the relationships between the ascomycetes. **Conclusion: heterogeneous.**

Nitschkeaceae: There are too few correlations reported to draw any firm conclusions. However *Oedemium* belongs to an outlying group whereas *Acremonium* and *Cladorrhinum* are in the "phialidic" core group which suggests that the correlations are varied. **Conclusion: heterogeneous.**

Sordariaceae: On balance there are more states of a "phialidic" nature than others and this casts doubt especially on the link of *Diplococcium* with *Helminthosphaeria* (*Sordariales* inc. sedis) . Such tretic conidiogenesis is more characteristic of the bitunicate group. **Conclusion: heterogeneous.**

DOTHIDEALES

Asterinaceae: There are two series of dominant conidiogenous events in these correlations, the solitary or sympodially proliferating conidiogenous cell, and the "phialide". However, anamorphs in these groups have either distinctive coelomycetous or hyphomycetous conidiomata. The correlations with tretic (*Pirozynskia*) and thallic (*Bahusakala*) genera are not in this pattern so must also be viewed with suspicion. **Conclusion: heterogeneous.**

Botryosphaeriaceae: The single genus, *Botryosphaeria*, is linked with a number of coelomycetous anamorphs which to a large extent agree in conidiogenous events. **Conclusion: homogeneous.**

Capnodiaceae: Information on conidiogenesis for several genera is not available, and synanamorphic states complicate the situation in many genera. However, the patterns emerging from the work of Hughes (1976) show a high degree of relatedness amongst fungi which are polymorphic. **Conclusion: homogeneous.**

Chaetothyriaceae: Only two correlations have been made and these involve very different anamorphs, one with simultaneously produced loci (*Kazulia*) and the other with thallic conidiogenesis (*Merismella*). **Conclusion: heterogeneous.**

Dothioraceae: With some correlations questionable (*Cytospora* with *Dothiora*, and *Haplosporella* with *Bagnisiella*), others dubious, and the known difficulty of distinguishing between *Aureobasidium* and *Hormonema*, this is a mixture of relationships. **Conclusion: heterogeneous.**

Englerulaceae: The few correlations for three genera in the family are in accord. **Conclusion: homogeneous.**

Hysteriaceae: The several correlations cover a wide range of conidiogenous events but ones in which monopodial or sympodial hologenous chains are the usual pattern. However, *Septocyta* and *Hysteropycnis* differ in having coelomycetous conidiomata and their conidia are solitary. **Conclusion: heterogeneous.**

Leptopeltidaceae: Only two correlations have been made, of the same anamorph genus, *Leptothyrium*, with two different ascomycete genera, *Leptopeltopsis* and *Pycnothyrium.*

Microthyriaceae: The only correlations to have been made involve *Dothidella* and for this genus no less than five different anamorph genera have been linked. These are quite disparate and include distinctive hyphomycetes and coelomycetes. **Conclusion: heterogeneous.**

Mycosphaerellaceae: *Guignardia* with fourteen, *Mycosphaerella* with twenty seven, and most other genera with several different anamorph generic correlations account for an immense diversity of relationships in the family. Of the 43 combinations of conidiogenous events at least 16 have been reported for the *Mycosphaerellaceae*. For a more detailed assessment of *Mycosphaerella* (see below). **Conclusion: heterogeneous.**

Myriangiaceae: The correlation between *Elsinoe* and *Sphaceloma* is consistent and although *Tubercularia* is indeed similar in respect of conidiogenous events, the correlation would seem to be dubious on account of the links between this genus and *Nectria* in the Hypocreales. There are no data on *Articularia.* **Conclusion: ? homogeneous.**

Parodiopsidaceae: There are several correlations here and all conform to a pattern of conidiogenous events with either solitary conidia formed on each conidiogenous cell or conidiogenous cells showing annellations (a form of "phialide"). The single exception is the outlier tretic development in *Tretospora* linked with *Balladyna.* **Conclusion: ? homogeneous.**

Pleosporaceae: The family is large and the number of correlations is extensive so overall the links are quite varied. However, there are some well-defined patterns in individual genera, such as the links between *Drechslera, Bipolaris, Curvularia* and *Exserohilum,* all of which are tretic and their teleomorphs in this family. Elsewhere there are a number of one to one relationships which are verified, such as *Eudarluca* and *Sphaerellopsis, Gemmamyces* and *Megaloseptoria, Lidophia* and *Dilophospora, Magnaporthe* and *Nakataea, Tubeufia* and *Helicosporium.* However, there are many more ascomycete genera here in which several anamorphic genera of dubious identity or with unrelated conidiogenous events have been correlated and this combined with the presence of both macroconidial and microconidial states in some species confuses the situation. **Conclusion: heterogeneous.**

Polystomellaceae: Only two correlations have been made and these are one to one relationships between ascomycetes and anamorph genera. Details of conidiogenesis are missing in one of these so it is difficult to comment.

Schizothyriaceae: Too few correlations have been made in this family. Those that are known are very different - *Leptothyrium* and *Zygophiala* have been linked with *Schizothyrium* and represent quite distinct types of conidiogenous development. **Conclusion: heterogeneous.**

Venturiaceae: As with the *Pleosporaceae* there are some individual links which are consistent, such as *Venturia* with a closely related group of hyphomycetes that includes *Fusicladium, Pollaccia* and *Spilocaea.* These share similar conidiogenous events. In *Didymella* however there are several anamorphs which show little relationship in this respect. Since there are also some correlations involving outdated anamorph generic names and others which are of dubious application the situation is unclear. The correlation of *Helminthosporium* with *Gibbera* which involves the outlying tretic mode of conidiogenesis seems to be quite incorrect. **Conclusion: ? homogeneous.**

SACCHAROMYCETALES

For the families in this order it is not practicable to attempt determination of relationships based on morphology and ontogeny when the traditional systematics are physiologically and biochemically orientated. From the available data on conidiogenesis the relationships are very mixed.

XYLARIALES

Amphisphaeriaceae: There is a wide range of seemingly unrelated conidiogenous events reported for anamorphs in the family. The correlation of *Dendryphiopsis* with *Amphisphaeria* is at variance with all other correlations in the family because the anamorph is one of the outliers (conidiogenesis is tretic). The situation has been resolved by the transfer of *A. incrustans* to *Kirschsteiniothelia* by Hawksworth (1985), who places it in the *Dothideales*, *Pleosporaceae* where similar anamorphs are known. Also the reliable correlation of *Arthrinium*, another outlier but with basauxically proliferating conidiogenous cells, is reliably reported for 3 different genera in the family. The anamorph of *Blogiascospora*, *Seiridium*, is closely related to some anamorphs of other genera in the *Amphisphaeriaceae*, such as *Pestalotiopsis* and *Seimatosporium*, supporting the inclusion of the *Amphisphaerellaceae*. This suggests that the systematics of the family need attention, especially *Physalospora* which is reported correlated with 14 different anamorphic genera. **Conclusion: heterogeneous.**

Xylariaceae: The pattern of correlations is so consistent in terms of conidiogenous events that isolated reports involving "phialidic" genera such as *Cryptocline* and *Scopulariopsis* linked to genera of the family are certainly incorrect. **Conclusions: homogeneous.**

FAMILIES INCERTAE SEDIS

Amorphothecaeae: *Hormoconis* linked with *Amorphotheca* and *Paecilomyces/Polypaecilum* with *Thermoascus* have quite different unrelated types of conidiogenous activity. **Conclusion: heterogeneous.**

Cephalothecaceae: Only *Acremonium* has been correlated with genera in this family.

Cryptomycetaceae: The number of correlations is not high and only involve two ascomycete genera, *Potebniamyces* and *Darkera*. All anamorphs show variations on the "phialidic" theme so are related, even though the seven correlations known for *Potebniamyces* involve four different anamorph genera. **Conclusion: ? homogeneous.**

Pseudeurotiaceae: The correlations are very mixed, being a combination of anamorphs belonging to the core combinations of conidiogenesis and others which are outliers. *Sporothrix, Chalara* and *Acremonium* are all quite distinct. **Conclusion: heterogeneous.**

Saccardiaceae, Seuratiaceae: There is a mixture of anamorphs, one for which there is no reliable information on conidiogenesis, and two which have one to one relationships with teleomorphs. *Seuratia* linked with *Atichia* does not have immediately recognisable counterparts in either ascomycetes or mitosporic fungi. There appears to be no underlying pattern. **Conclusion: heterogeneous.**

CONNECTIONS IN *SCLEROTINIACEAE* AND *PEZIZACEAE*

Progressive and parallel refinement of the taxonomies of conidial anamorphs and the corresponding members of the *Sclerotiniaceae* are examples of the interactive development of fungal taxonomy to reach some homogeneity in teleomorph-anamorph connections as an expression of monophyletic taxa.

The generic name *Botrytis* was introduced by Micheli (1729) and validated by Persoon (1801). *Botrytis cinerea* was designated as lectotype species by Gregory (1949).

The genus was enlarged very soon and thereafter spasmodically after Persoon's description by the addition of more than 280 species, and submitted to periodic revisions tending to segregate from it new genera with a botryose conidial apparatus. Nees (1817) introduced the generic names *Virgaria* and *Cladobotryum* to accommodate some *Botrytis* species proposed by Link (1809). De Bary (1863) introduced *Phytophthora* based on *Botrytis infestans* and excluded from *Botrytis* a number of *Peronosporales* (Hennebert, 1973).

De Bary (1884) revised the characters of *Botrytis* and illustrated the ampulliform conidiogenous cells bearing synchronous conidia that Fresenius (1850) had quite accurately described. He also established the organic connection of the sclerotial anamorph *Sclerotium durum* with the apothecial teleomorph that he named *Peziza fuckeliana*. De Bary (1869, 1884) confirmed the characters of the species, completing them with a description of the spermatial state (Hennebert, 1973).

The same situation occurred in the genus *Peziza*. Fuckel (1870) introduced the generic name *Sclerotinia* for *Peziza sclerotiorum* (as *Sclerotinia libertiana*) and *Peziza fuckeliana* (as *Sclerotinia fuckeliana*). In a similar way to *Botrytis,* the genus *Sclerotinia* was amplified by the addition of numerous species and revised thereafter. Honey (1928) described the genus *Monilinia* to receive *Sclerotinia* species with a conidial state of the *Monilia* type to distinguish them from the typical *Sclerotinia* species which produce no conidial state but only a spermatial state. For the same reason, Whetzel (1937) introduced the genus *Septotinia*, distinct on account of its sympodial thalloblastic conidial production. Weiss (1940) erected the genus *Ovulinia* characterized by its thalloblastic solitary conidia. Whetzel (1945) recognized several new genera which he segregated from *Sclerotinia* on the basis of their anamorphic conidial or stromatic states. He described the genus *Botryotinia* for those *Sclerotinia* species producing botryose conidiophores "of the *Botrytis* type", the genus *Streptotinia* for those with "streptiform" *Botrytis*-like conidiophores (e.g. *Botrytis streptothrix*), both genera developing sclerotia incorporating some host tissues, and *Seaveriania* for *Sclerotinia geranii* producing "botryose conidiophores" with verrucose conidia and diffuse stromatized host tisues. He erected the genera *Ciborinia* and *Verpatinia* and accepted *Stromatinia*, all without closed or partly closed conidiomata but with distinct types of sclerotial anamorph.

Wilson et al. (1954) reclassified *Sclerotinia temulenta* in the new genus *Gloeotinia,* characterized by its presumably phialidic cylindrical to allantoid conidia (*Endoconidium* macroconidia) in addition to spermatia and discrete and diffuse intramatrical stromatic hyphae.

Buchwald (1949) in correspondance with the taxonomic development of the *Sclerotiniaceae*, proposed an amendment to *Botrytis*, excluding species producing elongate conidiogenous cells on dichotomous conidiophores and no sclerotia, such as *B. isabellina, B. spectabilis, B. epigea* and others. Those species were considered by Juel (1920) in reference to *Hyphelia*. In his amendment of *Botrytis* Buchwald introduced the subgenus *Verrucobotrys* to classify separately *Botrytis geranii*, the anamorph of *Seaverinia geranii*.

Hennebert (1973) treated the subgenus *Verrucobotrys* (nom. inval.) as a genus, *Verrucobotrys,* with *Verrucobotrys geranii* as type species. He erected the genus *Streptobotrys* for the streptiform botryose anamorphs of *Streptotinia* species. He also distinguished *Amphobotrys ricini* (the teleomorph still classified in *Botryotinia*) from *Botrytis* for dichotomous rather than alternate branching in the conidiophore.

Also Buchwald (1970), who was willing to provide a name for each different kind of anamorph in the *Sclerotiniaceae*, named the anamorph of *Septotinia* and that of *Ovulinia* under the generic names *Septotis* and *Ovulitis* respectively.

Hughes (1958) also contributed greatly to the revision of *Botrytis* by the re-examination and reclassification of many species connected or not connected with operculate discomycetes or pyrenomycetes.

Polyactis depraedens, classified in *Botrytis* subgenus *Cristularia* by Saccardo (1886), has been reclassified by von Höhnel (1916) in the new genus *Cristulariella.* Recently Harada and Noro (1988) described another anamorph, *Cristulariella pruni,* the teleomorph of which belongs to their new genus of the *Sclerotiniaceae*, *Grovesinia.* The globose or pyramidal botryose *Cristulariella* conidiomata consist of conidiophores covered by short blastic conidia (Niedbalski et al., 1979).

After the publication of *Valdensinia* in connection with the anamorph *Valdensia* and of *Pycnopeziza* in connection with the anamorph *Varicosporium,* Korf (1973) published a general key to the *Sclerotiniaceae*. All members of the family are characterized by a spermatial state of the *Myrioconium* type and by some sort of diffuse or sclerotial stroma. The key further divides the family according to the presence and type or absence of a conidial anamorph. The conidial anamorphs become therefore essential diagnostic characters of the holomorphs.

Successive amendments of *Botrytis* to a strictly restricted sense and a homogeneous circumscription has therefore contributed to establishing a refined parallel taxonomy of the *Sclerotiniaceae*, thereby attaining univocal anamorph-teleomorph connections.

The genus *Botrytis sensu lato* has also been the repository of a number of other conidial anamorphs that are being connected to species of the Pezizales (Hennebert and Bellemère, 1979).

In the *Pezizaceae*, tribe Pezizeae, anamorphs in *Oedocephalum* Preuss have been recognized in *Iodophanus* and in several species of *Peziza* (*P. repanda, P. cerea, P. vesiculosa*) (Berthet, 1964; Korf, 1972; Paden, 1972; Stalpers, 1974). *Chromelosporium*

Table 2. Teleomorph-anamorph connections in the *Leotiales* and *Pezizales*

LEOTIALES		
Sclerotiniaceae		
spermatial state phialidic *Myrioconium*		
apothecia brownish, stalked, ascus pore blue in iodine, arising from		
*a sclerotium		
♦ conidial state present and spermatial state		
▲ synchronous and botryose		
► conidiophore straight		
◄ alternate branching	*Botrytis*	*Botryotinia*
◄ dichotomous	*Amphobotrys*	?*Botryotinia*
► conidiophore twisted		
◄ alternate branching	*Streptobotrys*	*Streptotinia*
♦ no conidial state, but spermatial state		*Sclerotinia*
*stromatized host tissues		
♦ conidial state present and spermatial state		
▲ synchronous and botryose		
◄ alternate branching	*Verrucobotrys*	*Seaverinia*
▲ acropetal chains	*Monilia*	*Monilinia*
▲ sympodial multiseptate conidia		
◄ free conidiophores	*Valdensia*	*Valdensinia*
◄ in acervuli	*Septotis*	*Septotinia*
◄ on cupulate stroma	*Acarosporium*	*Pycnopeziza*
▲ basipetal conidia	*Ovulitis*	*Ovulinia*
▲ slimy cylindric conidia	*Endoconidium*	*Gloeotinia*
▲ dry allantoid conidia	unnamed	*Scleromitrula*
♦ no conidial state but spermatial state		*Stromatinia, Ciboria* etc.
PEZIZALES		
Pezizaceae		
Tribe Pezizeae		
Botrytis-like	*Oedocephalum*	*Peziza,* *Iodophanus*
Botrytis fulva	*Chromelosporium*	*Peziza,* *Plicaria,* *Muciturbo*
Pyronemataceae		
Botrytis spectabilis	*Dichobotrys*	*Trichophaea,* *Sphaerosporella,* *Pyropyxis*
Botrytis Dichobotrys	*Lauprospora*	
Sarcosomataceae		
Botrytis geniculata	*Conoplea*	*Urnula,* *Plectania,* *Korfiella*
Morchellaceae		
Botrytis terrestris	*Constantinella*	*Morchella*

has been rehabilitated for *Ostracoderma*-like or *Hyphelia*-like conidiomata previously classified as *Botrytis* or *Polyactis* species (Hennebert, 1973). *Botrytis crystallina, B. spectabilis, B. fulva,* and *B. luteobrunnea* are synonyms of *Chromelosporium fulvum*, the anamorph of *Peziza ostracoderma* (Hennebert and Korf, 1975). Paden (1972) described other *Chromelosporium* anamorphs in *Plicaria trachycarpa* (*Lamprospora trachycarpa*, anamorph *Chromelosporium trachycarpum*), *Plicaria endocarpoides* and others. A *Chromelosporium*-like anamorph is described in *Muciturbo reticulatus* by Warcup and Talbot (1989).

In the *Pyronemataceae*, *Botrytis*-like anamorphs of *Trichophaea* species have been placed in the new genus *Dichobotrys* (1973), which differs from *Botrytis* in having dichotomous branching with furcate ampulliform conidiogenous cells bearing simultaneous blastic conidia (Hennebert, 1973).

In the *Sarcosomataceae*, *Conoplea* is known to be the anamorph of several species in the very closely related genera *Urnula, Korfiella* and *Plectania* (Korf, 1972; Paden, 1972). *Conoplea geniculata*, originally named *Botrytis geniculata*, is characterized by geniculate branched conidiophores bearing simple conidia (Hennebert, 1973).

In the *Morchellaceae*, *Constantinella terrestris*, originally named *Botrytis terrestris*, is reputed to be the anamorph of a *Morchella* species (Molliard, 1904; Paden, 1972). *Constantinella* accommodates several *Botrytis*-like species having verticillate conidiomata with terminal corymbiform multilocal conidiogenous cells producing a succession of solitary globose blastic conidia.

The conclusion is that the nearly one to one correlation between teleomorphs and anamorphs obtained in the *Sclerotiniaceae* does not appear to occur in the present state of the taxonomy of the *Pezizales*. Similar anamorphs are found in species of one to three presently distinguished teleomorphic genera. A possibility exists that such similar anamorphs deserve closer examination in the light of their connected teleomorphs and require generic division. Alternatively, the teleomorphic genera with similar anamorphs might need reconsideration. *Plicaria* may be closer to some species of *Peziza* than those species are to the rest of *Peziza* with a different anamorph. *Sphaerosporella* may be merged with *Trichophaea* and perhaps *Urnula, Korfiella* and *Plectania* may be placed in synonymy. It might, however be acceptable phylogenetically to have more than one teleomorph genus associated with a single type of anamorph. Here we have an example of an interactive taxonomy at work, where the anamorphs may inform the work on teleomorphs and *vice versa*.

MYCOSPHAERELLA AND ITS ANAMORPHS

Mycosphaerella (*Dothideales*, *Dothideaceae*) and its segregates have been the subject of various attempts to relate anamorphs with their teleomorphs. Klebahn (1918) in fact split *Mycosphaerella* into genera segregated on the basis of the anamorphs, recognizing *Septorisphaerella* for species with *Septoria* anamorphs, *Ramularisphaerella* for those with *Ramularia* anamorphs and *Cercosphaerella* for those with *Cercospora* anamorphs. Later Laibach (1922) changed the name of *Ramularisphaerella* to *Ramosphaerella* and added another segregate, *Ovosphaerella,* for species with *Ovularia* anamorphs. Although these genera have not met with general acceptance, they demonstrate that at least 70 years ago there were mycologists who felt that in *Mycosphaerella* there was some justification for rationalizing systematics of the genus on the basis of its anamorphic relationships. More recently Barr (1972) however, commented:

> "that knowledge of the conidial states of fungi belonging to the *Dothideales* is limited as relatively few connections have been authenticated. No generalizations can be made on the value of this character at present."

Mycosphaerella has at one time or another been reported with at least 27 different correlated anamorph genera (Kendrick and DiCosmo, 1979). Some anamorph genera can be eliminated from consideration in this analysis. *Cercoseptoria*, although initially maintained by Deighton (1983), was later sunk into synonymy with *Pseudocercospora* by Deighton (1987), a decision endorsed by Sutton et al. (1987) and Sutton and Pascoe (1988). *Cylindrosporella* was shown by Sutton (1980) to be a synonym of *Asteroma*. The correlation of *Marssonina* was considered by Kendrick and DiCosmo (1979) to be doubtful, a conclusion endorsed by the fact that most *Marssonina* species are more correctly correlated with *Diplocarpon* (Sutton, 1980) or *Drepanopeziza* (Rimpau, 1961). *Ovularia* is generally regarded as a synonym of *Ramularia* (Hughes, 1949) and this conclusion was endorsed by Sutton and Waller (1988). *Phyllosticta* sensu stricto is now used for what used to be known as *Phyllostictina,* the teleomorphs of which belong exclusively in *Guignardia* (van der Aa, 1973). Alternatively *Phyllosticta* was interpreted in the past as a foliicolous *Phoma,* in which case the relationships could be not only with *Mycosphaerella* but also with 21 other genera for which the genus has been reported as correlated. *Placosphaeria* and *Septocylindrium* are of doubtful application because they have not been recently revised or redescribed. *Septogloeum* sensu stricto contains only 3 species which have not been reported with any teleomorphs (Sutton and Pollack, 1974; Sutton and Webster, 1984).

Von Arx (1983) dealt specifically with the anamorphs of *Mycosphaerella.* He accepted and keyed out 23 genera which consisted of anamorph genera which had been correlated with *Mycosphaerella* and a number of other conidial genera which he considered to be related. His treatment differed from the compilation of Kendrick and DiCosmo (1979), especially in suggesting new synonymies and rejecting old ones. *Phloeospora* was included as a synonym of *Septoria. Thedgonia* was placed as a synonym of *Cercoseptoria* which was itself maintained as distinct from *Pseudocercospora. Cercosporidium* and *Fusicladium* were placed as synonyms of *Passalora* and *Stenella* was sunk into synonymy with *Cladosporium,* but *Heterosporium* kept as distinct. *Septocylindrium* was placed in synonymy with *Ramularia* but *Ovularia* was also kept as distinct. This list of correlations has been rationalized with that of Kendrick and DiCosmo (1979) by the following modifications: *Cercoseptoria* is accepted as a synonym of *Pseudocercospora. Heterosporium* is a synonym of *Cladosporium* and *Ovularia* is a synonym of *Ramularia. Microdochium* is reliably correlated with *Monographella* by von Arx (1984). *Asperisporium, Gloeocercospora, Mastigosporium, Mycocentrospora, Mycovellosiella,* and *Ramulispora* have not been correlated with any *Mycosphaerella* species and are omitted from further consideration.

Combining these revised lists of Kendrick and DiCosmo (1979) and von Arx (1983) gives a total of 23 correlated genera. These are arranged according to their general conidiogenous events and some patterns begin to emerge (Table 3).

Correlated anamorph genera with solitary conidia are a mixed group. *Fusicladiella* differs from *Miuraea* and *Toxosporium* in having thickened conidiogenous loci and realistically belongs with genera in the group containing *Cercospora. Miuraea* may well have more affinity with the group containing *Phloeospora*, whilst *Toxosporium* is an oddity. The versicolored conidial morphology is at variance with all other *Mycosphaerella* anamorphs, to the extent that the validity of its correlation must be questioned.

Anamorphs with "phialidic" or "annellidic" anamorphs all belong to the coelomycetes, except for *Stigmina.* But sporodochial conidiomata are not fundamentally different from erumpent acervular conidiomata so *Stigmina* is not that distinct from *Lecanosticta.* "Phialides" and "annellides" are known to be the extremes of a continuum so this makes a very natural group based on conidiogenesis, although it must not be forgotten that *Asteroma* is a microconidial rather than a macroconidial state. The "phialidic" genera are characterized by hyaline conidia whereas the "annellidic" group have pigmented conidia. There may well be two series here.

All the sympodial anamorphs are hyphomycetes except for *Septoria* and the circumstances of this exception are not easily resolved. Firstly the genus *Septoria* is heterogeneous (Sutton, 1980; Constantinescu, 1984; Sutton and Pascoe, 1987, 1989; Farr, 1991) and although the type species is sympodial there are many taxa which are not, showing either "phialidic", "annellidic" or solitary types of development. None of these has the thickened scars which are so characteristic of the other sympodial genera. So, although there is another recognizable group with sympodial development and thickened scars, *Septoria* is perhaps a separate problem owing to its very heterogeneity, maybe linked in to *Phloeospora* and related genera. The other members of this group have thickened scars both on loci and on conidia. In this feature they are heterogeneous. David (1993) has demonstrated at least three different types of scar structure, the *Cladosporium, Cercospora,* and *Stenella* types. Deighton (1973, 1979) erected *Cercosporella, Cercosporidium,* and *Paracercospora* on the basis of differences in scar structure. Von Arx (1983) also separated the anamorphs of *Mycosphaerella* and related conidial fungi by scar structure, recognizing four groups. Much more ultrastructural work is required to explore and document the differences but they are there and it is conceivable that they will provide a better insight to separating groups within *Mycosphaerella* than heretofore.

Pseudocercospora is to some extent relevant here for it combines repeatedly percurrent conidiogenous cell growth with sympodial growth associated with conidial production. The fact that the scars are unthickened places it closer to the "annellidic " group of coelomycetes than the sympodial group containing *Cercospora*. There is some variation in conidiomatal structure amongst species of *Pseudocercospora*, inasmuch as in many species they are sporodochial/acervular rather than hyphal and therefore close to *Phloeospora*, even though pigmentation is quite different.

Table 3. Combined list of accepted anamorphs for *Mycosphaerella* (from Kendrick and DiCosmo, 1979; von Arx, 1983) arranged according to conidiogenous events

	Conidiogenous event	**Conidiomatal type (coding from Hawksworth et al., 1983)**
**Fusicladiella*	1	1
Miuraea	1	1
Toxosporium	1	3
Ascochyta	15	4
Asteroma	15	6
Asteromella	15	4
Phoma	15	4
Lecanosticta	19	6
Phloeospora	19	6
Stigmina	19	1/3
Pseudocercospora	21	1/3
**Cercospora*	31	1
**Cercosporella*	10	1
**Cercosporidium*	10	1
**Cladosporium*	3	1
**Passalora*	10/31	1
**Phaeoisariopsis*	10/31	2
**Phaeoramularia*	3	1
**Polythrincium*	10	1
**Ramularia*	3/10	1
Septoria	3	4
**Stenella*	3/10	1
Thedgonia	39	1

* thickened scars/conidiogenous loci

Thedgonia is an oddity. Von Arx (1983) included the genus as a synonym of *Cercoseptoria*, itself a synonym of *Pseudocercospora*. The thallic development combined with sympodial proliferation is comparable with what is seen in *Geotrichum* and is some distance from the situation in *Pseudocercospora*. The presence of a *Mycosphaerella* teleomorph is inexplicable.

Barr (1972) has proposed a subgeneric, sectional and series arrangement for species of *Mycosphaerella*. The taxa studied were exclusively from North America and consequently not all the species with known anamorphs were included. Unfortunately none in the *Cercospora*-like group were listed so this precludes other than the most rudimentary of conclusions. *Septoria* is found only in subgen. *Mycosphaerella*, but there it occurs in four out of five different sections. *Ramularia* is also confined to subgen. *Mycosphaerella* and is reported in two of the five sections. Anamorphs for subgen. *Didymellina* are a mixed batch with *Polythrincium, Stigmina* and *Cladosporium*. The latter is confined to this subgenus and arises in two of the five sections.

The indications based on this very small and incomplete sample are that anamorphs contribute little to supporting the subgeneric and sectional organization in *Mycosphaerella* as proposed by Barr (1972). However, this is only part of the picture and until all reliably correlated species have been assessed in this respect it would be premature to assume that there are no patterns whatsoever. The grouping of *Mycosphaerella* anamorphs in the manner described above shows that there are series that may be worth exploring in relation to the concepts of Klebahn (1918) where genera were segregated from *Mycosphaerella* on the basis of their anamorphs. Although the separation may not be at generic level there might be merit in using the data at some subgeneric rank.

CONCLUSIONS

Our knowledge of ascomycetes and the value of anamorphs in systematic syntheses has barely changed in the intervening years since previous attempts in the late 1970s were concluded to be mostly unsuccessful. The published information is still not wholly reliable. There are insufficient well-documented correlations linking accurately identified ascomycetes with accurately identified anamorphs. The caveats enumerated earlier in this paper continue to apply when considering the value of conidiogenesis in ascomycete relationships. Taking conidiogenesis as the sole criterion will only give a rough assessment of relationships. Here we have been able to indicate three categories:

(1) **homogeneous** - for correlations in which conidiogenesis is similar in genera of anamorphs and therefore can be rationalised with the taxonomic arrangement for the ascomycetes

(2) **? homogeneous** - where most correlations are as in the first category but there are some exceptions which obfuscate the relationships and for various reasons can be discounted. Additional work is needed to resolve these.

(3) **heterogeneous** - where conidiogenesis in correlations shows no identifiable pattern and where attempts to eliminate erroneous or dubious records fail to rectify the situation. This indicates that more basic work is required.

Conidiogenesis should be used in conjunction with a range of other conidial parameters, as has been demonstrated in *Botrytis* and its teleomorphs and in *Mycosphaerella* and its anamorphs. Thus applied, unitary parameters of conidiogenesis in anamorphs could well have an influence on how ascomycetes are assessed at the ordinal and family level. They will perhaps make a greater contribution to understanding at the generic and infrageneric level. Many more descriptive and correlative studies are required, preferably undertaken by ascomycetologists working closely with mycologists specialising in conidial fungi. The results need to be dovetailed into information derived from molecular techniques before predictive phylogenetic arrangements can be postulated for most genera of ascomycetes. This is particularly important because conidial fungi have been involved in evolutionary processes separate from sexual ascomycetes. Such classificatory schemes that we have for them may not realistically be superimposed over ascomycete classifications and be expected to match with anything but the most basic of parameters.

REFERENCES

Aa, H.A., van der, 1973, Studies in *Phyllosticta* I, *Studies in Mycology, Baarn* 5: 1-110.

von Arx, J.A., 1983, *Mycosphaerella* and its anamorphs, *Proceedings Koniklije Nederlandse akademie van wetenschappen*, ser. C, 86: 15-54.

von Arx, J.A., 1984, Notes on *Monographella* and *Microdochium*, *Transactions of the British Mycological Society* 82: 373-374.

Barr, M.E., 1972, Preliminary studies on the *Dothideales* in temperate North America, *Contributions to the University of Michigan Herbarium* 9: 523-638.
Berthet, P., 1964, Formes conidiennes de divers Discomycètes, *Bulletin de la Société mycologique de France* 80: 125-149.
Buchwald, N.F., 1949, Studies in the *Sclerotiniaceae*. I. Taxonomy of the *Sclerotiniaceae*, *Kongelige Veterinaer-og Landbohoiokoles Aursskrift* 1949: 75-191.
Buchwald, N.F., 1970, *Septotis* g. nov. and *Ovulitis* g. nov. two new form-genera of *Sclerotiniaceae*, *Friesia* 9: 326-329.
Cole, G.T., and R.A. Samson, 1979, *Patterns of Development in Conidial Fungi*, Pitman, London.
Constantin, J., 1888, *Les Mucédinées Simples*, Paul Klincksieck, Paris.
Constantinescu, O., 1984, Taxonomic revision of *Septoria*-like fungi parasitic on *Betulaceae*, *Transactions of the British Mycological Society* 83: 383-398.
David, J.C., 1993, *Heterosporium and its relationship to Cladosporium*. PhD thesis, University of Reading.
de Bary, A., 1863, Recherches sur le développement de quelques champignons parasites, *Annales des sciences naturelles, Botanique, sér.* IV, 20: 5-148.
de Bary, A., 1864, Beiträge zur Morphologie und Physiologie der Pilze, *Abhandlungen Senckenbergischen naturforsalenden Gessellschaft* 5: 137.
de Bary, A., 1869, Ueber Schimmel und Hefe, *In*: *Sammlung gemeinverständlicher wissenschaftlicher Vorträge* (R. Virchow and F. von Holzendorf, eds): IV, 87-88: 551-630.
de Bary, A., 1884, *Vergleichende Morphologie und Biologie der Pilze, Mycetozoen und Bacterien*, W.Engelmann, Leipzig.
de Hoog, G.S., 1979, Deductive classification - worked examples using anamorph and teleomorph data in the ascomycetes, *In*: *The Whole Fungus* (W.B. Kendrick, ed.) 1: 215-239. National Museum of Natural Sciences, Ottawa.
Deighton, F.C., 1973, Studies on *Cercospora* and allied genera. IV. *Cercosporella* Sacc., *Pseudocercosporella* gen. nov. and *Pseudocercosporidium* gen. nov., *Mycological Papers* 133: 1-62.
Deighton, F.C., 1979, Studies on *Cercospora* and allied genera. VII. New species and redispositions, *Mycological Papers* 144: 1-56.
Deighton, F.C., 1983, Studies on *Cercospora* and allied genera. VIII. Further notes on *Cercoseptoria* and some new species and redispositions, *Mycological Papers* 151: 1-13.
Deighton, F.C., 1987, New species of *Pseudocercospora* and *Mycovellosiella,* and new combinations into *Pseudocercospora* and *Phaeoramularia*, *Transactions of the British Mycological Society* 88: 365-391.
Eriksson, O.E. and D.L. Hawksworth, 1993, Outline of the ascomycetes - 1993, *Systema Ascomycetum* 12: 51-257.
Farr, D.F., 1991, *Septoria* species on *Cornus*, *Mycologia* 83: 611-623.
Fresenius, J.B.G.W., 1850, *Beiträge zur Mykologie*, Heinrich Ludwig Bönner, Frankfurt.
Fuckel, L., 1870, Symbolae Mycologicae. Beiträge zur Kenntniss der Rheinischen Pilze, *Jahrbuch Nassaurschen Vereins Naturkunde* 23-24: 1-459.
Fuckel, L., 1873, Symbolae Mycologicae. Beiträge zur Kenntniss der Rheinischen Pilze Zoreiter Nächtrag, *Jahrbuch Nassaurschen Vereins Naturkunde* 27-28: 1-99.
Gregory, P.H., (1949), Studies on *Sclerotinia* and *Botrytis*. II. De Bary's description and specimens of *Peziza fuckeliana*, *Transactions of the British Mycological Society* 32: 1-10.
Harada, Y., and S.-I. Noro, 1988, *Grovesinia pruni* sp. nov., the teleomorph of a new zonate leaf spot fungus on *Prunus* in Japan, *Transactions of the Mycological Society of Japan* 29: 85-92.
Hawksworth, D.L., 1985, *Kirschsteiniothelia*, a new genus for the *Microthelia incrustans*-group (*Dothideales*), *Botanical Journal of the Linnean Society* 91: 181-202.
Hawksworth, D.L., B.C. Sutton, and G.C. Ainsworth, 1983, *Ainsworth & Bisby's Dictionary of the Fungi*. 7th edn. Commonwealth Agricultural Bureaux, Slough.
Hennebert, G.L., 1973, *Botrytis* and *Botrytis*-like genera, *Persoonia* 7: 183-204.
Hennebert, G.L., and A. Bellemère, 1979, Les formes conidiennes des Discomycètes. Essai taxonomique, *Revue des Mycologie* 43: 259-315.
Hennebert, G.L., and R.P. Korf, 1975, The peat mould, *Chromelosporium ollare*, conidial state of *Peziza ostracoderma*, and its misapplied names, *Botrytis crystallina, Botrytis spectabilis, Ostracoderma epigaeum* and *Peziza atrovinosa*, *Mycologia* 67: 214-240.
Höhnel,F., von, 1916, Fragmente zur Mykologie 997. Über *Illosporium Diedickeanum* Saccardo, *Sitzungsberichte Akademie Wissenschaften Wien*, Abt. 1, 125: 96-98.
Höhnel, F., von, 1923, System der Fungi Imperfecti Fuckel, *Mycologisches Untersuchungen* 1: 301-369.
Honey, E.E., 1928, The monilioid species of *Sclerotinia*, *Mycologia* 20: 127-157.

Hughes, S.J., 1949, Studies on some diseases on sainfoin (*Onobrychis sativa*) II. The life history of *Ramularia onobrychidis* Allescher, *Transactions of the British Mycological Society* 32: 34-59.

Hughes, S.J., 1953, Conidiophores, conidia, and classification, *Canadian Journal of Botany* 31: 577-659.

Hughes, S.J., 1958, Revisiones hyphomycetum aliquot cum appendice de nominibus rejiciendis, *Canadian Journal of Botany* 36: 727-838.

Hughes, S.J., 1976, Sooty moulds, *Mycologia* 68: 693-820.

Juel, H.O., 1920, Uber *Hyphelia* und *Ostracoderma*, zwei von Fries aufgestellte Pilzgattungen, *Svensk botanisk tidskrift* 14: 212-222.

Kendrick, W.B. (ed.), 1971, *Taxonomy of Fungi Imperfecti*, University of Toronto Press, Toronto.

Kendrick, W.B. (ed.), 1979, *The Whole Fungus*. 2 vols. National Museum of Natural Sciences, Ottawa.

Kendrick, W.B., 1981, The systematics of hyphomycetes, *In: Biology of Conidial Fungi* (G.T. Cole and W.B. Kendrick, eds) 1: 21-42. Academic Press, London and New York.

Kendrick, W.B., and F. DiCosmo, 1979, Teleomorph-anamorph connections in ascomycetes, *In*: *The Whole Fungus* (W.B. Kendrick, ed.) 1: 283-410. National Museum of Natural Sciences, Ottawa.

Klebahn, H., 1918, *Haupt- und Nebenfruchtformen der Askomyzeten*, Gebrüder Borntrager, Leipzig.

Korf, R.P., 1972, Synoptic key to the genera of *Pezizales*, *Mycologia* 64: 937-994.

Korf, R.P., 1973, Discomycetes and Tuberales, *In*: *The Fungi* (G.C. Ainsworth, F.K. Sparrow and A.S. Sussman eds) IVA: 249-319. Academic Press, London, New York.

Laibach, F., 1922, Untersuchungen uber einige *Ramularia*- und *Ovularia*-Arten und ihre Beziehungen zur Askomyzetengattung *Mycosphaerella*, *Zentralblatt Bakteriologie, Parasitnekunde, Infektionskrankheiten Hygiene* 55: 284-293.

Link, H.F., 1809, Observationes in Ordines Plantarum naturales. Dissertatio I, *Magazin Besellschaft Naturforschender Freunde zu Berlin* 3: 3-42.

Luttrell, E.S., 1979, Deuteromycetes and their relationships, *In*: *The Whole Fungus* (W.B. Kendrick, ed.) 1: 241-264. National Museum of Natural Sciences, Ottawa.

Mason, E.W., 1937, Annotated aacount of fungi received at the Imperial Mycological Institute, List II (Fasc. 3 - General Part), *Mycological Papers* 4: 68-99.

Micheli, P.A., 1729, *Nova Plantarum Genera*, Bernard Paperini, Florence.

Minter, D.W., P.M. Kirk, and B.C. Sutton, 1982, Holoblastic phialides, *Transactions of the British Mycological Society* 79: 75-93.

Minter, D.W., P.M. Kirk, and B.C. Sutton, 1983a, Thallic phialides, *Transactions of the British Mycological Society* 80 : 39-66.

Minter, D.W., Sutton, B.C. and Brady, B.L., 1983b, What are phialides anyway?, *Transactions of the British Mycological Society* 81: 109-120.

Molliard, M., 1904, Forme conidienne et sclérotes de *Morchella esculenta* Pers., *Revue générale des botanique* 16: 209-218.

Müller, E., 1971, Imperfect-perfect connections in ascomycetes, *In*: *Taxonomy of Fungi Imperfecti* (W.B. Kendrick, ed.): 184-201. University of Toronto Press, Toronto.

Nag Raj, T.R., 1981, Coelomycete systematics, *In*: *The Biology of Conidial Fungi* (G.T. Cole and W.B. Kendrick, eds) 1: 43-84. Academic Press, London and New York.

Nees, C.G.D., 1817, *Das System der Pilze und Schwämme*, Stachelschen Buchhandlung, Würzburg.

Niedbalski, M., J.L. Crane, and D. Neely, 1979, Illinois fungi 10. Development, morphology and taxonomy of *Cristulariella pyramidalis*, *Mycologia* 71: 722-730.

Paden, J.W., 1972, Imperfect states and the taxonomy of the *Pezizales*, *Persoonia* 6: 405-414.

Persoon, C.H., 1801, *Synopsis Methodica Fungorum*, Gottingen.

Reynolds, D.R., and J.W. Taylor (eds), 1993, *The Fungal Holomorph: mitotic, meiotic and pleomorphic speciation in fungal systematics*. CAB International, Wallingford.

Rimpau, R.H., 1961, Untersuchungen über die Gattung *Drepanopeziza* (Kleb.) v.Höhn., *Phytopathologische Zeitschrifte* 43: 257-306.

Saccardo, P.A., 1884, *Sylloge Fungorum omnium hucusque cognitorum*. Vol. 3. Padova.

Saccardo, P.A., 1886, *Sylloge Fungorum omnium hucusque cognitorum*. Vol. 6. Padova.

Stalpers, J.A., 1974, Revision of the genus *Oedocephalum* (Fungi Imperfecti), *Proceedings Koniklije Nederlandse akademie van wetenschappen*, C, 77: 402-407.

Subramanian, C.V., 1971, *Hyphomycetes. An account of Indian species, except Cercosporae*, Indian Council of Agricultural Research, New Delhi.

Subramanian, C.V., 1983, *The Hyphomycetes. Taxonomy and biology,* Academic Press, London and New York.

Sutton, B.C., 1973, Coelomycetes, *In*: *The Fungi* (G.C. Ainsworth, F.K. Sparrow and A.S. Sussman, eds) IVA: 513-582. Academic Press, London and New York.

Sutton, B.C., 1980, *The Coelomycetes*, Commonwealth Agricultural Bureaux, Slough.

Sutton, B.C., 1993, Mitosporic Fungi (deuteromycetes) in the *Dictionary of the Fungi*, *In*: *The Fungal Holomorph: mitotic, meiotic and pleomorphic speciation in fungal systematics* (D.R. Reynolds and J.W. Taylor, eds): 27-55. CAB International, Wallingford.
Sutton, B.C., and I.G. Pascoe, 1987, *Septoria* spoecies on *Acacia*, *Transactions of the British Mycological Society* 89: 521-632.
Sutton, B.C., and I.G. Pascoe, 1988, *Pseudocercospora correicola* sp. nov., another leaf pathogen of *Correa* species from Australia, *Australian Systematic Botany* 1: 87-94.
Sutton, B.C., and I.G. Pascoe, 1989, Some *Septoria* species on native Australian plants, *Studies in Mycology, Baarn* 31: 177-186.
Sutton, B.C., I.G. Pascoe, and I.K. Sharma, 1987, *Pseudocercospora correae* sp. nov., a leaf pathogen of *Correa* species from Australia, *Australian Journal of Botany* 35: 227-234.
Sutton, B.C., and F.G. Pollack, 1974, Microfungi on *Cercocarpus*, *Mycopathologia et Mycologia applicata* 52: 331-351.
Sutton, B.C., and J.M. Waller, 1988, Taxonomy of *Ophiocladium hordei*, causing leaf leasions on *Triticale* and other *Gramineae*, *Transactions of the British Mycological Society* 90: 55-61.
Sutton, B.C., and J. Webster, 1984, *Septogloeum japonicum* and *Marssonina pakistanica* spp. nov., coelomycetes with *Tricellula*-like conidia, *Transactions of the British Mycological Society* 83: 59-64.
Tubaki, K., 1958, Studies on the Japanese hyphomycetes V. Leaf and stem group with a discussion of the classification of hyphomycetes and their perfect states, *Journal of the Hattori Botany Laboratory* 20: 142-244.
Tubaki, K., 1981, *Hyphomycetes- their perfect-imperfect connexions*, J.Cramer, Vaduz.
Tulasne, L.R., and C. Tulasne, 1861-1865, *Selecta Carpologia Fungorum*, 3 vols. Imperial Press, Paris.
Vobis, G., 1980, Bau und Entwicklung der Flechten-Pyknidien und ihrer conidien, *Bibliotheca Lichenologica* 14: 1-141.
Vobis, G., and D.L. Hawksworth, 1981, Conidial lichen-forming fungi, *In*: *Biology of Conidial Fungi* (G.T. Cole and W.B. Kendrick, eds) 1: 245-273. Academic Press, London and New York.
Vuillemin, P., 1910a, Matériaux pour une classification rationelle des Fungi imperfecti, *Compte Rendu Hebdomadaire Séances de l'Academie des Sciences Paris* 150: 882-884.
Vuillemin, P., 1910b, Les conidiosporés, *Bulletin de la Société des Sciences de Nancy* 2: 129-172.
Vuillemin, P., 1911, Les aleuriosporés, *Bulletin de la Société des Sciences de Nancy* 3: 151-172.
Warcup, J.H., and P.H.B. Talbot, 1989, *Muciturbo*: a new genus of hypogeous ectomycorrhizal ascomycetes, *Mycological Research* 92: 95-100.
Weiss, F., 1940, *Ovulinia*, a new generic segregate from *Sclerotiia*, *Phytopathology* 30: 236-244.
Whetzel, H.H., 1937, *Septotinia*, a new genus of the *Ciborioideae*, *Mycologia* 29: 128-146.
Whetzel, H.H., 1945, A synopsis of the genera and species of the *Sclerotiniaceae*, a family of stromatic inoperculate discomycetes, *Mycologia* 37: 648-714.
Wilson, M., M. Noble, and E. Gray, 1954, *Gloeotinia* - a new genus of the *Sclerotiniaceae*, *Transactions of the British Mycological Society* 37: 29-32.

DISCUSSION

Korf: Some mycologists use the term anamorph only to refer to a sporulation object, but for me a sclerotium and a hypha are both anamorphs. To say that a fungus has no anamorphs therefore drives me up the wall !

Hennebert: The term anamorph was originally coined by Weresub and myself for sporulating structures, however we came to appreciate that some spore forms such as chlamydospores were not easily dispersed and, therefore, bulbils and microsclerotia were also included as they had a survival value and can be dispersed by some means. Hyphae may also be dispersed by invertebrates (e.g. mycangia). I now think that the anamorph can be conceived as a large part of the fungus, without which ascomata could not arise. Anamorph is thus not synonymous with mitosporic fungi. The eight ascospores also arise by mitosis (following previous meiosis), as do structures in the ascomata up the gametangial stage. Anamorph and teleomorph were introduced as mutually exclusive terms to facilitate the understanding and application of dual nomenclature in the Code.

Kimbrough: If you have *Trichophaea* in culture along with its *Dichobotrys* anamorph, chlamydospores or other propagules, you would in essence have the holomorph?

Hennebert: Not necessarily, because other states may be found in natural environments. The term holomorph is an abstraction for us but exists in nature.

Kimbrough: Then we will never know the holomorph.

Taylor, J.: The discussion suggests that the terms anamorph/ teleomorph/ holomorph are to be preferred because they do not make any assumptions about mitosis/ meiosis, but these terms do not convey much biological information. I want to know whether a species is clonal or can recombine in nature, something, unfortunately, that cannot be known without substantial work. However, if we use the mitosporic/ meiosporic terms we are at least thinking along those lines and in a manner that can be understood by other biologists. In teaching, I find the mitotic/ meiotic terminology is immediately understood, whereas after I have taken some minutes to explain the anamorph terminology, the students have lost interest. The erection of special terms for fungi makes it harder for others to appreciate what fungi are capable of.

Hennebert: Anamorph and teleomorph are of Greek origin and I appreciate this language is rarely taught today. However, if you know: *ana-* = secondary, accessory; *teleo-* = final; and *holo-* = entire, that is easily understood and more meaningful than the ambiguous imperfect/ perfect that mycologists used to employ. The rapid adoption of the anamorph terminology after its introduction was proof of its ability to satisfy a real need.

Hawksworth: With regard to the word anamorph, it may not be generally appreciated that it has a non-mycological meaning in English. It is applied to deformed figures which appear correctly proportioned when rightly viewed (e.g. elongated letters painted on road surfaces to facilitate legibility to drivers viewing at an angle) and in studies of perception. If a student encounters this word in a mycological text, and then refers to a standard English dictionary, confusion and frustration will result ! It is unfortunate that this usage was not considered prior to the introduction of the term into mycology. Anything that makes communication difficult is counter-productive for our subject.

Hennebert: As there are already many cases where the same word has dissimilar meanings in different fields I do not think that precludes its use in mycology. The word was originally coined by Donk.

Galloway: Acharius (*Methodus Lichenum,* 1803), introduced a new terminology for lichenology. At the time he was roundly condemned by his contemporaries for introducing too many Greek terms into the purity of botanical Latin, yet almost two centuries on these are still in use.

Gams: Hennebert has summarized the advantages of the anamorph/ teleomorph terminology. Why then switch in this case and also from conidia to mitospore ? Now this terminology has been established since 1971, surely it is more cumbersome to consider a change.

Rossman: In *Hypocreales,* connecting teleomorphs and anamorphs has been tremendously exciting. Samuels and myself have cultured anamorphs from hypocrealean teleomorphs, and through the years we have correlated many morphs by such methods. I would like to dispel the myth that *Nectria* has 19 anamorph genera. *Nectria* needs to be more narrowly circumscribed and when this is done it will have a single anamorph. It has been especially exciting to add molecular data to morphological hypothesis and see how they correlate. It is definitely worth trying to collect fresh ascomata and grow the ascospores.

Korf: Many fungi produce spermatia, which are not conidia in the sense that they do not normally germinate (there is one report in *Botryotinia*) and are an integral part of the sexual process. They are certainly mitospores, and to me are also anamorphs even though they do not have a role in dispersal.

Poelt: I do not think spermatia can be termed anamorphs as they cannot form an independent mycelium. I would like to give two warnings with respect to this discussion: (1) never try to define concepts in biology to 100 %; and (2) introduce general concepts for students so that they do not become overloaded. The future of mycology is our students.

Lodha: All terms, even that of *Fungi*, is not entirely convenient. Every term will have some defects. If changes are made in well-known terms, many books will continue to employ others for some time causing continuing problems for students. Remember that mycology is not only for specialists but also for students.

Blackwell: To me a fungus is one morph. It is a single mycelium with one DNA capable of producing diaspores that can be of different types. This is an easier and more exciting way for students to view the situation.

Hennebert: This is why we used to speak of botanical vs. anatomical nomenclatures in relation to these terms. For the former we need only one name for the fungus as it would be nonsensical to allocate separate names to each organ of a flowering plant. We call the organs in fungi morphs in an anatomical system.

Blackwell: But the whole flowering plant is a sporophyte, something lacking in the true fungi. We have a single type of mycelium that may or may not be capable of producing different types of diaspores.

Gams: I would agree with Blackwell that expressions such as "the anamorph of *Venturia* is pathogenic and the teleomorph not" are incorrect. Rather, the fungus *is* pathogenic and produces the anamorph during the pathogenic phase. The morph terminology is heavy, but as it has been fixed in the Code since 1981 I do not feel a change should be enforced now.

Hawksworth: The Code does not dictate practices, but is designed to meet the needs of its users. A provision introduced by mycologists as recently as 1981 can easily be removed at a subsequent Congress if that is what mycologists wish.

CAN WE RECOGNIZE MONOPHYLETIC GROUPS AMONG HYPHOMYCETES?

W.B. Kendrick and G. Murase

Biology Department
University of Waterloo
Waterloo, Ontario
Canada N2L 3G1

SUMMARY

Although evolutionary pressures have produced an extremely wide range of phenotypes among conidial fungi, morphology may sometimes be sufficiently distinctive or conservative as to suggest either monophyly or strong convergence in certain groups of anamorph-genera. A number of these groups are discussed and illustrated, with confirmatory or confounding evidence from the known teleomorphs.

INTRODUCTION

To what extent do phenotypic similarities reflect actual relationships? This has always been a crucial question in fungal taxonomy, because until the advent of molecular techniques, morphological and, more recently, developmental characters provided almost the entire basis for our comparisons of most fungal taxa. This has been particularly true of the conidial fungi, which in most cases still remain unconnected to a teleomorph, leaving their relationships and ancestry in doubt (Kendrick, 1981). In addition, the fungi are replete with examples of close similarities produced by convergent evolution, which must inject a cautionary note into these discussions. For example, the group of morphologically similar fungi placed in the order *Ophiostomatales* has now been shown to be genetically heterogeneous, and to include several independently evolved lines (see Wingfield *et al.*, this volume). Various members of the anamorph genus *Acremonium* have teleomorphs in at least 29 different holomorph genera representing nine different orders of ascomycetes (Murase and Kendrick, unpubl.): this is because the structures seen in *Acremonium* are considered to constitute a simple, rather generalized anamorph which has almost certainly evolved on several occasions. Species of *Chrysosporium*, another rather generalized anamorph-genus, have been connected with at least 22 ascomycetous holomorph genera of 3 orders (Murase and Kendrick, unpubl.), and with one basidiomycetous holomorph (*Phanerochaete chrysosporium*; Kendrick and Watling, 1979; Carmichael, 1962).

However, hope springs eternal, and the coexistence of several unusual features in the 13 groups of hyphomycetous anamorph genera assembled for this paper provide some grounds for suggesting that at least some of these groups are internally homogeneous or truly related. In those cases where connections have been established between these anamorphs and their teleomorphs, these provide a good test of that homogeneity. It is instructive to see if the holomorphs of each group cluster within individual holomorphic orders or not, thus reinforcing or weakening the prediction of relationship.

Table 1. Anamorph genera grouped by morphology and development

Case number & Features	Anamorph genus	Development	Holomorph: Genus ; (no. of connections)	Order
1. Conidia lunate, amero or didymo or phragmo, colourless, usually with one appendage at each end (bisetulate); conidiogenous cells phialidic.	*Codinaeopsis*	blastic-phialidic	-	
	Dictyochaeta	blastic-phialidic	*Chaetosphaeria* (4)	Trichosphaeriales
	"	" "	*Striatosphaeria* (2)	Trichosphaeriales
	Menidochium	blastic-phialidic	-	
	Menispora	blastic-phialidic	*Chaetosphaeria* (2)	Trichosphaeriales
	Menisporopsis	blastic-phialidic	-	
	Thozetella	blastic-phialidic	-	
	Venustusynnema	blastic-phialidic	-	
2. Conidia biconic, pigmented with colourless equatorial band; separating cells; conidiophores dark, with lobed base.	*Beltrania*	blastic-sympodial	-	
	Beltraniella	blastic-sympodial	*Pseudomassaria* (2)	Sphaeriales
	Beltraniopsis	blastic-sympodial	-	
	Pseudobeltrania	blastic-sympodial	-	
	Rhombostilbella	blastic-sympodial	-	
3. Conidia with colourless germ-slit; conidiophore septa dark, thick; conidiogenous axis arises from phialide-like mother cell.	*Arthrinium*	basauxic	*Apiospora* (4)	Sphaeriales
	"	"	*Physalospora* (1)	Sphaeriales
	"	"	*Pseudoguignardia* (1)	Sphaeriales
	"	"	[?]*Rhopographus* (1)	Dothideales
	Cordella	basauxic	-	
	Dictyoarthrinium	basauxic	-	
	Pteroconium	basauxic	*Apiospora* (1)	Sphaeriales
4. Conidia spheroidal, loosely constructed by branching, involuted, anastomosing hyphae, exterior often lattice- or net-like.	*Clathroconium*	branched hyphal system	-	
	Clathrosphaerina	branched hyphal system	*Hyaloscypha* (1)	Leotiales
	"	" " "	*Mollisia* (1)	Leotiales
	Clathrosporium	branched hyphal system	-	
	Spirosphaera	branched hyphal system	-	
	Strumella	branched hyphal system	-	
5. Conidia colourless, usually branched, without septa but with regularly-spaced constrictions (isthmi).	*Isthmolongispora*	blastic-sympodial	-	
	Phalangispora	blastic-sympodial	-	
	Ramaraomyces	blastic-sympodial	-	
	Speiropsis	blastic-sympodial	-	
	Wiesneriomyces	blastic-sympodial	-	
6. Conidia phragmo, monilioid, pigmented; arising from darker mother cells.	*Bahusaganda*	blastic-acropetal	-	
	Dwayabeeja	blastic-acropetal	-	
	Torula	blastic-acropetal	-	
7. Conidia highly condensed branching systems, lobed.	*Dendrosporium*	branched hyphal system	-	
	Desmidiospora	branched hyphal system	-	
	Riessia	branched hyphal system	-	

8. Conidia stauro, with large, rounded, mamillate or dome-like peripheral processes.	*Pyramidospora*	branched hyphal system	-	
	Sopagraha	branched hyphal system	-	
	Stephanoma	branched hyphal system	*Hypomyces*	Hypocreales
	Uberispora	branched hyphal system	-	
	Uvarispora	branched hyphal system	-	
9. Conidia angular, with long apical setulae; conidiophores pigmented.	*Chalarodes*	blastic-phialidic	-	
	Nawawia	blastic-phialidic	-	
	Phialosporostilbe	blastic-phialidic	-	
10. Conidia retrogressive, amero or didymo, colourless, in false chains; conidiogenous cells becoming shorter during conidiation.	*Basipetospora*	blastic-retrogressive	*Monascus* (3)	Eurotiales
	"	"	*Xeromyces* (1)	Eurotiales
	Cladobotryum	blastic-retrogressive	[?] *Coniochaetidium* (1)	Sordariales
	"	"	*Hypomyces* (17)	Hypocreales
	Trichothecium	blastic-retrogressive	*Heleococcum* (1)	Hypocreales
	"	"	*Hypomyces* (3)	Hypocreales
11. Conidia amero, arising synchronously on ampullae; conidiophores often dichotomously branched.	*Amphobotrys*	blastic-synchronous	*Botryotinia* (2)	Leotiales
	Botrytis	blastic-synchronous	*Botryotinia* (15)	Leotiales
	Streptobotrys	blastic-synchronous	*Streptotinia* (2)	Leotiales
	Verrucobotrys	blastic-synchronous	*Seaverinia* (1)	Leotiales
	"	"	*Sclerotinia* (1)	Leotiales
	Chromelosporium	blastic-synchronous	*Marcelleina* (1)	Pezizales
	"	"	*Muciturbo* (1)	Pezizales
	"	"	*Peziza* (5)	Pezizales
	Dichobotrys	blastic-synchronous	*Pyropyxis* (1)	Pezizales
	"	"	*Sphaerosporella* (1)	Pezizales
	"	"	*Trichophaea* (5)	Pezizales
	Padixonia	blastic-synchronous	*Xylaria* (2)	Xylariales
	"	"	*Xylosphaera* (1)	Xylariales
	Botryosporium	blastic-synchronous	-	
	Glischroderma	blastic-synchronous	-	
	Gonatobotrys	blastic-synchronous	-	
	Gonatobotryum	blastic-synchronous	-	
	Nematogonum	blastic-synchronous	-	
	Ostracodermidium	blastic-synchronous	-	
	Phymatotrichopsis	blastic-synchronous	-	
	Pulchromyces	blastic-synchronous	-	
12. Conidia cheiro, constructed of many separate but parallel hyphae.	*Cryptocoryneopsis*	branched hyphal system	-	
	Cryptocoryneum	branched hyphal system	-	
	Nidulispora	branched hyphal system	-	
	Petrakiopsis	branched hyphal system	-	
13. Conidia developing bilaterally from conidiogenous locus, attachment appears median.	*Ancoraspora*	monoblastic	-	
	Ceratosporella	blastic-percurrent	-	
	Latericonis	blastic-sympodial	-	
	Weufia	blastic-sympodial	-	

MATERIALS AND METHODS

Assembly of groups. We began by sifting through published illustrations of more than 1200 genera of hyphomycetes. The majority are illustrated in Carmichael *et al.* (1980), and the illustrations of newer genera are derived from the original publications. We proceeded to establish informal groups for those with several shared features. In this way we established 13 groups (Table 1). More groups could have been recognized, but those discussed here are adequate to test our hypothesis that at least some monophyletic groups can be recognized visually. Where it was available we added information on connections established with ascomycetous holomorphs, and on the orders in which those holomorphs have been provisionally placed (Murase and Kendrick, unpubl.).

RESULTS AND DISCUSSION

Members of most of our groups of anamorph genera share strikingly unusual or rare morphological features. Within most of our groups the genera also exhibit similar developmental processes, especially those involved in conidium ontogeny; in two examples (case 3, Fig. 1 [lower]; and case 10, Fig. 2 [lower]) the developmental process is also rare. On reflection, such a conjunction of morphological and/or developmental rarity or uniqueness will be seen to be a prerequisite in the kind of analysis we are attempting.

We note, however, that several of the methods by which conidia develop may reasonably be assumed to have evolved on more than one occasion (because there are only so many ways in which a conidium can arise from a parent cell, and because they are associated with such diverse holomorph taxa). In addition, some methods of conidium ontogeny are rather plastic, and a single anamorph may either switch from one method to another, and sometimes back again, during conidiation. Alternatively, what is often regarded as a single anamorph may produce conidia by two different methods at the same time. Therefore, we did not feel it necessary to restrict membership of some of our groups to fungi with identical developmental processes.

Testing of Groups

One way of discovering if conidial fungi are indeed related is to examine the dikaryomycotan holomorphs, where these are known. If the known holomorphs are all closely related, the likelihood of relatedness/monophyly is increased. One reasonably accessible measure of holomorph relationship is the ordinal placement of the holomorphs. If they are in the same order, a relatively close relationship seems likely. This criterion is generally met in our cases 1 and 3 (Fig. 1 [upper and lower, respectively]): the holomorphic exception (disposed in the holomorph order *Dothideales*) in case 3 is isolated in the face of four other congruent connections (all in the holomorph order *Sphaeriales*), and should be re-examined.

In case 10 (Fig. 2 [lower]), and case 11 (Fig. 3 [upper]) the known holomorphs are distributed among three different orders, suggesting that the anamorphic facies in question may have arisen more than once. In cases 2, 4 and 8, only one or two connections have been established, so no generalization is yet possible, although the discovery of any further connections would clearly provide a test for our hypotheses. In cases 5, 6, 7, 9, 12 and 13, nothing is yet known about the holomorphs, so our predictions remain untested at that level.

Homogeneous Groups

It is worth examining our successful groupings more closely: for example, the mode of conidium development which has been termed "basauxic" (case 3, Fig. 1 [lower]), is relatively rare. When considered in conjunction with other unusual features such as the thick, darkly pigmented septa of the conidiogenous axis, and the germ slits in the conidia, characters shared by most of the fungi exhibiting this kind of conidium development, there appears to be a strong *prima facie* case for monophyly.

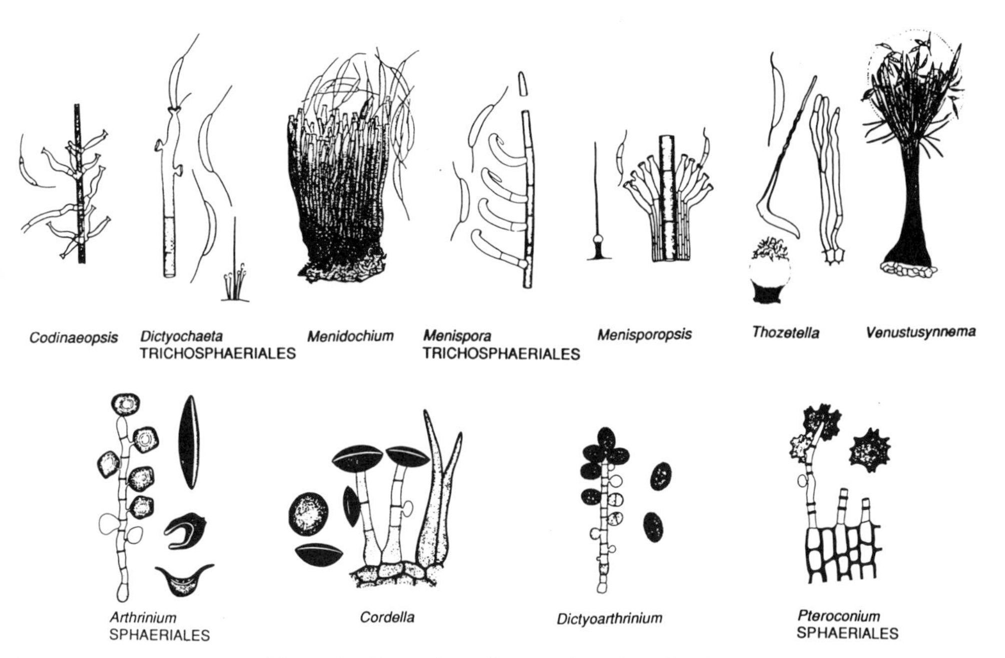

Figure 1. Illustrations of cases 1 (upper), and 3 (lower).

From the point of view of rarity, we might suggest that what has been called "blastic-retrogressive" conidium development (case 10, Fig. 2 [lower]) should find a confirmatory reflection in the known holomorphs. Yet here we are confronted with a mixture of three holomorphic orders, and we must either admit that even such a rare method of conidium development has evolved more than once, or we must suggest, rather boldly, that the existence of similar but rare anamorphs in the three orders may indicate ancestral connections among them. Only molecular biology can test this hypothesis.

Unfortunately, none of our groups has been connected in its entirety to dikaryomycotan holomorphs, and because of the strong probability that many conidial anamorphs have lost their ability to form teleomorphs (i.e. that they are anamorphic holomorphs), we realise that the recognition of holomorphic orders is unlikely to provide the final answers, and that our putative groupings may all ultimately have to be confirmed or rejected by molecular techniques. We believe that many of our groupings constitute appropriate starting points for such investigations.

Although morphology and development are consistent within most of our groups, the anamorph genera in case 13 (Fig. 3 [lower]) do not share the same kind of conidiogenesis, though all do have the same basic shape, orientation and connection of the conidium to the conidiophore. We readily admit that this group is more speculative than most of the others.

We regard some of the cases listed in Table 1 as not only monophyletic, but so closely related that they should probably be considered congeneric as well (an approach suggested by Kendrick, 1980). Among these are the genera of case 4: *Clathroconium*, *Clathrosphaerina*, *Clathrosporium*, *Spirosphaera*, and *Strumella*, with basically similar spheroidal propagules produced by branching, involuted and anastomosing hyphal systems. The genera of case 12: *Chalarodes, Nawawia*, and *Phialosporostilbe* are also very similar — the unusual blastic-phialidic, angular conidia with long apical setulae suggest a close relationship.

In a test-case for "condensation of genera" (Kendrick, 1981) a group of genera almost identical to the present case 2 (*Beltrania, Beltraniella, Beltraniopsis, Pseudobeltra-*

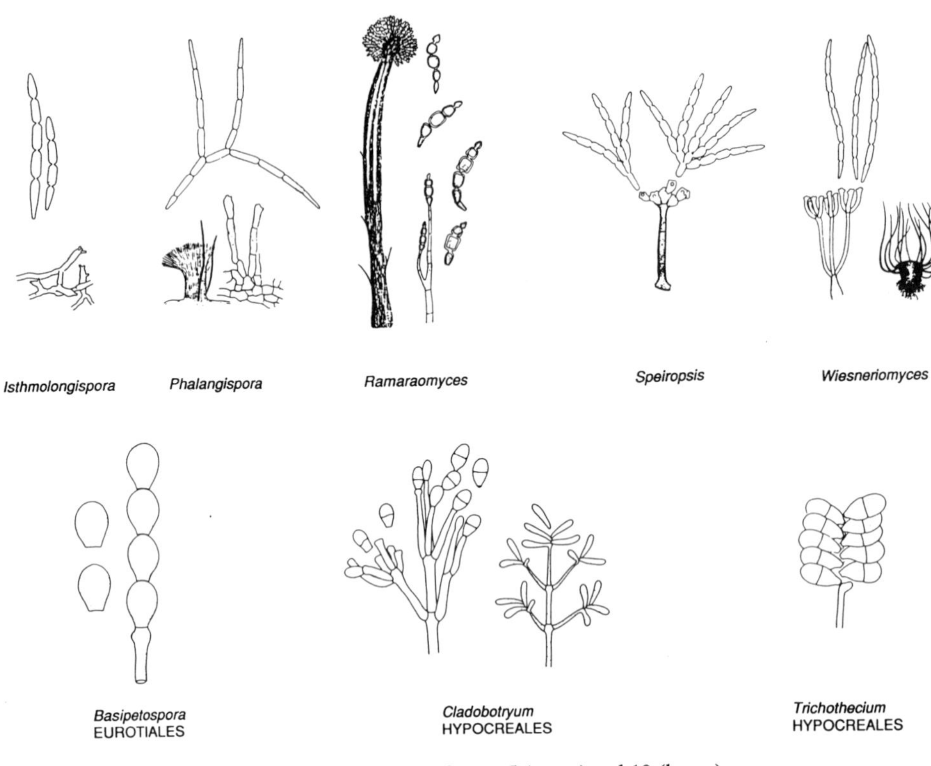

Figure 2. Illustrations of cases 5 (upper) and 10 (lower).

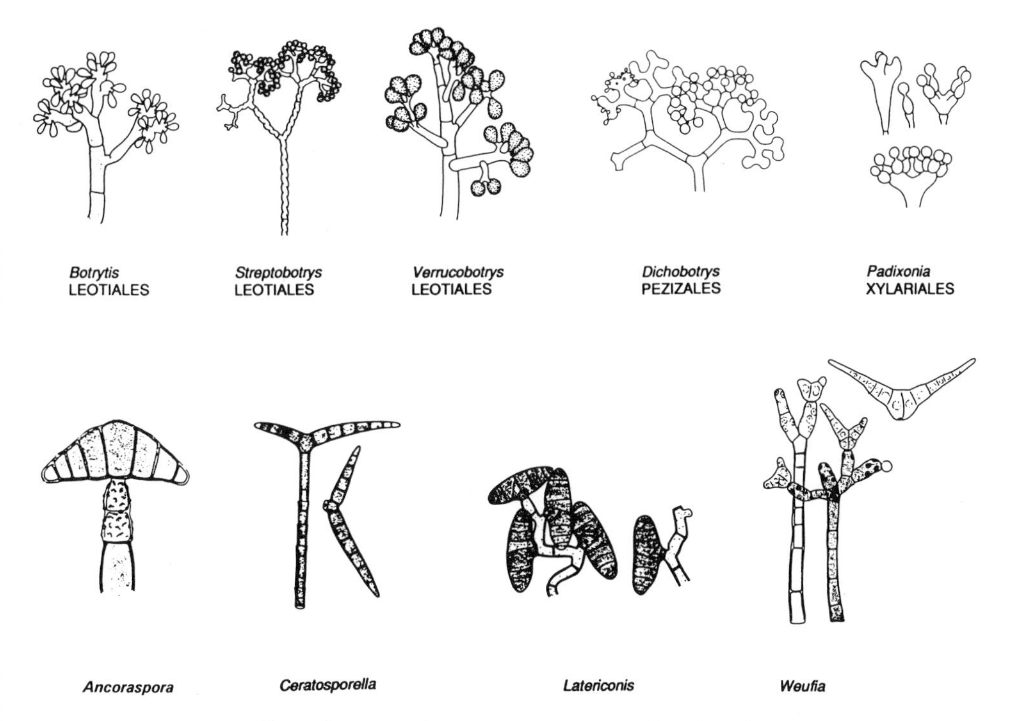

Figure 3. Illustrations of cases 5 (upper) and 10 (lower).

nia, Rhombostilbella) was proposed for amalgamation. Although this step has not yet been taken, it may be one way of rationalising the rapidly increasing number of anamorph genera.

ACKNOWLEDGMENTS

The authors acknowledge financial support from the Natural Sciences and Engineering Research Council of Canada in the form of an operating grant to Professor Kendrick.

REFERENCES

Bhat, D.J. and Sutton, B.C., 1983, New and interesting hyphomycetes from Ethiopia, *Transactions of the British Mycological Society* 85: 107-122.

Carmichael, J.W., 1962, *Chrysosporium* and some other aleuriosporic hyphomycetes, *Canadian Journal of Botany* 40: 1137-1174.

Carmichael, J.W., W.B. Kendrick, I.L. Conners, and L. Sigler, 1980, *Genera of Hyphomycetes*, University of Alberta Press, Edmonton.

Castañeda Ruiz, R.F. and Kendrick, B., 1990, Conidial fungi from Cuba:1, *University of Biology, ser.* 32: 45-48.

Eriksson, O.E., and D.L. Hawksworth, 1991, Outline of the ascomycetes - 1990, *Systema Ascomycetum* 9: 39-271.

Hawksworth, D.L., B.C. Sutton, and G.C. Ainsworth, 1983, *Ainsworth & Bisby's Dictionary of the Fungi*, 7th edn, Commonwealth Mycological Institute, Kew.

Kendrick, B., 1980, The generic concept in hyphomycetes — a reappraisal, *Mycotaxon* 11: 339-364.

Kendrick, B., 1981, The systematics of Hyphomycetes. *In*: *Biology of Conidial Fungi* (G.T. Role and B. Kendrick, eds) 1: 21-42. Academic Press, New York.

Kendrick, B., and R. Watling, 1979, Mitospores in Basidiomycetes, *In*: *The Whole Fungus* (B. Kendrick, ed): 473-545. National Museums of Canada, Ottawa.

Marvanová, L., 1980, New or noteworthy aquatic hyphomycetes, *Clavatospora, Heliscella, Nawawia* and *Heliscina*, *Transactions of the British Mycological Society* 75: 226-227.

McKenzie, E.H.C., 1991, Dematiaceous hyphomycetes on *Freycinetia* (*Pandanaceae*). 3. *Chalarodes* gen. nov., *Mycotaxon* 42: 89-93.

Mercado Sierra, A., and J. Mena Portales, 1905, Neuvo género de hifomicete fialídico de Cuba, *Revista del Jardin Botanico Nacional, La Habana* 6: 59-60.

Murase, G., and B. Kendrick (in preparation), *Anamorph-Teleomorph Connections in Ascomycetes*. Mycologue Publications, Waterloo.

Nawawi, A., and A.J. Kuthubutheen, 1987, *Clathrosporium intricatum* gen. et sp.nov. an aeroaquatic hyphomycete, *Transactions of the British Mycological Society* 89: 407-411.

Nawawi, A., and J. Webster, 1982, *Phalangispora constricta* gen. et sp.nov., a sporodochial hyphomycete with branched conidia, *Transactions of the British Mycological Society* 79: 45-64.

Rao, N.K., C. Manoharachary, and R.D. Goos, 1989, Forest litter hyphomycetes from Andhra Pradesh, India. IV. A new genus of synnematous hyphomycetes, *Mycologia* 81: 790-793.

Rao, V., K.A. Reddy, and G.S. de Hoog, 1984, *Latericonis*, a new genus of dematiaceous hyphomycetes, *Mycotaxon* 19: 409-412.

Rodrigues, N., 1981, *Ancoraspora*, un genéro nuevo de hyphomycetes de la fumagina, *Revista del Jardin Botanico Nacional, La Habana* 2: 19-27.

Samson R.A., and H.C. Evans, 1982, *Clathroconium*, a new helicosporous hyphomycete genus from spiders, *Canadian Journal of Botany* 60: 1577-1580.

Ultrastructure

ASCI AND ASCOSPORES IN ASCOMYCETE SYSTEMATICS

A. Bellemère

53 jardins Boïeldieu
F-92800 Puteaux
France

SUMMARY

Ascus characters are only of secondary importance in ascomycete systematics. This results in part from rather difficult technical problems in interpreting observations made only by light microscopy. The contribution to ascomycete systematics from ascus characters are illustrated in different groups, highlighting recent progress. Difficulties resulting from superficial or imperfect use of these criteria are also mentioned.

The proper systematic value of different ascus characters is considered in several groups of ascomycetes lichenized and including ascus formation, morphological characters and their variations during ascus development, structure of the epiplasm, ascus wall structure, apical wall structure, ascus dehiscence, ascosporogenesis, and the fine structure of the ascospore wall.

The systematic consequences resulting from a more important and more coherent utilization of ascus characteristics are considered. The relative constancy in several types of ascus seems to indicate that they can be used to define taxa of a rather high level. Conversely, ascospore wall structure is probably more dependent on external variations and has to be used only to define lower systematic levels.

Data relating to ascus characters are frequently insufficient in the systematic literature. If asci are precisely described, and if their variations during ascus development are carefully reported, important progress will be made in systematics and in our appreciation of evolution in ascomycete fungi.

INTRODUCTION

The ascus, characteristic cell of ascomycetes, is quite remarkable not only by its origin from the fusion of dikaryotic cells with the formation of a hook, but also by the successive realization of a series of major biological functions: karyogamy, meiosis, ascosporogenis, and ascospore liberation. Asci are surely a result of a long evolution during which these functions have been progressively concentrated in the same cell. The ascus is thus a very significative cell and its variations are surely of special systematic value. In this contribution only the characters of the walls of asci (lateral and apical) and of the ascospores are considered.

ASCI AND ASCOSPORES IN ASCOMYCETE SYSTEMATICS

Asci and Ascospores in Classical Systems

It is well-known that in classical classifications the major groups of ascomycetes (i.e. *Discomycetes*, *Pyrenomycetes*, *Loculoascomycetes*) are essentially defined by characters of the structure and development of the ascomata, with which ascus characters are secondarily associated (i.e. bitunicate asci, with "jack in the box" dehiscence in *Locu-*

Ascomycete Systematics: Problems and Perspectives in the Nineties
Edited by D.L. Hawksworth, Plenum Press, New York, 1994

loascomycetes, unitunicate asci in *Disco-* or *Pyrenomycetes*). In these groups several ascus characters are sometimes specifically used at a lower level to distinguish secondary groups (i.e. ascus dehiscence in *Discomycetes*: operculate *vs.* inoperculate). But generally these characters are associated with other criteria in the subdivision of the major groups (i.e. archeasceous asci and lichenization to define *Lecanorales*; apothecial structure, ascus and ascospore morphology to define *Ostropales* and *Rhytismatales*). At the generic or species level ascospore morphology has been abundantly used by classical authors (i.e. Rehm, Saccardo, von Höhnel, Petrak), but it has also been strongly underlined that this character is of a limited value (Nannfeldt, 1932).

Difficulties Encountered in the Use of Ascus Characteristics in Systematics

The study of asci in the LM (= light microscope) is relatively difficult for several reasons:

(1) Asci have to be correctly extracted from the ascoma, they are small, and aspects of fresh material may considerably differ from herbaria or fixed material (Baral, 1992), ascus morphology and structure vary during ascus development, and controversies exist in relation to reagents in use (Baral, 1987b). The scanning electron microscope (SEM) does not give internal details. Studies with the transmission electron microscope (TEM) are long, specialized, difficult, and may lead to artefacts and figures which do not always exactly correlate with LM data.

(2) Moreover, precise ascus descriptions are frequently neglected by authors (ancient and also a few modern ones) and suitable drawings are frequently lacking. Consequently, data concerning asci are generally insufficient to facilitate sound systematic work.

However, in spite of these difficulties, recent studies by LM, SEM or TEM have provided numerous positive items of data which have led to important modifications in ascomycete systematics.

Recent Progresses Using Ascus Characters

In recent years development of TEM studies correlated with SEM ones, has brought many fine data sets on ascus and ascospore structure which are very useful as systematic tools. First, the TEM confirmed the reality of the ascus apical structures, the observation of which was difficult in the LM. Conversely, some which previously appeared more or less aberrant have been interpreted by means of the TEM as variants of a general scheme (i.e. asci of *Lecanorales*). Fine details unsuspected by the LM have been revealed by the TEM (i.e. a large hyaline perispore or very fine ornamentation of the wall). The assimilation of TEM data makes a better interpretation of observations in the LM possible.

One of the most significant examples of recent progresses is a new delimitation, in the lichens, of the long-established families *Lecanoraceae* s.lat. and *Lecideaceae* s.lat. (Hafellner, 1984) where genera were more sharply defined by consideration of their ascus apices. This led to a restriction in the extent of these families, the definition a many new families and genera, and to a change in the systematic position of numerous taxa. More recently, new systematic concepts have been brought into other lichenized families through consideration of ascus apical structure, as in *Cetrariaceae* (Kärnefelt et al., 1992), *Alectoriaceae* and *Parmeliaceae* (Kärnefelt and Thell, 1992), *Lichinaceae* (Moreno and Egea, 1991), and *Pyrenulaceae* (Aptroot, 1991). Some families have been suppressed, their members being included in others which may be new (i.e. *Asterothyriaceae* placed in *Thelotremataceae*, *Gomphillaceae* and also *Solorinellaceae*; Vezda and Poelt, 1990).

At the ordinal level, it was recognized that the bitunicate fissitunicate asci of *Arthoniales* and *Opegraphales* have a similar apex construction (with a ring frequently reacting with KOH+I), and the fusion of the two orders was proposed (Tehler, 1990).

In operculate *Discomycetes*, the classification was reinforced by a consideration of the number of nuclei in the ascospores (i.e. *Helvellaceae* 4, *Morchellaceae* n, other families 1-2; Berthet, 1964). This criterion had its part in the dividing of the *Tuberaceae* (*Tuberales*) and the redisposition of their constitutive genera in different families of *Pez-*

izales (Trappe, 1979). Fine structure of the asci and ascospores indicate a similarity between *Tuber* and *Terfezia* (Janex-Favre and Parguey-Leduc, 1985), and also led to the inclusion of *Discina* in *Gyromitra* (Kimbrough et al., 1990). It also confirms the homogeneity of some genera (e.g. *Thelebolus*; Samuelson and Kimbrough, 1978). Conversely, *Sarcoscyphaceae* were distinguished from *Sarcosomataceae* by fine ascus apical structure (Bellemère et al., 1990). New genera were also separated on the basis of this criterion (e.g. *Pseudascozonus*, Brummelen, 1987; *Donadinia*, Bellemère et al., 1990). However, a few studies in which several types of fine structure of asci or ascospores were revealed have not yet led to formal systematic changes.

In inoperculate discomycetes, the new family *Dactylosporaceae* was recognized because of a peculiar ascus structure and dehiscence (Bellemère and Hafellner, 1982). In the *Eupyrenomycetes*, genera have been clearly separated from others in the *Halosphaeriales* as they differ in the constitution of their ascospore appendages (Johnson, 1980). In the *Coryneliaceae* (*Coryneliales*) a special type of ascus was observed in *Corynelia*, but not in *Coryneliopsis* (Johnston and Minter, 1989). Similarities in ascus apex lead some authors to fuse some orders.

In different groups of ascomycetes, several types of ascus structure and dehiscence have been frequently recognized by authors, often by TEM studies, but with no formal systematic changes proposed (i.e. Baral, 1987a; Bellemère, 1977; Bellemère et al., 1987; Eriksson, 1981; Merkus, 1973-76; Parguey-Leduc and Janex-Favre, 1984; Reynolds, 1987, 1989b; Brummelen, 1978; Verkley, 1992). Attention was drawn to possible convergence in ascospore wall structure in several genera (Bellemère et al., 1992).

Discrepancies in the Use of Ascus Characters in Systematics

In numerous families, asci of several genera clearly differ from those of the type genus. For instance, in lichenized families, this is the case in *Lichinaceae*, *Gyalectaceae* (only a few genera have asci of the *Gyalecta*-type), *Thelotremataceae* (*Nadvornikia*), and *Graphidaceae* (*Helminthocarpon*). In inoperculate *Discomycetes*, *Leotiaceae*, *Sclerotiniaceae*, and *Odontotremataceae* are heterogenous. In operculate *Discomycetes*, asci of *Cyttariales* are of the inoperculate *Bulgaria*-type (Mengoni, 1986). The new genus *Reddellomyces* differs from other *Pyronemataceae* in the ascus wall becoming thinner at the apex. Sometimes, ascus types differ in different families in the same order, as for instance in *Asterothyriaceae* and *Thelotremataceae* of *Graphidales*, *Lichinaceae* and *Peltulaceae* of *Lichinales*, and some families in *Dothideales*.

It must also be mentioned that sometimes systematic difficulties result from the utilization of ascus characters. A few genera showing the main characters of a family may, however, differ in ascus structure (i.e. *Auriculora* and *Bacidiaceae*; Henssen and Titze, 1990) and so have to be placed in an "*incertae sedis*" position. In other cases, ascus characters of genera are intermediate between those of two families (i.e. *Llimonea* between *Opegraphaceae* and *Roccellaceae*; Torrente and Egea, 1992). Uncertainty often arises because the characteristics of asci are imperfectly known: i.e. an ascus wall is mentioned as amyloid or not according to the authors, asci of several *Pyrenomycetes* s.lat. are described either as unitunicate or bitunicate-fissitunicate and subsequently placed either into *Eupyrenomycetes* or *Loculoascomycetes* (i.e. *Trichodelitschia*, Barr, 1990; *Trabutia*, Barr, 1987; von Arx, 1987). Further, authors often neglect to take ascus characters into account when proposing systematic modifications or describing new genera.

ASCUS INDIVIDUALIZATION

The origin of the ascus differs in major groups, and this is generally considered as a fundamental character (in relation to ascoma development) and is taken into account in the present system. *Saccharomycetaceae*, *Taphrinaceae*, and *Protomycetaceae* are separated from *Plectomycetes* with chains of asci and from other ascomycetes with typical ascogenous hyphae.

Early but precise studies have shown that the sporophytic apparatus is complex and that ascogenous hyphae may be of different types (Chadefaud, 1943). The systematic value of this criterion is still unclear (Wong and Chien, 1986; Wu and Kimbrough, 1990). Typical croziers are encountered in many families, but differences may exist in genera of the same family (*Leotiaceae*) or in species of the same genus (*Lachnum* s.lat.).

Croziers of a special type are well-known in *Orbiliaceae*, and observed also elsewhere (*Coryneliopsis*). Ascogenous hyphae form early or late, are more or less developed, and are spread-out or erect; these variations are probably correlated with ascomatal development. More data are needed on all these topics.

ASCUS MORPHOLOGY

It is well-known that asci largely differ in their general morphology and ascospore disposition. In my opinion, this criterion is probably significant in systematics and should have been considered more fully, but as morphological changes appear during ascus development, only similar and well-defined stages can be taken into account for comparisons in the description. The special elongation of protruding asci (*Ascobolus*), or of the subterminal part of asci in *Stictidaceae* (Bellemère, 1960), is probably an adaptative character of secondary systematic signification.

ASCUS LATERAL WALL STRUCTURE

The composition of the ascus wall is surely an important systematic criterion, but, for the present, data are insufficient for systematic application. The cellulosic wall mentioned in the *Protomycetaceae* is of interest. The IK + KOH reaction of the wall, or part of the wall, has been frequently used in systematics (i.e. *Lecanorales*; Nannfeldt, 1932).

With LM, two parts are generally recognized in the ascus wall: the exo- and the endoascus. The exoascus may be externally lined with a gelin (= gelatinized fuzzy coat; = periascus resulting from modifications in the external part of the *a* layer of the ascus). If the exo- and endoascus are thin, homogenous, and fused, the ascus is known as "unitunicate". Usually it is said to be "bitunicate" (s.lat.) if the endoascus is thick and more or less sublayered (and eventually separates during ascus dehiscence). Using TEM, it frequently appears that the wall structure of developing asci is more complex. In my opinion, four layers, *a*, *b*, *c*, *d*, may be discerned in a thick ascus wall (Fig. 1.I). The *a*, *b*, and *c* layers which are present throughout the development show only limited variation. The *a* layer is reactive and always thin; its covering ascus gelin, more or less important, reactive or not, frequently forms a loose net. This gelin is generally well-developed in lichenized asci where it sometimes spreads out at maturity but, in non-lichenized ascomycetes, it is generally reduced, non-amyloid, and so often neglected by mycologists. Under the *a* layer, the thin clear lamella of the *b* layer does not contain carbohydrates; it is only discernible in the TEM when the wall is orthogonally cut. It is probably a proteinaceous film of fundamental importance in the control of the exchange between asci and the environment. Note that in Bellemère (1971), the *b* layer is figured as a rather thick layer because the *c*1 sublayer and sometimes a part of the *c*2, was erroneously included in the *b* layer. The *c* layer, usually well-developed and moderately reactive, has a more or less granular structure; it frequently shows three sublayers *c*1, *c*2, and *c*3; the *c*2 being clear. Its thickness changes slightly during ascus development, weakly thickening, its three sublayers remaining distinct. The *d* layer is more or less developed according to species, and may be absent; it is generally made of more or less parallel fibrils which are moderately reactive. In any very young ascus no *d* layer is distinct in the periplasmic space between the plasmalemma and ascus wall. If this status is maintained throughout ascus development, no *d* layer forms. But frequently a *d* layer is deposited more or less early-on inside the *c* layer; it may becomes important and, in some cases, may be composed of two distinct sublayers, *d*1 and *d*2. In many taxa the *d* layer only develops in the upper part of the ascus. During ascus development the *d* layer is modified. In the beginning it is thin and formed of parallel fibrils, but then it thickens and its structure may become more or less lamellate (giving a tangentially banded pattern in TEM; Reynolds, 1971), or sometimes disperses and eventually assumes a network-like aspect. Later it becomes thinner and its internal structure often becomes indistinct. Frequently, the *c* and the *d* layers, as well as the *d*1 and *d*2 sublayers, are separated by a very thin clear lamella.

It is especially important to note that the *d* layer is not an initial (or primary) layer of the ascus, but rather a sort of secondary layer which only differentiates in the course of

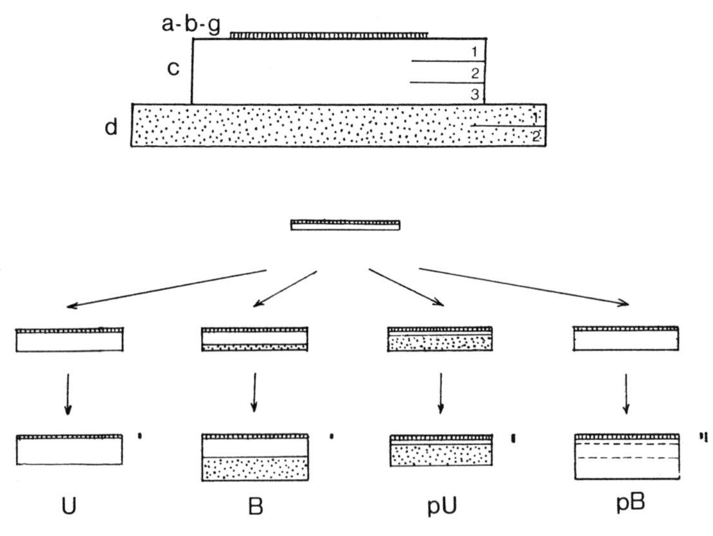

Figure 1.I. Ascus wall structure, see the text for explanation; *a* = *a* layer; *b* = *b* layer; *c* = *c* layer (with the 1, 2, and 3 sublayers); *d* = *d* layer (with the 1, and 2 sublayers); *g*, gelin. **II**, Development of convergent aspects in ascus wall structure from a young stage (layers marked as in **I**), see the text for explanation; *B* = bitunicate wall; *pB* = pseudobitunicate wall; *pU* = pseudounitunicate wall; *U* = unitunicate wall (lateral mark = "exoascus").

ascus development. This is clearly apparent when studying the development of ascus wall layers, but it is also clear in some characteristic figures (i.e. the overlapping of the basal pore plug by secondary wall material [= *d* layer] in the ascus of *Gyromitra perlata*; Kimbrough et al., 1991: 423). As the ascus *d* layer forms after a sort of hiatus or crisis in ascus development, it may be consequently considered as an adaptive acquisition. The *d* layer probably exists in all groups (*Endomycetales*, Curry, 1985; *Pezizales*, Samuelson and Kimbrough, 1978).

The problem of "unitunicate" or "bitunicate" asci has to be discussed in connection with the question of the development of the *d* layer. If no *d* layer exists, asci are clearly unitunicate (Fig. 1, II U); then the *c* layer forms the endoascus. When the *d* layer is present, and if the *d* and *c* layers are important, asci are clearly bitunicate (s.lat.) (Fig.1, II B; many *Loculoascomycetes*, numbers of *Pyrenomycetes*, *Coryneliales*, Johnston and Minter, 1989; and some *Discomycetes*, *Cookeina*). There the endoascus is probably made of the $c + d$ layers and the exoascus of the $a + b$ layers. If the *d* layer is developed and the *c* layer is especially thin, the exoascus may be interpreted as formed by the $a + b + c$ layers; the *d* layer forming the endoascus, the ascus may be then considered as "pseudo-unitunicate" (Fig.1, II pU). That is perhaps the case of typical *Sordariales*, *Xylariales* and some *Diatrypales*, where the loose "endoascus" structure is aberrant for typical unitunicates, and recalls bitunicates (Parguey-Leduc and Janex-Favre, 1984). This should have to be proved by a precise examination of the ascus wall development. If there is no *d* layer, and a very much developed *c* layer with well differentiated *c*1, *c*2 and *c*3 sublayers, then the exoascus is formed by $a + b + c1$ and the endoascus by $c2 + c3$ and seems bilayered. This recalls the structure of a bitunicate (Fig.1.II p B), and such asci could be named "pseudo-bitunicate". They are perhaps encountered in *Phyllachorales* (Cannon, 1991) or in taxa whose position is in discussion between *Loculo-* and *Pyrenomycetes*.

In conclusion, the descriptive terms exo- and endoscus have no structural value as the exoascus may be either $a + b$, or $a + b + c1$, or $a + b + c$. Moreover, convergent aspects may exist in ascus wall structure giving an unitunicate or a bitunicate aspect. This could be of systematic importance, but the use of ascus wall structure as a systematic criterion needs precise developmental studies.

ASCUS APICAL WALL STRUCTURE

During ascus development, the apical wall of the ascus generally modifies and an apical apparatus results. It has been extensively used in systematics (Chadefaud, 1973). This apparatus becomes discernible when the fusion nucleus divides (or somewhat earlier or later); then it generally regresses when ascospores are formed, but remains identifiable until ascospore liberation. In this contribution it is admitted that a continuity of the different wall layers exists at the ascus apex, and the apical apparatus is considered to result from changes in these layers - variations in thickness and differentiation. Synchronism in these modifications does not necessarily exist.

Modification of the Apical Wall Thickness

Variations in the *a* layer thickness are very reduced; but the gelin can form a cap which is more or less developed (i.e. *Dactylospora*). As the very thin *b* layer is unchanged, the modifications in wall thickness are essentially due to the two other layers (*c* + *d*). First they are considered here as a whole, but later, variations of *c* and *d* will be examined separately.

Thickening of the apical wall. The apical wall thickening is called the "tholus" (= apical dome, = dome apical; Fig. 2.I). In some cases the upward thickening is at first gradual and only above does it become abruptly more important. So a "bourrelet"

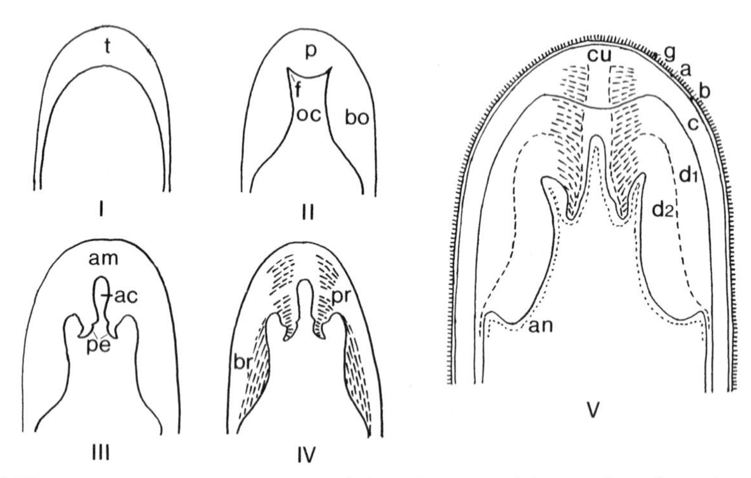

Figure 2.I-IV. Ascus apex components. *ac* = axial canal; *am* = axial mass; *bo* = bourrelet; *br* = ring in the bourrelet; *f* = furrow; *oc* = ocular chamber; *p* = plug; *pe* = pendant; *pr* = rings in the plug and pendant; *t* = tholus. **V**, Ascus apex structure. *a* = *a* layer; *an* = apical nasse; *b* = *b* layer; *c* = *c* layer; *cu* = cushion; *d*1 and *d*2 = sublayers of the *d* layer.

(= bourrelet sous-apical = subapical thickening) can be discerned under a plug (Fig. 2.II); a more or less well-marked furrow (= sinus) separates these two parts of the thickening. The ocular chamber (= chambre oculaire, = oculus, = apical chamber) is the part of the epiplasm surrounded by the bourrelet. Frequently, the apical wall thickening develops abruptly near the ascus top. In other cases the thickening is only made of the bourrelet and there is either no plug or a very narrow one. Many variants are known. The bourrelet may be more or less extended laterally, sometimes reaching the ascal foot. The plug is occasionaly very developed (e.g. *Clavicipitales*) or restricted. Its base may be prominent, flat, or concave. The plug is usually partially hollowed upwards from its base by an axial canal (= canal axial) extending at the top of the ocular chamber (Fig. 2.III). Above the axial canal, the very axial part of the thickening is known as the "axial mass" (= masse axiale, = axial body, = corps axial, = pièce axiale, = corps ombiliqué, = pseudo manubrium, = central mass, = central cylinder). The axial canal can be subcylindrical, but often its section is more or less stellate because it is laterally festooned.

In LM observations a peculiar refringent aspect results called the "apical nasse" (= nasse apicale; Chadefaud, 1942, 1973). This "nasse" (Fig. 2.V) spreads downwards along the bourrelet showing ramifications and anastomosis. An annular protrusion frequently extends downwards in the ocular chamber from the base of the plug forming a more or less important "pendant" (= pendentif).

Absence of thickening of the apical wall in maturing asci is observed in different taxa (i.e. *Gyalectaceae* and *Lichinaceae*) where the ascus wall is generally thin. Such asci are known as "prototunicate".

Reduction of wall thickness at the ascus apex. In several taxa the ascus apical wall does not become as thick as the lateral wall during ascus maturation (e.g. *Lasiosphaeriopsis stereocaulicola*; Eriksson and Santesson, 1986).

Variation in Thickness of *c* and *d* Layers

Thickening of the apical wall. Apical thickening of the *c* layer always seems limited to the plug and forms the cushion (= coussinet apical) which is sometimes refractive (= pulvillus); the *d* layer eventually forms the inferior part of the plug. If an axial canal developed before the deposition of the *d* layer, the whole axial mass is entirely formed by the cushion; if not the cushion only forms its superior part. The bourrelet always results from the thickening of the *d* layer. The *d2* sublayer may be less extended downwards than the *d1*.

Unchanged or thinner apical wall thickness. In such cases a convergent aspect may be realized at the ascus apex. For instance, if only the *c* or the *d* layer thickens at the apex, and if the other layer becomes thinner, then the wall thickness may appear unchanged at the apex. However, the wall structure is nevertheless modified. Similarly, when the wall becomes thinner at the apex, either the two layers *c* and *d*, or only *c* (or *d*) may be concerned.

Apical Differentiation in the Wall Layers

A characteristic feature of ascus apex differentiation is the formation of ring structures in the plug, and also in the bourrelet (Fig. 2.IV). These rings are frequently amyloïd but may also be "chitinoïd" (Congo Red +), or merely be somewhat refractive. In LM they generally appear as gently layered, or exceptionally net-like (*Hapsidascus*; Kohlmeyer and Volkmann-Kohlmeyer, 1991). In TEM, ring sites appear as more reactive parts of the wall and contain more or less reactive fibrils mixed with transparent granules or globules. Their extension may be somewhat different from LM observations, and a ring may be observable in the TEM whilst no ring is discernible in the LM. Moreover, there is no strict correlation between amyloïdy of the rings and reactivity with the Patag technique. Consequently, one must be cautious in the use of rings as a systematic criterion.

In the plug, rings may develop in the *c* as well as in the *d* layer (Fig. 2.V). They frequently show two levels (*c* + *d*), but sometimes several levels are distinct as independent rings may develop in either the *d1* or *d2* sublayers (and eventually in a part of the *c* sublayers). Generally, the lower ring is involved in the pendent. The internal side of ring structures is usually clearly delimited and separated from the axial canal by an annular space (espace annulaire) devoid of differentiation. On their external side, the delimitation of rings is frequently not so distinct because bundles of fibrils may extend more or less laterally. In the bourrelet the aspect and structure or the ring-like internal differentiation of the *d1* and *d2* sublayers may differ (*Stictidaceae*, *Teloschistaceae*).

The above considerations relate to asci with a typically rounded tip. In asci with other tip morphologies similar structural constituents are generally recognized. In some cases difficulties arise. For instance asci with a conical tip (as many *Rhytismataceae*) show very reduced apical structures which are not easily recognized even in the TEM, but the apical structure nevertheless seems classical. In most *Pezizales*, the asci are large, provided with an apical (or subapical) operculum, and frequently, the *d* layer is present (e.g. *Thelebolus*; Samuelson and Kimbrough, 1978). The operculum is surrounded by an annular differentiation which I consider is equivalent to the ring; thus the operculum is composed of the enlarged axial mass covered by the cushion. A bourrelet may exist. In some cases lateral special differentiations occur outside the wall (shoulder = épaulement, e.g. *Pseudopythiella*; Donadini et al., 1989); they probably act as sphincters during ascospore liberation. The operculum of the *Pezizales* probably results from an adaptation to dehiscence of a peculiar type of apex where the axial mass is enlarged ("opercular" type). This "opercular" type of ascus probably exists outside the *Pezizales* (*Propolis*, ? = *Propolomyces*, *Orbilia*) where it does not necessarily imply an operculate dehiscence. If this interpretation is of some value, the consequences would be considerable for discomycete systematics.

Many types of ascus apex are known, differing in the morphology, structure and differentiation of their wall layers (Baral, 1987a; Hertel and Rambold, 1988; Pietschmann, 1990; Rambold and Triebel, 1992; Roux et al., 1987; Spooner, 1987; Brummelen, 1978). Some of the more frequently encountered have been recognized for a long time (operculate, bitunicate, annellascés, archéascés; Luttrell, 1951, Chadefaud, 1973) and used in systematics to define major taxa. They probably correspond better to an adaptation to peculiar environmental conditions. But intermediate types are also known whose systematic interpretation causes problems (Eriksson, 1981; Sherwood, 1981; Hawksworth, 1990; Pascoe, 1990). Precise developmental studies of some of these, comparative to other taxa, have led to the view that a progressive diversification of the apical apparatus took place along several different evolutionary lines. These probably diverged from an initial type of apical apparatus. This perhaps resulted from modifications parallel to a sort of miscarriage in a process of division of the ascal wall. Suitable data are still insufficient to provide a clear view of ascus evolution.

ASCUS DEHISCENCE

Predehiscence Phase

Various changes may occur in asci (Fig. 3.I) before ascospore liberation, and these have not to be neglected for systematic purposes:

Ascus form modification: generally an elongation carrying the ascus tip to the level of the hymenium surface or above: protruding asci (i.e. *Ascobolus*, where ascus enlargement is also important).

Changes in ascus lateral walls: the lateral ascus wall frequently becomes thinner (i.e. fissitunicate asci where the internal layer sometimes becomes more or less evanescent; Bellemère et al., 1986a).

Changes in apical structures: height of the tholus is frequently reduced; the importance, texture and reactivity of the rings and axial body are more or less affected; in operculate asci, an annular differentiation appears, limiting the future operculum.

Ascus liberation: sometimes asci separate from the subhymenium before they dehisce and become free in the ascoma cavity (e.g. *Ceratostomella*).

Dehiscence Phase: Ascospore Liberation

The dehiscence phase (Fig. 3.II) has been used in systematics for a long time (inoperculés et operculés, Boudier, 1879; bitunicate, Luttrell, 1951).

Complete dissociation of the wall: i.e. evanescent asci. This takes place in several scattered orders: *Eurotiales*, *Microascales*, *Ophiostomatales*, *Meliolales*, *Caliciales*, *Lichinales*, and *Tuberales*. It generally affects prototunicate asci, but it is not directly correlated with the absence of apical structures (i.e. *Hydnotrya cerebriformis* has a non-functional apical apparatus persisting until the dissociation of the ascus wall; Zhuang, 1991). The precise mechanism is poorly known.

Lateral wall rupture: A limited circular dissociation appears laterally in the ascus wall either near the base of the ascus (e.g. *Coryneliales*), or in a subterminal part

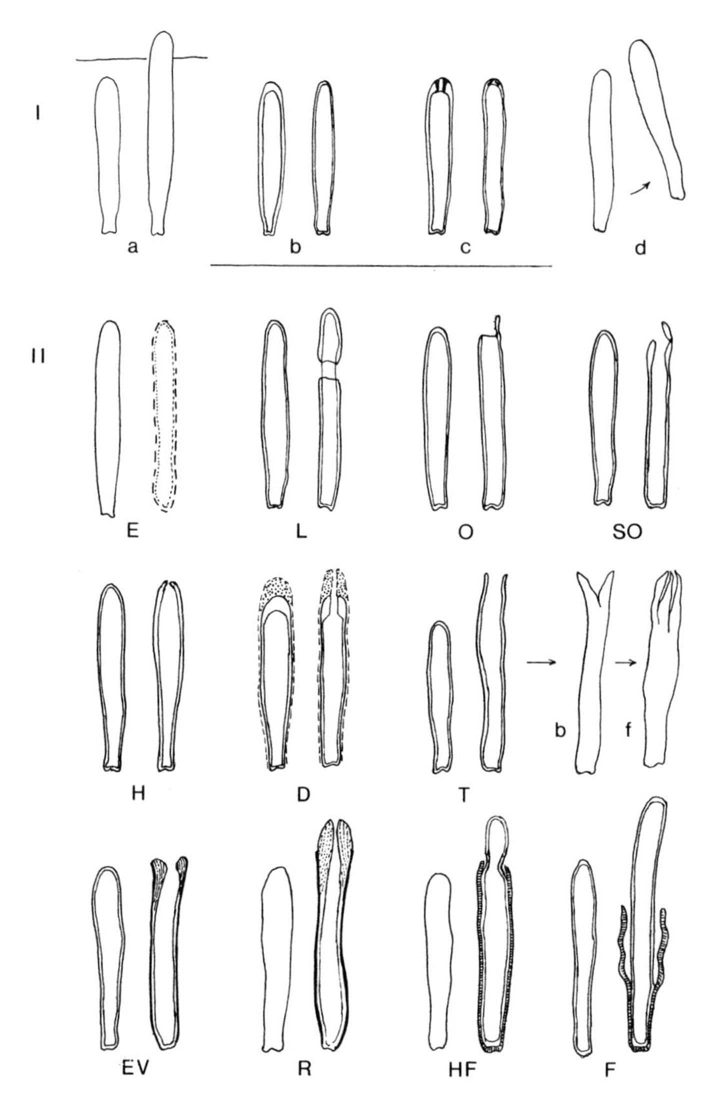

Figure 3.I. Predehiscence stage of asci. *a* = protruding ascus; *b* = ascus wall becoming thinner; *c* = change in apical structure; *d* = ascus liberation. **II**, Dehiscence stage of asci, see text for explanation; evanescent ascus (*E*); rupture of lateral wall (*L*); subapical rupture (*O*, operculate and *SO*, suboperculate dehiscence); rupture by apical wall without extrusion (*H*, pore-like dehiscence); *D*, *Dactylospora*-type; *T*, *Teloschistes*-type = extenditunicate (*b* = bivalve, *f* = fissurate variants); rupture with extrusion (*EV*, eversion; *R*, rostrate; *HF*, hemifissitunicate; *F*, fissitunicate).

(*Arthrorhaphis*; Bellemère, unpubl.). Possibly, the different wall layers are not simultaneously nor similarly involved.

Subapical rupture: A subapical circular aperture delimits a hole covered by an operculum which is laterally rejected on ascospore liberation, and then persists fixed by a hinge (numerous *Pezizaceae*). In several *Sarcosomataceae* s.lat. the *d* layer thickening forms a sort of chimney under the operculum (suboperculate asci; i.e. *Sarcoscypha*, *Pseudopythiella*).

Apical rupture: In the beginning the aperture is a simple hole, whose formation simultaneously affects all the wall layers. Two major types exist. In the first there is no bursting out of the internal part of the wall through the aperture. The dehiscence may be simple and typically pore-like (inoperculate discomycetes, *Trapelia*, *Heppia*, *Coccocarpia*), or it may be more complex as in the *Dactylospora*-type (with liberation of the ascospores by the "gelin"; Bellemère and Hafellner, 1982), or in the *Teloschistes*-type (Bellemère and Letrouit-Galinou, 1987), similar to the extenditunicate type (Reynolds, 1989b) where the apical pore forms after a predehiscent elongation. In this first type of apical rupture there is frequently a subsequent fissure-like splitting at the apical part of the wall; bivalve or fissurate dehiscence (Minter and Cannon, 1984). In a second type of apical rupture, there is a bursting out of the internal part of the wall. In a first infratype, all the wall layers remain joined in the bursting out; there is only an eversion if this is limited (eversion type, i.e. *Sclerotiniaceae*, *Dipodascopsis*; Curry, 1985), but if the bursting out is rather important, a rostrum forms (rostrum type, i.e. *Lecanoraceae* et al.). In a second infratype, the wall composition is modified and the internal layer may slide along the external one; in the hemifissitunicate type (e.g. *Rhizocarpon*) sliding is reduced; in the fissitunicate type (= bitunicate s.str., Luttrell, 1951; Reynolds, 1989a; = "jack in the box", Ingold, 1933) there is an important sliding (i.e. numerous *Dothideales*).

Several of the many types of ascus dehiscence which have been described are frequently associated with a particular type of ascus structure (i.e. evanescent with prototunicate, pore-like with apical annulus, rostrate with *Lecanora*-type of apex, fissitunicate with bitunicate wall). However, there is no strict correlation between apex structure and dehiscence type. So, according to the taxon, asci with apical ring may show either a pore-like, an evanescent, or a fissitunicate type of dehiscence. Consequently, ascus types cannot be defined only by terms characterizing either their apical structure or dehiscence; so, for instance, one must speaks not only of bitunicate asci, but of bitunicate-fissitunicate. One referring to structure, and the other to function.

SYSTEMATIC VALUE OF DEHISCENCE

Ascus dehiscence may differ among the taxa of a systematic group founded on non-ascal criteria. For example, in the *Dothideales* the asci are not exclusively fissitunicate (Reynolds, 1987, 1989b), and dehiscence of a non-operculate type is encountered in several *Pezizales*. Conversely, the same type of dehiscence may be encountered in various orders or families (i.e. fissitunicate in *Loculoascomycetes* and *Patellariales*; evanescent in *Lichinales*, *Tuberales*, *Ophiostomatales*, *Halosphaeriales* and *Diatrypales*). On the other hand, it is sometimes difficult to make a distinction between similar types of dehiscence (i.e. a long rostrum, hemifissitunicate, or fissitunicate). Moreover, convergent aspects may exist; fissitunicate dehiscence may perhaps sometimes result from a sliding between *d*1 and *d*2 sublayers; mazaedium-like structures have been encountered in *Pyrenulales* (Aptroot, 1991). The reality of several dehiscence types is questioned by some authors who claim, for instance, that artefacts are produced in LM observations made in water. The systematic value of dehiscence is uncertain. Dehiscence types are probably the result of a functional adaptation of structures to various conditions of the environment (with violent discharge or not), and should be considered as a secondary systematic criterion in comparison to ascus wall structure.

ASCOSPOROGENESIS

Ascospore Delimitation

Initial TEM studies revealed that ascospore delimitation proceeds from the fragmentation of an ascus vesicle made of two sheets. It has been established that this is a general process in ascomycetes, except for several taxa so far considered as systematically distinct from other ascomycetes, where ascospores are individually delimited by a double sheet of elements of endoplasmic reticulum (e.g. *Saccharomyces*, *Taphrina*). However, as no ascus vesicle has been observed in *Tuber* and allied genera, where ascospores are delimited by a single sheet issued from the plasmalemma (Parguey-Leduc et al., 1990: 47-68), the disposition of these fungi in the *Pezizales* (Trappe, 1979) remains questionable (Janex-Favre and Parguey-Leduc 1985).

Ascospore Wall Development

It has been shown by TEM that in the beginning the ascospore wall is reduced to a thin clear lamella compressed between two double-membranes. Later several layers with a particular differentiation develop in place of this lamella. Data relating to ascospore wall development shows that the process is similar in the whole of the ascomycetes. The ascospore wall structure of any taxon may be interpreted by the same basic scheme (Fig. 4.I). The observed diversity only results from variations in importance, differentiation, and chronological development of the constitutive layers (Fig. 4.II). Unfortunately, no consensus has been yet established between researchers for the denomination of these layers. This is a first step in ensuring difficulties in systematic implications. A second step results from differences in the interpretation of sequences in ascospore wall development. According to many authors, a primary wall develops first, and then a secondary wall, distinct from the preceding, appears outside it. However, the thickening of the primary wall is probably continuous with no hiatus between the development of the two walls, the secondary wall being only a differentiation appearing along the primary wall. The primary wall grows centrifugally (depositing material from the sporoplasm), but its differentiation moves centripetally (with deposition of complementary material from the epiplasm) and stops inside the wall by the construction of a thin pellicula which has an important physiological role in spore exchanges. This theory leads us to subdivide the wall into a proper wall (= the primary wall = épispore) and a perispore (= the secondary wall). The thin pellicula is the intermediary wall (more or less equivalent to the exospore). This is generally difficult to distinguish, even in the TEM, and may differentiate into sublayers which are only discernible in large ascospores with thick walls (i.e. operculate discomycetes; Bellemère and Meléndez-Howell, 1976).

The perispore is limited outside by the investing membrane (= ectospore), which is a thin but more or less differentiated and reactive, according to the taxon. The perispore may remain very thin or become thick; it is frequently unequally developed, sometimes forming apical obturators or apicules, or also apical or lateral appendages. Rarely the perispore of different spores fuse into a spore ball. The perispore is frequently composed of three sublayers (sometimes more), the median sublayer being generally well-developed and reactive, sometimes persisting as an ascospore ornamentation. This shows a great diversity of aspects. Diversity in perispore is of a reduced systematic value as its derivatives may either differ between species in the same genera or be similar in taxa of different orders including *Schistomatales* (Rosing, 1985; van Wyk and Wingfield, 1991), *Lecanorales* (*Pannaria* sp.), *Endomycetales* (*Saccharomyces rouxii*, *Debaryomyces*; Kreger van Rij and Veenhuis, 1975; Kreger van Rij, 1979). As it has been shown that transparent areas in the median sublayer of the perispore may develop by mutation in a single genus (Meléndez-Howell et al., 1987), perispore characters probably result from recent adaptations and do not characterize high level taxa, being only of interest at low levels.

The proper wall has a flat base; it is frequently composed of three feebly differentiated sublayers. In some cases a spore wall ornamentation may result from differentiation of the proper wall, whilst the perispore has no role (e.g. *Gelasinospora* , *Neurospora*; Bellemère et al., 1992; Fig.4.III, N,G), and probably also in *Persiciospora* (Krug, 1988), and *Chaetomiopsis* (Moustafa and Abdul-Wahid, 1990); the systematic value of such an ornamentation remains to be defined. In some cases terminal appendages observed in septate ascospores by LM look like perisporal appendages but differ in their structure as they are essentially built by the proper wall of the terminal cells (i.e. *Halosphaeriales*).

The endospore is a layer which forms rather late underneath the proper wall; it is frequently more or less layered, and usually its base is somewhat undulate. The endospore, which may be absent but exists in all groups, shows a wide diversity of aspect or importance. It can be unequally developed along the spore (polarilocular ascospores; Bellemère and Letrouit-Galinou, 1982; *Megalotremis*, Aptroot, 1991) or sometimes be only present at the poles of the spores (*Calicium*, Tibell, 1990). In some cases the endospore differentiates into a hard layer taking a part in ascospore ornamentation (Bellemère et al., 1992). In fact the endospore is probably originally an adaptative layer, produced later by the sporoplasm in response to external stress; in the course of evolution, its production has been more or less anticipated. The systematic value of endospore structures is consequently also reduced.

A spore envelope (= sac sporale) has sometimes been observed encircling maturing ascospores (e.g. *Tuber*, Janex-Favre and Parguey-Leduc, 1976; *Pseudascozonus*, Brummelen, 1987).

For the present, discrepancies may be encountered in relation to ascospore wall structure in the same family (i.e. convergent structures in *Lasiosphaeriaceae*; Bellemère et al., 1992; Fig. 4.III, P,A).

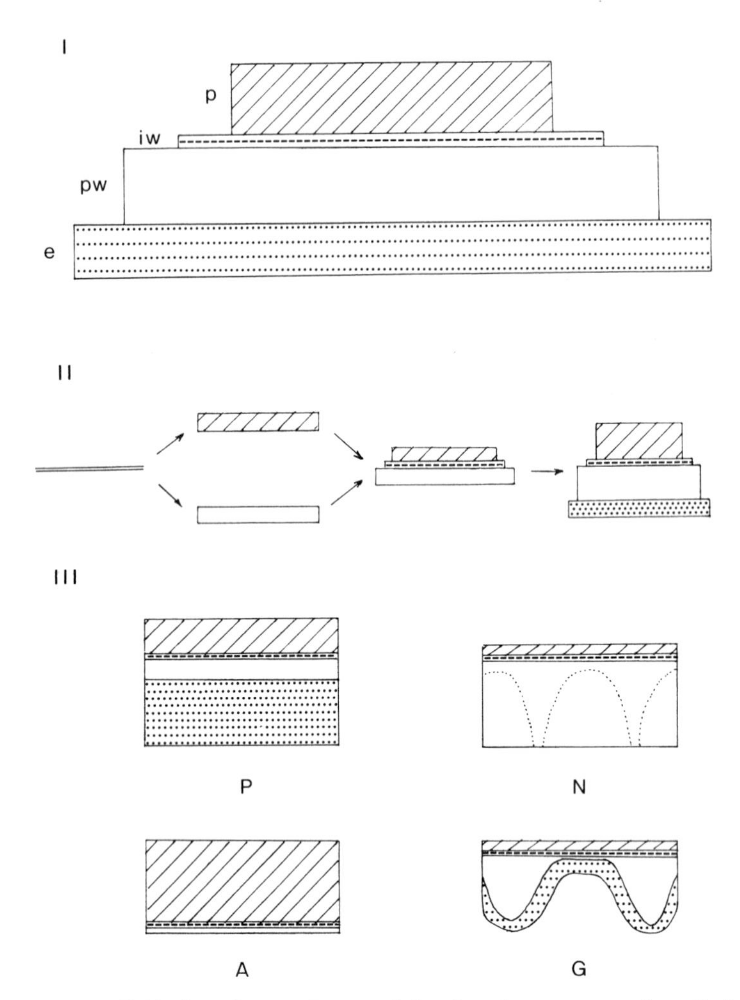

Figure 4. Ascospore wall (the investing membrane and the plasmalemma are not figured). **I,** Structure of the ascospore wall; *iw* = intermediary wall; *p* = perispore; *pw* = proper wall; *e* = endospore. **II**, Development of ascospore wall layers (layers marked as in **I**); the primary wall may be either the perispore or the proper wall. **III**, Differences in ascospore wall structure in *Lasiosphæriaceae* (*P*, *Podospora*; *A*, *Apiosordaria*) and *Sordariaceae* (*N*, *Neurospora*; *G*, *Gelasinospora*); layers marked as in **I.**

In conclusion, ascospore wall structure is a criterion to be used at a low level in systematics. Careful developmental studies of ascospore walls are, however, necessary to avoid misinterpretations leading to erroneous systematic conclusions.

Septation and Pigmentation of Ascospores

The septation of ascospores has been carefully analyzed by Chadefaud (1969), whose observations have been confirmed in the TEM. It appears that the scheme of ascospore septation is always the same (including polarilocular and muriform spores) and does not differ from hyphal septation. As mentioned above, this is a criterion of reduced importance in systematics. Studies by TEM revealed that the pigmentation of ascospores is generally a late metabolic process, independent of the wall-layered structure as pigmentation granules display randomly in the wall, and sometimes limited to two layers or sublayers (*Catolechia*, Bellemère and Hafellner, 1983; *Abrothallus*, Bellemère et al., 1986b). This criterion has to be used only to separate lower taxa. In other cases (i.e. *Ascobolus* and allied genera) the pigment is deposited in a special sublayer of the ascospore wall, but as its production is controlled by a single pigmentation gene; here also the systematic value remains limited.

CONCLUSIONS

It seems that developmental modalities of structural and morphological characters of the ascus wall, especially at the ascus apex, are the most valuable ascus criteria to be considered in the definition of higher systematic groups. The differentiations taking place in the constitutive layers of ascus or ascospore walls are probably more recent acquisitions whose systematic value has to be limited to the separation of lower taxa. Data relating to ascus characters are frequently insufficient in systematic literature concerning ascomycetes. If asci are precisely described, and if their variations during ascus development are carefully reported in diagnoses, then important progress will be made in systematics and in our appreciation of evolution in ascomycete fungi.

REFERENCES

Aptroot, A., 1991, A monograph of the *Pyrenulaceae* (excluding *Anthracothecium* and *Pyrenula*) and the *Requienellaceae*, with notes on the *Pleomassariaceae*, the *Trypetheliaceae* and *Mycomicrothelia* (lichenized and non lichenized ascomycetes), *Bibliotheca Lichenologica* 44: 1-78.

Arx, J.A. von, 1987, On *Trabutia* and its anamorphs, *Systema Ascomycetum* 6: 213-215.

Baral, H.O., 1987a, Der apical apparat der *Leotiales* - Eine lichtmikroskopische studie über Arten mit Amyloidring, *Zeitschrift für Mykologie* 53: 119-135.

Baral, H.O., 1987b, Lugol's solution IKI versus Melzer's reagent: hemiamyloïdy, a universal feature of the ascus wall, *Mycotaxon* 29: 399-450.

Baral, H.O., 1992, Vital versus herbarium taxonomy: morphological differences between living and dead cells of ascomycetes and their taxonomical implications, *Mycotaxon* 44: 333-390.

Barr, M.E., 1987, The genus *Trabutia*, *Mycologia* 79: 188-192.

Barr, M.E., 1990, Prodromus to non lichenized, pyrenomycetous members of class *Hymenoascomycetes*, *Mycotaxon* 39: 43-184.

Bellemère, A., 1960, Remarques sur le développement des asques du *Schizoxylon berkeleyanum* (Dur. et Lév.) Fuck. (Discomycéte inoperculé: Ostropale), *Comptes Rendu Hebdomadaire des Séances de l'Académie des Sciences Paris* 251: 2569-2571.

Bellemère, A., 1971, Les asques et les apothécies des Discomycètes bituniqués, *Annales des Sciences Naturelles, Botanique et Biologie végétale* 12: 429-464.

Bellemère, A., 1977, L'appareil apical de l'asque chez quelques Discomycètes: étude ultrastructurale comparative, *Revue de Mycologie* 41: 233-264.

Bellemère, A., and J. Hafellner, 1982, L'ultrastructure des asques du genre *Dactylospora* (Discomycètes) et son intérét taxonomique, *Cryptogamie, Mycologie* 3: 71-93.

Bellemère, A., and J. Hafellner, 1983, L'appareil apical des asques et la paroi des ascospores du *Catolechia wahlenbergii* (Ach.) Flotow ex Koerber et de l'*Epilichen scabrosus* (Ach.) Clem. ex Haf. (lichens, *Lecanorales*: étude ultrastructurale, *Cryptogamie, Bryologie et Lichénologie* 4: 1-36.

Bellemère, A., M.-C. Janex-Favre, L.-M. Meléndez-Howell, and A. Parguey-Leduc, 1992, Diversité ultrastructurale de la paroi ascosporale chez quelques Eupyrénomycètes, *Cryptogamie, Mycologie* 13: 215-246.

Bellemère, A., and M.-A. Letrouit-Galinou, 1982, Le développement des ascospores chez le *Caloplaca marina* Wedd. et chez quelques lichens de la famille des *Teloschistaceae* (*Caloplaca, Fulgensia, Xanthoria*) : étude ultrastructurale, *Cryptogamie, Bryologie et Lichénologie* 3: 95-137.

Bellemère, A., and M.-A. Letrouit-Galinou, 1987, Differentiation of lichen asci including dehiscence and sporogenesis: an ultrastructural survey, *Bibliotheca Lichenologica* 25: 137-161.

Bellemère, A., M.-C. Malherbe, H. Chacun, and J. Hafellner, 1986a, Les asques bituniqués du *Lecanidion atratum* (Hedw.) Rabh. (= *Patellaria atrata* (Hedw.) Fr.) (*Lecanidiaceae*): étude ultrastructurale de la paroi au cours du développement et à la déhiscence, *Cryptogamie, Mycologie* 7: 113-147.

Bellemère, A., M.-C. Malherbe, H. Chacun, and J. Hafellner, 1986b, Etude ultrastructurale des asques et des ascospores chez les espèces lichénicoles non lichénisées *Abrothallus bertianus* de Not. et *A. parmeliarum* (Sommerf.) Nyl., *Cryptogamie, Mycologie* 7: 47-85.

Bellemère, A., M.C. Malherbe, H. Chacun, and L.M. Meléndez-Howell, 1990, L'étude ultrastructurale des asques et des ascospores de l'*Urnula helvelloides* Donadini, Berthet et Astier et les concepts d'asque suboperculé et de *Sarcosomataceae*, *Cryptogamie, Mycologie* 11: 203-238.

Bellemère, A., and L.M. Meléndez-Howell, 1976, Etude ultrastructurale comparée de l'ornementation externe de la paroi des ascospores de deux *Pezizales*: *Peziza fortini* n. sp. récoltée au Mexique, et *Aleuria aurantia* (Ced. ex. Fr.) Fuck., *Revue de Mycologie* 40: 3-19.

Berthet, P., 1964, *Essai biotaxonomique sur les Discomycètes*. Thèse, Université de Lyon.

Boudier, E., 1879, On the importance that should be attached to the dehiscence of asci in classification of the discomycetes, *Grevillea* 8: 45-49.

Brummelen, J. van, 1978, The operculate ascus and allied forms, *Persoonia* 10: 113-128.

Brummelen, J. van, 1987, Ultrastructure of the ascus and ascospores in *Pseudoascozonus* (*Pezizales*, *Ascomycotina*), *Persoonia* 13: 369- 377.

Cannon, P.F., 1991, A revision of *Phyllachora* and some similar genera on the host family *Leguminosae*, *Mycological Papers* 163: 1-302.

Chadefaud, M., 1942, Etudes d'asques, II. Structure et anatomie comparée de l'appareil apical des asques chez divers Discomycètes et Pyrénomycètes, *Revue de Mycologie* 7: 57-88.

Chadefaud, M., 1943, Sur les divers types d'éléments dangeardiens des Ascomycètes et sur la formation des asques chez la Pezize *Pustularia catinus*, *Revue Scientifique* 81 (2): 70-78.

Chadefaud, M., 1969, Une interprétation de la paroi des ascospores septées notamment celle des *Aglaospora* et des *Pleospora*, *Bulletin de la Société Mycologique de France* 85: 145-157.

Chadefaud, M., 1973, Les asques et la systématique des Ascomycètes, *Bulletin de la Société Mycologique de France* 89: 127-170.

Curry, K.J., 1985, Ascosporogenesis in *Dipodascopsis tothii* (*Hemiascomycetidae*), *Mycologia* 77: 401-411.

Donadini, J-C., H. Chacun, M.C. Malherbe, and A. Bellemère, 1989, Ultrastructure des asques et des ascospores du *Pseudopithyella minuscula*. (Ascomycetes, *Pezizales*, *Sarcoscomataceae*), *Cryptogamie, Mycologie* 10 : 283-304.

Eriksson, O., 1981, The families of bitunicate ascomycetes, *Opera botanica* 60: 1-220.

Eriksson, O., and R. Santesson, 1986, *Lasiosphæriopsis stereocaulicola*, *Mycotaxon* 25: 569-580.

Hafellner, J., 1984, Studien in Richtung einer natürlichen Gliederung der Sammelfamilien *Lecanoraceae* und *Lecideaceae*, *Beiheft zur Nova Hedwigia* 79: 241-371.

Hawksworth, D.L., 1990, *Globosphaeria*, a remarkable new pyrenomycete on *Normandina* from Tasmania, *Lichenologist* 22: 301-308.

Henssen, A., and A. Titze, 1990, *Auriculora byssomorpha*, a tropical lichen with a remarkable developmental morphology, *Botanica Acta* 101: 131-139.

Hertel, H., and G. Rambold, 1988, Cephalodiate Arten der Gattung *Lecidea* sensus lato (Ascomycetes lichenisati), *Plant Systematics and Evolution* 158: 289-312.

Ingold, C.T., 1933, Spore discharge in the ascomycetes I, Pyrenomycetes, *New Phytologist* 32: 175-196.

Janex-Favre, M.-C., and A. Parguey-Leduc, 1976, La formation des ascospores chez deux truffes: *Tuber rufum* Pico et *Tuber aestivum* Vitt. (Tubéracées). *Compte Rendu Hebdomadaire des Séances de l'Académie des Sciences, Paris*, sér. D, 283: 1173-1175.

Janex-Favre, M.-C., and A. Parguey-Leduc, 1985, Les asques et les ascospores du *Terfezia claveryi* Ch. (Tubérales), *Cryptogamie, Mycologie* 6: 87-99.

Johnson, R.G., 1980, Ultrastructure of ascospore appendages of marine ascomycetes, *Botanica Marina* 23: 501-527.

Johnston, P.R., and D.W. Minter, 1989, Structure and taxonomic significance of the ascus in the *Coryneliaceae*, *Mycological Research* 92: 422-430.

Kärnefelt, I., and A. Thell, 1992, The evaluation of characters in lichenized families exemplified with the alectorioïd and some parmelioïd genera, *Plant Systematics and Evolution* 180: 181-204.

Kärnefelt, I., J.E. Mattson, and A. Thell 1992, Evolution and phylogeny of cetrarioid lichens, *Plant Systematics and Evolution* 183: 113-160.

Kimbrough, J.W., C.G. Wu, and J.L. Gibson, 1990, Ultrastructural observations on *Helvellaceae* (*Pezizales*, ascomycetes) - IV - Ascospore ontogeny in selected species of *Gyromitra* subgenus *Discina*, *Canadian Journal of Botany* 68: 317-328.

Kohlmeyer, J., and B. Volkmann-Kohlmeyer, 1991, *Hapsidascus hadrus* gen. and sp. nov. (*Ascomycotina*) from mangroves in the Caribbean. *Systema Ascomycetum* 10: 113-120.
Kreger van Rij, N.J.W., 1979, A comparative ultrastructural study of the ascospores of some *Saccharomyces* and *Kluyveromyces* species, *Archiv für Microbiologie* 121: 53-59.
Kreger van Rij, N.J.W., and M. Veenhuis, 1975, Electron microscopy of ascus formation in the yeast *Debaryomyces hansenii*, *Journal of General Microbiology* 89: 256-264.
Krug, J.C., 1988, A new species of *Persiciospora* from African soil, *Mycologia* 80: 414-417.
Luttrell, E.S., 1951, Taxonomy of the *Pyrenomycetes*, *University of Missouri Studies* 24: 1-120.
Meléndez-Howell, L.M., A. Bellemère, and J.L. Rossignol, 1987, Remarques à propos de l'ultrastructure d'ascospores "albinos" ou "granuleuses" de mutants d'*Ascobolus immersus* Pers. (gène b8), *Cryptogamie, Mycologie* 8: 269-288.
Mengoni, T., 1986, El aparato apical del asco de *Cyttaria harioti* (Ascomycetes, *Cyttariales*) con microscopia fotonica y electronica, *Boletin de la Sociedad Argentina de Botanica* 24: 393-401.
Merkus, E., 1973-1976, Ultrastructure of the ascospore wall in *Pezizales* (Ascomycetes) I, II, III, IV, *Persoonia* 7: 351-356; 8: 1-22, 227-247; 9: 1-38.
Minter, D.W., and P.F. Cannon, 1984, Ascospore discharge in some members of the *Rhytismataceae*, *Transactions of the British Mycological Society* 83: 65-92.
Moreno, P.P., and J.M. Egea, 1991, Biologia y taxonomia de la familia Lichinaceae con especial referencia a las especies del S.E. espanol y Norte de Africa, *Universidad de Murcia, Colección Bianca* 19: 1-87.
Moustafa, A.F., and O.A. Abdul-Wahid, 1990, A new perithecial ascomycete genus from Egyptian soils, *Mycologia* 82: 129-131.
Nannfeldt, J.A., 1932, Studien über die Morphologie und Systematik der nicht-lichenizierten Inoperculaten Discomyceten, *Nova Acta Regiae Societatis Scientarum Upsaliensis*, ser. IV, 8(2): 1-368.
Parguey-Leduc, A., and M.-C. Janex-Favre, 1984, La paroi des asques chez les Pyrénomycètes. Étude ultrastructurale II- Les asques unituniqués, *Cryptogamie, Mycologie* 5: 171-187.
Parguey-Leduc, A., M.-C. Janex-Favre, and C. Montant, 1990, L'appareil sporophytique et les asques du *Tuber melanosporum* Vitt. (Truffe noire du Périgord, Discomycètes), *Cryptogamie, Mycologie* 11: 47-68.
Pascoe, I.G., 1990, Observations on ascus structure of *Plectosphæria eucalypti* (*Phyllachoraceae*), *Mycological Research* 94: 675-684.
Pietschmann, M., 1990, Morphometrics of tubiform apical apparatus in *Lecideaceae*, *Micareaceae*, *Porpidiaceae* and allied families (lichenized ascomycetes, *Lecanorales*): limitations and perspectives of statistical inference, *Nova Hedwigia* 51: 521-549.
Rambold, G., and D. Triebel, 1992, The inter-lecanoralean associations, *Bibliotheca Lichenologica* 48: 3-201.
Reynolds, R., 1971, Wall structure of a bitunicate ascus, *Planta* 98: 244-257.
Reynolds, D.R., 1987, A non bitunicate ascus in the ascotromatic genus *Asterina*, *Cryptogamie, Mycologie* 8: 251-268.
Reynolds, D.R., 1989a, The bitunicate ascus paradigm, *Botanical Reviews* 55: 1-52.
Reynolds, D.R., 1989b, An extenditunicate ascus in the ascostromatic genus *Meliolina*, *Cryptogamie, Mycologie* 10: 305-320.
Rosing, W.C., 1985, Fine structure of cleistothecia, asci and ascospores of *Myxotrichum deflexum*, *Mycologia* 77: 920-926.
Roux, C., A. Bellemère, J.-C. Boissière, J. Esnault, M.-C. Janex-Favre, M.-A. Letrouit-Galinou, J. Wagner, 1987 ["1986"], Les bases de la systématique moderne des lichens, *Bulletin de la Société Botanique de France, Actualités Botaniques* 133: 7-40.
Samuelson, A., and J.W. Kimbrough, 1978, Asci of the *Pezizales*. IV- The apical apparatus of *Thelebolus*, *Botanical Gazette* 139: 346-361.
Sherwood, M.A., 1981, Convergent evolution in discomycetes from bark and wood, *Botanical Journal of the Linnean Society* 82: 15-34.
Spooner, B.M., 1987, *Helotiales* of Australasia: *Geoglossaceae*, *Orbiliaceae*, *Sclerotiniaceae*, *Hyaloscyphaceae*, *Bibliotheca Mycologica* 116: 1-711.
Tehler, A., 1990, A new approach to the phylogeny of *Euascomycetes* with a cladistic outline of *Arthoniales* focusing on *Roccellaceae*, *Canadian Journal of Botany* 68: 2458-2492.
Tibell, L., 1984, A reappraisal of *Caliciales*, *Beiheft zur Nova Hedwigia* 79: 587-713.
Torrente, P., and J.M. Egea, 1992, *Llimonaea* a new genus of lichenized fungi in the order *Opegraphales* (*Ascomycotina*), *Nova Hedwigia* 52: 239-245.
Trappe, J.M., 1979, The orders, families and genera of hypogeous *Ascomycotina* (truffles and their relatives), *Mycotaxon* 9: 297-340.

Verkley, J.M., 1992, Ultrastructure of the apical apparatus of asci in *Ombrophila violacea*, *Neobulgaria pura* and *Bulgaria inquinans*, *Persoonia* 15: 3-22.

Vezda, A., and J. Poelt, 1990, *Solorinellaceae* eine neue Flechtenfamilie, *Phyton* 30: 47-55.

Wong, H.C., and C.Y. Chien, 1986, Ultrastructure of sexual reproduction of *Monascus purpureus*, *Mycologia* 78: 713-721.

Wu, C.G., and J.M. Kimbrough, 1990, Ultrastuctural studies on the cleistothecial development of *Emericellopsis microspora* (*Eurotiales*, Ascomycetes), *Canadian Journal of Botany* 68: 1877-1888.

Wyk, P.W.J. van, and M.J. Wingfield, 1991, Ultrastructure of ascosporogenesis in *Ophiostoma davidsonii*, *Mycological Research* 95: 725-730.

Zhang, B.C., 1991, Morphology, cytology and taxonomy of *Hydnotrya cerebriformis Pezizales*, *Mycotaxon* 42: 155-162.

SEPTAL ULTRASTRUCTURE AND ASCOMYCETE SYSTEMATICS

J.W. Kimbrough

Plant Pathology Department
University of Florida
Gainesville, Florida 32611, USA

SUMMARY

Some of the most revolutionary changes in the systematics of various divisions of algae and certain classes of fungi were the results of unique differences found in the ultrastructure of septa. Among the fungi, septal ultrastructure has played a significant role in helping to delineate classes and subclasses of basidiomycetes. The application of septal ultrastructure to ascomycete systematics has been limited. Yet, there are strikingly different types of septal pore organelles in various groups of ascomycetes. This presentation demonstrates that septal structures may be useful in delimiting various groups of ascomycetes. The major focus is on septal structures associated with asci, ascogenous hyphae, and excipular cells of families of *Pezizales*. The use of these characters to show phylogenetic linkages between the epigeous and hypogeous tuberalean *Pezizales* is demonstrated. Septa as taxonomic criteria in other groups of ascomycetes are also discussed.

INTRODUCTION

In one of the initial studies of fungal ultrastructure, it became apparent that septa in the ascomycetes and basidiomycetes were strikingly different (Moore and McAlear, 1962). Septal structures have had a great impact on the taxonomy of basidiomycetes (Moore, 1978; Khan and Kimbrough, 1982; Oberwinkler, 1985; Boehm and McLaughlin, 1989), but even a greater impact among the algae (Pickett-Heaps and Marchant, 1972; Pueschel and Role, 1982). This has not been true of ascomycetes, however, and until the work of Curry and Kimbrough (1983), little attention had been given to the potential value of septal structures in the systematics of ascomycetes.

A number of very distinct organelles have been shown to be associated with septa in various ascomycetes. Shatkin and Tatum (1959) found a characteristic type of pore plugging in the vegetative cells of *Neurospora*, the ontogeny of which was studied in detail by Trinci and Collinge (1973). In this type, electron translucent forked projections extended from the septal rim into the dense pore plug. Curry and Kimbrough (1983) noted that the *Neurospora*-type septum was found in a number of *Plectomycetes*, *Pyrenomycetes*, and their anamorphs. Schrantz (1964) was the first to report a peculiar granular lamellate mass in the pores of vegetative cells of *Aleuria aurantia*. These "lamellate structures" appear to be unique features of taxa in the *Pezizales* (Kimbrough, 1991). Saito (1974) found in septal pores of sclerotial cells in *Sclerotinia sclerotiorum* a pore plugging structure similar to the *Neurospora*-type. This type of pore plugging is found in a variety of ascomycetes.

The most complex types of pore plugs are found within septal pores of asci and ascogenous hyphae. Carroll (1967) recorded very elaborate, hemispherical pore structures on the ascus side of the basal pore in *Ascodesmis sphaerospora* and *Saccobolus kerverni*.

Ascomycete Systematics: Problems and Perspectives in the Nineties
Edited by D.L. Hawksworth, Plenum Press, New York, 1994

Somewhat similar septal pore plugs were also found in *Ascobolus stercorarius* (Wells, 1972). Peculiar convex bands were observed across one or both sides of pores in asci and ascogenous hyphae in species of *Peziza* (as *Galactinia*; Schrantz, 1970). Subsequently, a variety of septal pore organelles has been discovered in the ascogenous hyphae and asci of all species of *Pezizales* studied thusfar (Kimbrough, 1991). I became especially interested in septa when during the study of the cytological development and ultrastructure of the ascogenous system and asci of a number of species of *Thelebolus* (Kimbrough, 1981), a consistent pattern of pore plugging was discovered in the ascogenous hyphae leading to ascus formation.

In light of the distinctive nature of septa in respective groups of basidiomycetes, the question arose as to whether specific septa might characterize various taxa of *Pezizales*. For more than a decade my students and I have been examining representative taxa of various families of *Pezizales*, and potentially related taxa. This survey was conducted to determine if there are unique types of septal pore structures that may be useful in resolving some of the taxonomic problems that exist within genera and families of the *Pezizales*, and that would link epigeous taxa (*Pezizales*) with supposedly related hypogeous taxa (*Tuberales*). In the following pages I describe some of the types of septal organelles found in the families of *Pezizales*, how these septal types are being used to establish taxonomic limits of genera and families, to establish linkages of epigeous *Pezizales* to traditional groups of *Tuberales*, and how septal data may be useful in the systematics and phylogeny of other groups of ascomycetes.

SEPTAL STRUCTURES IN PEZIZALES

Ascobolaceae

Wells (1972) was the first to point out the absence of Woronin bodies in the ascus of ascomycetes. He noted that in the diploid phase, both septal pores between the ascus and the ultimate and antipenultimate cells of the crozier became plugged with dome-shaped structures within the ascus. He did not show, however, the types of septal structures in other apothecial cells. Kimbrough and Curry (1985) examined septal structures in asci, ascogenous hyphae, paraphyses and excipular cells of six species of *Ascobolus*, two of *Saccobolus*, one of *Thecotheus*, and two of *Iodophanus*. An electron-opaque, hemispherical structure, similar to that found by Wells (1972), was recognized as the "ascoboloid septal type" (Fig. 1) and was found in all species of *Ascobolus*, *Saccobolus*, and *Thecotheus*. A biconvex band described later that we recognized as the "pezizoid septal type" was found in the septal pores of asci and ascogenous hyphae of all species of *Iodophanus* studied. Septal pores of vegetative cells possessed lamellate structures described earlier, although they were reduced and somewhat indistinct in some species of *Ascobolus* (Fig. 13).

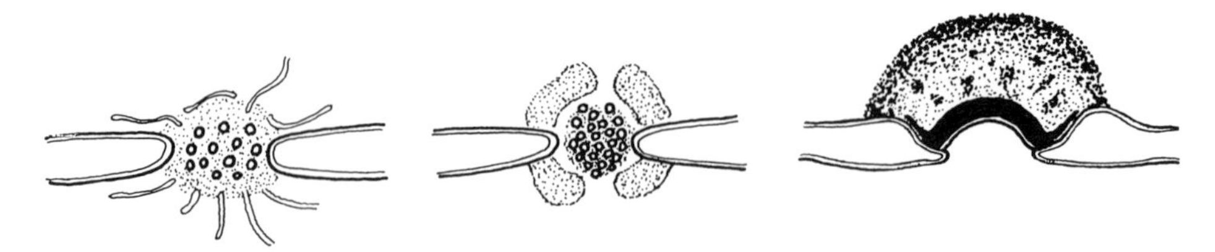

Figure 1. Schematic representation of septal pore structures in asci of *Ascobolaceae*.

The family position of *Iodophanus* and *Thecotheus* in *Pezizales* has been in question. Some retain both within the *Ascobolaceae* (Brummelen, 1967; Korf, 1973; Dennis, 1978), while others consider them among the *Pezizaceae* (Kimbrough and Korf, 1967; Rifai, 1968; Kimbrough et al., 1969). The biconvex bands in septal pores of asci and ascogenous hyphae and callous pectic spore ornaments of *Iodophanus*, which are similar to those of *Peziza*, suggest that *Iodophanus* belongs in the *Pezizaceae*. A number of features of *Thecotheus* point to a closer relationship with *Ascobolaceae*. These include the distinct dome-shaped ascal pore plugs, spores that remain relatively thick-walled throughout development, the lack of isolated spore ornaments, and an irregular secondary wall layer commonly having apiculi. The asci of species of *Thecotheus* are also very similar to those of *Ascobolus* (Samuelson, 1978).

Ascodesmidiaceae

Because of the pigmented ascospores and coprophilous habit, species of *Ascodesmis* were for many years placed within the *Ascobolaceae* (Brummelen, 1967; Rifai, 1968; Eckblad, 1968). Brummelen (1981) felt that *Ascodesmis* stood apart from other families of *Pezizales* and proposed that it be the sole genus of the *Ascodesmidiaceae*. It was in a species of *Ascodesmis* that Carroll (1967) first demonstrated elaborate pore plugging in the basal pores of asci in which a highly differentiated dome-shaped arrays of radiating tubular elements was found inside the ascus (Fig. 2). Brenner and Carroll (1967) also showed laminar and striate structures associated with the septal pore rim in vegetative cells; these were referred to as lamellate structures by Curry and Kimbrough (1983).

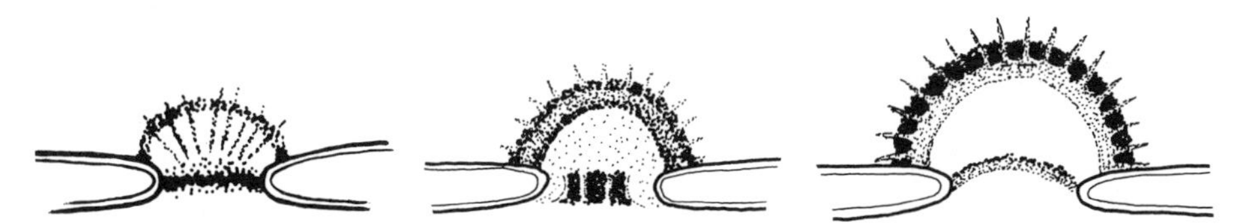

Figure 2. Schematic representation of septal pore structures in asci of *Ascodesmidiaceae*.

The "*Ascodesmis*-type" of ascal septal pore plug has subsequently been found in species of *Eleutherascus* (Steffens and Jones, 1983; Brummelen, 1989), and *Coprotus* (Kimbrough, unpubl.). Because these taxa form small, gymnohymenial apothecia which develop from clustered pairs of ascogonia and antheridia, combined with the unique ascal pore plug and other characters, Kimbrough (1989) proposed that the *Ascodesmidiaceae* be considered alongside the *Pyronemataceae* in a new suborder *Pyronemineae*.

Helvellaceae

The ultrastructure of septa of asci, ascogenous hyphae, paraphyses, and excipular cells of selected species of *Helvella* were described by Kimbrough and Gibson (1989). They found that electron-opaque, hemispherical structures occur in basal pores of young asci and with age become cone- to dumbbell-shape with "V"-shaped striations (Fig. 3). An electron-translucent torus separated the pore plug from the septal pore border.

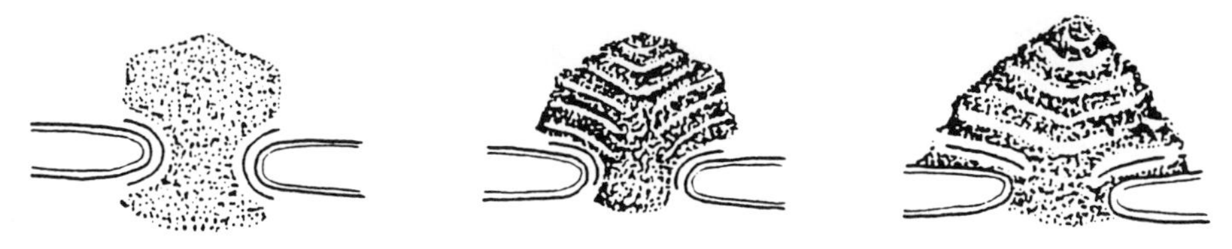

Figure 3. Schematic representation of septal pore structures in asci of *Helvellaceae*.

Septal pores of paraphyses and excipular cells possess the lamellate structure characteristic of other taxa of *Pezizales*. Woronin bodies were somewhat small and were globular to hexagonal (Fig. 13F). In a later study of five species of *Gyromitra* (Kimbrough, 1991), ascal pore plugs identical to those of *Helvella* were found in all species studied. It was in septal pores of vegetative cells, however, that in addition to lamellate structures, extremely elongate Woronin bodies were found, sometimes surrounding the pore (Fig. 13G), and at other times lodged within the pore. Data from these and related studies on ascosporogenesis suggest that the *Helvellaceae* should also encompass the *Rhizinaceae* and that perhaps only two tribes should be recognized, the *Helvelleae* and *Gyromitreae*.

Morchellaceae

In current studies of septa in the *Morchellaceae*, we have found that the septal pore plug at the base of asci is essentially the same as that found in members of the *Helvellaceae* (Fig. 4). *Morchella esculenta*, *M. elata*, and *M. semilibera* have a consistent pattern of pore-plugging in asci with a dome-shaped structure having "V"-shaped striations. Lamellate structures are found within septal pores of most vegetative cells. These are often pushed aside by extremely elongate Woronin bodies identical to those of species of *Gyromitra* (including *Discina*). In a previous study of *Morchella conica* (Kamaletdinowa and Vassilyev, 1982), both hexagonal and elongate retangular Woronin bodies were found (Fig. 13G). While inclusions, nuclear number, and ornamentation of spores are strikingly different in the two families, the unique septal structures, hymenial configurations, and other characters point to a close phylogenetic relationship of the *Helvellaceae* and *Morchellaceae*.

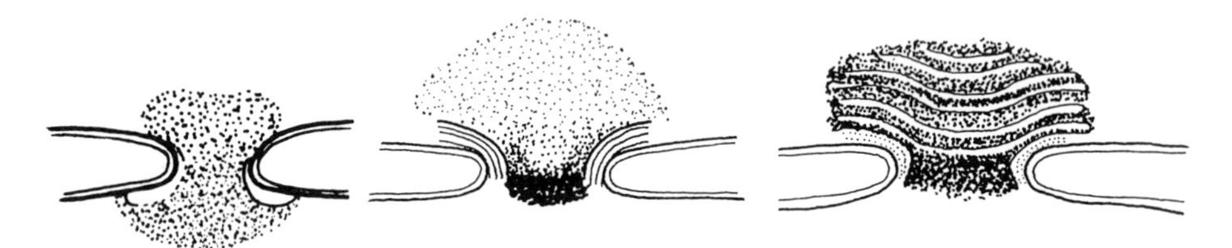

Figure 4. Schematic representation of septal pore structures in asci of *Morchellaceae*.

Otideaceae

The *Otideaceae* is the largest and most complicated family of *Pezizales*. Members of the family were originally described in the *Humariaceae* (see Rifai, 1968; Dennis, 1978), but may currently be found in the *Otideaceae* or *Pyronemataceae* (Eckblad, 1968), the *Aleuriaceae* and *Otideaceae* (Arpin, 1968; Kimbrough, 1970), the *Pyronemataceae* (Korf, 1973), and the *Otideaceae* (Korf and Zhuang, 1991; Eriksson and Hawksworth, 1991). The heterogeniety of this family is reflected also in the types of septal pore organelles found in various taxa. In a series of studies (Kimbrough and Curry, 1986a, 1986b; Wu, 1991), four patterns of ascal septal plugging were found among various genera:

(1) The "aleurioid-type", found in species of *Aleuria*, *Leucoscypha*, *Mycolachnea*, and *Trichophaea*, is characterized by a granular, opaque matrix which borders both the ascal and ascogenous hyphal sides of the pore (Fig. 5). In older asci the matrix appears in the ascal cell leaving a fan-shaped plug with a lamellate electron-translucent torus adjacent to the pore rim.

(2) The "otideoid-type", found in species of *Acervus*, *Anthracobia*, *Caloscypha*, *Octospora*, *Otidea*, and *Sphaerosporella*, is characterized by double translucent bands in the granular, opaque pore matrix. In mature asci, the pore plug is differentiated into two zones, an inner dense zone and an outer less opaque zone (Fig. 6). The double translucent bands adjacent to the pore rim in young asci are indistinct or disappears in mature asci.

(3) The "pulvinuloid-type", found in species of *Geopyxis* and *Pulvinula*, is strikingly similar to the early stages of the ascal pore plugs in the *Helvellaceae* (Fig. 7), although by maturity the "V"-shaped striations are fewer and less prominent.

(4) The "scutellinioid-type", found in *Cheilymenia*, *Coprobia*, and *Scutellinia*, is characterized by large hemispherical septal pore plugs that become zonate (Fig. 8), the inner zone adjacent to the pore lumen is electron-opaque and the outer, thicker zone of less dense material. As noted by Wu (1991), this type of ascal pore plug is remarkedly similar to that found in the *Ascobolaceae*.

Lamellate structures are found in septal pores of the cells of paraphyses and the exciple of all species of the *Otideaceae*. Woronin bodies tend to be small, largely globular,

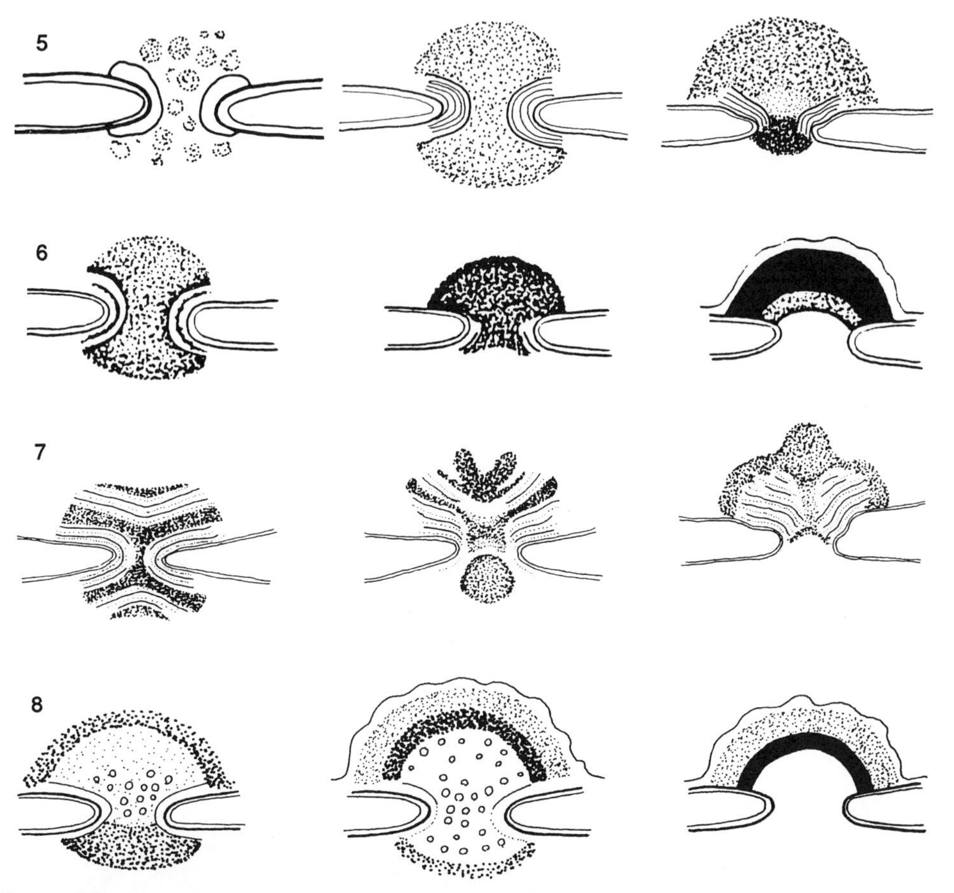

Figures 5-8. Schematic representations of septal pore structures in asci of the *Otideaceae*. **Fig. 5.** The "aleurioid" group. **Fig. 6.** Tthe "otidioid" group. **Fig. 7.** The "pulvinuloid" group. **Fig. 8.** The "scutelliniod" group.

but often hexagonal (Fig. 13E) in the "aleurioid", "pulvinuloid", and "otideoid" genera, while those of the "scutellinioid" genera are larger and globoid (Fig. 13B).

Pezizaceae

Septal structures of asci, ascogenous hyphae, paraphyses, and excipular cells were examined by Curry and Kimbrough (1983). In eight species of *Peziza* and one of *Plicaria*, lamellate structures (Fig. 13A), similar to those described by Schrantz (1964), were found in septa of vegetative cells of all apothecia examined. Large globose Woronin bodies were always found in the vicinity of septal pores. Electron-opaque convex and biconvex bands span the pores of asci in all species (Fig. 9). These bands often become covered with electron-opaque amorphous material or by additional secondary wall. Septal pore plugging in the ascogenous hyphae was similar to that of asci and occasional globose Woronin bodies were found associated with the convex band.

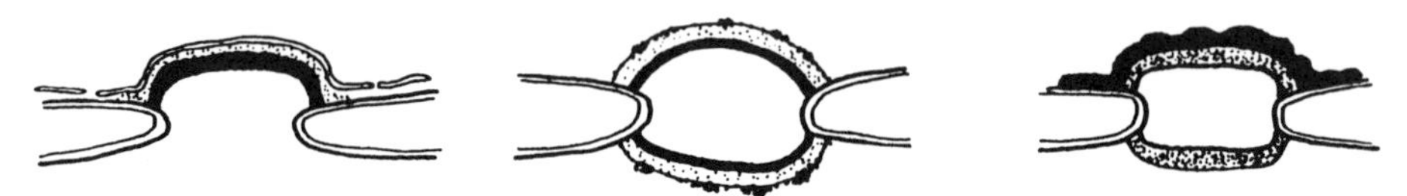

Figure 9. Schematic representation of septal pore structures in asci of *Pezizaceae.*

Convex and biconvex bands in pores of the ascal base appear to be a characteristic feature of the *Pezizaceae* in that they have subsequently been found in species of *Iodophanus* (Kimbrough and Curry, 1985), *Sarcosphaera*, and *Pachyella* (Kimbrough, unpubl.).

Pyronemataceae

Septal pore plugging in gymnohymenial species of *Pezizales* appear to be distinct from that found in other members of the order (Kimbrough, 1989). Septal pore plugs in asci and ascogenous hyphae of *Pyronema* are morphologically similar to those of *Ascobolaceae* (Fig. 10). The intercalary band adjacent to the central cytoplasmic core, however, is of a different electron density and small, radiating tubular elements similar to those of *Ascodesmis* extend away from the band. The tubular bands of *Pyronema* are smaller, fewer, and extend only a short distance into the hemispherical structure. An ascal pore plug identical to that found in *Ascodesmis* has also been found in species of *Coprotus* (Kimbrough, unpubl.).

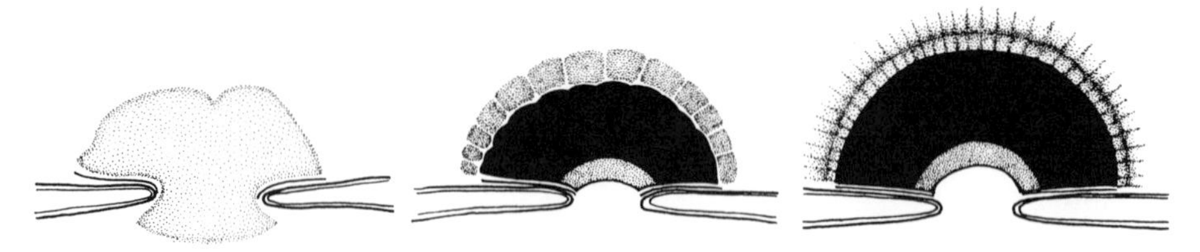

Figure 10. Schematic representation of septal pore structues in asci of *Pyronemataceae.*

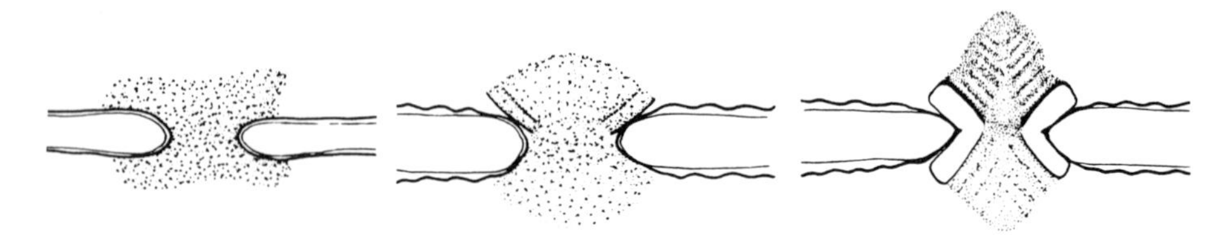

Figure 11. Schematic representation of septal pore structures in asci of *Sarcoscyphineae.*

Sarcoscyphaceae/Sarcosomataceae

There have been numerous ultrastructural studies on various genera of the *Sarcoscyphaceae* and *Sarsomataceae*, but they have been largely restricted to ascus structure and dehiscence (Brummelen, 1975; Samuelson, 1975; Samuelson et al., 1980). Kamaletdenowa and Vassilyev (1982) showed typical globose electron-opaque Woronin bodies adjacent to pores in *Sarcoscypha coccinea*. In my laboratory, Ms Li-tzu Li is currently engaged in an ultrastructural study of septa and of ascosporogenesis in both families. Preliminary data show that in septal pores of vegetative tissues lamellate structures are poorly differentiated or missing. Globose and occasionally hexagonal Woronin bodies like those of the *Otideaceae* (Fig. 13C, E) are typical in the *Sarcoscyphineae*. To our surprize, however, septal plugs of asci in species of *Sarcoscypha*, *Wynnea*, and *Urnula* are morphologically identical to those found in *Geopyxis* (Fig. 11) and seem intermediate between those of the *Otideaceae* and *Helvellaceae*.

Thelebolaceae

We no longer consider the *Thelebolaceae* a valid family of the *Pezizales*. Studies on ascus structure and ascocarp development of *Thelebolus* (Samuelson and Kimbrough, 1978; Brummelen, 1978; Kimbrough, 1981) show that species of *Thelebolus* have bitunicate asci and that ascomatal development is similar to that of certain *Loculoascomycetes*. Cytochemical and ultrastructural data on other genera previously assigned to the *Thelebolaceae* suggest that they belong elsewhere (Korf and Zhaung, 1991).

TAXONOMIC CONSIDERATIONS WITHIN PEZIZALES

A number of general conclusions may be drawn from the nature of septal pore structures in the asci and ascogenous hyphae of *Pezizales*. Septal pore organelles in asci and ascogenous hyphae are distinctly different from those of vegetative cells. The consistency and electron density of pore plugs suggest that they are chemically different from Woronin bodies and lamellate structures of vegetative cells. Septal pore plugs in asci of *Helvella* poststained with silver proteinate showed a conspicuous absence of polysaccharide (Kimbrough and Gibson, 1989). Beckett (1981) came to the same conclusion in other taxa. A number of studies have concluded that spherical and hexagonol Woronin bodies are proteinaceous. Cytochemical evidence indicates that septal plugging structures are not proteinaceous, do not contain lipids, and therefore, probably not derived from Woronin bodies (Nakai and Ushiyama, 1976).

Carroll (1967) suggested that the occurrence of complex pore caps at the base of developing asci may be related to the process of spore production. Since septal pore organelles in the ascogenous hyphae are also more elaborate than in vegetative cells, one might speculate that once the dikaryon is established in the ascogonium, such pore plugs may occur to insure the dikaryon in maintained in the ascogenous system. A fundamental feature vital to sexual reproduction might also be a very conservative character and, therefore, useful in showing taxonomic and phylogenetic relationships.

Among the *Pezizales*, the most simplistic type of pore plugging of asci and ascogenous hyphae occurs within the *Pezizaceae* (Fig. 12). That the convex and biconvex bands have been found in all *Pezizaceae* examined thus far suggest that it is a good family character. The elaboration or transformation of the convex band into a hemispheric plug is seen in the *Otideaceae* (*Scutellineae* and *Otideae*), *Ascobolaceae*, *Pyronemataceae*, and *Ascodesmidaceae* (Fig. 12). Although species of a limited number of genera have been examined in the *Ascodesmidaceae* and *Pyronemataceae*, the unique ascal pore plugging appears consistent at the family level. This is not true, however, among the *Otideaceae* in which there is a less well-defined pore plug composed of amorphous, granular material. Within the *Otideae*, in addition to the zonate hemispherical plug, there is the immergence of an electron-translucent, banded torus adjacent to the pore rim. In taxa such as *Mycolachnea* and members of the *Aleurieae*, the banded torus is more elaborate and the outer hemispheric structure is less zonate and not as well organized (Fig. 12). In taxa with the "pulvinuloid" ascal pore structure, the torus becomes more elaborate with a distinct "V"-shaped banding pattern. Pore plugs composed of a rather loose, granular matrix with more elaborate "V"-shaped straitions are seen in *Geopyxis*, and in members of

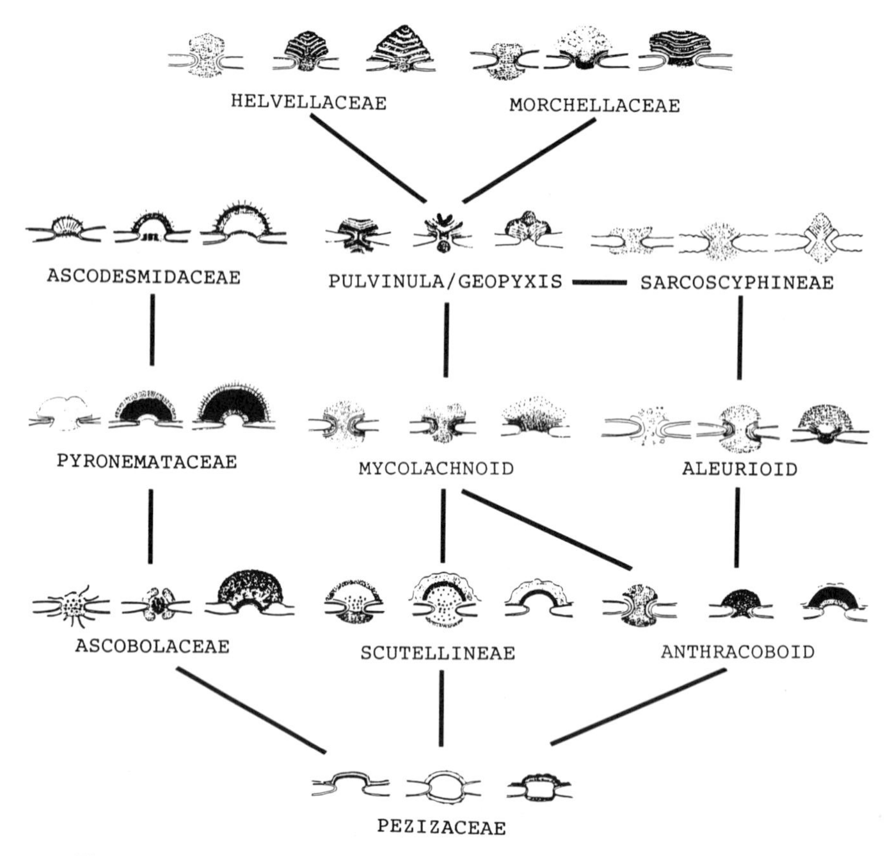

Figure 12. A phylogenetic scheme of *Pezizales* based upon septal structures in asci.

the *Sarcoscyphaceae* and *Sarcosomataceae* (Fig. 12). The *Helvellaceae* and *Morchellaceae* have the most elaborate of this type of pore plug. In taxa of these families, the septal pore at the base of asci is plugged with a large, cone-shaped structure with numerous "V"-shaped striations (Fig. 12).

When one contrast the types of lamellate structures and Woronin bodies associated with septal pores of vegetative cells (Fig. 13), there is an interesting correlation with the type of pore organelles found in asci and ascogenous hyphae of the various *Pezizales*. For example, large, globose Woronin bodies and elaborate lamellate structures are found in members of the *Scutellineae* and the *Pezizaceae* (Fig. 13A, B). Whereas, very small, globose to hexagonal Woronin bodies and poorly defined lamellate structures are found in the *Aleurieae*, *Otideae*, and the *Ascobolaceae* (Figs 13C, D, E). In the *Helvellaceae* and *Morchellaceae*, Woronin bodies range from occasionally globose to hexagonal or predominantly long and rectangular (Figs 13F, G).

Based upon morphological and cytological data, the *Helvellaceae* and *Morchellaceae* are considered to be highly evolved groups. The unique types of septal structures found in both vegetative and reproductive hyphae of these groups would tend to substantiate this idea. The next question is, to which groups of *Pezizales* are they most closely related? Septal structures would suggest members of the *Otideaceae*, especially certain taxa of the tribes *Mycolachneae* and *Aleurieae*. Because of the blueing of asci in iodine, members of the *Pezizaceae* and *Ascobolaceae* have also been considered closely related. Again, there is some similarity in the type of pore plugging found in asci of the two groups. In respect to septa, the *Scutellinieae* appear to be more similar to the *Ascobolaceae* than to other members of the *Otideaceae* to which they have traditionally been considered to be related. Septal structures would tend to reinforce the idea that the *Ascodesmidaceae* are related to the *Ascobolaceae*. Because of apothecial texture and peculiar ascus structure in the *Sarcoscyphaceae* and *Sarcosomataceae*, they tend to stand apart

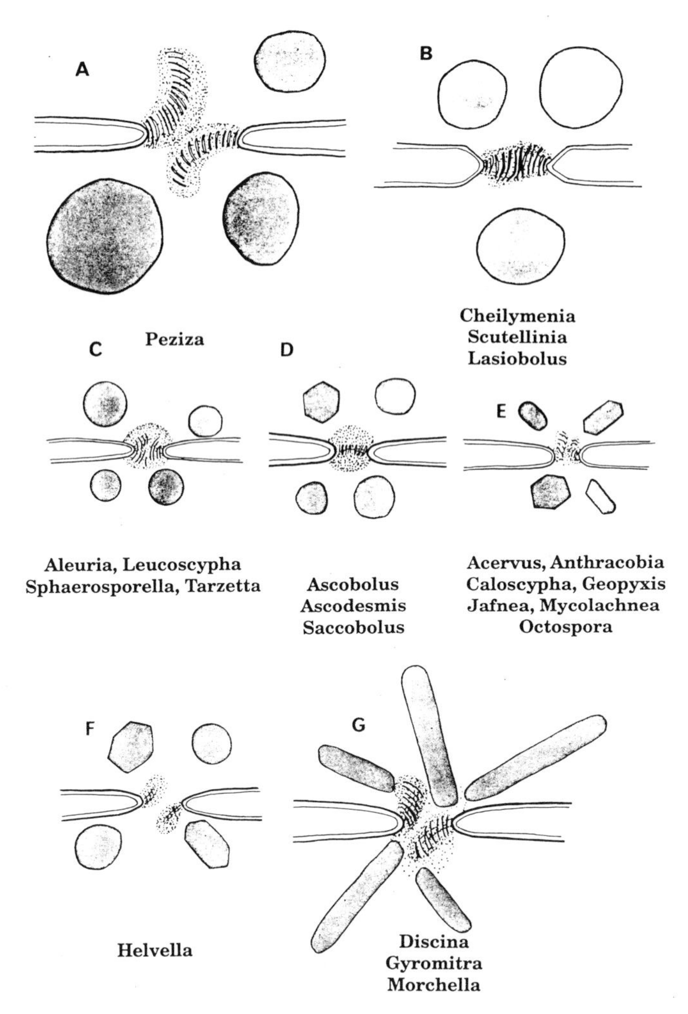

Figure 13. Schematic representation of septal pore structures and Woronin bodies in vegetative cells of various groups of *Pezizales*.

from other *Pezizales* and the question remains, to which groups of *Pezizales* might they be most closely related? Septal structures suggest that they may be related to members of the *Otideaceae*, especially the *Geopyxis-Pulvinula*-like groups.

Molecular studies of the *Pezizales* have been limited. Momol (1992) did a nuclear ribosomal DNA sequence analysis of selected members of the *Pezizales* (Fig. 14). The most parsimonious tree for the 5.8S sequences using *Neurospora* as the outgroup confirmed a close relationship of *Plicaria*, *Iodophanus*, and *Peziza* in the *Pezizaceae*, but was unable to delineate *Pyronema*, *Eleutheroascus*, *Ascodesmis*, *Saccobolus*, *Lamprospora*, and *Thelebolus*. Surprisingly, *Gyromitra* showed the closest affinity to *Otidea*, supporting our ultrastructural data that suggested a *Helvellaceae-Otideaceae* relationship. Since the 5.8S coding region was too conserved to resolve lower-level relationships of the eight grouped taxa, sequences from variable regions such as ITS were used. The ITS2 region was selected for this analysis because it was similar in length among the species tested. *Iodophanus* was used as the outgroup and a single most parsimonious tree was found (Fig. 15). Here, species of *Ascodesmis* and *Saccobolus* were clustered, *Pyronema* formed an unresolved line with *Eleutherascus* and *Lamprospora*, and *Thelebolus* stood apart as an independent line. These data agree with our septal data in bunching the *Ascobolaceae* and *Ascodesmitaceae* but left unresolved the exact position of the *Pyronemataceae*.

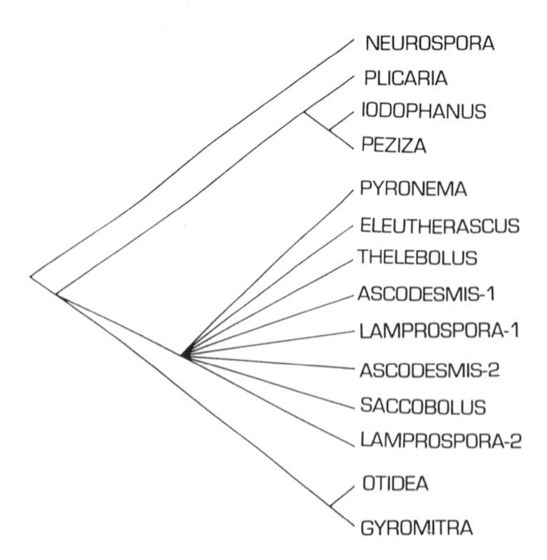

Figure 14. The most parsimonious tree for the 5.8S rDNA sequences using *Neurospora* as an outgroup (from Momol, 1992).

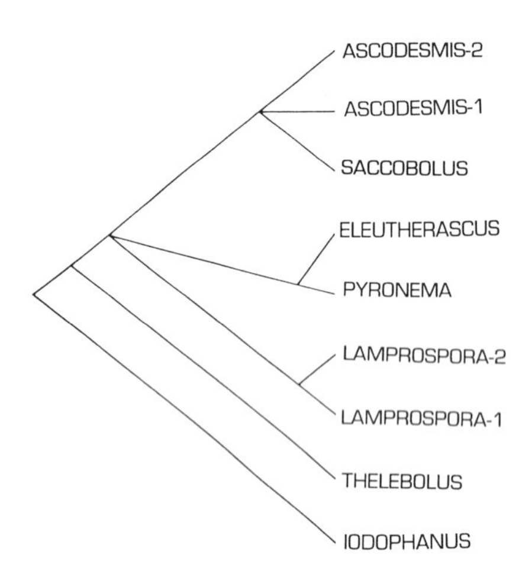

Figure 15. The most parsimonious tree for the ITS2 sequences using *Iodophanus* as an outgroup (from Momol, 1992).

SEPTAL ULTRASTRUCTURE AND PEZIZALES/TUBERALES LINKAGES

From the middle of the nineteenth century to the present, there has been the general recognition that *Pezizales* and *Tuberales* were closely related and the distinction between the two groups is not always clear (Burdsall, 1968; Eckblad, 1968). Trappe (1979), likewise, was unable to clearly distinguish epigeous and hypogeous taxa of the *Pezizales/Tuberales* complex and proposed abandoning the *Tuberales*. Using morphological, anatomical, and cytochemical data he placed a number of hypogeous taxa in established epigeous families of *Pezizales*. Others he left in their respective "truffle" families but placed those families within the *Pezizales*. Although there have been a number of ultrastructural studies of taxa of *Tuberales*, Kimbrough et al. (1991) were the first to use ultrastructural data, especially those of septal structures, to establish taxonomic linkages between the two groups. Cytological and ultrastructural features of asci, septa, and spore wall ontogeny of *Hydnobolites* revealed a number of characters that linked the genus to the *Pezizaceae*. These included the presence of electron-dense, biconvex bands in septal pores of asci, dextrinoid and weak blueing reaction of asci in iodine, and a type of spore wall deposition similar to members of the *Pezizaceae*.

Research currently in progress has determined that species of the "truffle" genus *Barssia* have tetranucleate ascospores with a large lipid droplet, a cardinal feature of the *Helvellaceae*. Septal pore plugs were also found to be of the *Helvella*-type with a large cone-shaped matrix having prominent "V"-shaped striations. These data combined with anatomical data clearly places *Barssia* within the *Helvellaceae*. Similar ultrastructural studies of *Genea gardenerii* reveal a type of septal plugging and spore wall ontogeny identical to that of *Mycolachnea* and other members of the *Otideaceae*. In earlier light microscope studies, Pfister (1984) also concluded that *Genea* belonged to the *Pyronemataceae* (*Otideaceae sensu* Korf and Zhuang, 1991). Recognizing the polyphyletic nature of hypogeous ascomycetes, we can anticipate linking other "truffle" taxa to existing families of *Pezizales*. For example, lamellate structures found in the "truffle" *Terfezia leptoderma* (Janex-Favre et al., 1988) clearly establish a relationship of this genus to *Pezizales*. The electron dense convex band and globose to hexagonal Woronin bodies in some of the vegetative cells and the type of spore ontogeny suggest a possible linkage to the *Otideaceae*.

SEPTA AND THE SYSTEMATICS OF OTHER ASCOMYCETES

Unfortunately, there have been relatively few studies specifically on septa of other orders and families of ascomycetes. Much of the information available comes from incidental micrographs where the author was concentrating on other features. Yet, there are some very distinct types of septa found among the other ascomycetes. Briefly, they include the following:

The "*Neurospora*-type" Vegetative Septum

The "*Neurospora*-type" septum was first described by Shatkin and Tatum (1959) in the vegetative cells of *N. crassa*. Its structure and ontogeny was most clearly demonstrated by Trinci and Collinge (1973) who showed the origin and progressive development of translucent, finger-like projections from the pore rim into the electron-opaque pore plug (Fig. 16A). An identical septal pore plug was found in species of *Hypocrea* by Nakia and Ushiyama (1976), who also demonstrated that the pore plug was structurally and chemically different from Woronin bodies. Subsequently, this type of pore plug has been found in several other ascomycetes (Curry and Kimbrough, 1983).

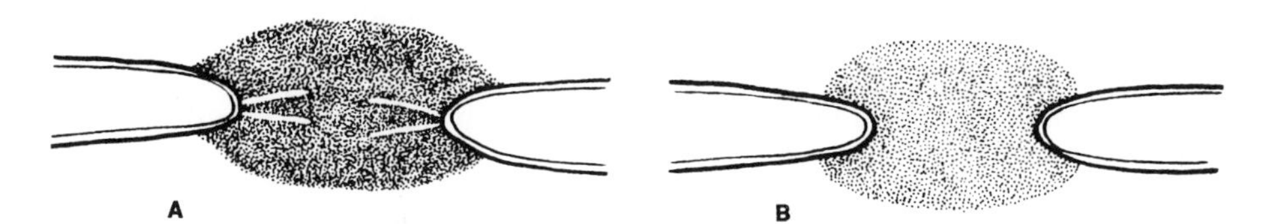

Figure 16. Schematic representation of the (A) "*Neurospora*-type" and (B) "*Sclerotinia*-type" septal pore structures in vegetative cells.

The "*Sclerotinia*-type" Vegetative Septum

The "*Sclerotinia*-type" septal pore plug was first demonstrated by Saitô (1974) in sclerotial cells of *Sclerotinia sclerotiorum*. It appears to have the same electron density and staining properties of the "*Neurospora*-type" but is lacking the electron-translucent finger-like projections (Fig. 16B). A number of workers have mistakenly interpreted this type of pore plug as a Woronin body. It has been clearly demonstrated that Woronin bodies can plug septal pores in vegetative cells of a number of ascomycetes (Curry and Kimbrough, 1983), however, in such cases Woronin bodies are membrane-bound but the pore plugs are not. The "*Sclerotinia*-type" septum has been found in various genera of ascomycetes including *Ampelomyces* (Hashioka and Nakai, 1980), and *Pseudopeziza* (Meyer and Luttrell, 1986).

Ascal and Ascogenous Septa of Other Ascomycetes

There is a great lack of information on septal pore organelles from the ascogenous hyphae and asci of other groups of ascomycetes. Furtado (1971) showed septal pore organelles in the ascogenous cells of *Sordaria fimicola* in which the rim of the septal pore was lined by a swollen torus from which tubular cisternae radiated into the adjacent cells. A narrow pore perforated the center of the ring. A similar type of septal structure has been found in *S. macrospora* (Beckett et al., 1974), and *S. humana* (Beckett, 1981), as well as in species of *Chaetomium* (Rosing, 1981), *Sporormia* (Blanchard, 1972) and *Leptosphaeria* (Parguey-Leduc et al., 1982). *Sporormia*, however, does not have the array of tubular cisternae radiating into the cytoplasm of adjacent cells. All of these genera are either *Pyrenomycetes* or *Loculoascomycetes*. Might this type of pore structure be characteristic of the *Pyrenomycetes* and *Loculoascomycetes*, or may there yet be other undiscovered septal types? I feel certain that other septa of this type will be uncovered with a deligent search of the literature and with additional EM research.

CONCLUSIONS

Our extensive research on septal ultrastructure in the *Pezizales* has demonstrated the usefulness of fine structural characters to resolve taxonomic problems within families and to reveal phylogenetic tendencies within the order, including both epigeous and hypogeous taxa. Lack of comparable data from other ascomycetes makes it difficult to use such data to determine phylogenetic relationships with other groups. Yet, septal data may be extremely useful to researchers in mycology, plant pathology, and mycorrhizas in which fungi refuse to sporulate and it is vital to know the group of fungi with which you are dealing.

It is in the study of mycorrhizas where septal ultrastructure has been of great practical use. A number of unknown mycorrhizal symbionts have been determined to be ascomycetes. They include, for example, species on the following hosts: *Pyrola* (Robertson and Robertson, 1985), *Picea* (Haug and Oberwinkler, 1987), *Abies* (Berndt and Oberwinkler, 1992), and boreal orchids (Currah and Sherburne, 1992). Berndt and Oberwinkler (1992) were able to determine that one of the mycorrhizas on silver fir (*Abies alba*) was not only an ascomycete but was a member of the *Pezizales* or *Tuberales*. From root morphology they concluded that it was most likely a species of *Tuber* and closely resembled *T. puberulum* which had been reported associated with the host.

It has not been possible to make an extensive survey of septa in non-pezizalean ascomycetes here, but it is clear to me that a wealth of useful information is hidden away in septal structures. In the use of rRNA sequencing in resolving broader level phylogenetic relationships within the basidiomycetes, Walker and Doolittle (1983) concluded that molecular data in various groups correlated more closely with septal data than to that of basidial and basidiocarp features. Preliminary studies noted above suggest that the same may be true of ascomycetes. If this is true, we may have been looking at the wrong end of the ascus to obtain some of the most conservative of phylogenetic characters.

ACKNOWLEDGEMENTS

I would like to thank the organizers of the "First International Workshop on Ascomycete Systematics" for the opportunity to present our research on ascomycete septa. I am grateful to the postdoctoral assistance of Dr K. Curry and Dr J. Gibson, graduate students Chi-guang Wu and Li-tzu Li, and my colleague Dr G.L. Benny in doing the extensive field work, laboratory preparations, and endless hours of electron microscopy necessary for these studies. I also appreciate the many years of support from the National Science Foundation. This manuscript belongs to the Florida Agricultural Experiment Station Journal Series.

REFERENCES

Arpin, N., 1968, *Les carotenoides de Discomycètes: essai chimiotaxonomique.* PhD thesis, Université de Lyon.

Beckett, A., 1981, The ultrastructure of septal pores and associated structures in ascogenous hyphae and asci of *Sordaria humana*, *Protoplasma* 107: 127-147.

Beckett, A., I.B. Heath, and D.J. McLaughlin, 1974, *An Atlas of Fungal Ultrastructure*, Longman Group, London.

Berndt, R. and F. Oberwinkler, 1992, Ultrastructure of septal pores of mycorrhiza-forming ascomycetes, *Mycologia* 84: 360-366.

Blanchard, R.O., 1972, Septa in *Sporormia australis*, *Mycologia* 64: 1330-1333.

Boehm, E.W.A. and D.J. McLaughlin, 1989, Phylogeny and ultrastructure in *Eocronartium muscicola*: meiosis and basidial development, *Mycologia* 81: 98-114.

Brenner, D.M. and G.C. Carroll, 1968, Fine structural correlates of growth in hyphae of *Ascodesmis sphaerospora*, *Journal of Bacteriology* 95: 658-671.

Brummelen, J. van, 1967, A world monograph of the genera *Ascobolus* and *Saccobolus* (ascomycetes, *Pezizales*), *Persoonia* (Suppl.) 1: 1-260.

Brummelen, J. van, 1975, Light and electron microscopic studies of the ascus top in *Sarcoscypha coccinea*, *Persoonia* 8: 259-271.

Brummelen, J. van, 1978, The operculate ascus and allied forms, *Persoonia* 10: 113-128.

Brummelen, J. van, 1981, The operculate ascus and allied forms. *In*: *Ascomycete Systematics*: *The Luttrellian Concept* (D.R. Reynolds, ed.): 27-48. Springer-Verlag, New York.

Brummelen, J. van, 1989, Ultrastructure of the ascus and the ascospore wall in *Eleutherascus* and *Ascodesmis* (*Ascomycotina*), *Persoonia* 14: 1-17.

Burdsall, H.H., 1968, A revision of the genus *Hydnocystis* (*Tuberales*) and of the hypogeous species of *Geopora* (*Pezizales*), *Mycologia* 60:496-525.

Carroll, G.C., 1967, The fine structdure of the ascus septum in *Ascodesmis sphaerospora* and *Saccobolus kerverni*, *Mycologia* 59: 527-532.

Currah, R.S. and R. Sherburne, 1992, Septal ultrastructure of some endophytes from boreal orchid mycorrhizas, *Mycological Research* 96: 583-587.

Curry, K.J. and J.W. Kimbrough, 1983, Septal structures in apothecial tissues of the *Pezizaceae* (*Pezizales*, ascomycetes), *Mycologia* 75: 781-794.

Dennis, R.W.G., 1978, *British Ascomycetes*, J. Cramer, Vaduz.

Eckblad, F.-E., 1968, The genera of the operculate *Discomycetes*. A reevaluation of their taxonomy, phylogeny, and nomenclature, *Nytt Magasin for Botanikk* 15: 1-191.

Eriksson, O.E. and D.L. Hawksworth, 1991, Notes on ascomycete systematics - Nos 1252-1293, *Systema Ascomycetum* 10: 135-149.

Furtado, J.S., 1971, The septal pore and other ultrastructural features of the pyrenomycete *Sordaria fimicola*, *Mycologia* 63: 104-113.

Hashioka, Y. and Y. Nakai, 1980, Ultrastructure of pycnidial development and mycoparasitism of *Ampelomyces quisequalis* parasitic on *Erysiphe*, *Transactions of the Mycoloogical Society of Japan* 21: 329-338.

Haug, I. and F. Oberwinkler, 1987, Some distinctive types of spruce mycorrhizae, *Trees* 1: 172-188.

Janex-Favre, M.-C., A. Parguey-Leduc, and L. Riousset, 1988, L'ascocarpe hypoge d'une terfez Francaise (*Terfezia leptoderma* Tul.) (Tubérales, Discomycètes), *Bulletin de la Société Mycologique de France* 104: 145-178.

Kamaletdinowa, F.I. and A.E. Vassilyev, 1982, *Cytology of Discomycetes*, Nauka Kaxakh SSR, Alma Ata.

Khan, S.R. and J.W. Kimbrough, 1982, A reevaluation of the basidiomycetes based upon septal and basidial structures, *Mycotaxon* 15: 103-120.

Kimbrough, J.W., 1970, Current trends in the classification of discomycetes, *Botanical Review* 36: 91-161.

Kimbrough, J.W., 1981, Cytology, ultrastructure, and the taxonomy of *Thelebolus* (ascomycetes), *Mycologia* 72: 1-27.

Kimbrough, J.W., 1989, Arguments towards restricting the limits of the *Pyronemataceae* (*Pezizales*, ascomycetes), *Memoirs of the New York Botanical Garden* 49: 326-345.

Kimbrough, J.W., 1991, Ultrastructural observations on *Helvellaceae* (*Pezizales*, ascomycetes). V. Septal structures in *Gyromitra*, *Mycological Research* 95: 421-426.

Kimbrough, J.W. and K.J. Curry, 1985, Septal ultrastructure in the *Ascobolaceae* (*Pezizales*, ascomycetes), *Mycologia* 77: 219-229.

Kimbrough, J.W. and K.J. Curry, 1986a, Septal structures in apothecial tissues of the tribe *Aleurieae* in the *Pyronemataceae* (*Pezizales*, ascomycetes), *Mycologia* 78: 407-417.

Kimbrough, J.W. and K.J. Curry, 1986b, Septal structures in apothecial tissues of taxa in the tribes *Scutellineae* and *Sowerbyelleae* (*Pyronemataceae, Pezizales*), *Mycologia* 78: 734-743.
Kimbrough, J.W. and J.L. Gibson, 1989, Ultrastructural observations on *Helvellaceae* (*Pezizales*: ascomycetes). III. Septal structures in *Helvella, Mycologia* 81: 914-920.
Kimbrough, J.W. and R.P. Korf, 1967, A synopsis of the genera and species of the tribe *Thelebeleae* (=*Pseudoascoboleae*), *American Journal of Botany* 54: 9-23.
Kimbrough, J.W., R. Luck-Allen, and R.F. Cain, 1969, *Iodophanus*, the *Pezizeae* segregate of *Ascophanus* (*Pezizales*), *American Journal of Botany* 56: 1187-1202.
Kimbrough, J.W., C.G. Wu, and J.L. Gibson, 1991, Ultrastructural evidence for a linkage of the truffle genus *Hydnobolites* to the *Pezizaceae* (*Pezizales*), *Botanical Gazette* 152: 408-420.
Korf, R.P., 1973, *Discomycetes* and *Tuberales*, *In*: *The Fungi. An advanced treatise* (G.C. Ainsworth, F.K. Sparrow, and A.S. Sussman, eds), 4A: 249-319. Academic Press, New York.
Korf, R.P. and W.-Y. Zhuang, 1991, Preliminary discomycete flora of Macronesia: Part 16, *Otideaceae, Scutellinioideae*, *Mycotaxon* 40: 79-106.
Meyer, S.F.L. and E.S. Luttrell, 1986, Ascoma morphology of *Pseudopeziza trifolii* forma specialis *medicaginis-sativae* (*Demateaceae*) on alfalfa, *Mycologia* 78: 529-542.
Momol, E., 1992, *Nuclear ribosomal DNA sequence analysis in molecular systematics of Pezizales (Ascomycetes)*. PhD dissertation, University of Florida, Gainesville.
Moore, R.T., 1978, Taxonomic significance of septal ultrastructure with particular reference to the jelly fungi, *Transactions of the British Mycological Society* 70: 1007-1024.
Moore, R.T. and J.H. McAlear, 1962, Fine structure of *Mycota*. 7. Observations on septa of ascomycetes and basidiomycetes, *American Journal of Botany* 49: 86-94.
Nakai, Y., and R. Ushiyama, 1976, Ultrastructure of septum pores in *Hypocrea schweinitzii*, *Transactions of the Mycological Society of Japan* 17: 401-408.
Oberwinkler, F., 1985, Ammerkungen zur Evolution und Systematik der Basidiomyceten, *Botamosches Jahrbücher* 107: 541-580.
Parguey-Leduc, A., M.C. Janex-Favre, S. Andrieu, L. Lacoste, F. Traore, 1982, Les peritheces et les asques du *Leptosphaeria* (?) *senegalensis* Segretain, Baylet, Darasse et Camain, *Annales de Parasitolegie Humaine et Comparee* 57: 179-195.
Pfister, D.H., 1984, *Genea-Jafneadelphus*-a tuberalean-pezizalean connection, *Mycologia* 76: 170-172.
Pickett-Heaps, J.D. and H.J. Marchant, 1972, The phylogeny of the green algae: a new proposal, *Cytobios* 6: 255-264.
Pueschel, C.M., and K.M. Cole, 1982, Rhodophycean pit plugs: an ultrastructural survey with taxonomic implications, *American Journal of Botany* 69: 703-720.
Rifai, M.A., 1968, The Australasian *Pezizales* in the Herbarium of the Royal Botanic Gardens Kew, *Verhandelingen der Koninklijke Nederlandsche Akademie van Wetenschappen, Afdeeling Natuurkunde, Tweede sect.* 57: 1-295.
Robertson, D.C. and J.A. Robertson, 1985, Ultrastructural aspects of *Pyrola* mycorrhizae, *Canadian Journal of Botany* 63: 1089-1098.
Rosing, W.C., 1981, Ultrastructure of septa in *Chaetomium brasiliense* (*Ascomycotina*), *Mycologia* 73: 1204-1207.
Saito, I., 1974, Ultrastructural aspects of the maturation of sclerotia of *Sclerotinia sclerotiorum* (Lib.) de Bary, *Transactions of the Mycological Society of Japan* 15: 384-400.
Samuelson, D.A., 1975, The apical apparatus of the suboperculate ascus, *Canadian Journal of Botany* 53: 2660-2679.
Samuelson, D.A., 1978, Asci of the *Pezizales*. I. The apical apparatus of the iodine-positive species, *Canadian Journal of Botany* 56: 1860-1875.
Samuelson, D.A. and J.W. Kimbrough, 1978, Asci of the *Pezizales*. IV. The apical apparauts of *Thelebolus*, *Botanical Gazette* 139: 346-361.
Samuelson, D.A., G.L. Benny, and J.W. Kimbrough, 1980, Asci of *Pezizales* VII. The apical apparatus of *Galiella rufa* and *Sarcosoma globosum*: reevaluation of the suboperculate ascus, *Canadian Journal of Botany* 58: 1235-1243.
Schrantz, J.P., 1964, Étude au microscope electronique des synapse de deux Discomycètes, *Compte Rendu Hebdomadaire des Séances del' Academie des Sciences sér. D* 258: 3342-3344.
Schrantz, J.P., 1970, Étude cytologique en microscopie optique et electronique, de quelques Ascomycètes. II. La paroi, *Revue de Cytologie et de Biologie Vegetale* 33: 111-168.
Shatkin, A.J. and E.L. Tatum, 1959, Electron microscopy of *Neurospora crassa* mycelia, *Journal of Biophyscis, Biochemistry and Cytology* 6: 423-426.
Steffens, W.L. and J.P. Jones, 1983, Ascus and ascospore development in *Eleutherascus peruvianus*. 2. Nuclear cytology and early ascus ontogeny, *Canadian Journal of Botany* 61: 1599-1617.

Trappe, J.M., 1979, The orders, families, and genera of hypogeous *Ascomycotina* (truffles and their allies), *Mycotaxon* 9: 297-340.

Trinci, A.P.J. and A.J. Collinge, 1973, Structure and plugging of septa of a wild type and spreading colonial mutant of *Neurospora crassa*, *Archiv für Mikrobiologie* 91: 355-364.

Walker, W.F., and W.F. Doolittle, 1983, 5S rRNA sequences from eight basidiomycetes and other fungi imperfecti, *Nucleic Acids Research* 11: 7625-7630.

Wells, K., 1972, Light and electron microscopic studies of *Ascobolus stercorarius* II. Ascus and ascospore ontogeny. *University of California, Publications in Botany* 62: 1-93.

Wu, C.G., 1991, *Ultrastructural investigations of Humariaceae (Pezizales, Ascomycetes): Spore ontogeny, septal structure, and phylogeny*. PhD dissertation, University of Florida, Gainesville.

Secondary Chemistry

SECONDARY METABOLITES OF SOME NON-LICHENIZED ASCOMYCETES

P.G. Mantle

Biochemistry Department
Imperial College of Science, Technology and Medicine
London SW7 2AY, UK

SUMMARY

Many secondary metabolites offer potential supplementary taxonomic criteria for ascomycete fungi, in that they are all derived from a few key intermediates of primary metabolism. They are therefore closely linked to some of the most fundamental enzymes of living processes. It is not yet clear whether all enzymes of so-called secondary metabolic pathways are unique to that role or are primary enzymes with multiple functions, but some secondary metabolites derived from several diverse precursors (e.g. indole-isoprenoids, prenylated diketopiperazines, benzodiazepines) may involve more DNA sequences than pathways leading to pigments. Specific end-products of secondary metabolic pathways are probably not good taxonomic criteria for dichotomous differentiation at the genus level, and even pathways predictably involving at least several enzymes may be too widely distributed to augment current taxonomic boundaries. For example, ochratoxin A, first found in *Aspergillus ochraceus*, is more consistently produced by *Penicillium verrucosum*. Taking a wider perspective, non-lichenized ascomycetes may elaborate a vast array of chemical structures but may only express these biosyntheses when the fungi are unrecognisable as having ascomycete sexuality. Elaboration of complex secondary metabolites, often of no perceived function for the organism, is usually incompatible with the metabolic demands of sexual fructification. Nevertheless, some secondary metabolites may have a role in rationalising taxonomic boundaries at the level of species as has been shown in recent detailed study of *Penicillium aurantiogriseum* and closely related species.

INTRODUCTION

First of all I would like to thank the organisers of this conference who have dared to invite me to speak. I can tell you, frankly, that I would never have had the gall to stand in front of a group of taxonomists if I hadn't been invited. It is, to use the French, "très, très formidable". I am not a taxonomist, but perhaps I can claim associate membership of the club having after about 30 years work recently described the teleomorph of a new *Claviceps* (Frederickson *et al.*, 1991). Taxonomy is something that my students hate like the plague, but I am finding that as the years go by it is more and more fascinating to look at order and to spend time defining order in biochemical terms.

SECONDARY METABOLITES AS TAXONOMIC CRITERIA

Twenty years ago secondary metabolites were hardly contemplated as possible tools within ascomycete taxonomy. Now that perhaps 8000 compounds are known to be produced by fungi, and many of these are produced by ascomycetes, taxonomists are more

Ascomycete Systematics: Problems and Perspectives in the Nineties
Edited by D.L. Hawksworth, Plenum Press, New York, 1994

interested in trying to include this type of parameter in the whole range of characters which might have value in defining order within the ascomycetes.

The present discussions will ignore philosophical considerations because it is possible to debate endlessly about the role of secondary metabolites and whether the fungus needs them. This particular philosophy is irrelevant in the present context because we are considering compounds which are end products of secondary metabolism. It is not so much the products that are important, although these can be extracted and observed, but it is the secondary metabolism we should consider (Bu'Lock, 1980) and that is a concept that is difficult to comprehend adequately. The end products of secondary metabolism may be recognised visually as colour in the whole organism or perhaps only by sophisticated analytical techniques such as TLC, HPLC, gas chromatography, mass spectrometry or UV spectroscopy - a whole range of physical techniques that now allow perception of compounds that are not coloured.

So, whether coloured or not, these secondary metabolites are merely an expression of a biosynthetic pathway, of few or many steps according to the complexity of structure, involving enzyme catalysed chemical reactions in an ordered sequence and with some regulatory mechanisms on a few steps. These enzymes relate to segments of the genome, the most fundamental source of taxonomic data, but so little is known about the enzymes. It is an open question whether they are all special to the pathways or are enzymes of primary metabolism performing additional functions. Most enzymes of biosynthetic pathways leading to secondary metabolites of non-lichenized ascomycetes are hypothetical. They are of course there, but it is only possible to deduce what type of enzyme each is by the function it performs in that type of step. However, usually the enzymes have not been isolated and therefore little is known about them.

A question arises: is there a highly specific distribution of particular secondary metabolites or secondary metabolic pathways in ascomycetes? The answer is, clearly, no. But the topic deserves more careful examination.

ASCOMYCETE SECONDARY METABOLISM

To assist consideration of the range of structural types of secondary metabolites, Table 1 lists a few of the major types of secondary metabolites produced by fungi. This list is not totally comprehensive because it is meant to be reasonably comprehensible. It shows the major groups of secondary metabolites, arising from key compounds of primary metabolism. A very large group of polyketides, biosynthetically analogous to some of the lichenized compounds illustrated in the following presentation, are derived from acetate, and amongst non-lichenized ascomycetes can be found types which elaborate anthraquinones, citrinin, aflatoxin and griseofulvin. These are secondary metabolites which are well-known pigments, or antibiotics or mycotoxins, the carbon skeletons of which are all derived directly and almost completely from acetate.

Another group also all come from acetate, though acetate which has first been diverted into the mevalonate pathway which normally leads to the production of sterols. Sterols are vital in the function of eukaryotic membranes. From this pathway, which is a primary metabolic one, there may be a diversion of mevalonate into a variety of complex polyisoprenoid compounds which are termed terpenoids. Ophiobolin, fusicoccin and gibberellic acid are just three curious examples which profoundly affect green plants.

Another rich source of secondary biosynthetic units is amino acids. Generally destined for protein synthesis, some are diverted to a range of different types of secondary metabolites. Some are derived by the condensation of two α-amino acids and are known as diketopiperazines. Table 1 lists two: both sporidesmin, produced by *Pithomyces chartarum*, and verruculogen, one of the tremorgenic mycotoxins produced by *Penicillium* spp., are formed by condensation of tryptophan and another amino acid. A variation on that theme occurs in benzodiazopines which are constructed from anthranilic acid and an α-amino acid. The first found example is cyclopenin.

More complex secondary metabolites derived from amino acids are the cyclic peptides; phomopsin A is an example.

The amino acid methionine is rarely involved as a complete amino acid in the biosynthesis of secondary metabolites, but it is the principal donor of one carbon units in elaborating the complexity of secondary metabolites. Citrinin is an example, because it is

Table 1. Principal groups of ascomycete secondary metabolites derived from acetate and/or amino acids, key intermediates of primary metabolism.

PRIMARY INTERMEDIATE	SECONDARY METABOLITE GROUP	ASCOMYCETE SECONDARY METABOLITES
ACETATE	POLYKETIDES	CITRININ AFLATOXIN B_1 GRISEOFULVIN
ACETATE ↓ MEVALONATE	ISOPRENOIDS (TERPENOIDS)	OPHIOBOLIN FUSICOCCIN GIBBERELLIC ACID
AMINO ACIDS	DIKETOPIPERAZINES (PAIRS OF α-AMINO ACIDS)	SPORIDESMIN VERRUCULOGEN
	BENZODIAZEPINES (ANTHRANILIC ACID & ANOTHER α-AMINO ACID)	CYCLOPENIN
	CYCLIC PEPTIDES	PHOMOPSIN A
(METHIONINE)	1 CARBON DONOR	CITRININ
COMBINATIONS OF THE ABOVE	STRUCTURALLY- COMPLEX BIOSYNTHETIC HYBRIDS	ASPERLICIN OCHRATOXIN A ERGOTAMINE

a polyketide in which there are three extra carbons derived from the methyl of methionine.

For further complexity of structure there can be polyketide, isoprenoid and/or amino acid-derived components substituted with one carbon units and all assembled by well-ordered enzymic mechanisms in a complex web of branching pathways. All necessarily involve many steps in their biosynthesis and the enzymes arise out of codings in the genome. Table 1 lists three examples of secondary metabolites derived from multiple precursors - asperlicin, ochratoxin A, and the ergot alkaloid ergotamine.

To illustrate how these groups of compounds vary structurally according to the biosynthetic precursors involved, consider the compound ochratoxin A, which is a kidney toxin and a carcinogen (Fig. 1). It is composed of a polyketide - actually a pentaketide derived from five acetate units folded to form an isocoumarin moiety. An additional carbon is donated by methionine and this becomes amide linked with the α-amino nitrogen of phenylalanine. There is also a chlorine substitution. Therefore, just to make that relatively small compound there have been several steps involved. Although first found as a metabolite of *Aspergillus ochraceus* the compound is probably more widely produced in the natural environment by *Penicillium verrucosum*. These two fungi are probably not taxonomically close.

Another example is gibberellic acid. There are several giberellins produced by *Gibberella fujikuroi*, but they are all derived from a four-isoprene chain condensed head to tail, folded in a certain way and then cyclized and subjected to further intramolecular modifications to give gibberellic acid. The whole carbon skeleton is formed from one type of five-carbon unit, but the initial condensation sequence later becomes significantly disturbed. Gibberellic acid is a rather special compound as a fungal secondary metabo-

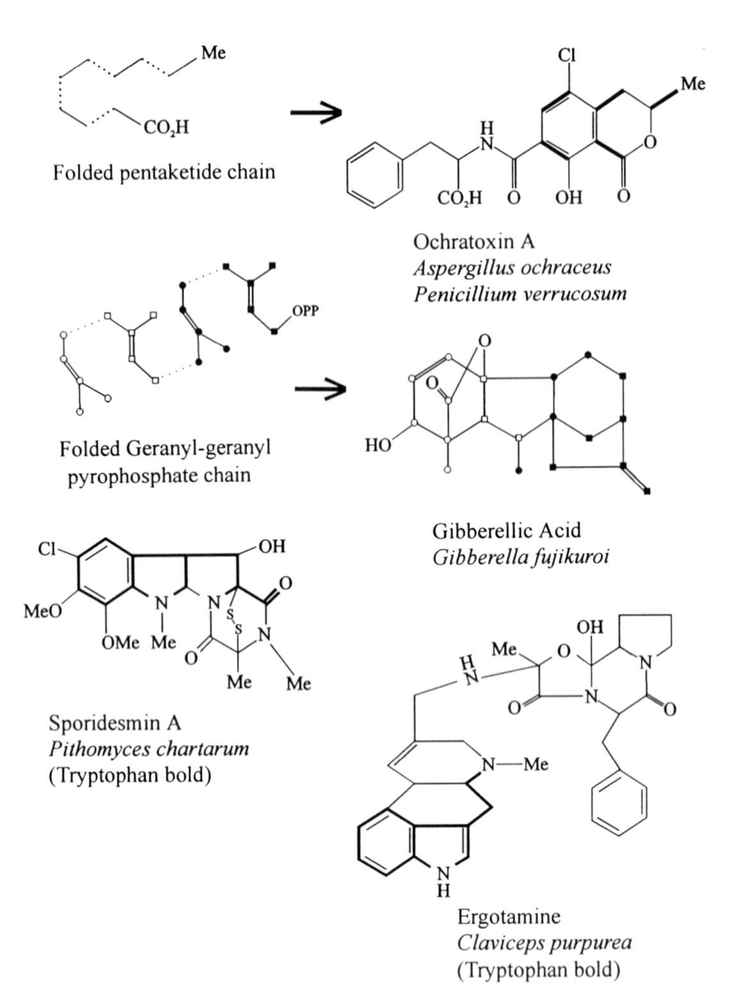

Figure 1. Structural involvement, in ascomycete and ascomycete-related fungi, of polyketide and isoprenoid components in ochratoxin A and gibberellic acid biosynthesis, respectively, and of tryptophan in the structures of sporidesmin and ergotamine.

lite, though commonly produced in green plants, and requires a complex biosynthetic process in the fungus that makes it.

Sporidesmin is the cause of facial eczema, a most distressing condition in sheep and cattle in New Zealand and some other parts of the world. Though uniquely the product of only one fungus there are close structural relatives which are products of other fungi. The principal structural unit is provided by tryptophan that is linked with alanine to form a piperazine ring, to which a one carbon group is donated from methionine, and a disulphide bridge is added for extra complexity. A further four one carbon units are substituted and chlorine is again involved. Forming this compound involves many biosynthetic steps. While sporidesmin is diagnostic for *Pithomyces chartarum*, closely related compounds, the gliotoxins, are metabolites of a mycologically rather different *Penicillium* species.

Ergotamine is also based on tryptophan, with an isoprene unit added from mevalonate. The tryptophan α-amino nitrogen is methylated and there follows a series of cyclisation steps leading to a tetracylic ergolene ring system. Linkage is then made with three amino acids condensed together into a cyclic tripeptide to form the ergotamine molecule. Even prior to the cyclic tripeptide linkage the molecule requires at least 9 or 10 proven steps and each will need an enzyme. Then there is the condensing of the cyclic tripeptide, and the overall construction of the whole molecule together in an ordered sequence.

Polyketides and isoprenoids provide many of the pigments which are well used taxonomic characters. Often these could be regarded as secondary metabolites, not being essential for growth processes. Being derived from acetate they come from a structurally simple part of the carbohydrate metabolic pool which is usually the richest aspect of the endogenous nutritional status of fungi in their natural environment. By comparison, the nitrogen-containing amino acids contribute to secondary metabolism in an apparently rather more sacrificial way, being the immediate building blocks of structural and enzymic proteins. Thus, alkaloids which are derived from an amino acid could be regarded as more special idiolytes. Those that are made from more than one are more special still. Even more special, and thus worthy of serious taxonomic attention, might be those derived from a combination of amino acids with moieties derived directly from acetate via the polyketide or isoprenoid pathways.

Thus it should be apparent that secondary metabolites, which might just seem to be for the amusement of organic chemists, are, for the organisms that produce them, demanding much metabolic organization, and expenditure of energy, arising from a significant part of the genome. The latter is perhaps the aspect that should be considered most here.

REGULATION OF METABOLITE BIOSYNTHESIS

It may be helpful to mention some of the major factors in the process of expression of secondary metabolism. It is one thing for an organism to seem just to make secondary metabolites occasionally, but this conceals the fact that such processes are determined rather precisely by the physiology of the organism. Contrary to the impression given in some text books, which present too simplistic an approach to microbial biochemistry, secondary metabolites are not always produced after growth. However, they quite often are. In laboratory batch fermentations of ascomycete fungi producing secondary metabolites there is inevitably a dynamic decline in nutrients. The production of secondary metabolites is frequently associated with a particular stage in the decline of the balance of nutrients that are available to the organism. This is rather different from the natural environment, which tends to be more like continuous culture, but scientists don't study the biosynthesis of these compounds in the natural environment. Note also that the appearance of a secondary metabolite in a culture of an organism does not necessarily coincide temporally with operation of the whole biosynthetic pathway. Strictly speaking, it only corresponds with the operation of the last step. That is the step which transforms an intermediate, that one may not know about or be able to recognise, into a product that one can. There are quite a number of instances of biosynthetic pathways to secondary metabolites in fungi where the early steps in the pathway may, in a batch culture, happen hours or days ahead of the first appearance of the end product. So, nutrients are very important for heterotrophs such as ascomycetes. The relative concentrations of carbon, nitrogen, and phosphorus sources, and those of a range of inorganic ions are powerful regulatory influences on secondary metabolism expression.

Differentiation may also be important. We are not dealing with organisms that have no way in which they can differentiate. Some produce several types of cell and some actually produce types of fructification. Differentiation of, for example, the plectenchymatic sclerotial cells in *Claviceps purpurea*, may be induced by certain amino acids in a fermentation process (Mantle and Nisbet, 1976), and penicillus formation in some *Penicillium* species in submerged culture is stimulated by the inclusion of relatively large amounts of calcium in the medium (Roncal *et al.*, 1993).

Consequently both the *Claviceps* and the *Penicillium* then produce metabolites *in vitro* which they otherwise only produce in a less artificial environment in which they freely differentiate special types of cell.

Further, using more biochemical terminology, it may be very important that the fungus is released from a regulatory mechanism termed catabolite repression. Catabolite repression may be due to the type of carbon source. An example is that penicillin is only formed in fermentation when glucose has been fully utilised and the fungus is forced to resort to a carbon source which is difficult to use; otherwise penicillin biosynthesis is repressed. The nitrogen source may be influential in regulation with respect to the gibberellins, which are only produced in fermentation as the glycine nitrogen source runs out (Bu'Lock *et al.*, 1974). Phosphate may also be involved in the regulation of ergot alkaloid biosynthesis.

Temperature is important, for example, in the production of the oestrogenic metabolite zearalenone by *Fusarium graminearum*; low temperature may stimulate this in storage after cereals have been moulded by the fungus in the field.

The physical circumstances in culture may also be influential. In fermentations the viscosity of the medium and the shear forces which are involved in a stirred fermenter will be extremely important in contriving a certain type of pellet form. That will in turn involve oxygen concentration dynamics throughout the pellet, so that at various levels within a pellet there will be the appropriate oxygen tension for the switching on of a particular type of metabolic pathway.

Manipulation of the chemical and physical environment will improve expression of secondary metabolite genome. Small changes in the organism's genome, induced in the laboratory and involving the generation of so called regulatory mutants, may alter the amino acid sequence of a protein with vital enzymic function for a secondary metabolic pathway. If this enzyme is normally subject to tight allosteric regulation by the end-product, even a change in one amino acid residue in an allosteric binding site may markedly change the conformation of the site, reduce the efficiency of regulation of enzyme function and allow freer expression of a compound which previously never accumulated in recognisable amount.

SOME DISTRIBUTION OF METABOLITES AMONGST ASCOMYCETES

Table 2 shows some structurally related secondary metabolites produced by ecologically obligate plant parasitic fungi which are either overtly ascomycetes in their sexual expression, or are presumed to be genetically ascomycete-like though, in practice only having an anamorphic fructification.

Claviceps paspali, a widely distributed pathogen of paspalum grass in Australia, New Zealand, the Americas, and northern Mediterranean countries, produces tremorgenic compounds called the paspalinines (Fig. 2) in parasitic sclerotia. Sclerotia quite readily give rise to the ascomycete teleomorph and the anamorph, which is a *Sphacelia*, is always prominent during early pathogenesis. Also requiring a host plant for proliferation of new biomass are the endophytes of grasses. Having very simple conidiogenesis they are classified as *Acremonium* species. They used to be regarded as rather inconsequential organisms, but are now known to produce indole-diterpenoid substances toxic to animals, and also other compounds which help confer protection of the plant against some insect pests. Accepted assignment to the *Albo-lanosa* section of *Acremonium* implies close affinity with *Epichlöe*, which is indisputably ascomycete, on the basis also of isoenzyme analysis. So here are similar secondary metabolites sharing an early part of the biosynthetic pathway in two genera within the *Hypocreales*.

Indole-diterpenoids which seem to be further branches of the same early pathway expressed in *Claviceps paspali* and *Acremonium lolii* are products of several common *Penicillium* species. All these fungi are saprophytes and have no known teleomorph. It is hard to imagine what the *Penicillium crustosum* sexual stage might have looked like millions of years ago, if it had one, or what someone might yet find, but it was probably not like *Claviceps* or *Epichlöe*. Therefore, there is in fungi a fairly wide range of occurrence of compounds of complex structure though with a paxilline-like core structure (Fig. 2). Structural variety is achieved mainly through substitution with two or three more isoprenes.

Biosynthesis of ergot alkaloids, which are structurally quite different from indole-diterpenoids, can also be superimposed across a similar range of fungi (Table 3). Complex cyclic tripeptide derivatives of the ergolene nucleus which are best recognised to occur in *Claviceps* also occur in *Epichlöe*. They also occur in at least two *Acremonium* species which are endophytes in grass so much as to affect ruminant health in the USA and Australasia. Simple ergot alkaloids have also been detected as metabolites of *Penicillium* and *Aspergillus in vitro*. The first 8 or 9 steps of this pathway are standard in most of these fungi. They also occur, in seeds of the convolvulaceous green plant *Ipomoea caerulea* so that it has been a mild source of "drugs of abuse". So the genome for coding for enzymes of ergot alkaloid biosynthesis is in a plant as well as in a range of ascomycete fungi. So widely is this secondary metabolism distributed that it is easy to conclude that there is very little close association between current expressions of taxonomy and incidence of these complex pathways amongst non-lichenised ascomycetes.

Table 2. Correlation, in the expression of indole-diterpenoid secondary metabolism, between plant parasitic hypocreacous ascomycetes of close affinity and a group of taxonomically disparate saprophytic *Penicillium* species. The form in bold type is particularly associated with metabolite biosynthesis.

ASCOMYCETE TELEOMORPH	ANAMORPH	INDOLE-DITERPENOID SECONDARY METABOLITES
ECOLOGICALLY OBLIGATE PLANT PARASITES - *HYPOCREALES*		
Claviceps paspali	*Sphacelia* sp.	Paspalinines
"*Epichlöe*-like" (Sect. *Albo-lanosa*)	***Acremonium lolii*** Paxilline	Lolitrems
SAPROPHYTES		
? But not like *Claviceps*	***Penicillium crustosum*** (Terverticillate, subgenus *Penicillium*)	Penitrems
or *Epichlöe*	***P. janczewskii*** ***P. janthinellum***	Penitrems Janthitrems
? which all ? have very simple		
? anamorph fructifications genus *Furcatum*)	***P. paxilli*** (Biverticillate, sub-	Paxilline

A point of view is that "ascomycete" is only the expression of a retained ability for a particular differentiation in certain fungi. Of course, mycologists see this as amazingly elegant biology and as embodying some of the most stable characters. Apart from colour, which may have taxonomic value, secondary metabolism is usually absent or sparse in ascomycete fructifications. Generally, secondary metabolism seems to be incompatible with sex. After all, the production of free-standing apothecia, perithecia in complex stromata, or even less well-organised cleisthothecia, is directed towards the metabolically-demanding objective of producing spores rich in nitrogen-containing nuclei and proteinaceous cytoplasm. Usually the spores have enough nutrient reserves to support germination and are often packaged in an explosive mechanism. It is, therefore, not surprising that when expressing the ascomycete characters, fungal secondary metabolism is economically excluded.

If, however, ascomycete characters are seen as only a brief expression of biological potential, taxonomists should also consider the same organism growing in vegetative form with or without anamorphic fructification in the laboratory on various nutrient media. It is then much more likely that secondary metabolism will become apparent in batch fermentation conditions. Philosophical considerations about whether some secondary metabolites are laboratory artefacts are irrelevant. Even if they are artefacts, they are still mainly expressions of genome. Taxonomists should be anxious to evaluate and possibly exploit biochemical evidence of order which arises from the fundamental codes of molecular biology.

In contrast to the situation expressed in lichenized ascomycetes where, if the special association between fungus and alga occurs, it can usually be seen on a rock or a tree but might have some difficulty growing and flourishing in the laboratory, non-lichenized ascomycetes can usually be grown readily in laboratory culture. Expression, then, of secondary metabolic potential may occur in suitable nutrient and physical environments, translated in biochemical language as activation or derepression of genes coding for particular enzymes of secondary metabolism.

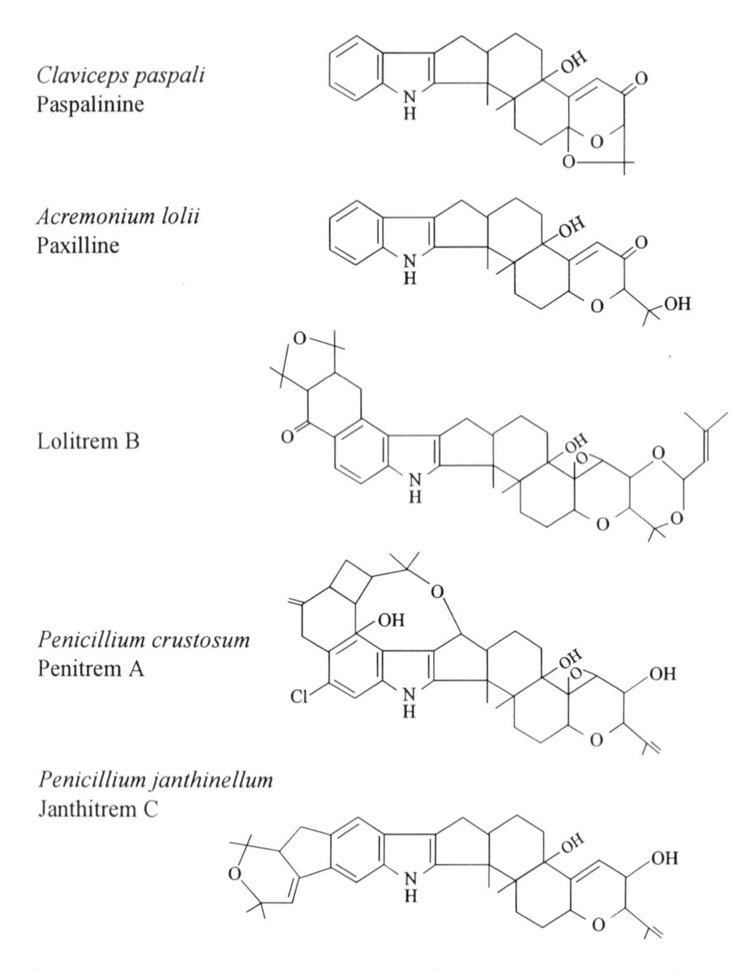

Figure 2. Indole-diterpenoid tremorgenic secondary metabolites of ascomycete and ascomycete-related fungi, alligned to show the paxilline structure in common and variety achieved mainly by isoprene substituents.

Earlier it was concluded that there is no highly specific distribution of secondary metabolites in ascomycetes. However, as increasingly sensitive analytical methods reveal wider incidence of secondary pathways in ascomycetes, and also in fungi which are predictably analogous though have just lost their sexuality, correlation patterns may appear. Already, secondary metabolites are proving to be of considerable value in redefining order amongst the many common fungi that are readily assigned to *Penicillium aurantiogriseum* or some closely similar fungi (Lund and Frisvad, 1994).

However, secondary metabolites may not be so reliable in making dichotomous keys at a high level in taxonomic organization where curious idiolyte genomes seem to be quite widely distributed. New knowledge may change this but much will depend on judgements on the relative contribution of genomic sequences to each expressed character. Some of the biosynthetic potential expressed only with great difficulty in axenic culture may be valuable to taxonomic authorities in redefining boundaries, but for purposes of quick and reliable identification only secondary metabolism that is readily expressed in simple laboratory culture will be useful.

Table 3. Distribution of ergot alkaloid biosynthesis in hypocreaceous ascomycetes, taxonomically distinct *Penicillium* species and a green plant.

ASCOMYCETE TELEOMORPH	ANAMORPH	ERGOT ALKALOID TYPE
Claviceps spp.	*Sphacelia* sp.	Complex cyclic tripeptide derivatives of ergolene nucleus
Epichlöe typhina	? *Ephelis* sp.	
"*Epichlöe*-like"	*Acremonium lolii*; *A. coenophialum*	
?	*Penicillium* spp.	simple ergolene derivatives
?	*Aspergillus* spp.	
Also *Ipomoea* spp. (convolvulaceous plants)		

It could be that the secondary metabolites recognized to date in lichenized ascomycetes are somewhat limited by the greater difficulty of laboratory culture and the consequent constraints on using radiolabelled putative precursors as probes for metabolites derived from such precursors. This may account for a heavy dominance of rather abundant polyketide metabolites within the distribution of which it is easier to see patterns of association with particular types of fungal symbiont.

REFERENCES

Bu'Lock, J.D., R.W. Detroy, Z. Hostalek and A. Munim-al-Shakarchi, 1974, Regulation of secondary biosynthesis in *Gibberella fujikuroi*, *Transactions of the British Mycological Society* 62: 377-389.

Bu'Lock, J.D., 1980, Mycotoxins as secondary metabolites, *In*: *The Biosynthesis of Mycotoxins* (P.S. Steyn, ed.): 1-16. Academic Press, New York.

Frederickson, D.E., P.G. Mantle and W.A.J. De Milliano, 1991, *Claviceps africana* sp.nov., the distinctive ergot pathogen of sorghum in Africa, *Mycological Research* 95: 1101-1107.

Lund, F. and J.C. Frisvad, 1994, Chemotaxonomy of *Penicillium aurantiogriseum* and related species, *Mycological Research* 98: in press.

Mantle, P.G. and L.J. Nisbet, 1976, Differentiation of *Claviceps purpurea* in axenic culture, *Journal of General Microbiology* 93: 321-334.

Roncal, R., U.O. Ugalde and A. Irastorza, 1993, Calcium-induced conidiation in *Penicillium cyclopium*: calcium triggers cytosolic alkalinization at the hyphal tip, *Journal of Bacteriology* 175: 879-886.

DISCUSSION

Whalley: To what extent can secondary metabolites be grouped into categories in relation to taxonomic position ? In the *Xylariaceae* three main types can be recognized: (1) metabolites with a high correlation with systematic position (e.g. dihydroisocumarins, butyrolactones); (2) metabolites which indicate taxonomic trends but seem more associated with habitat or life-style (e.g. cytochalasins); and (3) those which have no correlation with either.

Mantle: In well-researched families such as the *Xylariaceae* there is no doubt that secondary metabolites provide useful and systematically meaningful characters. However, we must be cautious in principle in accepting that secondary metabolites will be widely useful immediately, for example in dichotomous key construction. The mosaic occurrence of many compounds in different ascomycete groups mitigates against this as a general hypothesis. As more information is obtained, increasingly meaningful patterns may arise, especially as molecular data refines taxonomic boundaries.

The magnitude of the expression of secondary metabolites is not important, but rather the demonstration of the existence of particular biosynthetic pathways (and thus the implied enzymes and their coding genome). This could be achieved even if the incidence of an end product is very low in a culture by the use of ELISA techniques.

Rassing: With respect to the characterization of species by the production of metabolites, the pattern of metabolites might be a more useful tool in taxonomy than looking at the presence or absence of single metabolites.

Rossman: Were the *Ipomoea* seeds checked to see if any endophytic fungi could be producing the squalene?

Mantle: Probably not, but as a scientist one should keep an open mind to all possibilities until proved otherwise. I am not aware of endophytes in *Ipomoea*.

SECONDARY METABOLITES AS A TOOL IN ASCOMYCETE SYSTEMATICS: LICHENIZED FUNGI

William Louis Culberson and Chicita F. Culberson

Department of Botany
Duke University
Durham, North Carolina 27708, USA

SUMMARY

For many years, the secondary products called lichen substances have been intimately involved in the systematics of the lichen-forming ascomycetes. Today they take on a new importance as markers in genetic studies and for the analysis of gene flow and reproductive isolation. Lichen systematics has been much hampered in that artificial crosses are not possible. Refined methods of thin-layer chromatography and high-performance liquid chromatography now make it possible to chemotype single-spore cultures grown *in vitro* and consequently to analyze the progeny of maternal individuals from nature. Since lichen fungi are haploid, the appearance of a progeny chemotype different from the maternal one can detect and identify the source of gene flow. Examples from the *Cladonia chlorophaea* and *Ramalina siliquosa* species complexes are used to demonstrate this approach to the analysis of gene exchange in lichen fungi.

INTRODUCTION

The lichen-forming ascomycetes are famous for their secondary-product chemistry, which has been used commonly in systematics since the end of World War II. Chemistry was seized upon by some lichen systematists in search of new characters for a group of organisms where good characters were often in short supply. Although most well-defined lichen morphs were found to have a uniform secondary chemistry throughout their range, and chemistry became as much a key character as the morphology characterizing the species, other major morphs were found to be differentiated into two or more chemical races or chemotypes, a discovery that lead to extensive taxonomic controversy. What are lichen chemotypes? Originally, most workers saw them as genetically trivial polymorphisms unworthy of taxonomic recognition. A minority of workers, however, saw them as markers of sibling species because of their biogeographically significant ranges and frequent ecological distinctness. In many groups of organisms, this question would have been resolved immediately by the appropriate experimental crosses. Such crosses in the lichen fungi, however, are not yet possible for technical reasons. But the lichen substances may offer a solution to the very dilemma that they created. In the absence of the possibility of artificial crosses, lichen secondary products offer genetic markers for the indirect analysis of progeny from maternal individuals in nature. We shall discuss the first two genetic studies in which lichen substances have been used to assess gene flow and reproductive isolation in lichen-forming fungi.

The rationale for the analysis of lichen chemotypes is to make *in vitro* cultures of the offspring of maternal (apotheciate) individuals which in nature could have had the possibility of crossing with paternal (spermagoniate) individuals of the same morph but of different chemotype, in other words, to analyze the progeny of maternals that grew inti-

Ascomycete Systematics: Problems and Perspectives in the Nineties
Edited by D.L. Hawksworth, Plenum Press, New York, 1994

mately intermixed with at least one other chemotype. Most lichens are bisexual, but whether they are hetero- or homothallic is unknown. In lichens that grow densely packed, however, the possibility of crossing always exists because there is no morphological barrier to prevent a foreign spermatium from reaching the trichogyne. (There is no functional equivalent of the floral morphology that assures self-pollination in some angiosperms, for example.) Lichen fungi are haploid, and the appearance of one of the other chemotypes among the progeny of a maternal individual will at once detect outcrossing and identify the chemotype of the paternal parent. Qualitatively, this analysis should determine the ability or the inability of chemotypes to cross; quantitatively, it should give some measure of the actual extent of gene flow.

For this method to succeed, the sporelings must have the physiological capacity of making *in vitro* the characteristic secondary products that define them. Additionally, one must have the technical capacity of chemotyping the sporelings. Refinements of thin-layer chromatography (TLC) and high-performance liquid chromatography (HPLC) now make such chemical analyses possible. This indirect analysis of gene movement using secondary products as markers in populations is possible only because all of these conditions have now been met. We shall describe two analyses of breeding systems done by this technique, an American example involving the *Cladonia chlorophaea* group and a European one involving the *Ramalina siliquosa* group.

GENE FLOW IN THE *CLADONIA CHLOROPHAEA* COMPLEX

Our first genetic study (C. F. Culberson et al., 1988) was of the *Cladonia chlorophaea* complex, common lichens with goblet-shaped podetia familiar to all naturalists. Worldwide, this group is differentiated into 14 major chemotypes with extensively overlapping geographic ranges. In North Carolina, five chemotypes, characterized by their major compounds (Figs 1-5), grow in both pure and mixed populations. For

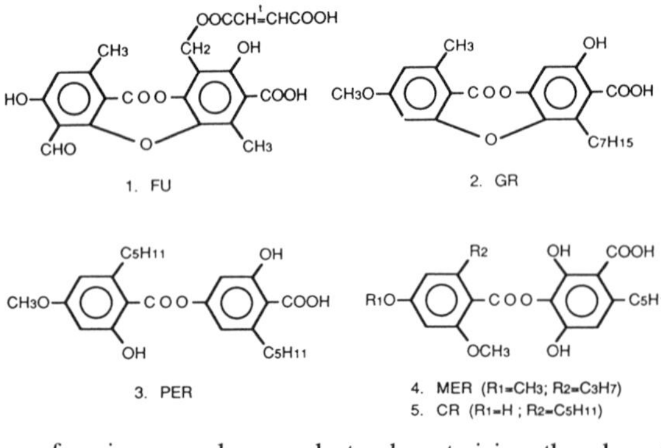

1. FU

2. GR

3. PER

4. MER (R_1=CH_3; R_2=C_3H_7)
5. CR (R_1=H ; R_2=C_5H_{11})

Figures 1-5. Structures of major secondary products characterizing the chemotypes of the *Cladonia chlorophaea* group in the Mountains and Coastal Plain of North Carolina. **Fig. 1.** Fumarprotocetraric acid. **Fig. 2.** Grayanic acid. **Fig. 3.** Perlatolic acid. **Fig. 4.** Merochlorophaeic acid. **Fig. 5.** Cryptochlorophaeic acid.

convenience we will refer to the chemotypes by their coded names (Table 1). We wanted to address the question of whether chemotypes interbreed when they grow in close proximity. The plan consisted of making single-spore isolates from appropriate maternals, lichenizing them with an alga (at that time believed to be required for the production of secondary products), and growing the progeny to sufficient size to be chemotyped. To this end, many soil mats of these lichens were collected in the Southern Appalachian Mountains and in the Coastal Plain, 1172 apotheciate podetia were removed and their positions in the mats marked (Fig. 6), apothecia were frozen, and sections of the podetia immediately adjoining them were chemotyped (TLC). These chemical data allowed the identification of the 431 apotheciate individuals that grew very close (within 20 mm; av-

Table 1. The chemotypes of the *Cladonia chlorophaea* group in North Carolina.

Chemotype	Symbol	As Taxonomic Species
cryptochlorophaeic acid	*cr*	*C. cryptochlorophaea*
fumarprotocetraric acid	*fu*	*C. chlorophaea*
grayanic acid	*gr*	*C. grayi*
merochlorophaeic acid	*mer*	*C. merochlorophaea*
perlatolic acid	*per*	*C. perlomera*

erage distance, 9.5 mm) to a podetium of different chemotype. The steps involved are summarized in Figs 7-8. From each spore family, 2-4 sporelings that looked most different were selected to increase the chance of detecting hybrid progeny. The 763 isolates were inoculated with a culture (kindly supplied by V. Ahmadjian) of a foreign alga, *Trebouxia erici* from the distantly related lichen *Cladonia cristatella*, to assure that algal genotype would be invariant. After 15-20 months, these cultures were analyzed (HPLC, TLC), and 167 (22%) proved to be chemotypeable and useable for the analysis of gene flow. Of these, 141 (84%) matched the chemotype of the maternal parent; 26 (16%) did not but, instead, matched the chemotype of a nearby individual.

In the Mountains, all chemotypeable progeny from *cr* (cryptochlorophaeic acid individuals) consistently matched the chemotype of the maternal (Fig. 9). In mixtures with the *gr* (grayanic acid individuals) and *mer* (merochlorophaeic acid individuals), there was no evidence of hybridization with these different chemotypes, and in the Mountains the *cr* chemotype appears to be a good biological species. In the Coastal Plain, however, *cr* interbreeds with *per*, the rare endemic chemotype that contains perlatolic acid in addition to the related 4-*O*-methylated *meta*-depsides, and *cr* and *per* are a polymorphism. The *cr* chemotype in North Carolina, however, is not uniform: In the Mountains, it usually produces fumarprotocetraric acid in addition to cryptochlorophaeic acid whereas in the

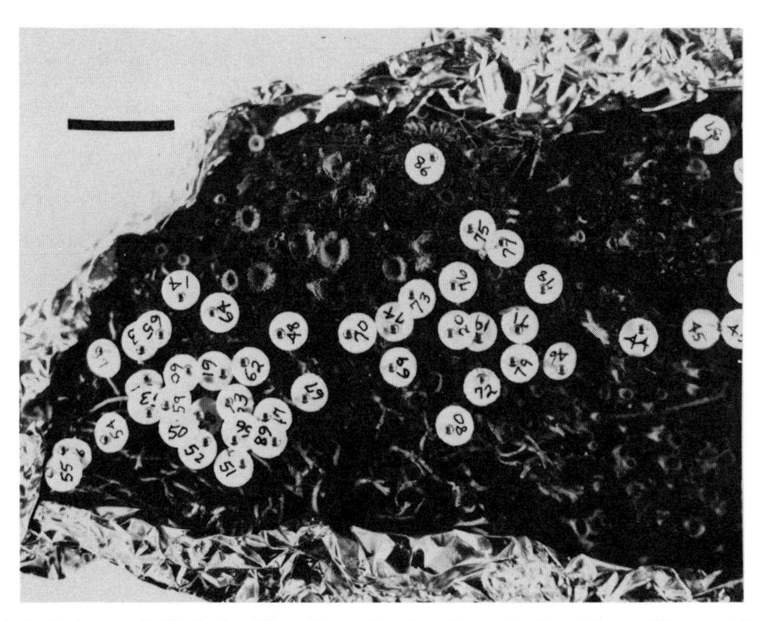

Figure 6. A typical mat of *Cladonia chlorophaea* showing the marked positions of harvested apotheciate podetia. The scale is 20 mm, the maximum distance permitted for different chemotypes to be considered as growing "closely together."

Coastal Plain it rarely does so (C. Culberson et al., 1977). Therefore, the *cr* chemotype in North Carolina may represent at least two distinctly different biological species. Proof of genetic variability within chemotypes has just come from a molecular-biological study of this group in the Southern Appalachian Mountains. DePriest and Been (1992) found a remarkable level of genetic variation within and between chemotypes, proof of genetic variability within and between chemotypes due in large part to group-I introns.

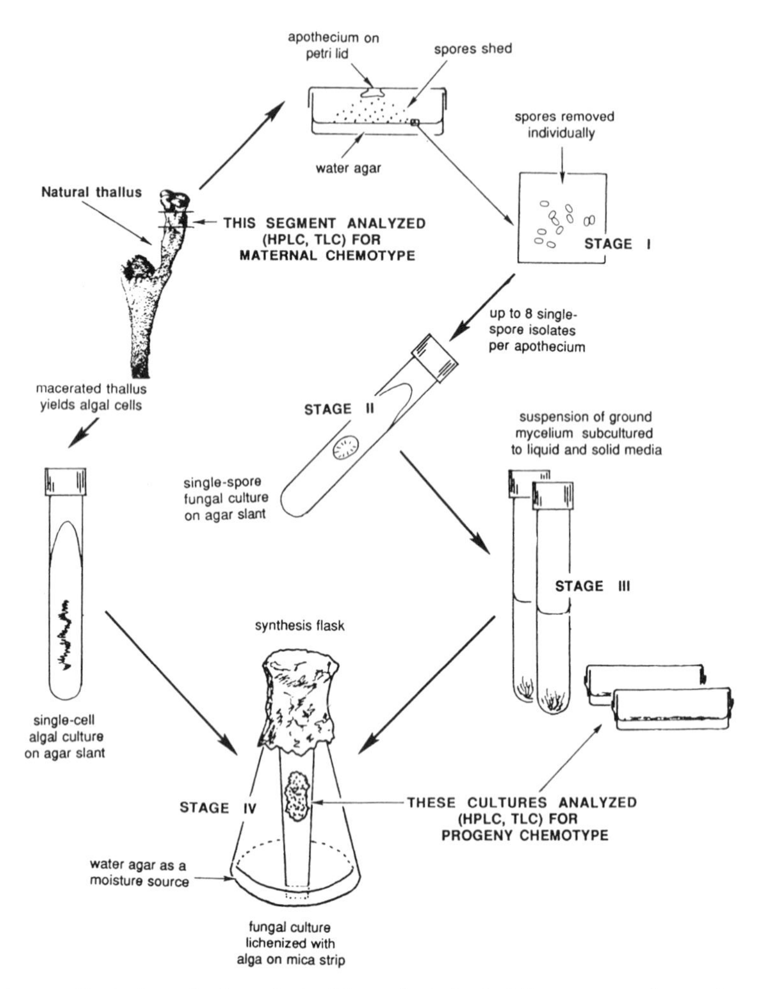

Figure 7. A generalized schematic flow chart of the methods for isolating spores and preparing cultures for the analysis of progeny chemotypes. In the *Cladonia chlorophaea* study, the fungus was grown in a synthesis flask by the method of Ahmadjian and Jacobs (1981), using a single-cell isolate of the alga from *C. cristatella*. For the progeny analysis of the *Ramalina siliquosa* complex, alga-free surface cultures in petri dishes were analyzed.

In contrast to the apparent reproductive isolation of *cr*, *gr* and *mer* both gave progeny including each other's chemotype when they grew in mixtures. When *gr* grew alone, all of the chemotypeable progeny produced grayanic acid. (The *mer* chemotype, here relatively infrequent and at the southern limit of its range in eastern North America, was never found growing alone.) Neither *mer* nor *gr* gave progeny of *cr* when they grew in mixtures with that chemotype. In terms of chemical similarity, the *mer* and *cr* chemotypes are the most closely related. Apparently hybridization cannot be predicted from the biogenetic relationships of the compounds. This study, using secondary products as genetic markers, was the first to demonstrate gene flow in lichens.

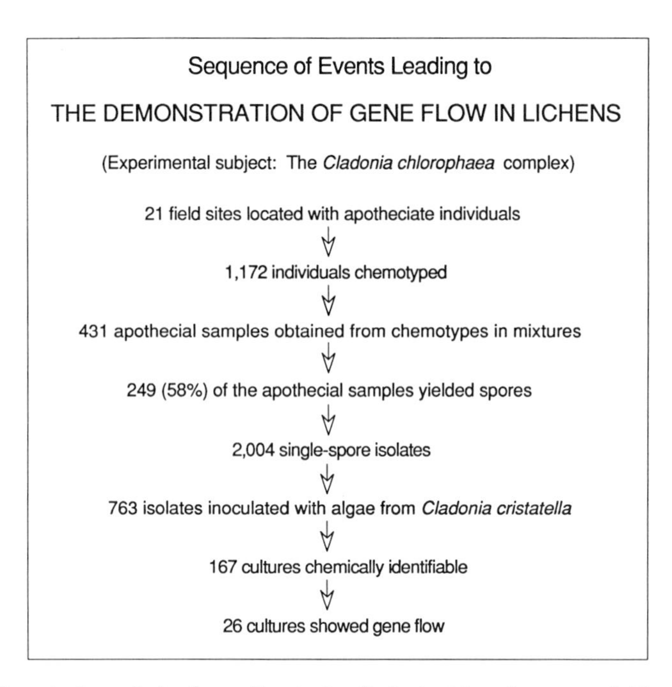

Figure 8. Steps in the analysis of gene flow in the *Cladonia chlorophaea* group in North Carolina.

PROGENY OF MATERNAL CHEMOTYPES (CLADONIA)*

	CHEMOTYPE(S) WITHIN 20 MM				
MATERNALS	*gr*	*mer*	*gr* *mer* *cr*	*cr*	*per*
Mountains:					
gr (67)	GR	GR 9 MER	GR 11 MER	GR	
mer (10)	MER 1 GR		MER 2 GR		
cr (31)	CR		CR	CR	
Coastal Plain:					
cr (7)					CR 1 PER
per (5)				PER 1 CR	

* Numbers refer to maternal apothecia.

Figure 9. Results of the study of the *Cladonia chlorophaea* group.

GENE FLOW IN THE *RAMALINA SILIQUOSA* COMPLEX

Our second genetic study (W. Culberson et al., 1993), which we have just completed, involved the chemotypes of the *Ramalina siliquosa* complex of the rocky maritime shores of western Europe. By evolving distinct chemotypes with different amplitudes of ecological tolerance, these lichens have successfully dominated all of the available space in the supralittoral zone, from above the level of high tide up to where vascular plants first appear (W. Culberson and C. Culberson, 1967). These chemotypes continue the sharp ecological zonation shown by the sessile invertebrate animals and benthic algae of the littoral zone below them (Lewis, 1964). Through a combination of genetic incompatibilities, spatial separations resulting from zonation, and short distances over which spermatia are functional, these chemotypes have extremely high levels of reproductive isolation.

The six chemotypes of the *siliquosa* lichens in northern Europe are characterized by a series of chemosyndromes identified by their major medullary products, ß-orcinol depsidones (Figs 10-14). One chemotype lacking depsidones is referred to as acid deficient (*ad*). All chemotypes, including *ad*, contain usnic acid (Fig. 15), apparently confined to apothecia and spermagonia (C. Culberson et al., 1993). The remarkable chemical and ecological differentiation shown by these lichens has lead to extensive studies (e.g., chemistry, C. Culberson, 1965; ecology, W. Culberson and Culberson, 1967; biogeography, W. Culberson, 1967; phenetics, Sheard, 1978a; and isozymes, Mattsson and Kärnefelt, 1986). Taxonomically, the six chemotypes are generally regarded as two morphological species each with three chemotypes (Sheard, 1978a; Purvis et al., 1992), but they have also been interpreted as distinct sibling species (W. Culberson, 1967, 1986).

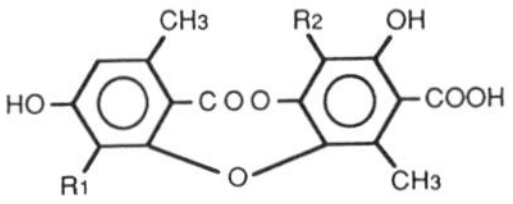

10. HYPO ($R_1=R_2=CH_3$)
11. PR ($R_1=CHO$; $R_2=CH_2OH$)
12. NST ($R_1=CHO$; $R_2=CH_3$; $R_3=H$) 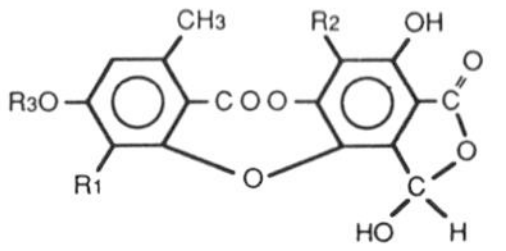
13. ST ($R_1=CHO$; $R_2=R_3=CH_3$)
14. SAL ($R_1=CHO$; $R_2=CH_2OH$; $R_3=H$)
15. US 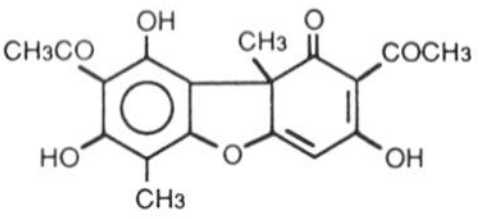

Figures 10-15. Structures of major secondary products characterizing the chemotypes of the *Ramalina siliquosa* group in Wales. **Fig. 10.** Hypoprotocetraric acid. **Fig. 11.** Protocetraric acid. **Fig. 12.** Norstictic acid. **Fig. 13.** Stictic acid. **Fig. 14.** Salazinic acid. **Fig. 15.** Usnic acid.

The breeding behaviors implied by the two-species and the six-species taxonomies are obviously very different. The two-species idea suggests intrinsic reproductive barriers between "morphological" species and freely hybridizing chemotypes within them, the zonation of the particular chemotypes then resulting from the constant elimination of genotypes unadapted to the "wrong" habitats. The six-species model, on the other hand, predicts reproductive isolation among sibling species (i.e., the chemotypes themselves) adapted and confined to successive zones. We thought that a progeny analysis along transects through the entire supralittoral zone, traversing pure zones of chemotypes and zones of overlap between them, would reveal the real breeding behaviour.

In 1988, we returned to the small promentory in Wales where we had originally found the *siliquosa* chemotypes zoned - from the harshest, most exposed habitat at the bottom of the supralittoral zone and facing the sea to the most protected habitat at the top and facing the land (W. Culberson and C. Culberson, 1967; Table 2). From four of the original transects, we collected 10 individuals per quadrat (35 cm square). Thalli were chemotyped, and 261 (68%) of the apothecia yielded spores. Of the 1004 single-spore isolates, 620 (62%) grew and were transferred to solid media for analysis. In this experiment, the secondary products characterizing these chemotypes were synthesized by the mycobiont alone, eliminating the tedious and time-consuming step of lichenization and the need to select an appropriate algal symbiont (C. Culberson et al., 1992). In fact, the percentage of chemotypeable *R. siliquosa* group mycobionts (62%) grown in the absence of the alga was much greater than the percentage of *C. chlorophaea* group mycobionts (22%) grown with an alga in synthesis flasks under conditions favouring lichenization.

The results of the progeny analysis showed a very high level of reproductive isolation among the chemotypes (Fig. 16). Of the 386 chemotypeable progeny, 96% matched the maternal thallus. Two common chemotypes, *hypo* and *nst*, one from each of the major "morphological" species, never gave progeny including any other chemotype. Furthermore, all chemotypes in unmixed quadrats invariably produced progeny like themselves. In mixed quadrats, there was no hybridization between any chemotypes belonging to the different "morphological" species (Table 2), confirming the clear distinction between these two major groups. However, within "morphological" species, we did not detect the random mating predicted by a two-species taxonomy.

PROGENY OF MATERNAL CHEMOTYPES (RAMALINA) *

TRANSECTS

From sheltered, facing the land ⟶ To exposed, facing the sea

CHEMOTYPES IN THE SAME QUADRATS WITH MATERNAL THALLUS

MATERNALS	*hypo*	*hypo* *sal* *nst*	*hypo* *nst* ±*ad* ±*st*	*nst* ±*ad*	*nst* *st*	*st*
hypo (45)	HYPO	HYPO	HYPO			
sal (2)		SAL 1 [HYPO]				
nst (119)		NST	NST	NST	NST	
ad (5)			2 [NST]	3 [NST]		
st (56)					ST 4 [NST]	ST

* Numbers refer to maternal apothecia.

Figure 16. Results of the study of the *Ramlina siliquosa* group.

Only two chemotypes emerged as a clear polymorphism, *ad* and *nst*, which always grow intermixed. Every *ad* maternal produced *nst* among its progeny. However, we could not detect true *ad*'s among the progeny of *nst* because *ad* has no medullary depsidone to serve as a marker, and every chemotype produced some progeny that failed to make their distinctive chemical products for whatever reason. The finding of this *nst*/*ad* polymorphism explains why *nst* and *ad* show no difference in zonation. At this site, however, only five *ad* maternals were included in our sample compared to 119 *nst* maternals, implying a more complicated genetics than these data suggest.

Hypo and *sal* occupy the sheltered habitat on these maritime cliffs, but the former is more abundant in the southern half of the group's range and the latter more common in the northern half (W. Culberson et al., 1977). At our locality, *hypo* dominated the sheltered site, and our sample included only two *sal* maternals. Surprisingly, in spite of the small sample number, one *sal* maternal produced a *hypo* sporeling. We found no *sal* progeny from *hypo* maternals, but of the *hypo* thalli in quadrats known to include *sal*'s, only one gave sporelings for analysis. Apparently these chemotypes can hybridize, but they must certainly be extensively reproductively isolated because they are so rarely sympatric.

Table 2. The chemotypes of the *Ramalina siliquosa* complex as two morphological species (Sheard, 1978a, b) or as six sibling species and their ecological zonation on seashore cliffs of western Europe.

Chemotype	Symbol	As morphological species	As sibling species	Ecological zonation
protocetraric[a]	-	*R. siliquosa*	*R. siliquosa*	very sheltered
hypoprotocetraric	*hypo*	"	*R. crassa*	sheltered
salazinic	*sal*	"	*R. incrassata*	sheltered
norstictic	*nst*	*R. cuspidata*	*R. stenoclada*	intermediate
acid deficient	*ad*	"	*R. atlantica*	intermediate
stictic	*st*	"	*R. cuspidata*	exposed

[a] Not encountered in our study in Wales (W. Culberson and C. Culberson, 1993).

Our results are best in showing the relationship between *st*, which occupies the most exposed zone, and *nst*, which grows just above it in the intermediate zone (Table 2). In unmixed quadrats, these chemotypes produce offspring invariably like themselves. In the relatively narrow zone of overlap, 24% of the *st* maternals produced *nst* sporelings among their progeny. Curiously, however, no *nst* maternal in the zone of overlap produced *st* progeny. Of the 31 progeny from *nst* maternals in these mixed quadrats, 30 (97%) were *nst* (and one was unchemotypeable). Gene flow here may be unidirectional.

The *nst* and *st* chemotypes offer an intriguing context in which to consider the old problem of allopatric *vs*. sympatric speciation. These chemotypes are ecologically differentiated populations that interbreed where their zonational bands overlap. Did they evolve allopatrically with little selective pressure for an intrinsic isolating mechanism or did they evolve sympatrically with intrinsic reproductive isolation not yet fully developed? In the allopatric model, the present ecologies and reproductive behavior would be the result of secondary contact. So-called "hybrid zones" between populations with incomplete reproductive isolation are currently a subject of much interest in population genetics (Hewitt, 1989; Harrison, 1990). Most discussions of this phenomenon revolve around some form of allopatric speciation.

Discussions of sympatric speciation are often theoretical (Levene, 1953; Maynard Smith, 1966; Dickinson and Antonovics, 1973; Caisse and Antonovics, 1978; Felsenstein, 1981). They involve a stable polymorphism confronted by a heterogeneous environment to which the two forms are differentially adapted. Assortative mating, which would result from the spatial isolation of habitat selection, would reinforce and accelerate differentiation, ultimately leading to speciation. If the *st* and *nst* chemotypes are sibling species that evolved sympatrically, their zonation is maintained by strong selection eliminating unadapted genotypes in the "wrong" habitat. This differentiation is sustained and accelerated by assortative mating in the large zones where they grow isolated from each other.

One would think that spermatia of *nst* would be washed down into the pure zone of *st*, resulting in *st* maternals there yielding *nst* offspring among their progeny. Our results, however, showed that this was not the case, and we concluded that spermatia here function only over short distances. This finding for gene flow in lichens agrees with the condition known for seed plants in which local pollen numerically overwhelms that of more distant origin (Levin and Kerster, 1967; Ehrlich and Raven, 1969).

These first experiments to assess gene flow and reproductive isolation in lichen fungi were possible because the unique secondary products of lichens can be used as genetic markers in the progeny analysis of maternals in natural populations. Not only are these compounds expressed *in vitro* by the fungus alone, they are also the only phenotypic characters from classical lichen taxonomy that are seen in such cultures. Lichen substances consequently have the potential of contributing to population-genetic studies just as effectively as they have for so long in biosystematics.

ACKNOWLEDGMENTS

Much of the research described here was done under grants BSR 85-07848, 88-06659, and 90-19702 from the National Science Foundation. We thank Janis Antonovics for helpful discussions and Anita Johnson for invaluable assistance over many years.

REFERENCES

Ahmadjian, V., and J.B. Jacobs, 1981, Relationship between fungus and alga in the lichen *Cladonia cristatella* Tuck., *Nature* 289: 169-172.

Caisse, M., and J. Antonovics, 1978, Evolution in closely adjacent plant populations. IX. Evolution of reproductive isolation in clinal populations, *Heredity* 40: 371-384.

Culberson, C.F., 1965, Some constituents of the lichen *Ramalina siliquosa*, *Phytochemistry* 4: 951-961.

Culberson, C.F., W.L. Culberson, and D.A. Arwood, 1977, Physiography and fumarprotocetraric acid production in the *Cladonia chlorophaea* group in North Carolina, *Bryologist* 80: 71-75.

Culberson, C.F., W.L. Culberson, and A. Johnson, 1988, Gene flow in lichens, *American Journal of Botany* 75: 1135-1139.

Culberson, C.F., W.L. Culberson, and A. Johnson, 1992, Characteristic lichen products in cultures of chemotypes of the *Ramalina siliquosa* complex, *Mycologia* 84: 705-714.
Culberson, C.F., W.L. Culberson, and A. Johnson, 1993, Occurrence and histological distribution of usnic acid in the *Ramalina siliquosa* species complex, *Bryologist* 96: 181-184.
Culberson, W.L., 1967, Analysis of chemical and morphological variation in the *Ramalina siliquosa* species complex, *Brittonia* 19: 333-352.
Culberson, W.L., 1986, Chemistry and sibling speciation in the lichen-forming fungi: ecological and biological considerations, *Bryologist* 89: 123-131.
Culberson, W.L., and C.F. Culberson, 1967, Habitat selection by chemically differentiated races of lichens, *Science* 158: 1195-1197.
Culberson, W.L., C.F. Culberson, and A. Johnson, 1977, Correlations between secondary-product chemistry and ecogeography in the *Ramalina siliquosa* group (lichens), *Plant Systematics and Evolution* 127: 191-200.
Culberson, W.L., C.F. Culberson, and A. Johnson, 1993, Speciation in the lichens of the *Ramalina siliquosa* complex (*Ascomycotina*, *Ramalinaceae*): gene flow and reproductive isolation, *American Journal of Botany* 80: 1472-1481.
DePriest, P.T., and M.D. Been, 1992, Numerous Group I introns with variable distributions in the ribosomal DNA of a lichen fungus, *Journal of Molecular Biology* 228: 315-321.
Dickinson, H., and J. Antonovics, 1973, Theoretical considerations of sympatric divergence, *American Naturalist* 107: 256-274.
Ehrlich, P.R., and P.H. Raven, 1969, Differentiation of populations, *Science* 165: 1228-1232.
Felsenstein, J., 1981, Skepticism towards Santa Rosalia, or why are so there so few kinds of animals?, *Evolution* 35: 124-138.
Harrison, R.G., 1990, Hybrid zones: windows on evolutionary process, *In*: *Oxford Surveys in Evolutionary Biology* (D. Futuyma and J. Antonovics, eds) 7: 69-128. Oxford University Press, New York.
Hewitt, G.M., 1989, The subdivision of species by hybrid zones, *In*: *Speciation and its Consequences* (D. Otte and J. A. Endler, eds): 85-110. Sinauer Associates, Sunderland.
Levene, H., 1953, Genetic equilibrium when more than one ecological niche is available, *American Naturalist* 87: 331-333.
Levin, D.A., and H.W. Kerster, 1967, Gene flow in seed plants, *Evolutionary Biology* 7: 139-220.
Lewis, J.R., 1964, *The Ecology of Rocky Shores*, English Universities Press, London.
Mattsson, J.-E., and J. Kärnefelt, 1986, Protein banding patterns in the *Ramalina siliquosa* group, *Lichenologist* 18: 231-240.
Maynard Smith, J., 1966, Sympatric speciation, *American Naturalist* 100: 637-650.
Purvis, O.W., B.J. Coppins, D.L. Hawksworth, P.W. James, and D.M. Moore, 1992, *The Lichen Flora of Great Britain and Ireland*, Natural History Museum Publications, London.
Sheard, J.W., 1978a, The taxonomy of the *Ramalina siliquosa* species aggregate (lichenized ascomycetes), *Canadian Journal of Botany* 56: 916-938.
Sheard, J.W., 1978b, The comparative ecology and distribution and within-species variation in the lichenized ascomycetes *Ramalina cuspidata* and *R. siliquosa* in the British Isles, *Canadian Journal of Botany* 56: 939-952.

Paleomycology, Cladistics, and Biogeography

THE FOSSIL HISTORY OF ASCOMYCETES

T.N. Taylor

Department of Plant Biology and
Byrd Polar Research Center
The Ohio State University
Columbus, Ohio 43210, USA

SUMMARY

Examples of fossil ascomycetes are described from throughout the geologic record with the oldest forms coming from rocks of Silurian age. During the Carboniferous there were several forms that morphologically resemble modern plectomycetes. Epiphyllous fungi are probably an ancient group that radiated extensively during the Tertiary. Both the fossil record and molecular sequence data suggest that the ascomycetes are older than previously thought.

INTRODUCTION

The geologic history of some fungal groups is well documented (Taylor, 1993). For example, it is now known that not only were chytrids highly diverse in the Lower Devonian, but many were also parasites of other organisms (e.g. Taylor et al., 1992a), including fungi (Hass et al., 1993). To date the oldest basidiomycetes come from the Upper Devonian. This is based on the presence of medallion clamp connections and mycelium in wood of the progymnosperm *Callixylon* (Stubblefield et al., 1985). Clamp connections (Dennis, 1970; Osborn et al., 1989), septate hyphae and symptoms that are today caused by wood rotting basidiomycetes (Stubblefield and Taylor, 1986; Stubblefield et al., 1985), further demonstrate that this group was highly diversified throughout the Paleozoic.

Despite the fact that ascomycetes represent the largest group of fungi today, the geologic history of this group continues to remain poorly understood. There are several reasons for this, including the difficulty of identifying diagnostic features in the fossils, and the generally held belief that the sac fungi did not evolve until much later, perhaps during the Cretaceous (Pirozynski and Weresub, 1979). Recent reports, however, suggest that ascomycetes existed during the Paleozoic, and are perhaps even older than the basidiomycetes. It is the intent of this paper to document the occurrence of various fossils believed to represent members of the *Ascomycotina*. No attempt has been made to document all fossil ascomycetes, but rather to illustrate several forms, ways in which they are preserved, and biological questions that relate to the geologic history of the sac fungi.

SILURIAN

To date the oldest evidence of possible ascomycetes comes from the mid-late Silurian (Ludlovian) of Gotland, Sweden (Sherwood-Pike and Gray, 1985). The fossils were recovered from both outcrop and drill core samples of the Burgsvik Sandstone, and

Ascomycete Systematics: Problems and Perspectives in the Nineties
Edited by D.L. Hawksworth, Plenum Press, New York, 1994

consist of a variety of fungal remains including both hyphae and spores. The most common fungal elements consist of large, ovoid-cylindrical transversely septate (1-9) spores up to 55 μm long and 25 μm wide (Fig. 1). The basis for suggesting that these spores were produced by ascomycetes is the presence of scars at the ends of some spores like those produced by certain conidial fungi. Hyphae with perforate septa are also present in the Gotland samples. Some possess a whorl of short branches that arise just below a septum (Fig. 2); the entire hyphal structure morphologically resembles the conidiophore of certain conidial fungi that produce phialides. Aggregations of hyphae in what are believed to represent fecal pellets provide an independent measure that these Silurian sites represent evidence of a terrestrial ecosystem, and thus capable of supporting ascomycetous fungi.

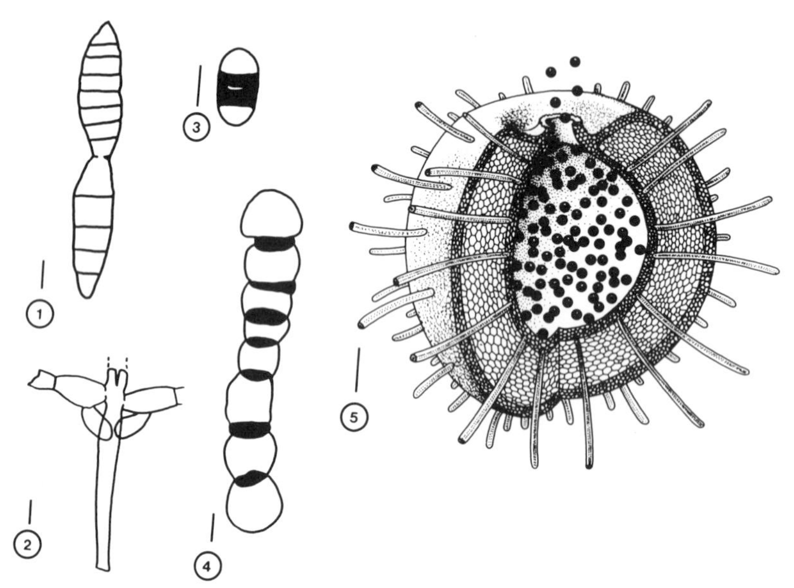

Figures 1-5. Fig. 1. Chain of two transversely septate conidia (Silurian; redrawn from Sherwood-Pike and Gray, 1985). Bar scale = 10 μm. **Fig. 2.** Hypha with perforate septa and phialide-like appendages (Silurian; redrawn from Sherwood-Pike and Gray, 1985). Bar scale = 10 μm. **Fig. 3.** Dicellate *Felixites* spore with thickened septum (Carboniferous; redrawn from Elsik, 1990). Bar scale = 20 μm. **Fig. 4.** Chain of *Reduviasporonites* spores (Permian; redrawn from Wilson, 1962). Bar scale = 10 μm. **Fig. 5.** *Endochaetophora antarctica* sporocarp (Triassic). Bar scale = 50 μm).

DEVONIAN

The occurrence of ascomycetes in the Devonian is based on the presence of round-elliptical structures on the cuticle of *Orestovia devonica*, an enigmatic Lower Devonian land plant from Siberia. Krassilov (1981) interprets these structures as thyriothecia that contain angular hypothecial cells, some with prominent hooked hyphae (Fig. 6). Present are asci (Fig. 7) and elongate cells that may represent paraphyses (Krassilov, 1981). Some cuticle preparations reveal structures that Krassilov interprets as hyphopodia. Based on the morphology of the thyriothecia it is believed the fungus has its closest affinities within the *Microthyriales*.

Another fossil that has been attributed to the ascomycetes is *Mycokidstonia sphaerialoides*, known from the Lower Devonian Rhynie chert (Pons and Locquin, 1981). What is described as a spherical ascoma is 175 μm diam with a central ostiole. The peridium is thought to be plectenchymatous, however, hyphae are not apparent on the micrographs. It is quite probable that *M. sphaerialoides* represents the zoosporangium of some chytridiaceous fungus, a group that is now known to have been well-represented in the Rhynie chert (Taylor et al., 1992b).

CARBONIFEROUS

Some of the most common fungi in Carboniferous rocks are conceptacle-like structures termed sporocarps. They occur singly or in groups of up to six; most are found in the peat matrix of Carboniferous coal balls, although they are also found in the vegetative and reproductive parts of vascular plants. Several genera are now recognized and include forms that range up to 1.0 mm diam with a wall composed of interlaced hyphae (Stubblefield et al., 1983). Some sporocarps possess an outer surface that is smooth, while in others like *Traquairia* (Fig. 10), the surface is ornamented by prominent radiating spines (Stubblefield and Taylor, 1983). Based on the morphology of the sporocarp wall these structures are interpreted as cleistothecia of some ascomycetous fungus.

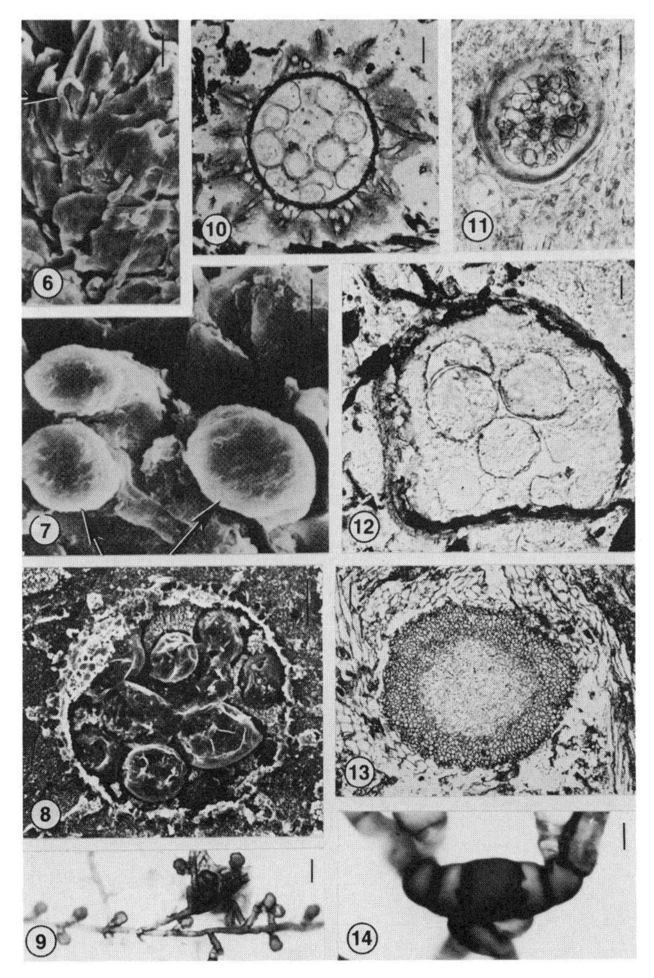

Figures 6-14. Fig. 6. Hooked hypha (arrow) in thyriothecium of *Orestovia* (Devonian; from Krassilov, 1981). Bar scale = 20 μm. **Fig. 7.** Asci (arrows) in thyriothecium of *Orestovia* (Devonian; from Krassilov, 1981). Bar scale = 10 μm. **Fig. 8.** Etched cleistothecium of *Traquairia* showing large spheres (asci) (Carboniferous). Bar scale = 50 μm. **Fig. 9.** Mycelium of *Meliolinites dilcheri* with capitate hyphopodia (Eocene). Bar scale = 5 μm. **Fig. 10.** Sporocarp of *Traquairia williamsonii* containing asci (Carboniferous). Bar scale = 50 μm. **Fig. 11.** Thick-walled fungal spore of *Palaeomyces* containing spheres (mycoparasites) (Devonian). Bar scale = 50 μm. **Fig. 12.** Possible ascus of *Coleocarpon* with ascospores (Carboniferous). Bar scale = 5 μm. **Fig. 13.** Sporocarp of *Palaeosclerotium pusillum* showing outer pseudoparenchymatous layer and inner zone of branched hyphae (Carboniferous). Bar scale = 50 μm. **Fig. 14.** Germinating bipolar multicellate spore of *Meliolinites dilcheri* (Eocene). Bar scale = 10 μm.

Perhaps the most difficult problem in interpreting these fungi as ascomycetes concerns the considerable variability found in the contents of the cleistothecium. Some contain numerous scattered spheres, while in others the lumen is empty. When contents are present they may be quite variable in size. For example, in *Sporocarpon cellulosum* the spheres are up to 80 μm diam, while in *Traquairia williamsonii* the size range of the spheres is 33-55 μm diam (Stubblefield and Taylor, 1983). There are several interpretations concerning the variation in cleistothecial contents in these Carboniferous sporocarps. The most likely explanation is that the large spheres represent asci (Fig. 12); the smaller spheres within these asci would then represent ascospores (Fig. 12). Cleistothecia with a large number of small spheres may indicate a developmental stage in which the asci have disintegrated prior to ascospore dispersal, so that only ascospores are present. A small number of large spores may represent a developmental stage prior to the formation of ascospores, or asci that have already shed their ascospores. Stubblefield and Taylor (1983) suggest that these Paleozoic sporocarps share the largest number of structural and morphological similarities with cleistothecia of the *Erysiphales* and *Eurotiales*.

Another hypothesis suggests that at least some of these sporocarps may have affinities with certain zygomycetes such as members of the *Endogonaceae* (Taylor and White, 1989). This idea is based on the absence of ascogonia, ascogenous hyphae and ascospores, as well as what is interpreted as the rigid wall of some of the large spheres (Fig. 8). Large thick-walled spores containing small spores are also known from the Lower Devonian Rhynie chert (Fig. 11). These thick-walled spheres represent some type of fungal spore; the small spheres inside represent mycoparasites (Hass et al., 1993).

Another interesting Carboniferous fungus that is believed to have some affinities with the ascomycetes is *Palaeosclerotium pusillum* (Rothwell, 1972). This Upper Carboniferous fungus is represented by spherical-ovoid cleistothecia approximately 1.0 mm diam (Fig. 13) that contain round bodies believed to represent asci and ascospores (Dennis, 1976). Asci are approximately 35 μm diam and contain ascospores in the 12 μm size range. The presence of dolipore-like septa and clamp connections on some hyphae has resulted in considerable controversy regarding the affinities of this fungus. McLaughlin (1976) suggests that *Palaeosclerotium* might be an ascomycetous cleistothecium parasitized by a basidiomycete, while Singer (1977) believes the affinities lie within ascomycetes, perhaps close to the modern *Eurotiales*. This is the same group of ascomycetes that share characters with the other Carboniferous sporocarps such as *Traquairia* and *Mycocarpon*. Finally, Pirozynski and Weresub (1979) offer the opinion that *Palaeosclerotium* is neither an independent ascomycete nor a basidiomycete, but rather a dikaryotic fungus that links the basidiomycetes with a group of extinct lichen-like forms.

There are a variety of Paleozoic fungal spores that occur routinely in palynological preparations that may have affinities with the ascomycetes (Elsik, 1992). Some of these may be solitary, while others, like those reported by Sherwood-Pike and Gray (1985), are multicellate (Fig. 1), or occur as chains of conidia (e.g. *Reduviasporonites*, Fig. 4; Wilson, 1962). Elsik (pers. comm.) suggests that *Felixites* (Elsik, 1990), a dicellate, inaperturate Carboniferous spore with a thick septum (Fig. 3) may belong to the ascomycetes. Because extant ascospores are so highly variable in size, shape, ornamentation and number of cells, it is unlikely that these palynomorphs will provide any reliable basis for distinguishing lineages of ascomycetes in the fossil record.

TRIASSIC

There are relatively few examples of structurally preserved ascomycetes from Mesozoic sediments since there are fewer permineralized plant bearing localities of this age than perhaps anywhere in geologic time. At localities where the paleoflora is permineralized, such as the early Middle Triassic of Antarctica, the mycoflora is quite diverse (White and Taylor, 1991; Taylor, 1993). Cleistothecia, like those reported from the Carboniferous, are also known from Triassic rocks (Taylor and White, 1989). For example, *Endochaetophora* is an ostiolate sporocarp up to 500 μm diam with numerous hyphal appendages ornamenting the outer surface (Fig. 5). A few specimens contain small (12-16 μm) spores, but none apparently contain asci. As is the case with the Paleozoic forms, the inability to demonstrate septate hyphae represents a continuous obstacle in classifying these fossil fungi based on the total complement of characters found in modern ascomycete analogues. To date there is no fossil evidence as to when and how hyphal septations evolved. Perhaps the Paleozoic and early Mesozoic sporocarps represent examples of fungal group(s) that are today extinct, but ancestral to the ascomycetes, basidiomycetes or both (White and Taylor, 1988).

CRETACEOUS

Microthyriaceous fungi are perhaps the best known of all fossil ascomycetes (Elsik, 1978). Leaf bearing shoot fragments from the Lower Cretaceous contain abundant microthyriaceous ascomata that are most similar to those of the modern *Micropeltaceae* (Alvin and Muir, 1970). The ostiolate ascoma of *Stomiopeltites cretacea* is dome-shaped and approximately 250 μm diam. Spores are not present, but what are interpreted as pycnidia are associated with the ascocarps. Another Cretaceous (Maastrichtian) microthyriaceous fungal fruiting structure is *Trichopeltinites* (Sweet and Kalgutkar, 1989). This fossil consists of a lobed stroma with ascomata of variable size. Nothing is known about the spores. A fossil member of the *Meliolaceae* is represented by *Meliolinites*, an ascomycete reported from angiosperm leaves of Lower Eocene age (Daghlian, 1978). Colonies of *M. dilcheri* are spherical and consist of two-celled short, lateral branches and capitate hyphopodia. All developmental stages are present including mature perithecia and 5-celled, tapered spores showing bipolar germination (Fig. 14).

TERTIARY

By Tertiary time there were many groups of epiphyllous fungi that were widely distributed geographically and comparable with most modern forms (Taylor, 1993). For example, Dilcher (1965) lists 18 genera associated with Middle Eocene angiosperm leaves, and Sheffy and Dilcher (1971) report 76 species of fungal spores from the same sediments. Although some fossil epiphyllous fungi have been useful as ecological indicators of paleoclimate and as biostratigraphic markers, little is known about the origin of major epiphyllous groups or their biogeographical patterns. The investigation of problems of this type will require the in-depth analysis of major fossil leaf collections and host/fungus relationships through time; studies that today are unfortunately receiving little attention. Another problem in elucidating the evolution of foliar borne fungi is that most epiphyllous parasites are preserved as impressions/compressions, and thus potentially important diagnostic features are often not preserved. Thus, although there are numerous reports of pre-Tertiary epiphyllous fungi, including some extending back to the Paleozoic (Dilcher, 1965; Pia, 1927; Tiffney and Barghoorn, 1974), their nature and affinities continue to remain problematic. The application of more modern imaging systems such as scanning electron microscopy to some of these specimens might prove rewarding if some of the original fossils can still be located.

Some authors (e.g. Pirozynski, 1976) discount all preCretaceous epiphyllous fungi, suggesting that the origin of the group was tightly tied to the origin and adaptative radiation of the angiosperms. Heath (1978) suggests that vascular plants prior to the Cretaceous may have had effective defence mechanisms that inhibited foliar parasites. However, since the phylloplane habitat has been in existence since at least Middle Devonian time, it is surprising that more epiphyllous fungi have not been documented from pre-Tertiary sediments. That parasitizing fungi are known as early as the Lower Devonian (Taylor et al., 1992a) lends support to the hypothesis that various epiphyllous fungal groups were in existence early in the terrestrialization of Earth. The thyriothecia reported on *Orestovia* from Lower Devonian specimens support this assumption. It would appear that the few documented reports of pre-Cretaceous epiphyllous fungi is one more example showing that major groups of fossil fungi have not been extensively studied to date.

One exceptionally well-preserved Eocene ascomycete is *Cryptocolax*, a spherical, permineralized cleistothecium (90 μm diam) recovered from dicotyledonous wood (Scott, 1956). Each of the several hundred cleistothecia examined contain 70-80 ascospores; asci are evanescent with ascogenous hyphae and paraphyses absent. Affinities are believed to lie within the *Eurotiales*, including the extant genera *Anixiopsis* and *Cephalotheca*.

CONCLUSIONS

Historically there have been two major ideas regarding the origin of the ascomycetes, neither based on any fossil evidence. In one of these the ascomycete lineage is directly related to the red algae through the loss of pigmentation. Proponents of this hypothesis (e.g. Demoulin, 1985) point to many similarities in the life history, biology, and biochemistry of ascomycetes and certain red algae, while opponents (e.g. Cavalier-Smith, 1987) cite an equally detailed list of ultrastructural differences. The second hypothesis views ascomycetes as evolving from chytridiaceous ancestors through the zygomycete group *Endogonales* (Caviler-Smith, 1987). He suggests that a shift from zygospore to ascospores and the evolution of ascogenous hyphae are all that is necessary to alter the subterranean endogonalean sporocarp into an ascocarp of some early plectomycete. Sherwood-Pike (1991) also speculates that some protozygomycete is ancestral to the ascomycetes. In recent years molecular data have been used to estimate the time of origin and radiations of fungal groups. The most recent of these is an analysis by Berbee and Taylor (1993) based on a comparison of 37 fungal species using 18S rRNA sequences. Based on molecular clock estimates, their results suggest that the ascomycetes evolved during the Carboniferous, with the ancestor to the sac fungi and basidiomycetes perhaps present as early as the Silurian.

Although the red algae extend back to the Upper Proterozoic (Butterfield et al., 1990), and are well represented in the Paleozoic, none of these fossils provides any information about the potential relationship with ascomycetes. Moreover, it is unlikely that the fossil record ever will. The possible relationship between zygomycetes and ascomycetes is, however, a different matter, since some of the reproductive structures preserved in Paleozoic sediments both morphologically and structurally resemble the cleistothecia of certain plectomycetes. Like the ascocarps of living plectomycetes, the Carboniferous sporocarps lack paraphyses and an ostiole. Asci are typically spherical and break down to release ascospores within the cleistothecium. As noted above, some of the fossil sporocarps contain spherical structures that may represent asci. Other specimens with only small spores may represent developmental stages in which asci have already broken down. Other Paleozoic and Mesozoic sporocarps contain a single, large spore. Some of these forms may represent zygomycete sporocarps (White and Taylor, 1989), while others may demonstrate examples of lineages involved in the shift from zygospore to ascospores, as suggested by Caviler-Smith (1987).

One of the major problems in attempting to document the geologic history of fungi concerns the inability to identify certain characters used to delimit major groups. In the case of the ascomycetes, one such character is the ascoma. In modern forms this feature exhibits extensive morphological diversity which is based at least in part on the reproductive strategies of the organisms (Hawksworth, 1987). Some fossil ascomycetes possess ascoma that are morphologically similar to modern forms, while others exhibit a combination of characters not present in modern analogues. It is probable that particular fungal characters may not have evolved at the same rate within different groups, as is true for other groups of organisms. If major lineages of fungi diverged approximately 1000 Ma ago as suggested by Knoll (1992), it is not surprising that by the Paleozoic various types of ascomycetes were already well established. Paleomycology is no different from other branches of paleobiology in that it is unlikely that the study of fossil fungi will ever provide the transitional phases leading to the establishment of major living groups. Nevertheless, despite the imperfections in the geologic history of fungal groups, fossils continue to provide the only method available to test the validity of evolutionary assumptions based on modern forms, and to provide constraints for molecular clock hypotheses.

ACKNOWLEDGEMENTS

This study was supported in part by funds from the National Science Foundation (DPP-9118314). I am indebted to Dr M.L. Berbee and Professor J.W. Taylor for providing a preprint of their paper on the evolutionary radiations of the true fungi.

REFERENCES

Alvin, K.L. and M.D. Muir, 1970, An epiphyllous fungus from the Lower Cretaceous, *Biological Journal of the Linnean Society* 2: 55-59.

Berbee, M.L. and J.W. Taylor, 1993, Dating the evolutionary radiations of the true fungi, *Canadian Journal of Botany*: in press.

Butterfield, N.J., A.H. Knoll, and K. Swett, 1990, A bangiophyte red alga from the Proterozoic of Arctic Canada, *Science* 250: 104-107.

Cavalier-Smith, T., 1987, The origin of fungi and pseudofungi, *In*: *Evolutionary Bbiology of The Fungi* (A.D.M. Rayner, C.M.Brasier and D. Moore, eds): 339-353. Cambridge University Press, Cambridge.

Daghlian, C.P., 1978, A new melioloid fungus from the Early Eocene of Texas, *Palaeontology* 21: 141-146.

Demoulin, V., 1985, The red algal-higher fungi phylogenetic link: the last ten years, *BioSystems* 18: 347-356.

Dennis, R.L., 1970, A Middle Pennsylvanian basidiomycete with clamp connections, *Mycologia* 62: 578-584.

Dennis, R.L., 1976, *Palaeosclerotium*, a Pennsylvanian age fungus combining features of modern ascomycetes and basidiomycetes, *Science* 192: 66-68.

Dilcher, D.L., 1965, Epiphyllous fungi from Eocene deposits in western Tennessee, U.S.A., *Palaeontographica* 116B: 1-54.

Elsik, W.C., 1978 ["1976-77"], Classification and geologic history of the microthyriaceous fungi, *Proceedings of the IVth International Palynological Conference Lucknow* 1: 331-342.

Elsik, W.C., 1990, The fungal morphotype *Felixites* n. gen., *Pollen et Spores* 31: 155-159.

Elsik, W.C., 1992, The morphology, taxonomy, classification and geologic occurrence of fungal palynomorphs, *American Association, Stratigraphic Palynology Shortcourse*. American Association of Stratigraphic Palynologists Foundation.

Hass, H., T.N. Taylor, and W. Remy, 1993, Fungi from the Lower Devonian Rhynie chert: mycoparasitism, *American Journal of Botany* 81: 29-37.

Hawksworth, D.L., 1987, The evolution and adaptation of sexual reproductive structures in the *Ascomycotina*, *In*: *Evolutionary Biology of The Fungi* (A.D.M. Rayner, C.M. Brasier and D. Moore, eds): 179-189. Cambridge University Press, Cambridge.

Heath, M.C., 1978, Evolution of parasitism in the fungi, *In*: *Evolutionary Biology of The Fungi* (A.D.M. Rayner, C.M. Brasier and D. Moore, eds): 149-160. Cambridge University Press, Cambridge.

Knoll, A.H., 1992, The early evolution of eukaryotes: a geological perspective, *Science* 256: 622-627.

Krassilov, V., 1981, *Orestovia* and the origin of vascular plants, *Lethaia* 14: 235-250.

McLaughlin, D.J., 1976, On *Palaeosclerotium* as a link between ascomycetes and basidiomycetes, *Science* 193: 602.

Osborn, J.M., T.N. Taylor, and J.F. White, jr., 1989, *Palaeofibulus* gen. nov., a clamp-bearing fungus from the Triassic of Antarctica, *Mycologia* 81: 622-626.

Pia, J., 1927, *Thallophyta*, *In*: *Handbuch der Paläobotanik*: 1-136. R. Oldenbourg, München und Berlin.

Pirozynski, K.A., 1976, Fungal spores in fossil record, *Biological Memoirs* 1: 104-120.

Pirozynski, K.A. and L.W. Weresub, 1979, The classification and nomenclature of fossil fungi, *In*: *The Whole Fungus* (B. Kendrick, ed.) 2: 653-688. National Museum of Natural Sciences, Ottawa.

Pons, D. and M.V. Locquin, 1981, *Mycokidstonia sphaerialoides* Pons & Locquin, gen et sp. nov., Ascomycète fossile Dévonien, *Cahiers de Micropaléontologie* 1: 101-104.

Rothwell, G.W., 1972, *Palaeosclerotium pusillum* gen. et sp. nov., a fossil eumycete from the Pennsylvanian of Illinois, *Canadian Journal of Botany* 50: 2353-2356.

Scott, R.A., 1956, *Cryptocolax*-a new genus of fungi (*Aspergillaceae*) from the Eocene of Oregon, *American Journal of Botany* 43: 589-593.

Sheffy, M.V. and D.L. Dilcher, 1971, Morphology and taxonomy of fungal spores, *Palaeontographica* 133B: 34-51.

Sherwood-Pike, M.A., 1991, Fossils as keys to evolution in fungi, *BioSystems* 25: 121-129.

Sherwood-Pike, M.A. and J. Gray, 1985, Silurian fungal remains: probable records of the class *Ascomycetes*, *Lethaia* 18: 1-20.

Singer, R., 1977, An interpretation of *Palaeosclerotium*, *Mycologia* 69: 850-854.

Stubblefield, S.P. and T.N. Taylor, 1983, Studies of Paleozoic fungi. I. The structure and organization of *Traquairia* (*Ascomycota*), *American Journal of Botany* 70: 387-399.

Stubblefield, S.P. and T.N. Taylor, 1986, Wood decay in silicified gymnosperms from Antarctica, *Botanical Gazette* 147: 116-125.

Stubblefield, S.P., T.N. Taylor, and C.B. Beck, 1985, Studies of Paleozoic fungi.V. Wood-decaying fungi in *Callixylon newberryi* from the Upper Devonian, *American Journal of Botany* 72: 1765-1774.

Stubblefield, S.P., T.N. Taylor, C.E. Miller, and G.T. Cole, 1983, Studies of Carboniferous fungi. II. The structure and organization of *Mycocarpon, Sporocarpon, Dubiocarpon,* and *Coleocarpon* (*Ascomycotina*), *American Journal of Botany* 70: 1482-1498.

Sweet, A.R. and R.M. Kalgutkar, 1989, *Trichopeltinites* Cookson from the Latest Maastrichtian of Canada, *Contributions to Paleontology, Geological Survey of Canada Bulletin* 396: 223-227.

Taylor, T.N., 1993, Fungi, *In: The Fossil Record* (H.J. Benton, ed) 2: 9-13. Chapman and Hall, London.

Taylor, T.N., H. Hass, and W. Remy, 1992a, Devonian fungi: interactions with the green alga *Palaeonitella, Mycologia* 84: 901-910.

Taylor, T.N., W. Remy, and H. Hass, 1992b, Fungi from the Lower Devonian Rhynie chert: *Chytridiomycetes, American Journal of Botany* 79: 1233-1241.

Taylor, T.N. and J.F. White, jr., 1989, Fossil fungi (*Endogonaceae*) from the Triassic of Antarctica, *American Journal of Botany* 76: 389-396.

Tiffney, B.H. and E.S. Barghoorn, 1974, The fossil record of the fungi, *Occasional Papers of the Farlow Herbarium* 7: 1-42.

White, J.F., jr. and T.N. Taylor, 1988, Triassic fungus from Antarctica with possible ascomycetous affinities, *American Journal of Botany* 75: 1495-1550.

White, J.F., jr. and T.N. Taylor, 1989, Triassic fungi with suggested affinities to the *Endogonales* (*Zygomycotina*), *Review of Palaeobotany and Palynology* 61: 53-61.

White, J.F., jr. and T.N. Taylor, 1991, Fungal sporocarps from Triassic peat deposits in Antarctica, *Review of Palaeobotany and Palynology* 67: 229-236.

Wilson, L.R., 1962, A Permian fungus spore type from the Flowerpot Formation of Oklahoma, *Oklahoma Geological Notes* 22: 91-96.

BIOGEOGRAPHY AND ANCESTRY OF LICHENS AND OTHER ASCOMYCETES

D.J. Galloway

Department of Botany
The Natural History Museum
Cromwell Road, London SW7 5BD, UK

SUMMARY

Biogeography is simply defined as the study of geographical distribution of organisms; the patterns of distribution of taxa, and the processes by which these observed distributions have come about. The application of chemical and biogeographical data to the detection of ancestry in the *Ascomycotina* is discussed.

INTRODUCTION

Fungi are an ancient group of organisms (Cain, 1972), with the earliest possible records of ascomycetous fungi being from late Silurian and early Devonian deposits and contemporaneous with the earliest land plants (Sherwood-Pike and Gray, 1985; Hawksworth, 1985, 1991; T.N. Taylor, this volume). Sherwood-Pike and Gray (1985) state:

> "There is much debate and little conclusive evidence relating to the subject of ascomycete phylogeny. The fossil record, as presently known, gives few clues to the resolution of such important questions as which, if any, of the surviving subclasses and orders of ascomycetes should be regarded as primitive or ancestral, and which if any, of the surviving phylum of protists is ancestral to the ascomycetes . . . .".

They regard both *Basidiomycota* and *Ascomycota* as being of Palaeozoic origin, with fossil evidence suggesting the *Ascomycota* to be the earlier. A view formerly widely held, though not unanimously, was that ascomycete fungi have a red-alga or red alga-like ancestry (Cain, 1972; Demoulin, 1974, 1985; Hawksworth, 1982, 1991; Tehler, 1988), though this is now much doubted.

Given the ancient origin of ascomycete fungi and of simple photosynthetic organisms such as cyanobacteria and green algae, an early emergence of symbiotic associations between potential photobionts and ascomycete mycobionts must have occurred with cyanobacteria as an ancient group (Schopf and Walter, 1982) being among the earliest potential photobionts available for emerging fungi (Hawksworth, 1988a, 1998b; Büdel, 1992). Symbiogenesis, the evolution of new life forms by integrating partners from different taxa resulting from prolonged physical association, was a principle of evolutionary thinking in Russia from late last century, but has only been more widely accepted in the west during the past two decades (Khakina, 1992).

Lichens are now commonly cited as among the most successful of terrestrial symbiotic associations, having had a long and diverse evolutionary history (Hawksworth, 1978, 1988a, 1988b; Kendrick, 1991; Honegger, 1991, 1992), resulting in basic patterns of primary metabolism with respect to carbon and nitrogen transfer between autotrophic and heterotrophic partners (Smith, 1978; Smith and Douglas, 1987; Feige and Jensen, 1992); particular reproductive strategies (Bowler and Rundel, 1975; Mattson and Lumbsch, 1989); well-developed and unique secondary chemistry in some groups but not others; a wide variety of symbiont interactions (Hawksworth, 1978, 1982, 1988a, 1988b; Ahmadjian, 1987, 1992; Kendrick, 1991; Armaleo and Clerc, 1991); and characteristic morphological and physiological adaptations to a staggering diversity of habitats worldwide.

With only rather few reports of lichen fossils in the literature (Hallbauer et al., 1977; Sherwood-Pike, 1985; Bermudes and Back, 1991) we must necessarily look for evidence of the antiquity of ascomycete fungi in present-day taxa, a good example being the order *Peltigerales*. A scheme correlating ascus morphology, photobionts and habitat ecology in the *Ascomycotina* gives the *Peltigerales* a central position (Dick and Hawksworth, 1985) reflecting the inclusion of characters of particular antiquity (Hawksworth, 1982, 1988a, 1988b). These include: (1) studies on ascus structure (Eriksson, 1981; Hawksworth, 1982); (2) the exceptional number of genera and species of obligately lichenicolous fungi occurring on members of the *Peltigerales*; (3) cyanobacterial symbionts are common in the order both as primary photobionts and as cephalodia (James and Henssen, 1976; Galloway, 1988b, 1991a, 1991c, 1992; White and James, 1988); (4) secondary chemistry; and (5) geographical distribution in the light of current theories of historical biogeography.

In this last context might be mentioned the occurrence of taxa in New Zealand and Patagonia which have shown a notable lack of differentiation over tens of millions of years. Hay (1992) has recently accepted species stasis over 100 million years for certain *Araceae*. Heads (1993) suggests that austral disjunction in *Hebe* is "a normal component of the standard New Zealand-Patagonia track seen in many organisms and attributed to Mesozoic evolution and the subsequent break-up of Gondwanaland". The lichens *Degelia gayana* (Jφrgensen and James, 1990) and *Leioderma pycnophorum* (Galloway and Jφrgensen, 1987) are two of many examples (Galloway, 1987, 1988a).

I will discuss some ideas on chemistry, and biogeography of lichenized ascomycete fungi as possible ways of looking at evolutionary processes in this group.

CHEMISTRY

Lichens produce a wide range of primary and secondary metabolites and these may provide important clues to both taxonomic and evolutionary relationships (Galloway, 1991a, 1991c). The importance of chemical characters in sibling speciation in lichen-forming fungi where chemical and ecological differentiation is observed in taxa having highly conservative morphologies (Culberson, 1986, this volume; Culberson et al., 1988; Holtan-Hartwig, 1993) underlines the view that chemistry is a major marker of evolutionary change in some lichens.

Some genera in the *Peltigerales* show exceptional chemical diversity, elaborating compounds from all three major pathways of secondary metabolism, *viz.* the acetate-polymalonate pathway, the shikimic acid pathway, and the mevalonic acid pathway. In the families *Peltigeraceae*, *Solorinaceae* and *Nephromataceae* compounds from both the acetate-polymalonate and mevalonic acid pathways are produced.

In the *Lobariaceae* an even more complex chemistry occurs with *Pseudocyphellaria* having the most richly diverse chemistry of any genus, with several compounds synthesized *via* the shikimic acid pathway in addition to contributions from the other biosynthetic pathways. Triterpenoids from several different series show a distinct evolution in chemical complexity, and triterpenoid signatures such as those in *Nephroma* (White and James, 1988), *Peltigera* (Holtan-Hartwig, 1993) and *Pseudocyphellaria* (Galloway, 1988b, 1991a, 1991c, 1992) may be of value in reconstructing evolutionary relationships (Galloway, 1991a, 1991c). Galloway (1991c) has discussed triterpenoid composition and distribution in lichens in a preliminary and qualitative way.

Hopane triterpenoids are those most widespread in lichens being known from the genera *Heterodermia*, *Lobaria*, *Nephroma*, *Peltigera*, *Physcia*, *Pseudocyphellaria*, and *Rinodina*. Hopane-like compounds are present in a number of prokaryotes including *Nostoc*, and it is thought that they are components of membranes, analogous to sterols in eukaryotic cells. They are thought to be primitive phylogenetic precursors of sterols, derived from the same precursor squalene, but cyclised in the absence of oxygen and possibly representing persistence of a chemical adaptation compatible with archaebiotic, prephotosynthetic conditions (Ourisson et al., 1979). The similarity of lichen hopanes which are produced in quantity in species of *Nephroma*, *Peltigera*, and *Pseudocyphellaria* in cool temperate habitats, and of hopane biomarkers (such as 17α 21ß hopane and 17ß 21α hopane) used by petroleum chemists when assaying oil deposits in various geological formations, suggests the possibility of relating chemical structures to geological formations of known age, which may have important implications in phylogenetic reconstructions in the *Peltigerales* (Galloway, 1991c). Recently, Holtan-Hartwig in a detailed study of *Peltigera* populations in Norway, has mapped the distribution of chemodemes (differing mainly in triterpenoid composition) and has shown that in *P. aphthosa*, *P. malacea* and *P. neopolydactyla* several chemodemes exist which are morphologically, ecologically, and distributionally definable (Holtan-Hartwig, 1993).

Other more advanced triterpenoid groups are found in the genus *Pseudocyphellaria* in yellow-medulla species containing shikimic acid pathway compounds such as terphenylquinones and pulvinic acid derivatives. These include stictanes, seco stictanes, lupanes, and fernenes (Galloway, 1991c). Fernenes are rearrangement products of hopanes and are found in ferns and in mosses, and it has been suggested that fernenes may constitute a phylogenetic link between lichens, mosses and vascular plants (Markham and Porter, 1978). Triterpenoids are certainly key compounds in the detection of biogeographical and evolutionary relationships between taxa in ascomycete fungi.

BIOGEOGRAPHY

All organisms evolve in both space and time, a view early articulated by Alfred Russel Wallace when he stated (Wallace, 1855):

> "It has now been shown, though most briefly and imperfectly, how the law that "Every species has come into existence coincident in both time and space with a pre-existing closely allied species," connects together and renders intelligible a vast number of independent and hitherto unexplained facts . . ."

Evolution in space and time gives rise to the formation of patterns of distribution of organisms, and the study of these patterns and of the processes by which they have come about is known as biogeography. Taxonomy underpins biogeography; as Humphries (1983) has pointed out:

> "biogeography is dependent on the theoretical foundations of systematics since it can only be as good as the taxonomy it uses to establish historical distribution. Biogeography must therefore be regarded as an integral part of systematics; indeed changes in biogeographical explanations follow changes in taxonomic theory".

In the nineteenth century, once lichen collections were made available from the great exploring expeditions to the Southern Hemisphere and to tropical regions (Galloway, 1991a, 1991b) and lichenologists were freed from a "European" view of the world, a cosmopolitanism pervaded accounts of the geographical distribution of lichens (Hooker and Taylor, 1844; Lindsay, 1856; Berkeley, 1857; Nylander, 1858; Smith, 1921), which, although recognizing endemic or local taxa in many areas, neverthless tended to overstate the case that lichens generally occur widely from polar to tropical regions. An example of this is seen in the introduction to the first account of antarctic and subantarctic lichens collected by Joseph Hooker on the Ross Expedition of 1839-1843. Hooker and his lichenologist co-author, Thomas Taylor wrote (Hooker and Taylor, 1844):

> ". . . The uniformity of rocks as they appear above the soil, in all parts of the earth, has been well ascertained by geologists. Atmospheric influences disintegrate their surface. The atmosphere and rocks being identical in all latitudes, so must the first layers of decomposing matter on the surface of the latter be. Now such being the precise places that Lichens select for their seat, it would follow that the geographical distribution of the species should be extended to wider limits than that of any other tribe of plants. . ."

150 years later, our knowledge of the taxonomy and distribution of lichens has increased dramatically and a number of introductory accounts of lichen biogeography are now available (Galloway, 1979, 1987, 1988a, 1991a, 1991b; Jørgensen, 1983; Hawksworth and Hill, 1984; Egan, 1986), and there are an increasing number of taxonomic revisions of lichenized genera which draw biogeographical conclusions from their studies (e.g. Sheard, 1977; Jørgensen, 1979; Kärnefelt, 1979, 1989, 1990a, 1990b, 1991; Arvidsson and Galloway, 1981; Arvidsson, 1983; Galloway and Jørgensen, 1987; Tehler, 1983; Galloway, 1988b, 1992; Almborn, 1988, 1989, 1992; Thor, 1990; Stenroos, 1991; Stenroos and Ahti, 1992; Ahti, 1992; Kärnefelt et al., 1992; Karnefelt and Thell, 1992; Sipman, 1992; Breuss, 1993).

For non-lichenized ascomycete information on biogeography I cannot comment further beyond the excellent reviews of Pirozynski and Weresub (1979) and Pirozynski (1983), and discussions of the pan-austral *Nothofagus* parasite *Cyttaria* (Korf, 1983; Humphries et al., 1986) which show patterns of distribution similar to those found in lichens and other organisms. Ainsworth (1976) gives a concise account of early views on mycogeography.

Distribution patterns have been traditionally perceived as representing biogeographic affinities or elements in a particular biota, those most commonly perceived in ascomycete fungi being: (1) cosmopolitan, (2) pantropical, (3) palaeotropical, (4) austral, (5) australasian, (6) bipolar, and (7) endemic.

Explanations or models have been developed which help to explain and sometimes predict such patterns of geographical distribution of biota. Amongst biogeographers there are several schools of thought and method, *viz.* migrationist biogeography, vicariance biogeography and panbiogeography in historical approaches. Endler (1982a) states:

> ". . . as a result of the newness of quantitative techniques and ideas . . . we too often see large bodies of data analyzed with respect to only one of several competing hypotheses. Too often a model is chosen apparently because it is familiar, and explanations have often been those of plausibility rather than a comparison of alternative hypotheses. Even those interested in testing hypotheses rarely consider alternative hypotheses . . . Biogeography has become divided into "schools" which largely ignore one-another".

Heads (1990) believes that:

> ". . . Much of twentieth century biogeography can be read as a detour, via specific concepts of "centre" and "origin", which has avoided a sustained examination of the possible significance of biogeographic analysis for understanding the evolutionary process and change in general . . . distribution patterns of organisms with widely different ecologies and means of dispersal, such as alpine plants and marine animals, can be shown to follow the same tracks, centres and nodes of distribution, which reflect tectonic history rather than aspects of particular groups . . . Approaches to biogeography have been based all too often on consideration of particular lineages, emphasizing purely theoretical ancestor-descendant relationships and have maintained a blind spot towards the general effect of phases of modernization on a landscape and its biota . . .".

A discussion of present-day ecological factors, and historical approaches in biogeography and the unfortunate dichotomy between the two approaches is given in Endler (1982b) and in Henderson (1991), and an interesting paper of Smith (1989) considers the philosophical core of biogeography.

The three main schools of historical biogeography are briefly outlined below, following the synopsis of Wilson (1991).

Migrationist Biogeography

The migrationist or dispersalist tradition has today a wide constituency among biogeographers and is the basis of most biogeographic textbooks (Wilson, 1991). It has its theoretical roots in the views of Wallace and Darwin who invoked a theory of chance migration from single centres of origin. In Darwin's words (Darwin, 1859):

> "We are thus brought to the question which has been largely discussed by naturalists, namely whether species have been created at one or more points of the earth's surface. Undoubtedly there are many cases of extreme difficulty in understanding how the same species could possi-

bly have migrated from one point to several distant and isolated points, where now found. Nevertheless the simplicity of the view that each species was first produced within a single region captivates the mind . . .".

Known occurrences of taxa are mapped as ranges with extremes regarded as limits to distribution. There is no generally agreed method of analysis. Centres of origin are usually suggested, as are routes and methods of migration which are assumed to be the major cause of presently observed distributions. Ecology is regarded as being a major factor in determining distribution. No special use of taxonomy is made in analysis using narrative migrationist models. Taxa ocurring in centres of origin are sometimes assumed to be primitive (or sometimes advanced). Very many examples of migrationist explanations of distribution patterns are found in the lichenological literature among which may be cited studies on *Placopsis* (Lamb, 1947), *Parmelia* (Hale, 1987), *Pseudocyphellaria* (Galloway, 1988b, 1992), *Degelia* (Jørgensen and James, 1990), and *Xanthoparmelia* (Hale, 1990).

Vicariance Biogeography

Vicariance or cladistic biogeographers use cladistics to generate a taxonomic cladogram for each group of taxa studied and then replace each taxon with the area in which it occurs producing an area cladogram. Area cladograms from different groups are compared to determine congruence. Distribution is mapped as presence/absence in area units. It is assumed that an ancestral species occurred over the whole area and that parts were isolated by major geological events such as plate tectonics, mountain building and various geomorphological processes, giving rise to allopatric speciation. For reviews of recent developments see Humphries and Parenti (1986), and Humphries et al. (1988). Examples of this approach are given in studies on *Dirina* and *Roccellina* (Tehler, 1983), *Cyttaria* (Humphries et al., 1986), and *Chiodecton* (Thor, 1990).

Panbiogeography

The panbiogeographic method was developed by Croizat (1958, 1964) whose important insight into distribution patterns was that they are highly repetitive. These repetitive patterns he called "generalized tracks", lines on a map which summarize distributions of many diverse individual taxa. He rejected chance dispersal as the cause of generalized tracks because the pattern is independent of the individual taxon's dispersal abilities (Michaux, 1989). The "main massing" is emphasized rather than the limits of distribution. Tracks applied to global patterns of distribution are found to have no correlation whatever with modern geography but extend across all the world's oceans and continents.

Distributions are generalized by identifying major tracks, nodes (track junctions, and centres of endemism), and gates (major nodes). A track is a line graph drawn on a map of the geographical distribution of a particular taxon and connecting the disjunct collection localities or distribution areas of that taxon. The simplest way to construct such a graph is to form a minimal spanning tree, that is a graph that connects all localities or distribution areas occupied by a taxon so that the sum of link lengths connecting each locality or distribution area, is the smallest possible. Next a hypothesis of the baseline (diagnostic character) is prepared for each track depending on the particular ocean or sea basin or major tectonic feature that the track crosses or circumscribes (Craw, 1989). Biological tracks correlate with fomer zones of disturbance such as geological fold or fracture zones, or coastlines and their environs, with much present-day biology being (Heads, 1990):

". . . that of old, relic, coastal communities left stranded inland on plains, hills and mountains, all by normal geological processes such as uplift and eustatic sealevel changes".

Panbiogeography avoids assumption of any particular method of dispersal, although its adherents emphasize geological history and de-emphasize migration for some taxa although it is accepted that certain "weedy" taxa are actively migrating. Terrane tectonics (Craw, 1985; Cooper, 1989; Heads, 1989; Burrett et al., 1991) are often of special interest and have been used to explain lichen distributions of taxa in *Anzia*, *Pannaria*, and *Xanthoparmelia* in New Zealand (Heads, 1989), and of general bipolar distributions in lichens and a wide range of other biota between the southwest and northeast Pacific (Chin et al., 1991). Recent developments of panbiogeographic theory and methods are found in Craw (1988, 1989), Grehan (1988, 1989, 1991), Henderson (1989), and Page (1989).

CONCLUSIONS

In this brief overview of approaches and possibilities in the biogeography and detection of ancestry in lichens and other ascomycetes, I hope I have shown something of the prospects that exist both for analysing presently available data on lichens particularly, as well as for future lines of investigation. It seems to me that lichens are a biogeographer's group, par excellence, a group of which the following, written 234 years ago still has relevance for us today (Watson, 1758):

> ". . . we have in this genus of plants [lichens] a convincing instance of the utility which may result from the study of natural science in general, and even of its minuter and hitherto most neglected branches . . . posterity will doubtless find the means of employing them to many valuable purposes in human life unknown to us . . . The hopes of discovering some latent property, which may turn out to the advantage of his fellow creatures, will animate the man, whose mind is truly formed for relishing the pleasures of natural science; and however the result may be, the inspection and contemplation of nature's productions will ever afford that satisfaction, which will amply repay him for his trouble. The minuter, and, as they are commonly estimated, the most abject and insignificant things are not beneath our notice; and an attentive mind will readily conceive how much farther, and more extensively useful, every branch of nature's kingdom may yet prove in the oeconomy of human life. The man, therefore, whom a genius and love for natural history has allured into its pursuits, and whose leisure permits his gratification in such researches, if he is not happy enough to be crowned with success, at least deserves it, and merits the thanks of his fellow-creatures for his application and diligence."

REFERENCES

Ahmadjian, V., 1987, Coevolution in lichens, *Annals of the New York Academy of Science* 503: 307-315.

Ahmadjian, V., 1992, Basic mechanisms of signal exchange, recognition, and regulation in lichens, *In*: *Algae and Symbioses: Plants, Animals, Fungi, Viruses, Interactions Explored* (W. Reisser, ed.): 675-697, Biopress, Bristol.

Ahti, T., 1992, Biogeographic aspects of *Cladoniaceae* in the paramos, *In*: *Paramo, An Andean Ecosystem under Human Influence* (H. Balsev and J.L. Luteyn, eds): 111-117, Academic Press, London.

Ainsworth, G.C., 1976, *Introduction to the History of Mycology*, Cambridge University Press, Cambridge.

Almborn, O., 1988, Some distribution patterns in the lichen flora of South Africa, *Monographs on Systematic Botany, Missouri Botanical Garden* 25: 429-432.

Almborn, O., 1989, Revision of the lichen genus *Teloschistes* in central and southern Africa, *Nordic Journal of Botany* 8: 521-537

Almborn, O., 1992, Some overlooked or misidentified species of *Teloschistes* from South America and a key to the South-American species, *Nordic Journal of Botany* 12: 361-364.

Armaleo, D., and P. Clerc, 1991, Lichen chimeras: DNA analysis suggests that one fungus forms two morphotypes, *Experimental Mycology* 15: 1-10.

Arvidsson, L., 1983, A monograph of the lichen genus *Coccocarpia*, *Opera Botanica* 67: 1-96.

Arvidsson, L., and D.J. Galloway, 1981, *Degelia*, a new lichen genus in the *Pannariaceae*, *Lichenologist* 13: 27-50.

Berkeley, M.J., 1857, *Introduction to Cryptogamic Botany*, Bailliere, London.

Bermudes, D., and R.C. Back, 1991, Symbiosis inferred from the fossil record, *In*: *Symbiosis as a Source of Evolutionary Innovation* (L. Margulis and R. Fester, eds): 72-99, MIT Press, Cambridge, Mass.

Bowler, P.A., and P.W. Rundel, 1975, Reproductive strategies in lichens, *Botanical Journal of the Linnean Society* 70: 325-340.

Breuss, O., 1993, *Catapyrenium* (*Verrucariaceae*) species from South America, *Plant Systematics and Evolution* 185: 17-33.

Büdel, B., 1992, Taxonomy of lichenized procaryotic blue-green algae, *In*: *Algae and Symbioses: Plants, Animals, Fungi, Viruses, Interactions Explored* (W. Reisser, ed.): 301-324, Biopress, Bristol.

Burrett, C., N. Duhig, R. Berry, and R. Varne, 1991, Asian and south-western Pacific continental terranes derived from Gondwana, and their biogeographic significance, *Australian Systematic Botany* 4: 13-24.

Cain, R.F., 1972, Evolution of the fungi, *Mycologia* 64: 1-14.

Chin, N.K.M., M.T. Brown, and M.J. Heads, 1991, The biogeography of *Lessoniaceae*, with special reference to *Macrocystis* C. Agardh (*Phaeophyta*: *Laminarales*), *Hydrobiologia* 215: 1-11.

Cooper, R.A., 1989, New Zealand tectonostratigraphic terranes and panbiogeography, *New Zealand Journal of Zoology* 16: 699-712.

Craw, R., 1985, Classic problems of southern hemisphere biogeography re-examined, panbiogeographic analysis of the New Zealand frog *Leiopelma*, the ratite birds, and *Nothofagus*, *Zeitschrift für Zoologische Systematik und Evolutionsforschung* 23: 1-10.

Craw, R., 1988, Panbiogeography: method and synthesis in biogeography, *In*: *Analytical Biogeography* (A.A. Myers, and P.S. Giller, eds): 405-435, Chapman and Hall, London.

Craw, R., 1989, Quantitative panbiogeography: introduction to methods, *New Zealand Journal of Zoology* 16: 485-494.

Croizat, L., 1952, *Manual of Phytogeography*, Junk, The Hague.

Croizat, L., 1958, *Panbiogeography*, L. Croizat, Caracas.

Culberson, C.F., W.L. Culberson, and A. Johnson, 1988, Gene flow in lichens, *American Journal of Botany* 75: 1135-1139.

Culberson, W.L., 1986, Chemistry and sibling speciation in the lichen-forming fungi: ecological and biological considerations, *Bryologist* 89: 123-131.

Darwin, C., 1859, *On the Origin of Species by means of Natural Selection*, John Murray, London.

Demoulin, V., 1974, The origin of ascomycetes and basidiomycetes. The case for red algal ancestry, *Botanical Review* 40: 315-345.

Demoulin, V., 1985, The red algal-higher fungi phylogenetic link: the last ten years, *BioSystems* 18: 347-356.

Dick, M.W., and D.L. Hawksworth, 1982, A synopsis of the biology of the *Ascomycotina*, *Botanical Journal of the Linnean Society* 91: 175-179.

Egan, R.S., 1986, Correlations and non-correlations of chemical variation patterns with lichen morphology and geography, *Bryologist* 89: 99-110.

Endler, J.A., 1982a, Alternative hypotheses in biogeography: introduction and synopsis of the symposium, *American Zoologist* 22: 349-354.

Endler, J.A., 1982b, Problems in distinguishing historical from ecological factors in biogeography, *American Zoologist* 22: 441-452.

Eriksson, O., 1981, The families of bitunicate ascomycetes, *Opera Botanica* 60: 1-220.

Feige, G.B., and M. Jensen, 1992, Basic carbon and nitrogen metabolism of lichens, *In*: *Algae and Symbioses: Plants, Animals, Fungi, Viruses, Interactions explored* (W. Reisser, ed.): 255-275. Biopress, Bristol.

Galloway, D.J., 1979, Biogeographical elements in the New Zealand lichen flora, *In*: *Plants and Islands* (D.J. Bramwell, ed.): 201-224, Academic Press, London.

Galloway, D.J., 1987, Austral lichen genera: some biogeographical problems, *Bibliotheca Lichenologica* 25: 385-399.

Galloway, D.J., 1988a, Plate tectonics and the distribution of cool temperate Southern Hemisphere macrolichens, *Botanical Journal of the Linnean Society* 96: 45-55.

Galloway, D.J., 1988b, Studies in *Pseudocyphellaria* (lichens) I. The New Zealand species. *Bulletin of the British Museum (Natural History), Botany* 17: 1-267.

Galloway, D.J., 1991a, Phytogeography of Southern Hemisphere lichens, *In*: *Advances in Quantitative Phytogeography* (P.L.Nimis and T. Crovello, eds): 233-262, Kluwer, Dordrecht.

Galloway, D.J., 1991b, Biogeographical relationships of Pacific tropical lichen floras, *In*: *Tropical Lichens: Their Systematics, Conservation, and Ecology* (D.J. Galloway, ed.): 1-16, Clarendon Press, Oxford.

Galloway, D.J., 1991c, Chemical evolution in the order *Peltigerales*: triterpenoids, *Symbiosis* 11: 327-344.

Galloway, D.J., 1992, Studies in *Pseudocyphellaria* (lichens) III. The South American species, *Bibliotheca Lichenologica* 46: 1-275.

Galloway, D.J., and P.M. Jϕrgensen, 1987, Studies in the family *Pannariaceae* II: The genus *Leioderma* Nyl., *Lichenologist* 19: 345-400.

Grehan, J.R., 1988, Panbiogeography: evolution in space and time, *Rivista Biologia* 81: 469-498.

Grehan, J.R., 1989, New Zealand panbiogeography: past, present, and future, *New Zealand Journal of Zoology* 16: 513-525.

Grehan, J.R., 1991. A panbiogeographic perspective for pre-Cretaceous angiosperm-lepidoptera coevolution, *Australian Systematic Botany* 4: 91-110.

Hale, M.E., 1987, A monograph of the lichen genus *Parmelia* Acharius sensu stricto (*Ascomycotina: Parmeliaceae*), *Smithsonian Contributions to Botany* 66: 1-55.
Hale. M.E., 1990, A synopsis of the lichen genus *Xanthoparmelia* (Vainio) Hale (*Ascomycotina, Parmeliaceae*), *Smithsonian Contributions to Botany* 74: 1-250.
Hallbauer, D.K., H.M. Jahns, and H.A. Beltmann, 1977, Morphological and anatomical observations on some Precambrian plants from the Witwatersrand, South Africa, *Geologische Rundschau* 66: 477-491.
Hawksworth, D.L., 1978, The taxonomy of lichen-forming fungi: reflections on some fundamental problems, *In*: *Essays in Plant Taxonomy* (H.E. Street, ed.): 211-243, Academic Press, London.
Hawksworth, D.L., 1982, Co-evolution and the detection of ancestry in lichens, *Journal of the Hattori Botanical Laboratory* 52: 323-329.
Hawksworth, D.L., 1985, Problems and prospects in the systematics of the *Ascomycotina*, *Proceedings of the Indian Academy of Science (Plant Science)* 94: 319-339.
Hawksworth, D.L., 1988a, The variety of fungal-algal symbioses, their evolutionary significance, and the nature of lichens, *Botanical Journal of the Linnean Society* 96: 3-20.
Hawksworth, D.L., 1988b, Co-evolution of fungi with algae and cyanobacteria in lichen symbioses, *In*: *Co-evolution of Fungi with Plants and Animals* (K.A. Pirozynski and D.L. Hawksworth, eds): 125-148, Academic Press, London.
Hawksworth, D.L., 1991, The fungal dimension of biodiversity: magnitude, significance, and conservation, *Mycological Research* 95: 641-655.
Hawksworth, D.L., and D.J. Hill, 1984, *The Lichen-Forming Fungi*, Blackie, Glasgow and London.
Hay, A., 1992, Tribal and subtribal delimitation and circumscription of the genera of *Araceae* tribe *Lasieae*, *Annals of the Missouri Botanical Garden* 79: 184-205.
Heads, M., 1989, Integrating earth and life sciences in New Zealand natural history: the parallel arcs model, *New Zealand Journal of Zoology* 16: 549-585.
Heads, M., 1990, Mesozoic tectonics and the deconstruction of biogeography: a new model of Australasian biology, *Journal of Biogeography* 17: 223-225.
Heads, M., 1993, Biogeography and biodiversity in *Hebe*, a South Pacific genus of *Scrophulariaceae*, *Candollea* 48: 19-60.
Henderson, I.M., 1989, Quantitative panbiogeography: an investigation into concepts and methods, *New Zealand Journal of Zoology* 16: 495-510.
Henderson, I., 1991, Biogeography without area?, *Australian Systematic Botany* 4: 59-71.
Holtan-Hartwig, J., 1993, The lichen genus *Peltigera*, exclusive of the *P. canina* group, in Norway, *Sommerfeltia* 15: 1-77.
Honegger, R., 1991, Fungal evolution: symbiosis and morphogenesis, *In*: *Symbiosis as a Source of Evolutionary Innovation* (L. Margulis and R. Fester, eds): 319-340, MIT Press, Cambridge, Mass.
Honegger, R., 1992, Lichens: mycobiont-photobiont relationships, *In*: *Algae and Symbioses: Plants, Animals, Fungi, Viruses, Interactions Explored* (W. Reisser, ed.): 255-275, Biopress, Bristol.
Hooker, J.D., and T. Taylor, 1844, Lichenes antarctici . . . ., *London Journal of Botany* 3: 634-658.
Humphries, C.J., 1983, Biogeographical explanations and the southern beeches, *In*: *Evolution, Time and Space: the emergence of the biosphere* (R.W. Sims, J.H. Price and P.E.S. Whalley, eds): 335-365, Academic Press, London.
Humphries, C.J., J.M. Cox, and E.S. Nielsen, 1986, *Nothofagus*, and its parasites: a cladisitic approach to coevolution, *In*: *Coevolution and Systematics* (A.R. Stone and D.L. Hawksworth, eds): 55-76, Clarendon Press, Oxford.
Humphries, C.J., P.Y. Ladiges, M. Roos, and M. Zandee, 1988, Cladistic biogeography, *In*: *Analytical Biogeography* (A.A. Myers and P.S. Giller, eds): 371-404, Chapman and Hall, London.
Humphries, C.J., and L. Parenti, 1986, *Cladistic Biogeography*, Clarendon Press, Oxford.
James, P.W., and A. Henssen, 1976, The morphological and taxonomic significance of cephalodia, *In*: *Lichenology: Progress and Problems* (D.H. Brown, D.L. Hawksworth, and R.H. Bailey, eds): 27-77, Academic Press, London.
Jφrgensen, P.M., 1979, The phytogeographical relationships of the lichen flora of Tristan da Cunha (excluding Gough Island), *Canadian Journal of Botany* 57: 2279-2282.
Jφrgensen, P.M., 1983, Distribution patterns of lichens in the Pacific region, *Australian Journal of Botany, Supplement* 10: 43-66.
Jφrgensen, P.M., and P.W. James, 1990, Studies in the lichen family *Pannariaceae* IV: The genus *Degelia*, *Bibliotheca Lichenologica* 38: 253-276.
Kärnefelt, I., 1979, The brown fruticose species of *Cetraria*, *Opera Botanica* 46: 1-150.
Kärnefelt, I., 1989, Morphology and phylogeny in the *Teloschistales*, *Cryptogamic Botany* 1: 147-203.
Kärnefelt, I., 1990a, Isidiate taxa in the *Teloschistales* and their ecological and evolutionary significance, *Lichenologist* 22: 307-320.

Kärnefelt, I., 1990b, Evidence of a slow evolutionary change in the speciation of lichens, *Bibliotheca Lichenologica* 38: 291-306.
Kärnefelt, I., 1991, Evolutionary rates in the *Teloschistaceae*, *In*: *Tropical Lichens: their systematics, conservation and ecology* (D.J. Galloway, ed.): 105-121, Clarendon Press, Oxford.
Kärnefelt, I., J.-E. Mattson, and A. Thell, 1992, Evolution and phylogeny of cetrarioid lichens, *Plant Systematics and Evolution* 183: 113-160.
Kärnefelt, I., and A. Thell, 1992, The evaluation of characters in lichenized families, exemplified with the alectorioid and some parmelioid genera, *Plant Systematics and Evolution* 180: 181-204.
Kendrick, B., 1991, Fungal symbioses and evolutionary innovations, *In*: *Symbiosis as a Source of Evolutionary Innovation* (L. Margulis and R. Fester, eds): 249-260, MIT Press, Cambridge, Mass.
Khakina, L.N., 1992, *Concepts of Symbiogenesis. A historical and critical study of the research of Russian botanists*. Yale University Press, New Haven.
Korf, R.P., 1983, *Cyttaria* (*Cyttariales*): coevolution with *Nothofagus* and evolutionary relationships to the *Boedijnopezizeae* (*Pezizales*, *Sarcoscyphaceae*), *Australian Journal of Botany Supplement* 10: 77-87.
Lamb, I.M., 1947, A monograph of the lichen genus *Placopsis* Nyl., *Lilloa* 13: 151-288.
Lindsay, W.L., 1856, *A Popular History of British Lichens*, Lovell Reeve, London.
Markham, K.R., and L.J. Porter, 1978, Chemical constituents of the bryophytes, *Progress in Phytochemistry* 5: 181-272.
Mattson, J.-E., and H.T. Lumbsch, 1989, The use of the species pair concept in lichen taxonomy, *Taxon* 38: 238-241.
Michaux, B., 1989, Generalized tracks and geology, *Systematic Zoology* 38: 390-398.
Nylander, W.L., 1858, *Synopsis Methodica Lichenum*. Vol. 1, Martinet, Paris.
Ourisson, G., M. Rohmer, and R. Anton, 1979, From terpenes to sterols: macroevolution and microevolution, *Recent Advances in Phytochemistry* 24: 283-327.
Page, R.D.M., 1989, New Zealand and the new biogeography, *New Zealand Journal of Zoology* 16: 471-483.
Pirozynski, K.A., 1983, Pacific mycogeography: an appraisal, *Australian Journal of Botany Supplement* 10: 137-159.
Pirozynski, K.A., and L.K. Weresub, 1979, A biogeographic view of the history of ascomycetes and the development of their pleomorphism, *In*: *The Whole Fungus* (B. Kendrick, ed.) 1: 93-125, National Museums of Canada, Ottawa.
Schopf, J.W., and M.R. Walter, 1982, Origin and early evolution of cyanobacteria: the geological evidence, *Botanical Monographs* 19: 543-564.
Sheard, J.W., 1977, Palaeogeography, chemistry and taxonomy of the lichenized ascomycetes *Dimelaeana* and *Thamnolia*, *Bryologist* 80: 100-118.
Sherwood-Pike, M.A., 1985, *Pelicothallus* Dilcher, an overlooked fossil lichen, *Lichenologist* 17: 114-115.
Sherwood-Pike, M.A., and J. Gray, 1985, Silurian fungal remains: probable records of the class *Ascomycetes*, *Lethaia* 18: 1-20,
Sipman, H.J.M., 1992, The origin of the lichen flora of the Colombian paramos, *In*: *Paramo, an Andean Ecosystem under Human Influence* (H. Balsev and J.L.Luteyn, eds): 95-109, Academic Press, London.
Smith, A.L., 1921, *Lichens*, Cambridge University Press, Cambridge.
Smith, C.H., 1989, Historical biogeography: geography as evolution, evolution as geography, *New Zealand Journal of Zoology* 16: 773-785.
Smith, D.C., 1978, What can lichens tell us about real fungi? *Mycologia* 70: 915-934.
Smith, D.C., and A.E. Douglas, 1987, *The Biology of Symbiosis*, Edward Arnold, London.
Stenroos, S., 1991, The lichen genera *Parmelia* and *Punctelia* in Tierra del Fuego, *Annales Botanici Fennici* 28: 241-245.
Stenroos, S., and T. Ahti, 1990, The lichen family *Cladoniaceae* in Tierra del Fuego: problematic or otherwise noteworthy taxa, *Annales Botanici Fennici* 27: 317- 327.
Tehler, A., 1983, The genera *Dirina* and *Roccellina* (*Roccellaceae*), *Opera Botanica* 70: 1-86.
Tehler, A., 1988, A cladistic outline of the *Eumycota*, *Cladistics* 4: 227-277.
Thor, G., 1990, The lichen genus *Chiodecton* and five allied genera, *Opera Botanica* 103: 1-92.
Wallace, A.R., 1855, On the law which has regulated the introduction of new species, *Annals and Magazine of Natural History, ser.* 2, 16: 184-196.
Watson, W., 1759, An historical memoir concerning a genus of plants called *Lichen* by Micheli, Haller, and Linnaeus; and comprehended by Dillenius under the terms *Usnea*, *Coralloides*, and *Lichenoides*: tending principally to illustrate their several uses, *Philosophical Transactions of the Royal Society* 50: 652-688.

White, F.J., and P.W. James, 1988, Studies on the genus *Nephroma* II. The southern temperate species, *Lichenologist* 20: 103-166.
Wilson, J.B., 1991, A comparison of biogeographic models: migration, vicariance and panbiogeography, *Global Ecology and Biogeography Letters* 1: 84-87.

CLADISTIC ANALYSIS IN ASCOMYCETE SYSTEMATICS: THEORY AND PRACTICE

A. Tehler

Botaniska Institutionen
Stockholms Universitet
S-106 91 Stockholm, Sweden

SUMMARY

The theory and principle of cladistic methodology are summarized, and exemplified by applied cladistics in ascomycete systematics and classifications. General topics touched upon are: competing systematic methods, monophyly, paraphyly, terminal taxa, character coding, ingroup, outgroup, character polarization, parsimony, character weighting, equally parsimonious trees, problems with tree instability, consensus trees, and classification.

INTRODUCTION

This contribution will not attempt to give a full coverage of modern cladistics. Nor will it report any new taxonomic or systematic results in ascomycete systematics. Rather it is an attempt to give a brief and plain overview of the basic principles and working procedures in practising cladistic methodology. In spite of many defeatist comments, cladistics is just as applicable in mycology as in any other discipline of systematics. My hope is to inspire mycological workers to adopt the methodology by showing that the basic idea of cladistics, i.e. grouping *only* on nested sets of derived characters, is intriguingly simple and totally convincing once realized. In this attempt I will use some of my previous results to exemplify the method and the reasoning behind it.

Fungal cladistics is still in its bud, although cladistic methods for reconstructing phylogenies has become normal practice in many other fields of systematics such as phanerogamic botany, bryology, zoology, paleozoology, and paleobotany. The first phylogenetic hypothesis of the *Eumycota* based on cladistic methods was proposed by myself (Tehler, 1988). On a morphological basis my outline is cladistically still unchallenged although five years have passed since its publication. However, on a molecular basis some outline phylogenies inferred from nuclear small subunit rRNA data have been published recently (Bruns et al., 1991; Bruns et al., 1992). A wide systematic amplitude of morphologically well-studied fungi is of course essential when inferring phylogenetic relationships at higher taxonomic levels. But the best systematic and evolutionary studies are those based on many types of data and one of the most powerful uses of molecular data is that it presents a way to test hypotheses based on morphological data. Congruence in two such data sets should give more reliable and well-corroborated hypotheses for the recognition of monophyletic taxa.

For example, the congruity of major clades is considerable between my (Tehler, 1988) morphological, most parsimonious tree, and corresponding molecular, parsimonious trees by Bruns et al. (Bruns et al., 1991; Bruns et al., 1992). The monophyly of the *Eumycota* as well as the succession of major fungal groups is supported by molecular data, following my earlier (Tehler, 1988) classification: *Blastocladiomycota* as sister group to the rest of the *Eumycota*; *Chytridiomycota* as sister group to the *Amastigomycota*; *Zygomycotina* as sister group to the *Dicaryomycotina*; and *Ascomycetes* as sister group to the *Basidiomycetes*. The most conspicuous incongruity between the morphological and molecular trees (apart from the oddly placed *Mucor* with *Blastocladiella* and *Endogone* with the *Chytridiomycetes*) is the inclusion of the yeasts in a monophyletic group of *Ascomycetes* (Bruns et al., 1992). That hypothesis, which is supported by most traditional morphologists, is contrary to my tree where the *Ascomycetes* are monophyletic only by exclusion of the yeasts. Yet, this may be a consequence of unsampled taxa in the *Taphrinaceae* and various hemiascomycete groups (Bruns et al., 1992). Within the *Euascomycetes* the conflicts between the morphological tree and the molecular trees become more apparent (Tehler, 1988; Tehler, 1990; Berbee and Taylor, 1992; Gargas and Taylor, 1992; Tehler, 1994). Yet, much work remains to be done here, both in the morphological and molecular fields.

SYSTEMATIC METHODS

Presently, there are three different systematic approaches: cladistics, phenetics, and evolutionary (traditional) systematics. It is important to realize that the basic taxonomic work with character analyses and circumscription of species is, or should be independent of systematic approach. This is obvious, but needs to be emphasized as there is a common misconception that workers using cladistics (including those using phenetics, numerical taxonomy) do not study species and do not do character analyses. The species, the terminal taxon, is usually the basic element and the building stone in systematics, irrespective of systematic approach.

Cladistics

Cladistics is simply a method to be used in the search for monophyletic groups. Some basic assumptions must be accepted in working with cladistics (*cf.* Forey et al., 1992): Organisms form patterns of groups that can be hierarchically arranged. This hierarchy can be discovered by grouping the organisms on shared derived characters (synapomorphies), through character state changes (homology). *Ad hoc* hypotheses of character transformation are minimized by using the parsimony criterion. Cladograms are diagrams that represent the hierarchy, and branching episodes are absolutely decisive, therefore, a suggested classification must be in accordance with the cladogram. This means that in Fig. 1, taxon C must be placed with D-G although C would be phenetically more similar to A and B.

Each character used in the analysis has to be explicitly accounted for and character state homology has to be carefully evaluated. Character states are polarized into one primitive (plesiomorphic) state and one or more derived (apomorphic) states. In contrast to phenograms produced in phenetics (see below), cladograms can be translated into phylogenetic hypotheses because cladistics, in contrast to phenetics, supports the notion of common descent. Consequently, a proposed, phylogeny based on cladistics can be explained by evolutionary processes. Still, it is very important to realize that cladistics itself, i.e. finding the pattern, is independent of *a priori* assumptions about evolutionary processes. Only after a pattern has been proposed, questions about evolution and evolutionary processes can be asked to explain the pattern. The process of evolution can only be discerned through character distribution. The cladistic method is scientific in the sense that it can be repeated and the hypothesis be tested (falsified) by other workers by the Popperian falsification model.

Phenetics

As in cladistics, characters in phenetic, numerical taxonomy, approaches have to be explicitly accounted for, but in this case no polarization of the character states is made. Accordingly, groups are based also on primitive characters or non-characters or both, i.e. on overall percentage similarity; hence characters and taxa are not associated. In phenograms branching episodes are not decisive because groups are based on both absences and presences, which means that taxon C (Fig. 1) would be grouped with A and B apart from D-G because C would be more "similar" to A and B than to any of D-G. Phenetics has no logical, phylogenetic basis and thus cannot be used to generate hypotheses about evolution and so is useless in a phylogenetic context. As with cladistics, numerical taxonomy is scientific in the sense that it can be tested by applying the Popperian falsification model.

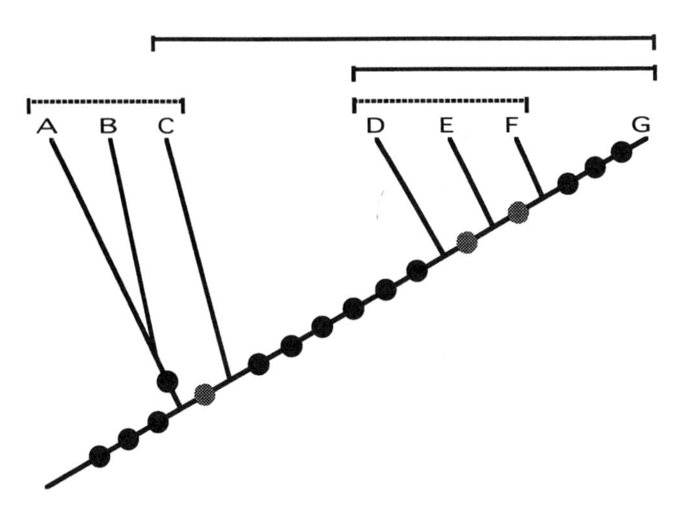

Figure 1. Created phylogeny illustrating monophyly, paraphyly, and polyphyly. Dots represent characters. Group C-G is monophyletic and group A-C is paraphyletic, although taxon C is phenetically more "similar" to A and B than to any of the taxa D-G. Likewise, Group D-G is monophyletic and D-F paraphyletic although the groups D-F are phenetically more "similar" to one another than any of them are to taxon G. Group C, D is polyphyletic.

Evolutionary (Traditional) Systematics

Phylogeny is reconstructed by successively incorporating evolutionary ideas during the process of systematic work. Only characters that are *a priori* considered evolutionary significant, or otherwise important, are selected by the worker on the group in question. The phylogenetic reconstruction is based on the authors' knowledge of the group and from which an evolutionary scenario is painted. As such it is based on intuition or *a priori* opinion of the phylogeny. It is unscientific in the sense that it cannot be tested by applying the Popperian falsification model.

Monophyly, Paraphyly and Polyphyly

A *monophyletic* group includes all members of a common ancestor as defined by a certain set of shared characters (synapomorphies). For example, groups A-B and C-G are both monophyletic (Fig. 1).

A *paraphyletic* group shares some characters (symplesiomorphies) of the monophyletic group but does not include all members. For example groups A-C and D-F are both paraphyletic.

A *polyphyletic* group is one based on analogous characters. A commonly used example to illustrate this is the presence of wings in both bats and birds.

CLADISTIC WORKING PROCEDURES

The idea of cladistics is simple and the practice is no more problematic than other systematic methods. However, for various reasons, many taxonomists have been deterred from cladistics. An explanation for this may be the technical language. Another explanation is perhaps the affected debate that took place some years ago. Fortunately, that debate is today more or less over and more and more workers turn to cladistics. In the following I will try to show the simplicity in cladistics, hoping it will inspire to, rather than deter from, further studies.

Study Group: The Normal Case

When beginning a cladistic study, an initial hypothesis of monophyly needs to be formulated for the study group. Such an hypothesis is based on one or more potential synapomorphies, and is normally called the "ingroup". The process of finding an ingroup is usually integrated with the choice of outgroup (see below). The two most common situations when starting a taxonomic or systematic work are:

(1) The study group is well-characterized, i.e. potentially monophyletic, but relationships within that group are unknown or poorly understood. Cladistic studies can start immediately.
(2) The study group is heterogeneous with no obvious monophyletic criteria. If this is the case, monophyletic units need to be searched for within the assemblage either by narrowing it down or enlarging it so that finally monophyly for a number of taxa can be postulated (*cf.* Tehler, 1988, 1990). For example, in the outset of my 1988 outline of the *Eumycota*, the obviously paraphyletic group "fungi" could be narrowed down to the proposedly monophyletic group chitinous fungi. Similarly, in my studies of the *Arthoniales* (Tehler, 1990) the larger group in which *Arthoniales* are included could be narrowed down to the potentially monophyletic but unresolved group, ascostromatic fungi. From here the *Arthoniales* could be postulated as being a monophyletic group on the basis of some potential synapomorphies, e.g. the hemiamyloid reaction of the endotunica and paraphysoidal hamathecium.

Terminal Taxa and Ingroup

As pointed out above, the definition and circumscription of taxa is a strictly taxonomic procedure and should be independent from any systematic approach. In cladistics, as in any systematic approach, the ideal terminal taxon, the working unit, is the species. Species are the building bricks in systematics because they can be conceived of as real and existing entities of nature; higher taxa are human abstractions.

Although not specifically a cladistic issue, the never-ending debate over the species concept will be touched upon briefly here. Despite the refined species concepts such as the biological and evolutionary species concepts, most botanists, and mycologists are in practice to a very large extent constrained to using a strictly morphological species concept. A very straight forward species concept is the one formulated by Nelson and Platnick (1981):

> "those samples that a biologist can distinguish, and tell others how to distinguish (diagnose), are called species"; and "species are simply the smallest detected samples of self-perpetuating organisms that have unique sets of characters."

The ultimate cladistic study would be to include "all" species and characters into one gigantic analysis. Yet, this is impracticable for several reasons, but principally because it is conceptually impossible to know when all species and characters are analyzed. Nevertheless, there is a need to bring monophyletic structure also to higher level systematics. To find higher level relationships we are compelled to select and to generalize characters and taxa, resulting in a phylogenetic experiment by using selected parts of a whole.

Characters and Character Coding

What is a character? A simple definition is the one by Nelson and Platnick (1981): any attribute of an organism by which we can distinguish samples. One of the most important parts of cladistic work is formulating hypotheses of character state homology. Any such homology statement requires an assumption of change, in which is implied evolutionary change. This is the only place in the cladistic process where such assumptions have to be made. Hence, in every character is included a hypotheses of character state change, and it is essential that each of these homology statements is carefully assessed. As a result of using a morphological species concept, hypotheses of homology are primarily based on observed similarity.

Binary Characters

Consider a character, *x*, with two character states: *a* <-> *b* (Fig. 2A). In the real world this could be translated into *x* = *spore septa*, *a* = *eseptate* and *b* = *1-septate*. The hypothesis is that the states *a* and *b* are homologous, i.e. *a* has transformed into *b* or *vice versa*. The direction of change is not known, but it is unnecessary at this stage. The character is simply coded as in the matrix (Fig. 3A), *eseptate* or *1-septate*, as applicable to the taxon. When binary characters are polarized transformation will be automatically established (Fig. 3C, D).

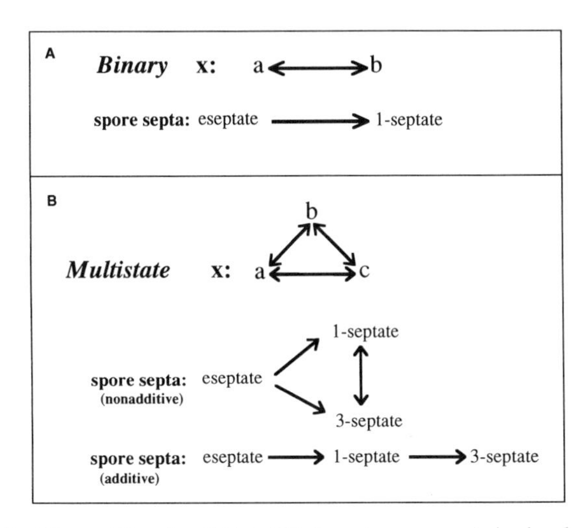

Figure 2. Binary (A) and multistate (B) characters, non-polarized and polarized.

Multistate Characters

Sometimes more than two character states are present in a character as for example the character describing *spore septa*, with the states *eseptate*, *1-septate* and *3-septate* (Fig. 2B). Such characters are called multistate characters. Sometimes the order can be determined by studying intermediate steps (ontogeny). Spore septum development in microcephalic spores, as in the example above, can be determined because each new septum divides a segment into equal parts. Consequently the character states can be included in the analysis in an additive (ordered), linear, transformation series, i.e. each character state includes the state that preceded it (Fig. 2B). But it is not necessary to know the ontogeny to set up an additive series. Consider for example the character *thalline exciple* and the transformation series: 0 *circular* → 1 *undulating* → 2 *undulating* and *verrucose* → 3 *undulating*, *verrucose* and *pruinose*. Here the transformation series is self-evident because each character state is obviously included in the state that preceded it. Additivity is a mathematical rationale and has nothing to do with evolution or *a priori* opinion about evolutionary direction. As with binary characters, additive series can be broken up and the character states be homoplasious. The resulting analysis will tell you if the series is additive or not.

Still, far from always can the order of transformation be determined. Consider the character *cortex structure* with the character states *anticlinally arranged*, *periclinally arranged*, or *interwovenly arranged* hyphae. Contrary to the previous examples, the character states in this case are not partitions of each other and must be left as non-additive (unordered), i.e. any state can go directly to any other state. Multistate characters can also be arranged in branching order.

		S. chloroleuca	S. leptothallina	S. glaucomoides	S. californica
1	thallus surface	rugose	even	verrucose	even
2	thalline margin cortex	poor or lacking	poor or lacking	well developed	well developed
3	medullary hyphae	smooth	granulose	granulose	smooth
4	ascomal shape	undulating	undulating	entire	entire
5	disc	flat	flat	convex	convex
6	thalline margin	in level	in level	in level	protruding
7	thalline margin algae	absent	present	present	present
8	epithecial hyphae	smooth	granulose	granulose	smooth

A

		S. chloroleuca	S. leptothallina	S. glaucomoides	S. californica
1	thallus surface	0	1	2	1
2	thalline margin cortex	0	0	1	1
3	medullary hyphae	1	0	0	1
4	ascomal shape	0	0	1	1
5	disc	0	0	1	1
6	thalline margin	1	1	1	0
7	thalline margin algae	0	1	1	1
8	epithecial hyphae	1	0	0	1

B

		Byssophoropsis	S. chloroleuca	S. leptothallina	S. glaucomoides	S. californica
1	thallus surface	rugose	rugose	even	verrucose	even
2	thalline margin cortex	poor or lacking	poor or lacking	poor or lacking	well developed	well developed
3	medullary hyphae	granulose	smooth	granulose	granulose	smooth
4	ascomal shape	undulating	undulating	undulating	entire	entire
5	disc	flat	flat	flat	convex	convex
6	thalline margin	protruding	in level	in level	in level	protruding
7	thalline margin algae	absent	absent	present	present	present
8	epithecial hyphae	granulose	smooth	granulose	granulose	smooth

C

		Byssophoropsis	S. chloroleuca	S. leptothallina	S. glaucomoides	S. californica
1	thallus surface	0	0	1	2	1
2	thalline margin cortex	0	0	0	1	1
3	medullary hyphae	0	1	0	0	1
4	ascomal shape	0	0	0	1	1
5	disc	0	0	0	1	1
6	thalline margin	0	1	1	1	0
7	thalline margin algae	0	0	1	1	1
8	epithecial hyphae	0	1	0	0	1

D

Figure 3. Character coding and outgroup selection (data from Tehler, 1993c). A, Matrix coded with character state names, when character state homology is determined for each character these are coded to the appropriate taxon, outgroup here excluded; B, same as A but character state names substituted for 0, 1 and 2; C, outgroup included in matrix with character state names; D, same as C but character state names substituted for 0, 1 and 2. By convention 0 is given to the plesiomorphic outgroup state and 1, 2, 3, etc. to apomorphic ingroup states.

Additive Binary Coding: The character states of a linear transformation series can be broken down to correlated binary characters, and then coded as independent characters.

Some characters may be inapplicable, i.e. a character state cannot be coded if the character itself is absent. For example no character state of the cortex structure can be coded if a cortex is not present.

When character state homology has been determined for each character, these are coded to the appropriate taxon into matrix form (Fig. 3A, B).

```
[                   180       190       200       210       220       230  ]
[                   .         .         .         .         .         .    ]
Saccharomyces  ACATGCTAAAAACCCCGACNNTTCGGAAGGGGTGTATTTATTAGATAAAAAACCAATGC
Arthonia       ACATGCGAAAAACCCCGACNNTTCGGAAGGGGTGTATTTATTAGATAAAAAGCCAACGC
Schismatomma   ACATGCTAAAAACCTCGACNNTTCGGAAGGGGTGTATTTATTAGATAAAAAGCCAATGC
Lecanactis     ACATGCCAAAAACCCCGACNNTTCGGAAGGGGTGTATTTATTAGATAAAAAGCCAACGC
Dendrographa   ACATGCGAAAAACCCCGACNNTTCGGAAGGGGTGTATTTATTAGATAAAAAGCCAACGC
Sphaerophorus  ACATGCTTAAAATCTCGACCCTTTGGAAGAGATGTATTTATTAGATAAAAAATCAATGT
Lecidea        ACATGCTAAAAACCCCGACNNTTCGGAAGGGGTGTATTTATTAGATTAAAAACCAATGC
Neurospora     ACATGCTAAAAACCTCGACNNTTCGGAAGGGGTGTATTTATTAGATAAAAAACCAATGC
```

Figure 4. Portion of a data matrix of nucleotide sequence data (data from Tehler, 1994).

Polarization

The determination of character polarity refers to the direction of character state change. The character is polarized when the state that preceded all the other states is determined (Fig. 3C). By far the most commonly used procedure for determining character state polarity is by outgroup comparison (Watrous and Wheeler, 1981). They define outgroup comparison as: "For a given character with two or more states within a group, the state occurring in related groups is assumed to be the plesiomorphic state". Hence, the outgroup state is by definition primitive. It is important to realize that both character states have to be present in the ingroup otherwise the character cannot be polarized. It is also important to note that, even if the character has been polarized, the direction of change in multistate character states is still undetermined (see multistate characters above). By convention, 0 is given to the plesiomorphic outgroup state and 1, 2, 3, etc. to apomorphic ingroup states (Fig. 3D).

In sequence data aligned positions are used as characters and each nucleotide, adenine, cytosine, guanine, thymine (A, C, G, T) as character states (Fig. 4).

Choice of Outgroup

The usual case when searching for an outgroup is that no phylogenetic (cladistic) hypotheses are available for the larger group in which the ingroup is included, and from which an outgroup can be chosen. Therefore, taxa that due to some potential synapomorphies can be considered related to the ingroup have to be considered. This can be exemplified by the choice of *Oomycota* as outgroup in my cladistic analysis of the *Eumycota* (Tehler, 1988), and the choice of *Cookella* for the *Arthoniales* analysis (Tehler, 1990).

Those analyses (Tehler, 1988, 1990) were a great help in forming a basis for further euascomycete and *Arthoniales* analyses (Tehler, 1993a, b, c; 1994). For example, my taxonomic studies of the formerly heterogeneous *Schismatomma* confirmed that it could to be divided into four monophyletic genera (Tehler, 1993a), but the relationships of these genera to other *Arthoniales* taxa were unclear. I did a reassessment of the *Arthoniales* analysis so that terminal taxa were represented by genera, and then included the four new genera into a new data set (Tehler, 1993a). Now, the choice of outgroup was clear, namely *Arthothelium*, the sister group from the earlier *Arthoniales* analysis (Tehler, 1990). In turn the resulting cladogram (Fig. 5) formed the basis for choosing outgroups when the genera *Schismatomma* and *Sigridea* were analyzed (Tehler, 1993b, c).

Parsimony

When establishing hypotheses of relationships, any systematist is inevitably confronted with contradictory features, so-called homoplasies. To maintain a proposed hypothesis of relationship, any such homoplasy has to be excused by inferring an *ad hoc* hypothesis, either with a parallelism or a reversal. *Ad hoc* hypotheses must be avoided because "Science requires that choice among theories be decided by evidence, and the effect of an *ad hoc* hypothesis is precisely to dispose of an observation that otherwise would provide evidence against a theory" (Farris, 1983). For any particular analysis, homoplasies are thus misinterpreted homologies. The objective with parsimony is to minimize *ad hoc* hypotheses, i.e. distribute informative characters in relation to a respective taxon so that as few as possible contradictions, homoplasies, are retained on the cladogram. Obviously, the shortest tree(s), i.e. the most parsimonious solution, is the simplest explanation and thus, is the hypothesis that must be accepted. A common objection heard against parsimony is that nature is not parsimonious. Yet, parsimony is not an evolutionary rationale, but a scientific criterion to choose among hypotheses.

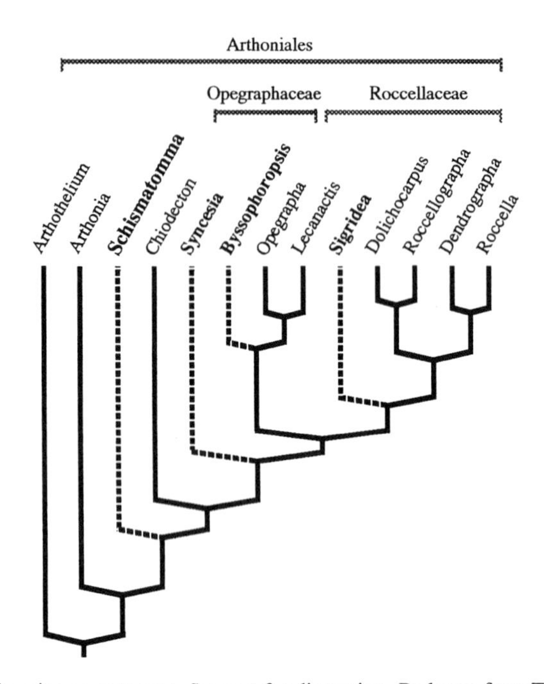

Figure 5. Choosing an outgroup. See text for discussion. Redrawn from Tehler (1993a).

Computers have to be used to process the data and when calculating the trees. The two most widely used desk-top computer programs for phylogenetic analysis are Hennig86 (Farris, 1988) for IBM compatibles, and PAUP (Swofford, 1991) for Macintosh.

The two most commonly used parsimony criteria are:

(1) *Wagner parsimony* (Farris, 1970) that uses additive (ordered) character state change through intervening character states in a transformation series.
(2) *Fitch parsimony* (Fitch, 1971) that uses non additive (unordered) character state changes and any state can transform directly into any other state.

Two other algorithms, not much used because the difficulty to argue for the constraints imposed, are: *Dollo parsimony* (Farris, 1977) in which parallelisms are not allowed meaning that each synapomorphy is uniquely derived and: *Camin-Sokal parsimony* (Camin and Sokal, 1965) in which character evolution is said to be irreversible, i.e. that character state change cannot go from a more derived state to a less derived state.

All characters are weighted, even though given equal weight. If a character for some reason is considered more important it can be given more weight than other characters. Still, it is very difficult to argue convincingly for an *a priori* weighting procedure. An *a posteriori* weighting procedure is *successive weighting* (Farris, 1988). Here, characters showing no or only little homoplasy from an initial analysis with characters equally weighted, are given more weight in successive analyses.

In searching for the most parsimonious tree the exact algorithm guarantees to find the shortest tree because all trees in the given data are considered. Unfortunately, it can handle only small matrices of around 10 taxa, maximally 20 depending on the structure of the data matrix. As the matrices become larger, the time needed to process the data becomes impractical. Therefore, a more generalized, heuristic, algorithm has to be employed for large data sets. All trees cannot be evaluated under this algorithm, which means that the shortest tree may not necessarily be found.

Consensus Trees

In cladistic analysis, multiple, equally parsimonious cladograms are often obtained from one data set. Therefore, there may be a problem with competing hypotheses since each cladogram is equally good as an hypothesis of the phylogeny of the investigated group. Carpenter (1988) argued that any of the competing cladograms should be preferred over a consensus tree because any one of them contains more information than a consensus tree. But, considering that there are no scientific bases for choosing among equally parsimonious trees, a consensus tree may nevertheless be proposed and used for classificatory purposes (Anderberg and Tehler, 1990). A consensus tree summarizes two or more equally parsimonious trees.

The most widely used consensus method is *strict consensus*. The strict consensus shows only the groups, components, that are constantly present in all the equally parsimonious trees. Inconsistent groups are collapsed into tri- or polychotomies. In *Combinable component consensus*, groups are included that are not in conflict with other groups. In the *Adams consensus* conflicting taxa are removed to the node common to all trees. The remaining group will then be included if it is present also in the other cladograms.

Tree Stability

To examine the stability and robustness of phylogenetic trees produced by the parsimony analyses, additional analyses and statistical treatments of the data sets have been developed. Some computer programs, PAUP (Swofford, 1991), MacClade (Maddison and Maddison, 1992), and Kara and Arn (Farris, unpubl. programs), may be used to calculate the data.

In the *Bremer support* (Källersjö et al., 1992), trees longer than the shortest are successively kept until groups are lost in consensus (Bremer, 1988; Fig. 6). Hence, the larger length of trees added before groups are lost the stronger evidence for support. Bremer's length difference is also referred to as the *decay test* or *decay index* (Donoghue et al., 1992).

Resampling of data by using bootstrap analysis (Felsenstein, 1985) is claimed to test tree stability. In a bootstrap analysis character rows in the real data set are usually randomly substituted with other character rows in the same data set. The new data set is analysed to give new trees. This procedure is repeated a number of times, and the percentage of the presence of a particular group in the sampled data is considered an index of support. Bootstrap is said to place confidence estimates on groups contained in the most parsimonious cladograms. However, it has been pointed out that this is not true confidence limits in a statistical sense (Forey et al., 1992).

Tree length distribution has been suggested to provide an indication of the presence of a phylogenetic signal (Hillis, 1991). The more left-skewed, that is the further away the most parsimonious tree is from the mean value of the tree lengths, the more phylogenetically structured is the data. But, the value of the skewness criterion as a phylogenetic marker has been questioned (Källersjö et al., 1992).

Permutation and total support tests. In the *permutation test*, following Faith and Cranston's (1991) permutation tail probability, the shortest tree obtained from real data is compared with a number of shortest trees obtained from permutated, randomized data of the same data set, to see if it is significantly shorter than most of the trees obtained from permutated data. *Total support* is the sum of group supports (Källersjö et al., 1992). The calculation procedure, and inference drawn from the result, is similar to that above but total support gives the support for the hierarchic structure in the data as a whole, *cf.* Bremer support above.

Congruence between data sets can be tested using the Mickevich and Farris (1981) method. Here the shortest tree from real data is calculated for each data set separately. Then the shortest tree from these two data sets combined is calculated. Thereafter, the characters in the data sets are shuffled and the same procedure is carried out again for a number of times. If the value obtained from real data is larger than that obtained from most of the allocated data, then the data sets are congruent.

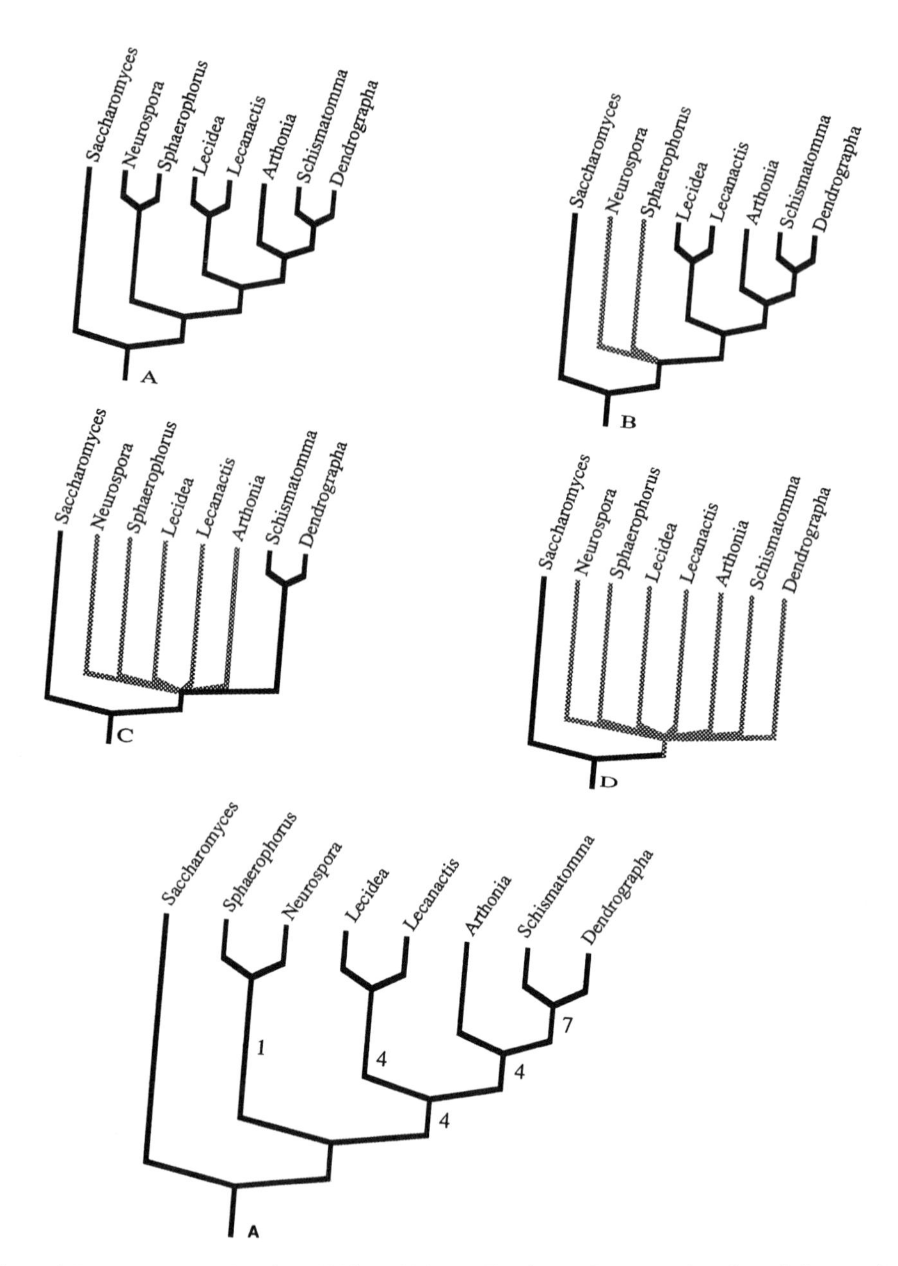

Figure 6. Bremer support (data from Tehler, 1994). A, Topology of most parsimonious cladogram, length of 131 steps; B, strict and combinable component consensus trees from two and three trees 132-134 steps and shorter; C, strict and combinable component consensus from nine trees 135-137 steps and shorter; D, all nodes collapse in trees starting from 138 steps; E, most parsimonious topology with Bremer support indices at each node.

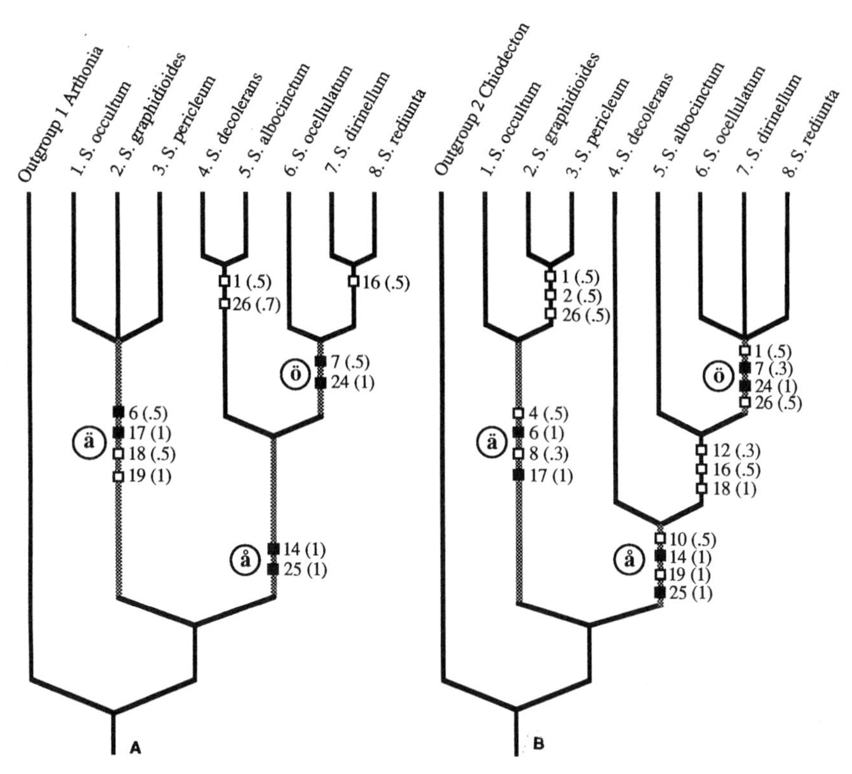

Figure 7. Two consensus trees based on the same data matrix but with different outgroups (data from Tehler, 1993b). Characters plotted are those that are common to nodes in all equally parsimonious cladograms in each analysis, respectively. A consistency index for each character is shown within brackets. Nodes á, ä and ö are common to both trees; A, Strict consensus tree from 3 equally parsimonious cladograms using *Arthonia radiata* as outgroup (characters symbolized by filled boxes are those that are common to the opposite consensus tree); B, strict consensus tree from 3 equally parsimonious cladograms using outgroup 2, *Chiodecton sphaerale* (characters symbolized by filled boxes are those that are common to the opposite consensus tree).

A different method was practised by myself recently (Tehler, 1993b; Fig. 7). I compared two consensus trees based on the same data matrix, but calculated using different outgroups. I plotted characters that were common to nodes in all equally parsimonious cladograms in each analysis respectively. I then checked for common groups in the two consensus trees as well as characters common to the nodes of these groups. If the same groups, characterized by the same features, appear in the two consensus trees this will be considered as evidence for support and that the phylogenetic hypothesis is well corroborated.

CLASSIFICATION

Classification is grouping and the naming of groups. It is hierarchical and should reflect phylogeny in terms of natural groups. There is little controversy over this. However, there are different opinions between cladists and evolutionary systematists on the definition of a natural group. Many evolutionary systematists are willing to accept a classification with unnatural groups even if they simultaneously present a tree showing otherwise (see Fig. 1). For example non-groups such as *Mastigomycotina* and *Monokaryomycota* are retained on a traditional basis. To a cladist this is unacceptable. Cladistically, the definition of a natural group is completely clear; namely equivalent to a monophyletic group (see above). The classification must follow the cladogram. The original suggestion by Hennig (1966) was that sister groups must have equal rank because otherwise paraphyletic groups are created.

CONCLUSION

The accomplishment of cladism was encapsulated quite well by Hull (1989):

". . . young workers, no matter how profoundly they may disagree with the old codgers who preceded them, now take for granted that they must master cladistic methods. More importantly, cladistics is winning because biologists other than systematists are finding the results of cladistic analysis useful in their own pursuits, for example, in disentangling the effects of history and of natural processes in the living world."

ACKNOWLEDGEMENT

I would like to thank Dr A. Anderberg (Naturhistoriska Riksmuseet) for his constructive and useful suggestions.

REFERENCES

Anderberg, A., and A. Tehler, 1990, Consensus trees, a taxonomic necessity, *Cladistics* 6: 399-492.

Berbee, M., and J.W. Taylor, 1992, Convergent ascospore dispersal among pyrenomycete fungi inferred from 18S ribosomal DNA sequence, *Molecular Phylogenetics and Evolution* 1: 274-284.

Bremer, K., 1988, The limits of amino-acid sequence data in angiosperm phylogenetic reconstruction, *Evolution* 42: 795-803.

Bruns, T.D., R. Vilgalys, S.M. Barns, D. Gonzalez, D.S. Hibett, J.D. Lane, L. Simon, S. Stickel, T.M. Szaro, W.G. Weisburg, and M.L. Sogin, 1992, Evolutionary relationships within the fungi: analyses of nuclear small subunit rRNA sequences, *Molecular Phylogenetics and Evolution* 1: 231-241.

Bruns, T.D., T.J. White, and J.W. Taylor, 1991, Fungal molecular systematics, *Annual Review of Ecology and Systematics* 22: 525-564.

Camin, J.H., and R.R. Sokal, 1965, A method for deducing branching sequences in phylogeny, *Evolution* 19: 311-326.

Carpenter, J.M., 1988, Choosing among multiple equally parsimonious cladograms, *Cladistics* 4: 291-296.

Donoghue, M.J., R.G. Olmstead, J.F. Smith, and J.D. Palmer, 1992, Phylogenetic relationships of *Dipsacales* based on rbcL. sequences, *Annals of the Missouri Botanical Garden*: in press.

Faith, D., and P. Cranston, 1991, Could a cladogram this short have arisen by chance alone?, *Cladistics* 7: 1-28.

Farris, J.S., 1970, Methods for computing Wagner trees, *Systematic Zoology* 19: 83-92.

Farris, J.S., 1977, Phylogenetic analysis under Dollo's law, *Systematic Zoology* 26: 77-88.

Farris, J.S., 1983, The logical basis for phylogenetic analysis. *In*: *Advances in Cladistics* (N. Platnick and V. Funk, eds). Colombia University Press, New York.

Farris, J.S., 1988, *Hennig86, Version 1.5. Computer Program and Reference Manual*, Stony Brook.

Felsenstein, J., 1985, Confidence limits on phylogenetics: an approach using the bootstrap, *Evolution* 39: 783-791.

Fitch, W.M., 1971, Toward defining the course of evolution: minimum change for a specified tree topology, *Systematic Zoology* 20: 406-416.

Forey, P.L., C.J. Humphries, I.J. Kitching, R.W. Scotland, D.J. Siebert, and D.M. Williams, 1992, *Cladistics. A Practical Course in Systematics*, Clarendon Press, Oxford.

Gargas, A., and J.W. Taylor, 1992, Molecular systematics of lichenized and non-lichenized fungi based on rDNA sequences. *In*: *Second International Symposium IAL 2. Abstracts* (I. Kärnefelt, ed.). Lunds Universitet, Lund.

Hennig, B., 1966, *Phylogenetic Systematics*. [transl. D.D. Davis and R. Zangerl.] University of Illinois Press, Urbana.

Hillis, D.M., 1991, Discriminating between phylogenetic signal and random noise in DNA sequences. *In*: *Phylogenetic Analysis of DNA Sequences* (M.M. Miyamoto and J. Cracraft, eds): 278-294. Oxford University Press, Oxford.

Hull, D.L., 1989, The evolution of phylogenetic systematics. *In*: *The Hierarchy of Life* (B. Fernholm, K. Bremer and H. Jörnvall, eds). Elsevier Science Publishers, Amsterdam.

Källersjö, M., S.M. Farris, A.G. Kluge, and C. Bult, 1992, Skewness and permutation, *Cladistics* 8: 275-287.

Maddison, W.P., and D.R. Maddison, 1992, *MacClade: Analysis of Phylogeny and Character Evolution. Version 3.0*, Sinauer Associates, Sunderland, Mass.

Mickevich, M.F., and J.S. Farris, 1981, The implications of congruence in *Menidia*, *Systematic Zoology* 30: 351-370.
Nelson, G., and N. Platnick, 1981, *Systematics and Biogeography, Cladistics and Vicariance*, Columbia University Press, New York.
Swofford, D.L., 1991, *PAUP: Phylogenetic Analysis Using Parsimony. Version 3.0*, Illinois Natural History Survey, Champaign.
Tehler, A., 1988, A cladistic outline of the *Eumycota*, *Cladistics* 4: 227-277.
Tehler, A., 1990, A new approach to the phylogeny of *Euascomycetes* with a cladistic outline of *Arthoniales* focussing on *Roccellaceae*, *Canadian Journal of Botany* 68: 2458-2492.
Tehler, A., 1993a, *Schismatomma* and three new or reinstated genera, a reassessment of generic relationships in *Arthoniales*, *Cryptogamic Botany* 3: in press.
Tehler, A., 1993b, The genus *Schismatomma* (*Arthoniales*, *Euascomycetidae*), *Opera Botanica*: in press.
Tehler, A., 1993c, The genus *Sigridea* (*Roccellaceae*, *Arthoniales*, *Euascomycetidae*), *Nova Hedwigia*: in press.
Tehler, A., 1994, *Arthoniales* phylogeny as indicated by morphological and rDNA sequence data. *Cryptogamic Botany*: in press.
Watrous, L.E., and Q.D. Wheeler, 1981, The outgroup comparison method of character analysis, *Systematic Zoology* 30: 1-11.

DISCUSSION

Taylor, J.: As a non-specialist, I find cladograms such as the ones shown very helpful. These provide a visual hypothesis that focuses our attention on what should be done next. I wish to encourage specialists to use cladistics to show all of us both characters and alternative trees.

Molecular Systematics

MOLECULAR EVOLUTION OF ASCOMYCETE FUNGI: PHYLOGENY AND CONFLICT

J.W. Taylor[1], E.C. Swann[2], and M.L. Berbee[3]

[1]Department of Plant Biology
111 Koshland Hall
University of California
Berkeley, California 94720, USA

[2]Department of Plant Biology
University of Minnesota
St Paul, Minnesota 55108, USA

[3]Department of Botany
University of British Columbia
Vancouver, British Columbia V6T 2B1, Canada

SUMMARY

Phylogenetic analysis of 18S ribosomal RNA genes puts *Chytridiomycota*, *Zygomycota*, *Ascomycota*, and *Basidiomycota* in the fungal kingdom. Fungi are part of the terminal eukaryotic radiation, and have animals plus choanogflagellates as closest relatives. Ascomycetes show an early radiation including *Taphrina* and *Schizosaccharomyces*, followed by a split leading to either budding yeasts or hyphal forms. Hyphal ascomycetes show a radiation of apothecial and loculoascomycete taxa together with the progenitors of two well-supported classes, *Pyrenomycetes* and *Plectomycetes*. *Pneumocystis carinii* provides an example of how conflicts between phylogenies inferred from morphological and molecular characters can be resolved. Molecular characters support its placement near the divergence of *Ascomycota* and *Basidiomycota* and its meiospore characters seem ascomycetous.

INTRODUCTION

The techniques of molecular biology, particularly DNA sequencing, combined with cladistics (Hennig, 1966), are having a profound effect on microbial phylogenetics. Having been first applied to the sequencing of prokaryotic ribosomal RNAs (cf. Woese et al., 1990), these technique were brought to fungi (e.g. Walker and Doolittle, 1982; Gottschalk and Blanz, 1985; Sogin, 1989), and became widespread with the advent of the polymerase chain reaction (White et al., 1990). So many data are now available that questions about the extent of the kingdom *Fungi* and the relation of this kingdom to other organisms can be addressed. These data can also be used to examine the validity of, and relationships among, higher taxa of fungi, in particular, classes of *Ascomycota* (Berbee and Taylor, 1992a). In this chapter ascomycete phylogeny will be discussed, as will methods for resolving conflicts that have arisen between phylogenetic histories inferred from morphological and molecular data.

Ascomycete Systematics: Problems and Perspectives in the Nineties
Edited by D.L. Hawksworth, Plenum Press, New York, 1994

THE PLACE OF FUNGI IN ALL BIOTA

Ecological considerations helped to move the fungi from part of the plant kingdom to their own kingdom (Whittaker, 1969). Phylogenetic analyses based on nucleotide sequence of nuclear small subunit ribosomal RNA genes (18S rDNA) have shown that the fungal kingdom is part of the terminal radiation of great eukaryotic groups (Fig. 1), which occurred about one billion years ago (Wainright et al., 1993; Sogin, 1989; Bruns et al., 1991). Among these kingdoms of eukaryotes, which include the fungi, animals, and plants, it now appears that the closest relatives to fungi are animals (Wainwright et al., 1993; Hasegawa et al., 1993). Wainwright et al. (1993) used 18S rDNA to show that fungi are the sister group to a clade comprising animals plus choanoflagellates. The branch supporting the clade of fungi and animals plus choanoflagellates is found in 80% to 85% of bootstrapped data sets; this figure rises to 98% if *Acanthamoeba* is omitted from the analysis. Hasegawa et al. (1993) also found that fungi and animals are sister taxa. They used amino acid sequence of elongation factor, a protein involved in the translation of messenger RNA. A link between fungi and animals was not unexpected, based on comparisons of carbohydrate storage molecules (both use glycogen), structural polysaccharide deposited outside of the plasma membrane (chitin), and the mitochondrial codon UGA specifying tryptophan instead of termination (Cavalier-Smith, 1987). The human fascination about which kingdom is closest to animals is understandable, as is the excitement among mycologists upon seeing their organisms made more prominent, but it would be a shame if these thoughts obscured the most exciting aspect of the eukaryotic tree - that a rapid radiation of extremely successful eukaryotes began one billion years ago.

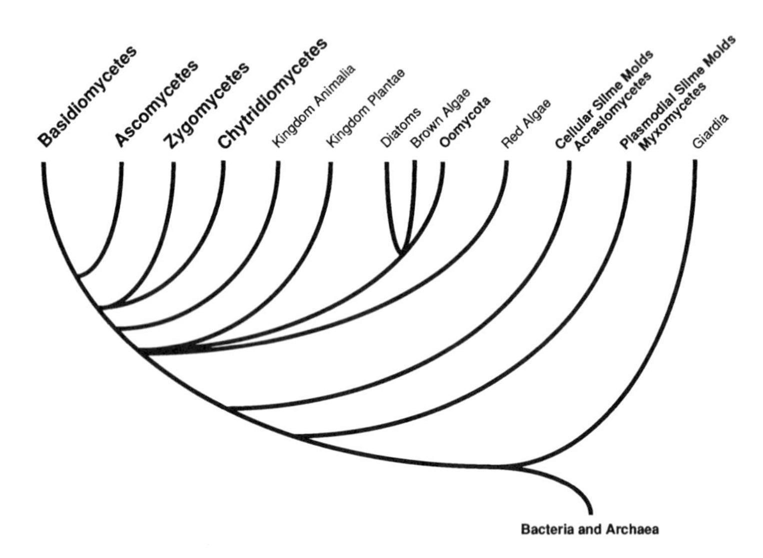

Figure 1. Phylogenetic tree showing relationships of eukaryotes, biased toward organisms traditionally studied by mycologists. This tree is a synthesis of trees based on 18S rDNA nucleotide sequences (Bruns et al., 1991, 1992; Bhattacharya et al., 1992; Sogin, 1989; Wainwright et al., 1993). The terminal radiation of eukaryotes includes organisms from the red algae to the fungi. *Giardia* represents an early diverging eukaryote (Sogin et al., 1989).

KINGDOM FUNGI

Sequence analysis of 18S rDNA has also helped to define the composition of the kingdom *Fungi* (Figs 1-2). Analysis of 18S rDNA sequence shows that four separate monophyletic groups include organisms claimed by mycologists: *Acrasiomycota*, *Myxomycota*, *Oomycota*, and *Fungi* (Bruns et al., 1991). In phylogenetic trees based on 18S rDNA, the two groups of slime molds diverged separately, prior to the terminal radiation of eukaryotes (Hasegawa et al., 1985; Sogin, 1989). This result is consistent with the many differences between slime molds and fungi in terms of form, function and life cycle. Similarly, 18S rDNA analysis of *Oomycota* shows these organisms to be part of the clade including brown algae, diatoms, and other algae containing chlorophylls a and c (Förster et al., 1990; Bhattacharya et al., 1992), again a result consistent with phenotypic characters (Barr, 1981).

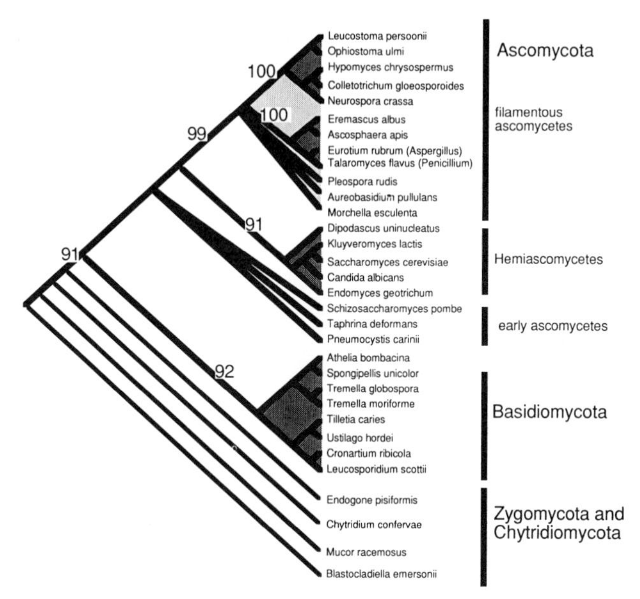

Figure 2. Phylogenetic tree based on parsimony analysis of 18S rDNA nucleotide sequence. The numbers above the branches are the percentage of bootstrapped data sets in which the branch appears; 95% and above is considered to be strong support (Felsenstein, 1985). Within the *Ascomycota*, there is an early divergence represented by the early ascomycetes and the progenators of the rest of the division. This divergence is followed by a bifurcation with branches leading to the *Hemiascomycetes* in one direction and filamentous ascomycetes with fruiting bodies in the other. In the filamentous ascomycetes, there is an early radiation of apothecial and fissitunicate ascomycetes plus the progenators of the *Plectomycetes* and *Pyrenomycetes*. Modified from Berbee and Taylor (1992b), where the sources of the 18S rDNA sequences can be found.

The monophyletic clade of fungi includes the *Chytridiomycota*, *Zygomycota*, *Ascomycota* and *Basidiomycota*. Inclusion of *Chytridiomycota* in the fungi has been controversial because these organisms possess flagella. On this basis, they have been included in the *Protoctista* (Margulis and Schwartz, 1988), but comparison of cell wall polysaccharide (Bartnicki-Garcia, 1970) and lysine synthesis (Vogel, 1964) linked them firmly with the *Zygomycota*, *Ascomycota* and *Basidiomycota*. Förster et al. (1990) showed that *Chytridiomycota* and *Oomycota* occupied separate clades. Bowman et al. (1992a) used neighbour joining analysis of bootstrapped 18S rDNA sequence data sets to show that a strongly supported branch separated (1) several chytridiomycetes and representatives of *Ascomycota* and *Basidiomycota* from (2) a ciliate protist and a plant (the outgroup). Subsequent analysis of many fungal 18S rDNA sequences (Bruns et al., 1992) showed that *Chytridiomycota* and *Zygomycota* are difficult to separate and are basal to the *Ascomycota* and *Basidiomycota*. Although *Chytridiomycota* form the basal branch in kingdom *Fungi*, flagella may have been lost more than once during the evolution of the ancestors to *Zygomycota*, *Ascomycota* and *Basidiomycota*. Among the nonflagellated fungi, the *Zygomycota* diverged first and the branch defining the *Ascomycota* and *Basidiomycota* as terminal sister groups is strongly supported (Bruns et al., 1992; Berbee and Taylor, 1992b).

ALTERNATIVE MOLECULAR PHYLOGENIES OF KINGDOM FUNGI

In contrast to analysis of 18S rDNA, comparison of nucleotide sequence of 5S rRNA (Hori and Osawa, 1987) or glyceraldehyde 3-phosphate dehydrogenase (Smith, 1989) results in a polyphyletic kingdom *Fungi*. With both of these molecules, it is likely that the variable positions are evolving too rapidly to be useful over the entire kingdom *Fungi*, as discussed in Bruns et al. (1991). Recently, Radford (1993) has compared the amino acid sequence of orotidine 5' monophosphate decarboxylase (OMPD) from representatives of *Zygomycota*, *Ascomycota* and *Basidiomycota*. Although the topology of the unrooted tree is congruent with that supported by 18S rDNA sequence, when Radford rooted the OMPD tree using animal sequence, the root fell within the *Ascomycota*. Morphologically, this is an unlikely place to root the fungal tree (cf. Tehler, 1988), and is probably due to the fact that OMPD amino acid substitution between animals and any fungus is *ca.* 50%. It is remarkable that the most parsimonious unrooted tree based on OMPD sequence is congruent with 18S rDNA sequence and morphology, given that sequence similarity of this molecule among *Ascomycota* is often less than 50%.

PHYLOGENETICS WITHIN THE ASCOMYCOTA

The recent trend in ascomycete classification has been to dispense with supraordinal taxa (Eriksson, 1982; Hawksworth et al., 1983). The traditional classes have been abandoned because the fruiting body characters used to define them are known to converge (Cain, 1972; Malloch, 1981) and can change state in different environments (von Arx, 1973). Although the traditional classes have fallen into disfavour, cladistic treatments of the *Ascomycota* (Tehler, 1990), and of all fungi (Tehler, 1988), have defined clades that bear a clear resemblance to some ascomycete classes (e.g. *Eurotiomycetidae*, *Saccharomycetaceae*, and bitunicate ascohymeniales plus ascolocular fungi are similar to *Plectomycetes*, *Hemiascomycetes* and *Loculoascomycetes*, respectively). This is not to say that these cladistic analyses have supported a traditional view of the *Ascomycota*, because they indicate that the *Ascomycota* is paraphyletic. In Tehler's analysis, the *Saccharomycetaceae* and *Taphrinaceae* are the sister group to the *Basidiomycota*, and the *Saccharomycetaceae*, *Taphrinaceae* and *Basidiomycota* together are the sister group to the rest of the *Ascomycota*.

Adding to this flux in ascomycete classification are nucleotide sequences of 18S rDNA from many taxa. From analysis of these sequences have come three well-supported clades of *Ascomycota* that closely resemble traditional classes (Fig. 2). The first is the *Hemiascomycetes*, fungi that typically grow as yeasts and produce unitunicate asci singly and without a fruiting body (Hendriks et al., 1992); the second is the *Plectomycetes*, hyphal fungi that typically produce unitunicate spherical asci that passively discharge their ascospores in closed fruiting bodies (Berbee and Taylor, 1992a); and the third is the *Pyrenomycetes*, hyphal fungi whose elongated unitunicate asci typically eject their spores through the open neck of the fruiting body (Berbee and Taylor, 1992a, c; Spatafora and Blackwell, 1992). The relationships among these three classes are also clear, the *Hemiascomycetes* diverge first, and a well-supported branch unites the two hyphal groups (Bruns et al., 1992; Berbee and Taylor, 1992b).

This simple picture is made more complex by adding fungi traditionally classified in the *Discomycetes* (apothecial ascomycetes) and *Loculoascomycetes* (herein called fissitunicate ascomycetes with the understanding that other types of asci may be found in this group), and by the discovery that some fungi thought to be *Hemiascomycetes* (*Schizosaccharomyces* and *Taphrina*) probably diverged very early in the history of ascomycetes, well before the divergence of the group now harbouring the hemiascomycete yeasts (Berbee and Taylor, 1992b; Bruns et al., 1992; Nishida and Sugiyama, 1993).

Early Diverging Ascomycetes

Although the branches uniting the *Ascomycota* and *Basidiomycota*, and the branch leading to the *Basidiomycota*, are strongly supported (as judged by parsimony analysis of bootstrapped 18S rDNA data sets; Bruns et al., 1992, Berbee and Taylor, 1992b), the branch leading to the *Ascomycota* is not (Fig. 2). This is due to the presence of *Taphrina*, *Pneumocystis* and *Schizosaccharomyces*, each of which has a most parsimonious position early on the ascomycete clade, but one without strong statistical support. With 18S rDNA sequence, statistical methods cannot be used to rule out placing these fungi at the base of the *Basidiomycota* or on the branch uniting *Ascomycota* and *Basidiomycota*. *Schizosaccharomyces* and *Taphrina* make ascospores in asci, which argues for their inclusion in the *Ascomycota*, but *Taphrina* does share character states with basidiomycetes (Savile, 1955; Tehler, 1988). No obvious shared derived (synapomorphic) phenotypic characters can be used to group *Schizosaccharomyces*, *Taphrina* and *Pneumocystis* into a single clade, nor can these fungi be united using 18S rDNA sequence. Because they represent deep divergences in the *Ascomycota*, they are equally distant from other ascomycetes; this feature has long been noted by developmental biologists working on *Saccharomyces* and *Schizosaccharomyces* (cf. Taylor et al., 1993). These fungi may be the surviving members of the first radiation of ascomycetes, a radiation that seems to have preceeded the divergence of the *Hemiascomycetes* and the filamentous ascomycetes.

Apothecial and Fissitunicate Ascomycetes

Analysis of 18S rDNA from representative apothecial species (Gargas, 1992) and two fissitunicate species (*Aureobasidium pullulans*, Illingworth et al., 1991 - we are assuming that this fungus is the asexual state of a fissitunicate ascomycete; and *Pleospora rudis*, Berbee and Taylor, 1992b), combined with that of other fungi, reveals a radiation of apothecial and fissitunicate forms that creates a polychotomy, which includes the branches leading to later radiations of *Pyrenomycetes* and *Plectomycetes* (Fig. 2). Although there is no support for a monophyletic clade of all fissitunicate ascomycetes, Spatafora et al. (1993) used nucleotide sequence of part of the 18S rDNA to show that several species of "black yeasts" (anamorphic taxa thought to have loculoascomycetous affinities) form a monophyletic clade similar to those formed by cleistothecial and perithecial ascomycetes.

In summary, the relationships of ascomycetes inferred from analysis of 18S rDNA suggest: (1) an early radiation represented by *Taphrina*, *Schizosaccharomyces* and *Pneumocystis*, followed shortly by (2) a bifurcation leading to *Hemiascomycetes* on one branch and filamentous fungi with fruiting bodies on the other. Among the hyphal fungi with fruiting bodies, there is (3) an early radiation comprising apothecial and fissitunicate forms and the progenitors of *Plectomycetes* and *Pyrenomycetes*, each of which subsequently radiated.

CONFLICTING PHYLOGENIES

Tehler brought cladistic methods to the phylogenetics of the *Ascomycota* (Tehler, 1990) and fungi as a whole (Tehler, 1988), building upon prior intuitive analyses (e.g. Bessey, 1950; Luttrell, 1955). In general, Tehler's results are in remarkable agreement with those of molecular studies, but it is the conflicts that are most interesting. These conflicts can be observed because 18S rDNA sequences have been determined for representatives of all of his ascomycetous clades except the *Dipodascaceae* (i.e. *Dipodascus albus*) and *Myxotrichum*. Two of Tehler's basal ascomycetous clades, *Endomycetaceae* (*Endomyces geotrichum*) and *Dipodascopsis* (*D. uninucleatus*), are part of the *Hemiascomycetes* clade in 18S rDNA analysis (Berbee and Taylor, 1993), and it seems likely that *D. albus* and *Myxotrichum* will be accommodated in *Hemiascomycetes* and *Plectomycetes*, respectively. Aside from these basal taxa, the main conflict concerns the placement of the *Hemiascomycetes*. Tehler's analysis makes them part of the sister group of the *Basidiomycota* while 18S rDNA analysis places them in the *Ascomycota*, subtended by a well-supported branch. Tehler uses two characters to unite *Hemiascomycetes* with *Basidiomycota*. The first character relies on a distinction between the sexual fusion of differentiated gametangia in *Chytridiomycota*, *Zygomycota* and *Ascomycota*, and of somatic cells in *Hemiascomycetes*, *Taphrina* and *Basidiomycota*. This character can be questioned on the grounds that gametangia, which are obvious in zygomycetes and filamentous ascomycetes (albeit with exceptions, i.e. *Neurospora tetrasperma*, Dodge, 1935), may be absent in yeasts simply as a consequence of the single cell growth form. Cells of *Saccharomyces cerevisiae* that are ready for sexual fusion have undergone profound cell biological changes following pheromone reception, and they have converted from vegetative cells to gametes in all ways except gross morphology (cf. Forsburg and

Nurse, 1991). The second character, direct budding of meiospores, remains as the sole character uniting *Hemiascomycetes*, early ascomycetes and *Basidiomycota*. It would require only one more step for this character to be ancestral to both the *Ascomycota* and *Basidiomycota*, and lost on the branch leading to filamentous ascomycetes with fruiting bodies.

Conflicts Among Molecular Characters

Conflicts between trees based on 18S rDNA sequence and those based on 5S rRNA or protein coding genes have already been discussed, and more can be expected when rapidly evolving molecules are used to address ancient divergences (Bruns et al., 1991). A recent example concerns the mitochondrial large subunit rRNA gene (mt 24S rDNA), which has been used to infer a close relationship between *Pneumocystis carinii* and the simple-septate, red basidiomycete yeasts (Wakefield et al., 1992). This result is in conflict with relationships inferred from analysis of 18S rDNA genes (van de Peer et al., 1992; Bruns et al., 1992; Berbee and Taylor, 1992b), which has led to an exchange of letters (Taylor and Bowman, 1993; Wakefield et al., 1993). This type of conflict is likely to become more commonplace as more molecules are employed for phylogenetic analysis, and a closer examination of how molecular and morphological features can be used to resolve these conflicts is appropriate.

The first step in conflict resolution is to use statistical methods to confirm that a conflict truly exists. Often, most parsimonious trees will have different topologies, but the data cannot be used to show that one is significantly more likely than the other. In the case of *P. carinii*, maximum likelihood analysis of the two conflicting trees (Fig. 3; one is based on 18S rDNA, the other on mt 24S rDNA) using the 18S rDNA sequence shows that the data are significantly more likely given the tree placing *P. carinii* at the base of the *Ascomycota* than they are given the tree placing *P. carinii* at the base of the clade of simple-septate, red basidiomycete yeasts. Thus, a conflict does exist. When an alignment of mt 24S rDNA is available that includes GenBank sequences for *Schizosaccharomyces pombe*, *Podospora anserina*, and *Saccharomyces cerevisiae* (Wakefield et al., 1993), it will be possible to extend the maximum likelihood analysis to this data set.

Resolution of the conflict requires examining the nucleotide sequence data to see if the amount of nucleotide substitution is appropriate for the phylogenetic question being addressed. A tenet of molecular evolution is that mutations increase over time so that the number of mutations seen between two taxa will indicate the relative time since their genomes diverged; i.e. closely related taxa should have fewer mutations than distantly related taxa. Unfortunately, as time passes, and mutations accrue, mutations are more likely to occur where a prior mutation occurred. These multiple hits become more common as time passes and cause the gap between the number of observed and true mutations to increase. Eventually, the gap becomes so large that the amount of observed mutation provides no information on the age of divergence. For rDNA molecules, the point where the linear relationship between mutation and time fails is about 30% dissimilarity (Hillis and Dixon, 1991). When 18S rDNA sequence from *Chytridiomycota*, *Zygomycota*, *Ascomycota* and *Basidiomycota* are aligned and compared, pair by pair, dissimilar sites (using the distance matrix option of PAUP 3.1, which adjusts for missing data) do not exceed 13%; addition of a ciliate to the data set still does not exceed 13% (Bowman et al., 1992a). Thus, in terms of pairwise dissimilarity, the 18S rDNA is appropriate for phylogenetic questions among all fungi. With mt 24S rDNA, dissimilar sites between two strains of *P. carinii* vary from 17% to 28%. The dissimilarity between *P. carinii* and other ascomycetes ranges from 51% to 69% and that between *P. carinii* and simple-septate basidiomycetes ranges from 38% to 44% (Wakefield et al., 1993). In light of the 30% dissimilarity threshold for phylogenetic analysis, the mt 24S rDNA is not appropriate for comparison between ascomycetes and basidiomycetes. Thus, the conflict is resolved: the mt 24S rDNA is evolving too rapidly to successfully analyze the relationships of fungi that diverged early in the evolution of the *Ascomycota* and *Basidiomycota*.

Morphological Characters and Conflict Resolution

As Wakefield et al. (1993) point out, phylogenetic conflicts may also be resolved by examining morphological characters, which they use to link *P. carinii* with simple-

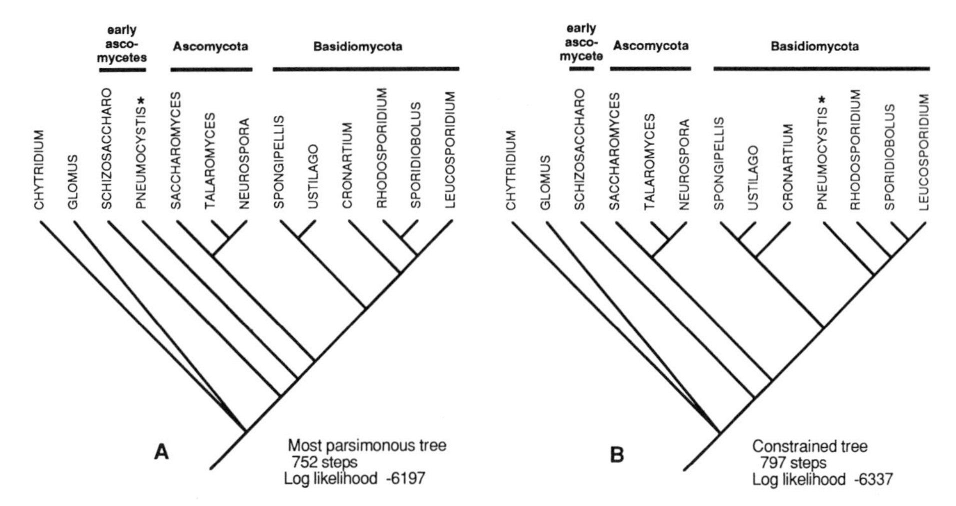

Figure 3. Two phylogenetic trees showing alternative hypotheses about the relationship of *Pneumocystis carinii* to other fungi:

A, One of two most parsimonous trees based on analysis (PAUP; Swofford, 1993) of 1589 unambiguously aligned nucleotides of 18S rDNA sequence. The two most parsimonous trees of 752 steps differ in the position of *Sporidiobolus*, which can be the sister taxon of either *Rhodosporidium* or *Leucosporidium*. The log likelihood of these data given this most parsimonous tree is -6197 (DNAML in PHYLIP, Felsenstein, 1991).

B, One of two most parsimonous tree based on 18S rDNA nucleotide sequence with the constraint that *Pneumocystis* must be on the clade of simple-septate basidiomycete yeasts. These two most parsimonous trees of 797 steps differ in the placement of *Schizosaccharomyces*, which may also be basal to the clade including *Saccharomyces*, *Talaromyces* and *Neurospora*. The log likelihood of the data given this tree was -6337, more than five standard deviations (24.8) lower than the log likelihood of the most parsimonous tree. When log likelihoods differ by more than two standard deviations, the data are significantly more likely given the tree with the higher log likelihood (Felsenstein, 1991). Therefore, *Pneumocystis* is unlikely to be a member of the clade of simple-septate basidiomycete yeasts. The log likelihoods of the trees not shown are: -6199 for the other most parsimonous tree, and -6336 with a standard deviation of 27.9 for the other constrained tree. Sources of 18S rDNA sequence are: *Chytridium* (Bowman et al., 1992a), *Glomus* (Bruns et al., 1992), *Schizosaccharomyes* (Bruns et al., 1992), *Pneumocystis* (Edman et al., 1988), *Saccharomyces* (Mankin et al., 1986), *Talaromyces* (Berbee and Taylor, 1992a), *Neurospora* (Sogin et al., 1986), *Spongipellis* (Bowman et al., 1992a), *Ustilago* (Berbee and Taylor, 1992b), *Cronartium* (Bruns et al., 1992), *Rhodosporidium* (van de Peer et al., 1992), *Sporidiobolus* (E.C. Swann, unpubl.), *Leucosporidium* (van de Peer et al, 1992).

septate, red basidiomycete yeasts. Morphological characters are likely to be cryptic in the case of *P. carinii* because they were not distinctive enough to prevent its being assigned to the *Protoctista* (cf. Cushion et al., 1991). Morphological studies have demonstrated its fungal affinities (Vavra and Kucera, 1970) and molecular studies have confirmed them (Edman et al., 1988; Stringer et al., 1989). In this way, *P. carinii* is similar to *Coccidioides immitis*, which was described as a protist, but has been shown to be an ascomycete using morphology (Sigler and Carmichael, 1976) and nucleic acid sequence (Bowman et al., 1992b). The two morphological characters important to the phylogenetic placement of *P. carinii* are cell walls and endospores (Wakefield et al., 1992, 1993).

Cell Walls. Kreger-van Rij and Veenhuis (1971) used electron microscopy of permanganate-fixed cell walls to show that basidiomycete yeasts have laminated walls and ascomycete yeasts do not. This feature is most clearly seen during budding, when the outer layers of the basidiomycete cell wall rupture. Permanganate-fixed cell walls of precysts and cysts of *P. carinii* do not show laminated walls (Vavra and Kucera, 1970; figs 30-31), however, ruptured walls cannot be seen because the rapidly dividing vegetative cells lack a thick wall and divide by fission (Richardson et al., 1989). More recent ultrastructural studies (Matsumoto and Yoshida, 1984; Yoshida, 1989) have used glutaraldehyde fixation. With glutaraldehyde, laminations may go undetected unless stains are used that highlight polysaccharides (e.g. Taylor and Wells, 1979; figs 26, 39). When such staining is used on glutaraldehyde fixed cell walls, laminated walls typical of basidiomycete yeasts are still not seen (Yoshida, 1989; fig. 13). Therefore, *P. carinii* cell walls have not been shown to have the diagnostic features of basidiomycete yeasts.

Endospores. *P. carinii* produces intracystic bodies, which have been regarded by Wakefield et al. (1993) as similar to the endospores produced by basidiomycete yeasts (Fell et al., 1973). Endospores, spores formed from within cells without involvement of the parental cell wall, are the rule in *Chytridiomycota* (zoospores; e.g. Taylor and Fuller, 1981), *Zygomycota* (sporangiospores; e.g. Bracker, 1968), and *Ascomycota* (ascospores; e.g. Carroll, 1967), and are found in some basidiomycetes. Thus, their presence is a shared ancestral character, or symplesiomorphy (albeit one that has been lost in most basidiomycetes), and therefore cannot be used to place *P. carinii* on the basidiomycete or ascomycete clade. Furthermore, the basidiomycete yeasts shown by Fell et al. to produce endospores, *Rhodosporidium bisporidiis*, *R. capitatum* and *R. infirmo-miniatum*, have been transferred to the genus *Cystofilobasidium* (Oberwinkler et al., 1983) and are not members of the clade of simple-septate, red pigmented basidiomycetes (Fell et al., 1992). The most convincing demonstration of endospores in a simple-septate, red pigmented basidiomycete is in *Sporidiobolus salmonicolor* (Bandoni, 1984; fig. 23). Here, yeast cells produce buds inside teliospores. Bandoni suggests that the yeast cells may represent meiospores. If ultrastructural and cytological studies should support this speculation, it would show that meiospores may be formed within the meiocyte both in *Ascomycota* and *Basidiomycota*; however, presence of the spores would still constitute a shared ancestral character for these two divisions.

Vavra and Kucera (1970) concluded from their thorough ultrastructural study of the *P. carinii* life cycle that: "Numerous fine structural analogies can be found between the mode of formation of intracystic bodies inside the cyst of *Pneumocystis* and the fine structure of yeast cells during ascus differentiation and ascospore formation." Their analogies refer to the membranes and walls that delimit the intracystic bodies: "... the formation of the membrane of each intracystic body within the cytoplasm is in agreement with the ascospore delineation mechanism in fungi.", "As in fungi, only a limited part of the cyst cytoplasm is included in the intracystic bodies.", and "The cell wall of the intracystic bodies grows thicker and consists, as in yeast ascospores, of a thick layer interposed between 2 membranes (Fig. 36)." In the past decade, Yoshida and colleagues (Matsumoto and Yoshida, 1984; Yoshida, 1989) have discovered synaptonemal complexes in the precyst of *P. carinii*, lending further support to the homology of cysts to asci and intracystic bodies to ascospores. They also provide evidence of a double-membrane delimiting each developing intracystic body (Matsumoto and Yoshida, 1984; fig. 15). If these membranes are double-ascospore-delimiting-membranes, as are known from ascomycetes, including *Saccharomyces cerevisiae* (Moens, 1971), *Schizosaccharomyces pombe* (Tanaka and Hirata, 1982), and *Ascobolus* (Wells, 1972), they would be a shared derived character linking *P. carinii* with the *Ascomycota*. Although the simple presence of endospores is a shared ancestral character, which cannot be used to place *P. carinii* in the *Ascomycota* or *Basidiomycota*, the formation of endospores following meiosis from walls deposited between double membranes provides a complex of shared derived characters that can be used to place *P. carinii* in the *Ascomycota*.

CONCLUSION

Phylogenetic analysis of 18S rDNA nucleotide sequence has been used to show that the kingdom *Fungi* comprises flagellated organisms (*Chytridiomycota*) as well as nonflagellated ones (*Zygomycota*, *Ascomycota* and *Basidiomycota*) and that the sister group to the kingdom *Fungi* is the animal kingdom plus choanoflagellates. These molecular analyses have also shown that some traditional classes of *Ascomycota* (*Hemiascomycetes*, *Plectomycetes*, and *Pyrenomycetes*), are represented by well-supported monophyletic clades. However, the cladistic state of apothecial or fissitunicate ascomycetes remains unresolved. *Schizosaccharomyces* and *Taphrina* (genera thought to be *Hemiascomycetes*) and *Pneumocystis carinii* are not part of the hemiascomycete clade, but may represent independent basal branches in the *Ascomycota*. Conflicts between morphological and molecular phylogenies are bound to arise; when they do they may be resolved by careful examination of the data which support the phylogenies, be they morphological or molecular. It is likely that *P. carinii* will be only one of many cases where morphological characters may help solve phylogenetic problems for which molecular data are inadequate or controversial.

ACKNOWLEDGMENTS

Preparation of this contribution was supported in part by grants from the National Institutes of Health (NIH) and the National Science Foundation (NSF). Austin Burt, Kathy LoBuglio, and Greg Saenz are thanked for reading drafts of this manuscript, and for many discussions of fungal evolution. Melanie Cushion and Chuck Staben are thanked for advice on *Pneumocystis* morphology and life history.

REFERENCES

von Arx, J.A., 1973, Ostiolate and nonostiolate pyrenomycetes, *Proceedings Koninklijke Nederlandse Akademie van Wetenschappen, sect. C*, 76: 289-296.

Bandoni, R.J., 1984, The *Tremellales* and *Auriculariales*: an alternative classification, *Transactions of the Mycological Society of Japan* 25: 489-530.

Barr, D.J.S., 1981, The phylogenetic and taxonomic implications of flagellar rootlet morphology among zoosporic fungi, *BioSystems* 14: 359-370.

Bartnicki-Garcia, S., 1970, Cell wall composition and other biochemical markers in fungal phylogeny, *In*: *Phytochemical Phylogeny* (J.P. Harborne, ed.): 81-104, Academic Press, New York.

Berbee, M.L., and J.W. Taylor, 1992a, Two ascomycete classes based on fruiting-body characters and ribosomal DNA sequence, *Molecular Biology and Evolution* 9: 278-284.

Berbee, M.L., and J.W. Taylor, 1992b, Detecting morphological convergence in true fungi, using 18S rRNA gene sequence data, *BioSystems* 28: 117-125.

Berbee, M.L., and J.W. Taylor, 1992c, Convergence in ascospore discharge mechanism among pyrenomycete fungi based on 18S ribosomal RNA gene sequence, *Molecular Phylogenetics and Evolution* 1: 59-71.

Berbee, M.L., and J.W. Taylor, 1993, Dating the evolutionary radiations of the true fungi, *Canadian Journal of Botany* 71: 1114-1127.

Bessey, E.A., 1950, *Morphology and Taxonomy of Fungi*, Blakiston, Philadelphia.

Bhattacharya, D., L. Medlin, P.O. Wainright, E.V. Ariztia, C. Bibeau, S.K. Stickel, and M.L. Sogin, 1992, Algae containing chlorophylls a + c are paraphyletic: molecular evolutionary analysis of the chromophyta, *Evolution* 46: 1801-1817.

Bowman, B.H., J.W. Taylor, A.G. Brownlee, J. Lee, S-D. Lu, and T.J. White, 1992a, Molecular evolution of the fungi: relationship of the basidiomycetes, ascomycetes, and chytridiomycetes, *Molecular Biology and Evolution* 9: 285-296.

Bowman, B., J.W. Taylor, and T.J. White. 1992b, Molecular evolution of the fungi: human pathogens, *Molecular Biology and Evolution* 9: 893-904.

Bracker, C.E. 1968, The ultrastructure and development of sporangia in *Gilbertella persicaria*, *Mycologia* 60: 1016-1067.

Bruns, T.D., T.J. White, and J.W. Taylor, 1991, Fungal molecular systematics, *Annual Review of Ecology and Systematics* 22: 525-564

Bruns, T.D., R. Vilgalys, S.M. Barns, D. Gonzalez, D.S. Hibbett, D.J. Lane, L. Simon, S. Stickel, T.M. Szaro, W.G. Weisburg, and M.L. Sogin, 1992, Evolutionary relationships within the fungi: analyses of nuclear small subunit rRNA sequences, *Molecular Phylogenetics and Evolution* 1: 231-241.

Cain, R.F., 1972, Evolution of the fungi, *Mycologia* 64: 1-14

Carroll, G.C., 1967, The ultrastructure of ascospore delimitation of *Saccobolus kerverni*, *Journal of Cell Biology* 33: 218-224.

Cavalier-Smith, T., 1987, The origin of *Fungi* and *Pseudofungi*, *In*: *Evolutionary Biology of the Fungi* (A.D.M. Rayner, C.M. Brasier, and D. Moore, eds): 339-353, Cambridge University Press, Cambridge.

Cushion, M.T., J.R. Stringer, and P.D. Walzer, 1991, Cellular and molecular biology of *Pneumocystis carinii*, *International Review of Cytology* 131: 59-107.

Dodge, B.O., 1935, The mechanics of sexual reproduction in *Neurospora*, *Mycologia* 27: 418-438.

Edman, J.C., J.A. Kovacs, H. Mansur, D.V. Santi, H.J. Elwood, and M.L. Sogin, 1988, Ribosomal RNA sequence shows *Pneumocystis carinii* to be a member of the fungi, *Nature* 334: 519-522.

Eriksson, O.[E.], 1982, Outline of the ascomycetes - 1982, *Mycotaxon* 15: 203-248.

Fell, J.W., I.L. Hunter, and A.S. Tallman, 1973, Marine basidiomycetous yeasts (*Rhodotorula* spp. n.) with tetrapolar and multiple allelic bipolar mating systems, *Canadian Journal of Microbiology* 19: 643-657.

Fell, J.W., A. Statzell-Tallman, M.J. Lutz, and C.P. Kurtzman, 1992, Partial rRNA sequences in marine yeasts: a model for identification of marine eukaryotes, *Molecular Marine Biology and Biotechnology* 1: 175-186.

Felsenstein, J., 1985, Confidence limits on phylogenies: an approach using the bootstrap, *Evolution* 39: 783-791.

Felsenstein, J., 1991. *Phylip 3.4*, Department of Genetics, University of Washington, Seattle.

Forsburg, S.L., and P. Nurse, 1991, Cell cycle regulation in the yeasts *Saccharomyces cerevisiae* and *Schizosaccharomyces pombe*, *Annual Review of Cell Biology* 7: 227-256.

Förster, H., M.D. Coffey, H. Elwood, and M.L. Sogin, 1990, Sequence analysis of the small subunit ribosomal RNAs of three zoosporic fungi and implications for fungal evolution, *Mycologia* 82: 306-312.

Gargas, A., 1992, Phylogeny of discomycetes and early radiations of the filamentous ascomycetes inferred from 18S rDNA sequence data. PhD Thesis, University of California, USA.

Gottschalk, M., and P.A. Blanz, 1985, Untersuchungen an 5S ribosomalen robonukleinsauren als beitrag zur Klarung von systematik und phylogenie der Basidiomyceten, *Zeitschrift für Mykologie* 51: 205-243.

Hasegawa, M., T. Hashimoto, J. Adachi, N. Iwabe, and T. Miyata, 1993, Early branchings in the evolution of eukaryotes: ancient divergence of *Entamoeba* that lacks mitochondria revealed by protein sequence data, *Journal of Molecular Evolution* 36: 380-388.

Hasegawa, M., Y. Iida, T. Yano, F. Takaiwa, and M. Iwabuchi, 1985, Phylogenetic relationships among eukaryotic kingdoms inferred from ribosomal RNA sequences, *Journal of Molecular Evolution* 22: 32-38

Hawksworth, D.L., B.C. Sutton, and G.C. Ainsworth, 1983, *Ainsworth & Bisby's Dictionary of the Fungi*. 7th edition, Commonwealth Agricultural Bureaux, Slough.

Hendriks, L., A. Goris, Y. van de Peer, J.-M. Neefs, M. Vancanneyt, K. Kersters, J.-F. Berny, G.L. Hennebert, R. R. De Wachter, 1992, Phylogenetic relationships among ascomycetes and ascomycete-like yeasts as deduced from small ribosomal subunit RNA sequences, *Systematic and Applied Microbiology* 15: 98-104.

Hennig, W., 1966, *Phylogenetic Systematics*, University of Illinois Press, Urbana.

Hillis, D.M., and M.T. Dixon, 1991, Ribosomal DNA: molecular evolution and phylogenetic inference, *Quarterly Review of Biology* 66: 411-453.

Hori, H., and S. Osawa, 1987, Origin and evolution of organisms as deduced from 5S ribosomal RNA sequences, *Molecular Biology and Evolution* 4: 445-472.

Illingworth, C.A., J.H. Andrews, C. Bibeau, and M.L. Sogin, 1991, Phylogenetic placement of *Athelia bombacina*, *Aureobasidium pullulans* and *Colletotrichum gloeosporoides* inferred from sequence comparisons of small-subunit ribosomal RNAs, *Experimental Mycology* 15: 65-75.

Kreger-van Rij, N.J.W., and M. Veenhuis, 1971, A comparative study of the cell wall structure of basidiomycetous and related yeasts, *Journal of Genneral Microbiology* 68: 87-95.

Luttrell, E.S., 1955, The ascostromatic ascomycetes, *Mycologia* 47: 511-532.

Malloch, D. 1981, The plectomycete centrum, *In*: *Ascomycete Systematics: The Luttrellian Concept* (D.R. Reynolds, ed.): 73-91, Springer, New York.

Mankin, A.S., K.G. Skryabin, and P.M. Rubtsov, 1986, Identification of ten additional nucleotides in the primary structure of yeast 18S rRNA, *Gene* 44: 143.

Margulis, L., and K.V. Schwartz, 1988, *Five Kingdoms*, W.H. Freeman, New York.

Matsumoto, Y., and Y. Yoshida, 1984, Sporogony in *Pneumocystis carinii*: synaptonemal complexes and meiotic nuclear divisions observed in precysts, *Journal of Protozoology* 31: 420-428.

Moens, P.B., 1971, Fine structure of ascospore development in the yeast *Saccharomyces cerevisiae*, *Canadian Journal of Microbiology* 17: 507-510.

Nishida, H., and J. Sugiyama, 1992, Phylogenetic relationships among *Taphrina*, *Saitoella*, and other higher fungi, *Molecular Biology and Evolution* 10: 431-436.

Oberwinkler, F., R. Bandoni, P. Blanz, and L. Kisimova-Horovitz, 1983, *Cystofilobasidium*: a new genus in the *Filobasidiaceae*, *Systematic and Applied Microbiology* 4: 114-122.

Radford, A., 1993, A fungal phylogeny based upon orotidine 5'-monophosphate decarboxylase, *Journal of Molecular Evolution* 36: 389-395.

Richardson, J.D., S.F. Queener, M. Bartlett, and J. Smith, 1989, Binary fission of *Pneumocystis carinii* trophozoites grown in vitro, *Journal of Protozoology* 36: 27S-29S.
Savile, D.B.O., 1955, A phylogeny of the basidiomycetes, *Canadian Journal of Botany* 33: 60-104.
Sigler, L., and J.W. Carmichael, 1976, Taxonomy of *Malbranchea* and some other hyphomycetes with arthroconidia, *Mycotaxon* 4: 349-488.
Smith, T.L., 1989, Disparate evolution of yeasts and filamentuous fungi indicated by phylogentic analysis of glyceraldehyde-3-phosphate dehyrogenase genes, *Proceedings of the National Academy of Sciences of the United States of America* 86: 7063-7066.
Sogin, M.L., 1989, Evolution of eukaryotic microorganisms and their small subunit ribosomal RNAs, *American Zoologist* 29: 487-499.
Sogin, M.L., J.H. Gunderson, H.J. Elwood, R.A. Alonso, and D.A. Peattie, 1989, Phylogenetic meaning of the kingdom concept: an unusual ribosomal RNA from *Giardia lamblia, Science* 243: 75-77.
Sogin, M.L., K. Miotto, and L. Miller, 1986, Primary structure of the *Neurospora crassa* small subunit ribosomal RNA coding region, *Nucleic Acids Research* 23: 9540.
Spatafora, J.W., and M. Blackwell, 1992, Monophyly and higher level systematics of ascomycetes, *Inoculum* 43 (1,2,3): 50.
Spatafora, J.W., R. Vilgalys, and T.G. Mitchell, 1993, Phylogenetic placement of the "black yeasts" (*Ascomycota*), *Inoculum* 44 (2): 57.
Stringer, S.L., K. Hudson, M.A. Blase, P.D. Walzer, M.T. Cushion, and J.R. Stringer, 1989, Sequence from ribosomal RNA of *Pneumocystis carinii* compared to those of four fungi suggests an ascomycetous affinity, *Journal of Protozoology* 36: 14S-16S.
Swofford, D.L., 1993, *PAUP: Phylogenetic Analysis Using Parsimony, Version 3.1*, Illinois Natural History Survey, Champaign, Illinois.
Tanaka, K., and A. Hirata, 1982, Ascospore development in the fission yeasts *S. pombe* and *S. japonicus*, *Journal of Cell Science* 56: 263-279.
Taylor, J.W., B. Bowman, M.L. Berbee, and T.J. White, 1993, Fungal model organisms: phylogenetics of *Saccharomyces*, *Aspergillus* and *Neurospora*, *Systematic Biology* 42: 440-457.
Taylor, J.W., and B.H. Bowman, 1993, *Pneumocystis carinii* and the ustomycetous red yeast fungi, *Molecular Microbiology* 8: 425-426.
Taylor, J.W., and M.S. Fuller, 1981, The golgi apparatus, zoosporogenesis, and development of the zoospore discharge apparatus of *Chytridium confervae. Experimental Mycology* 5: 35-59.
Taylor, J.W., and K. Wells, 1979, A light and electron microscopic study of mitosis in *Bullera alba* and the histochemistry of some cytoplasmic substances, *Protoplasma* 98: 31-62.
Tehler, A., 1988, A cladistic outline of the *Eumycota*, *Cladistics* 4: 227-277.
Tehler, A., 1989, A new approach to the phylogeny of *Euascomycetes* with a cladistic outline of *Arthoniales* focussing on *Roccellaceae*, *Canadian Journal of Botany* 68: 2458-2492.
van de Peer, Y., L. Hendriks, A. Goris, J.-M. Neefs, M. Vancanneyt, K. Kersters, J.-F. Berny, G.L. Hennebert, and R. De Wachter, 1992, Evolution of basidiomycetous yeasts as deduced from small ribosome subunit RNA sequences, *Systematic and Applied Microbiology* 15: 250-258.
Vavra, J., and K. Kucera, 1970, *Pneumocystis carinii* Delanoë, its ultrastructure and ultrastructural affinities, *Journal of Protozoology* 17:463-483.
Vogel, H.J., 1964, Distribution of lysine pathways among fungi: evolutionary implications, *American Naturalist* 98: 435-446.
Wainright, P.O., G. Hinkle, M.L. Sogin, and S.K. Stickel, 1993, Monophyletic origins of the metazoa: an evolutionary link with fungi, *Science* 260: 340-342.
Wakefield, A.E., J.M. Hopkin, P.D. Bridge, and D.L. Hawksworth, 1993, *Pneumocystis carinii* and the ustomycetous red yeast fungi, *Molecular Microbiology* 8: 426-427.
Wakefield, A.E., S.E. Peters, B. Suneale, P.D. Bridge, G.S. Hall, D.L. Hawksworth, L.A. Guiver, A.G. Allen, and J.M. Hopkin, 1992, *Pneumocystis carinii* shows DNA homology with the ustomycetous red yeast fungi, *Molecular Microbiology* 6: 1903-1911.
Walker, W.F., and W.F. Doolittle, 1982, Redividing the basidiomycetes on the basis of 5S rRNA sequences, *Nature* 299: 723-724.
Wells, K., 1972, Light and electron microscopic studies on *Ascobolus stercorarius*. II. Ascus and ascospore ontogeny, *University of California Publications in Botany* 62: 1-93.
White, T.J., T.D. Bruns, S. Lee, and J.W. Taylor, 1990, Amplification and direct sequencing of fungal ribosomal RNA genes for phylogenetics, *In*: *PCR Protocols* (M. Innis, D. Gelfand, J. Sninsky and T. White, eds): 315-322, Academic Press, San Diego.
Whittaker, R.H., 1969, New concepts of kingdoms of organisms, *Science* 163: 150-160.
Woese, C.R., O. Kandler, and M.L. Wheelis, 1990, Towards a natural system of organisms: proposal for the domains archaea, bacteria, and eucarya, *Proceedings of the National Academy of Sciences, United States of America* 87: 4576-4579
Yoshida, Y., 1989, Ultrastructural studies of *Pneumocystis carinii*, *Journal of Protozoology* 36: 53-60.

DISCUSSION

Tehler: Was the Sogin *et al.* figure you showed a cladogram ?

Taylor, J.: The figure from Sogin's laboratory is in Wainwright *et al.* (*Science* 260, 340-342, 1993). It was a cladogram and the paper included a consensus tree from boot-strapped datasets. The support for animals and fungi being sister groups was 80-85 % depending on the analysis. In the absence of *Acanthamoeba* the value rose to 98 %. The most solid aspect of this work was the demonstration of a crown-like radiation of higher eukaryotes (including fungi, animals, plants, brown algae, diatoms, *Oomycota*, and red algae). Determining which of these groups are closest relatives is more difficult to do in a statistically confident manner.

Tehler: The position of *Saccharomycetales* and *Taphrinales* you presented is interesting because it is in conflict with morphological phylogenetic trees. In my own studies (Tehler, *Cladistics* 4, 227, 1988) I was surprised that *Saccharomyces* grouped together with the basidiomycetes. The 18S data appears to constantly group these two orders with the ascomycetes, but morphological data suggest that they constitute a monophyletic group between the basidiomycetes and ascomycetes. Bootstrap data were missing for some lower nodes and you indicated that support for some of these branches is not strong; it appears there is a polychotomy.

Taylor, J.: You are correct. Based on 18S data, yeasts and hyphal ascomycetes are found on the same monophyletic clade, and *Taphrina* is not in the *Saccharomycetales* (but is position at the base of the *Ascomycota* is not strongly supported). One morphological character used to unite yeasts, *Taphrina* and basidiomycetes in one clade is the initiation of sex by the fusion of apparently vegetative cells; this charcater is not shared by the filamentous ascomycetes. However, yeast cells capable of sexual fusion are physiologically distinct from budding cells, and share with filamentous ascomycetes and basidiomycetes features such as cell cycle arrest and oligopeptide pheromone production. Characters such as peptide pheromones or dikaryotic cells may unite ascomycetes (including *Schizosaccharomyces* and *Taphrina*) with basidiomyctes to the exclusion of zygomycetes and chytridiomycetes.

Tehler: I am pleased to see more characters to support the dikaryotic fungi coming to light.

Brygoo: With respect to the position of *Taphrina*, I do not know if there are introns in the ß-tubulin genes, but that would be a simple criterion to assist in the placement of *Taphrina* with ascomycete or basidiomycete fungi.

Eriksson: Certain areas of the 18S gene are more variable and others are more conserved. Some stem loops seen in the secondary structure are characteristic for certain ascomycete groups. I compared the secondary structure of *Leucosporidium* with *Pneumocystis* and six ascomycetes, many positions were identical, but in 109 positions *Pneumocystis* was similar to the ascomycetes but different from *Leucosporidium*. In two places *Pneumocystis* was similar to *Leucosporidium* but different from the other ascomycetes considered.

18S RIBOSOMAL DNA SEQUENCE DATA AND DATING, CLASSIFYING, AND RANKING THE FUNGI

M.L. Berbee[1] and J.W. Taylor[2]

[1]Department of Botany
6270 University Boulevard
Vancouver, B.C., Canada V6T 1Z4

[2]Koshland Hall Rm 111
Department of Plant Biology
University of California
Berkeley, CA 94720, USA

SUMMARY

How could information from 18S ribosomal DNA sequences be incorporated into a classification system? Molecular systematic studies are yielding a wealth of new fungal taxonomic characters. In some cases, DNA sequence data support traditional taxonomic groups. In other cases, among ascomycetous yeasts for example, sequence data suggest new, unfamiliar groupings that require revision of the classification. Whether from molecular or traditional approaches, increased resolution of taxonomic relationships is leading to situations in which the number of ranks available for classification is inadequate for complete description of relationships. Possible solutions to the problem of proliferating ranks include accepting paraphyletic groups or eliminating the traditional ranks like family, class, order, and so on. Hennig proposed that taxa that diverged at about the same geological time should be given the same ranking. With a combination of evidence from sequence data and from the fossil record, we have estimated the timing of some fungal divergences, and we explore the consequences of basing rankings on geological age of origins.

INTRODUCTION

Over the past ten years, increasing numbers of fungal ribosomal DNA sequences have been accumulating in international public databases. As evidence of this accumulation, Table 1 shows the GenBank data base accession numbers for rDNA sequences of the fungi considered in this paper. Many other sequences, including sequences from other genes, are readily available through GenBank or EMBL to anyone with access to a university mainframe computer.

Sequence data from ribosomal DNA clearly do not resolve all questions about the relationships among fungi. However, the sequence data do support some fungal groupings very strongly. Several groupings in the sequence-based tree in Fig. 1 correspond to familiar fungal taxa. The ascomycetes are a traditional group; the basidiomycetes are a traditional group, and both ascomycetes and basidiomycetes receive support in phylogenetic trees inferred from sequence data.

Other fungi that group together based on their sequences do not correspond to traditional taxa. Groups without names are difficult to discuss. For example, I must describe, rather than name, the "ascomycetes plus the basidiomycetes", the sister group to

Table 1. Source of sequence data

Fungus species	GenBank accession
Ascomycotina	
Ascosphaera apis	M83264
Aureobasidium pullulans	M55639
Candida albicans	X53497
Colletotrichum gloeosporoides	M55640
Dipodascopsis uninucleata	U00969
Endomyces geotrichum	U00974
Eremascus albus	M83258
Eurotium rubrum	U00970
Hypomyces chrysospermus	M89782
Kluyveromyces lactis	X51830
Leucostoma persoonii	M83259
Morchella esculenta	A. Gargas, unpubl. data
Neurospora crassa	X04971
Ophiostoma ulmi	M83261
Pleospora rudis	U00975
Pneumocystis carinii	X12708
Saccharomyces cerevisiae	J01353
Schizosaccharomyces pombe	X54866
Taphrina deformans	U00971
Talaromyces flavus	M83262
Basidiomycotina	
Athelia bombacina	M55638
Cronartium ribicola	M94338
Leucosporidium scottii	X53499
Spongipellis unicolor	M59760
Tilletia caries	U00972
Tremella globospora	U00976
Tremella moriformis	U00977
Ustilago hordii	U00973
Zygomycetes	
Glomus intraradices	X58725
Chytridiomycetes	
Caecomyces (Shaeromonas) communis	M62707
Chytridium confervae	M59758
Neocallimastix sp.	M59761
Neocallimastix frontalis	M62704
Neocallimastix joynii	M62705
Piromyces (Pirinibas) communis	M62706
Spizellomyces acuminatus	M59759

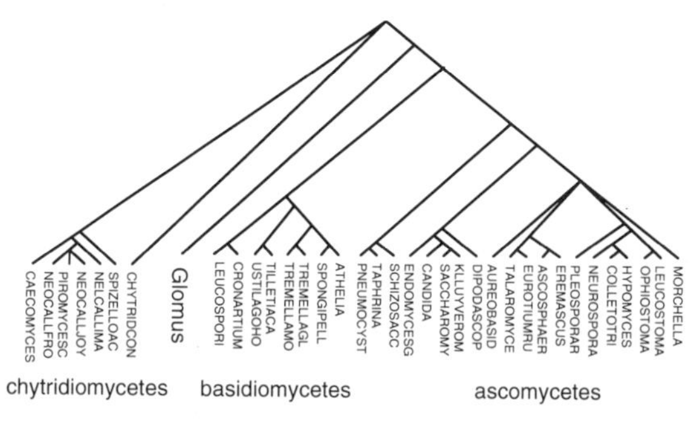

Figure 1. Consensus of six most parsimonious trees inferred from 18S ribosomal DNA of 36 fungal species using the phylogeny inference program PAUP 3.0s (Swofford, 1991). Full names of taxa are given in Table 1.

the *"Glomus"* group. From sequence data, we have inferred that the ascomycetous yeasts usually united in *Hemiascomycetes* or the *Endomycetales* are not a monophyletic group. Yeasts in the genera *Taphrina* and *Schizosaccharomyces* diverged early from the remaining ascomycetes. "True yeasts", including *Saccharomyces* and *Dipodascopsis*, are more closely related to each other and to the filamentous ascomycetes than they are to *Taphrina*. The monophyletic group consisting of filamentous ascomycetes plus the true yeasts is not named or ranked in current classifications.

CLASSIFICATION AND RANKING

As sequence data contribute to our knowledge of phylogenetic groupings, how should classification (arranging taxa in a formal nested hierarchy) and ranking (deciding on taxonomic levels) follow? Here are some of the possibilities:

(1) We could use one, widely known historical classification system, ignoring input from new sources of data. Each new generation of mycological systematic techniques has revealed new relationships; each new generation of mycologists has classified the fungi differently. Assuming the rate of change will slow, we could wait to change our classifications until our understanding of the relationships between fungi stabilizes. To take this approach would mean that we would have to learn and teach taxa we believe to be artificial. We would then have to learn and teach phylogenetic relationships among the fungi that do not correspond to taxon names. Even in the short run, such a system would prove awkward.

(2) We could accept the phylogenetic tree from sequence data or from a combination of sequence data and morphological data as the sufficient formal classification for taxa. The tree would replace the designation of ranks, eliminating the need for classes, orders, families, and so on. By this approach, new or historical names could be cross-referenced to trees (Queiroz and Gauthier, 1990; Vilgalys and Hibbit, 1993). Hierarchical positions can also be presented in an indented list, as given below, or they can be described with words (Queiroz and Gauthier, 1990). The ascomycetes, for example, are the group derived from a fungus with an ascus, or alternatively they are the fungi including all the descendants of the most recent common ancestor of *Taphrina deformans* and *Neurospora crassa*. Groups which are not monophyletic could still be formally defined (Queiroz and Gauthier, 1990). For example, the *Hemiascomycetes* sensu Ainsworth et al. (1973) could be defined as the descendants of the most recent common ancestor of *Taphrina deformans* and *Neurospora crassa* including *Eremascus* species but excluding the other descendants of filamentous ascomycetes with fruiting bodies. This system is flexible, it readily accommodates changes in information and it circumvents the difficulties involved in establishing fixed ranks. It permits clear communication among taxonomists who disagree. However, lots of words are required to communicate the same information that is implicit in, for example, a class name.

(3) We could incorporate information about fungal groupings from sequence data into a cladistic classification system, maintaining traditional ranks (Wiley, 1981). This option is probably closest to current standard taxonomic practice.

How might fungal classifications change with the incorporation of cladistic thinking and molecular data? According to Hennig (1966), sister groups should have the same rank. Hennig's logic was that two lineages that diverged from a common ancestor are evolutionary equals, and classification should reflect this equality. The partial classification below, based on the phylogeny from sequence data in Fig. 1, fulfills the requirement that sister taxa have equivalent hierarchical positions without including any rank assignments; ranks are evident from the indentations. More than two taxa may appear at the hierarchical position, indicating that their relationships were not fully resolved with available rDNA sequence data:

Chytrids
{**Unnamed** group, Zygomycetes in part, Ascomycetes, Basidiomycetes
    Glomus group
    {**Unnamed** group, Ascomycetes, Basidiomycetes}
        Basidiomycete
            {**Unnamed** group, smuts, jelly fungi, holobasidiomycetes}
                Group including *Tilletia* and *Ustilago* species on grasses.
                {**Unnamed** group, *Tremella* + holobasidiomycetes}
            {**Unnamed** group, rusts + *Leucosporidium*}
        Ascomycete
            {**Unnamed** group, evidence for monophyly weak, includes *Schizosaccharomyces, Taphrina, Pneumocystis*}
            {**Unnamed** group, yeasts like *Saccharomyces* + filamentous ascomycetes like *Neurospora*, + common ancestor to both}
                Endomycetes, monophyletic group including *Dipodascopsis* and *Endomyces geotrichum*
                    Dipodascopsis
                    {**Unnamed** group, *Endomyces geotrichum* + *Candida albicans* + *Saccharomyces cerviseae*}
                        Endomyces geotrichum
                        {*Candida albicans, Saccharomyces cerevisiae, Kluyveromyces Kluyveromyces lactis*}
                            Candida albicans
                            S. cerevisiae, K. lactis
                euascomycete
                    discomycetes
                    loculoascomycetes
                    plectomycetes
                        Trichocomaceae
                        {**Unnamed** group, includes *Ascosphaera apis* and *Eremascus albus*}
                    pyrenomycete
                        {**Unnamed** group, includes *Ophiostoma* and *Leucostoma*}
                        {**Unnamed** group, includes *Neurospora* and *Hypomyces*}

The number of unnamed groups in the indented classification above is symptomatic of a common consequence of following Hennig, and giving sister groups equal ranks (Wiley, 1981). Naming and ranking all sister groups may, depending on the branching pattern for the taxa, require more hierarchical groupings than the classes, orders and families that mycologists traditionally recognize. The number of names of higher taxa needed in a completely resolved phylogeny depends on branching patterns, but is at least equal to the total number of species in the classification minus one. The number of ranks or levels of nesting also increases with increasing numbers of species. If the phylogenetic tree is pectinate, with a succession of single species arising from a radiating lineage (Fig. 2), then the total number of ranks needed in a group is equal to the number of species. If the phylogenetic tree shows an equal, dichotomizing branching pattern, then taking the number of taxa, increasing the number of taxa to the number divisible by the nearest whole power of two, and then taking the logarithm to the base 2 of the nearest whole power of the number gives the number of ranks required. This means that to classify the 36 fungi in our cladogram, we would need at least 35 names in addition to the species names. To assign equal ranks to all sister groups in the cladogram, we would need a maximum of 35 ranks, if the branching were completely pectinate. The minimum number of ranks we can calculate as:

$35 = 2^5+3.$

2^6 is the nearest whole power of two.

$\log_2 2^{(6)} = 6$ ranks.

Based on the number of levels of branching in the trees we have reconstructed so far, it is not obvious that the traditional taxonomic system involving assigning ranks of division, class, order, family and tribe will involve an impossible number of "sub" and "super" categories. However, our trees from sequence data include only a minute fraction of ascomycete species. According to the Hawksworth et al. (1983), there are 28 650 species of ascomycetes. If the branching pattern were consistently dichotomizing, 15 ranks would suffice for complete, fully resolved classification. However, if the historical branching pattern for these fungi were pectinate, 28 649 ranks would be required. If a completely resolved tree of all ascomycetes included a high proportion of pectinate branches, then the number of ranks needed for complete classification would become cumbersome and confusing.

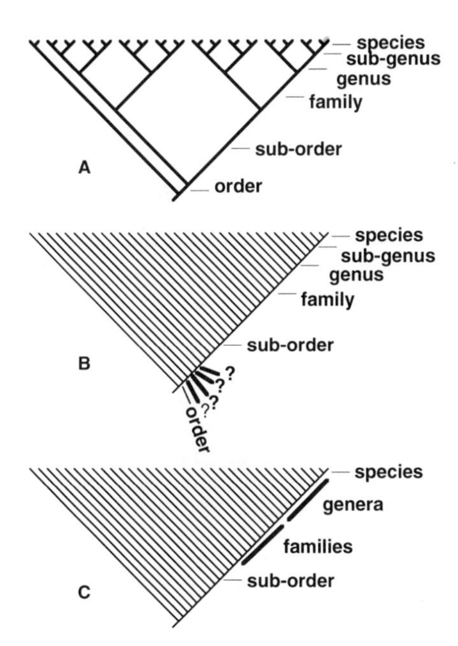

Figure 2. The number of ranks (i.e. species, genus, family, and so on) required for complete classification of a group of organisms depends on the branching relationships among the taxa. If, as in tree A, the branching is mainly dichotomous, then a relatively small number of ranks suffices for completely resolved classification. If as in trees B and C, however, the branching is pectinate, then the number of ranks needed for a completely resolved classification exceeds the number available in a traditional taxonomic system. One solution to the problem of excess ranks is to relax the requirement that each branching receive a separate rank as in tree C.

The problem of excess ranks could be dealt with by relaxing the requirement that each taxonomic level have a rank. Some clusters of genera could go unnamed; others, because they are interesting to human beings, would receive a rank such as subfamily or family. Wiley (1981) advocated adopting a phylogenetic "sequencing convention" which would indicate that species within a single rank are at different taxonomic levels by listing the species sequentially in their order of divergence (Fig. 2).

Placing taxa in paraphyletic groups starts to seem appealing when the branching pattern is pectinate and the groups delimited by each divergence of a single species lack interesting unifying characters. Invoking paraphyletic groups often involves accepting the amount of morphological similarity between groups in addition to phylogeny as a criterion for defining groups (Estabrook, 1986).

In phylogenetic trees based on sequence data, not all relationships are resolved with confidence. The pyrenomycete group, the plectomycete group, discomycetes like *Morchella esculenta*, and loculoascomycetes *Pleospora rudis*, and *Aureobasidium pullulans* are clearly members of the filamentous ascomycetes. However, our data set of 18S ribosomal DNA sequences does not resolve the branching order of these taxa within the filamentous ascomycetes. Wiley (1981) suggests designating taxa of "interchangeable taxonomic position" with the term "*sedis mutabilis*," rather than the term "*incertae sedis*," to indicate that ambiguity in branching order, even though membership in the higher taxonomic group is well-established.

RANKING AND GEOLOGICAL TIME

Three major lineages of multicellular organisms, plants, animals, and fungi, diverged from one another at about the same time, in the Precambrian, about 1000 million years ago (Sogin, 1989). Since they diverged from a common ancestor, plants and animals have evolved from unicells to redwood trees and human beings. Have fungi undergone comparable divergences over the same time period?

Hennig (1966) suggested using geological time as the criterion for determining ranks for non-sister taxa in a phylogenetic tree. Just as sister lineages are equivalent units because they are equal in age, other lineages that diverged at about the same time are also equivalent units, and should also receive equal ranks. Lineages that diverged between the Cambrian and the Carboniferous are in different classes; lineages that diverged between the coal age and the bottom of the Triassic are in different orders; lineages that diverged between the Triassic and the Tertiary are in different families and so on (Hennig, 1966).

Sibley and Ahlquist (1990) followed Hennig's suggestion. They assumed that time is proportional to the amount of change in melting temperatures in DNA hybridization studies. If DNA from two species is mixed and the melting temperature of the mixed, heteroduplex DNA is 10°C lower than the melting temperature for unmixed, homoduplex DNA, then the two species involved should be in separate families. If the difference in melting temperature were from 20-22°C, then the species should be in different orders.

Otherwise, a time-based ranking system has not been applied widely. Biologists have long accepted that an order of frogs may not encompass the same amount of diversity as an order of ascomycetes. Insisting that rank be linked to geological time would lead to changes in rank for familiar taxa. When our estimates of time of origin of groups change, all taxa in the groups would need to be re-ranked A geologically-based ranking system cannot reasonably apply to species definitions

Along with the substantial disadvantages, some advantages may be gained by ranking taxa based on the geological ages of groups. If the majority of features of the biology of plants, animals, and fungi result from clock-like accumulations of neutral mutations, then ranking based on time would provide a substantial amount of information. Genes from members of families (ranked by geological time) of fungi and animals would show similar levels of divergence for taxa with the same ranks. A molecular biologist who found that a mammalian gene worked in different genera within a family of mammals could predict that the homologous fungal gene would work in different genera of a family of fungi. Biochemists sometimes seek the most functionally essential parts of a gene by looking for the parts of the gene that are conserved across different taxa. Taxonomic ranks based on a geological scale would help biochemists choose taxa with appropriate levels of divergence. In addition, adding a time component would be helpful in assigning ranks to asexual lineages.

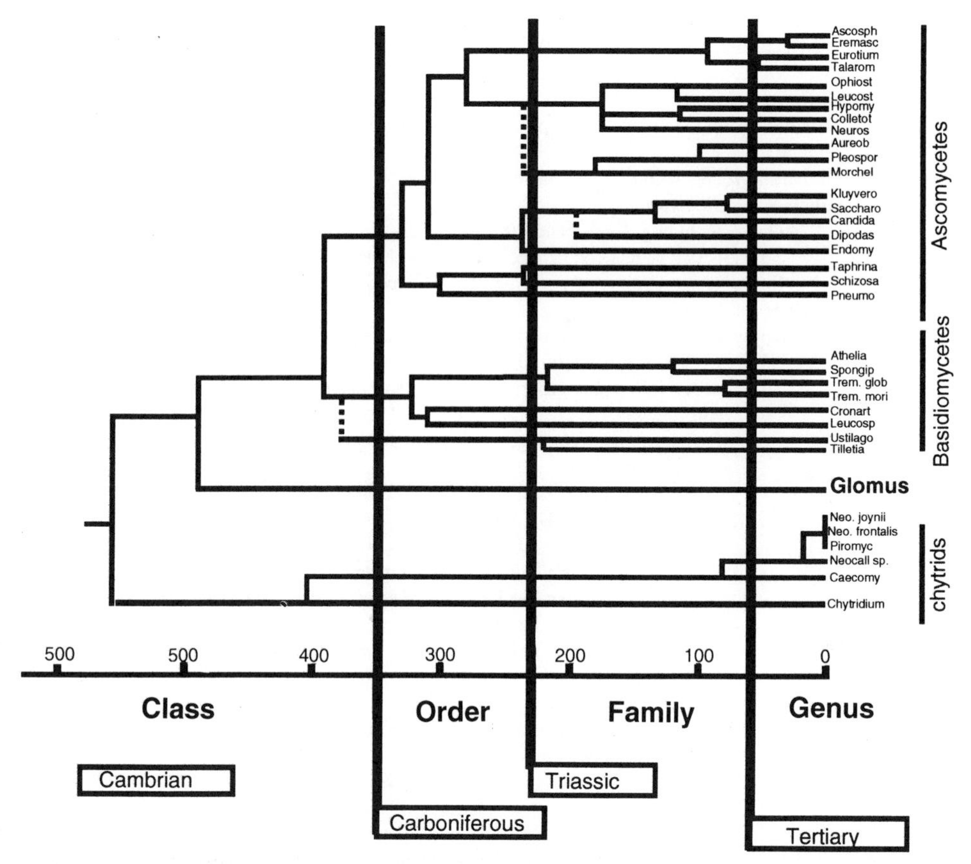

Figure 3. Hennig (1966) proposed that ranks should be assigned to taxonomic groups depending on their geological age. For example, the Cambrian to the Carboniferous would be the time of origin of class-level taxa. In this figure, Hennig's geologically-based rankings are superimposed on our estimates of the timing of fungal divergences.

To examine the effect of using our estimates of geological time of origin to rank fungal groups, we took the our estimate of the fungal time scale (Berbee and Taylor, 1993a, b), and superimposed ranks on divergences based on the timing of the divergences (Fig. 3). With the passage of time, nucleotide substitutions increase. Two lineages that have been separated for a long while differ by a higher percentage of nucleotide substitution than two recently-separated lineages. Taking advantage of the relationship between percent substitution and time, Ochman and Wilson (1987) established a time scale for bacterial evolution; Wolfe et al. (1989) reconstructed the time of divergence of monocot and dicot plants, and Simon et al. (1993) estimated the timing of radiation of endomycorrhizal fungi. We have used the percentage of nucleotide substitution that lineages have accumulated since divergence from a common ancestor to estimate the relative timing of origin of fungal lineages (Berbee and Taylor, 1993a, b). Using calibration points from the fossil record such as the appearance of fossilized fungal clamp connections, we estimated the absolute timing of origin of fungal groups. From our estimates, the chytrids split from a lineage of terrestrial fungi that gave rise to the zygomycetes, ascomycetes, and basidiomycetes about 550 Ma, the ascomycetes split from the basidiomycetes about 400 Ma, and major lineages of ascomycetes including the loculoascomycetes, the pyrenomycetes and plectomycetes were established through the Mesozoic (Berbee and Taylor, 1993b).

The list below shows the consequences of insisting that sister groups and taxa of equal age are of equal rank (Hennig, 1966):

Super class Chytrids
Super class {Unnamed group, Zygomycetes in part, Ascomycetes, Basidiomycetes}
  Class {*Glomus*}
  Class {Unnamed group, Ascomycetes, Basidiomycetes}
    Subclass Basidiomycetes
      Group including *Tilletia* and *Ustilago* species on grasses, *sedis mutabilis.*
      Super-order {Unnamed group, *Tremella* + holobasidiomycetes}
      Super-order {Unnamed group, rusts + *Leucosporidium*}
    Subclass Ascomycetes
      Super-order {Unnamed group, *Schizosaccharomyces, Taphrina, Pneumocystis*}
        {Relationships within this group are not clear}
      Super-order {Unnamed group, yeasts like *Saccharomyces* + filamentous euascomycetes}
        Order {monophyletic group including *Dipodascopsis*, *Candida*, and *Endomyces geotrichum*
          Sub-order {Unnamed group, *Dipodascopsis*}
          Sub-order {Unnamed group, *Endomyces geotrichum* + *Candida albicans* + *Saccharomyces cerviseae*}
            Super-family *E. geotrichum*
            Super-family {*C. albicans, S. cerevisiae, Kluyveromyces lactis*}
              Family *Candida albicans*
              Family {*S. cerevisiae, K. lactis*}
        Order euascomycete
          discomycetes, *sedis mutabilis*
          loculoascomycetes, *sedis mutabilis*
          plectomycetes, *sedis mutabilis*
                Sub-family *Ascosphaera apis* + *Eremascus albus*
                Sub-family *Trichocomaceae*
          pyrenomycetes, *sedis mutabilis*
            Super-family {Unnamed group, includes *Ophiostoma* and *Leucostoma*}
            Super-family {Unnamed group, includes *Neurospora* and *Hypomyces*}

Clearly some of the rankings based on a time-scale are unusual. However, in part because the rankings of higher taxa of fungi vary greatly in various historical classification systems, fungal groups ranked by the time-scale approach appear generally familiar.

Can results from sequence data be incorporated into a phylogenetic system for the fungi? I think that the answer is that they can and will be. The resulting system of classification will be substantially familiar but will include new taxa and re-defined taxa. How the incorporation should proceed remains open to discussion.

ACKNOWLEDGMENTS

We thank Dr Andrea Gargas for the unpublished sequence of *Morchella esculenta*, Martin Adamson for critical comments, and David Carmean for helpful discussions and aid in obtaining library materials. This work was supported in part by National Sciences and Engineering Research Council of Canada Operating Grant OGP0138427 and by National Institutes of Health grant NIH R01 AI28545.

REFERENCES

Ainsworth, G.L., F.K. Sparrow, and A.S. Sussman, eds, 1973, *The Fungi - An advanced treatise*. Vol. 4A. Academic Press, New York.

Berbee, M.L., and J.W. Taylor, 1993a, Ascomycete relationships: dating the origin of asexual lineages with 18S ribosomal RNA gene sequence data, *In: The Fungal Holomorph: Mitotic, Meiotic and Pleomorphic Speciation in Fungal Systematics* (D.R. Reynolds and J.W.Taylor, eds): 67-78. CAB International, Wallingford.

Berbee, M.L., and J.W. Taylor, 1993b, Dating the evolutionary radiations of the true fungi, *Canadian Journal of Botany* 71: 1114-1127.

Bruns, T.D., R. Vilgalys, S.M. Barns, D. Gonzales, D.S. Hibbett, D.J. Lane, L. Simon, S. Stickel, T.M. Szaro, W.G. Weisburg, and M.L. Sogin, 1992, Evolutionary relationsips within the Fungi: analyses of nuclear small subunit rRNA sequences, *Molecular Phylogenetics and Evolution*, 1: 231-241.

Estabrook, G.F. 1986, Evolutionary classification using convex phenetics, *Systematic Zoology* 35: 560-570.

Hawksworth, D.L., B.C. Sutton, and G.C. Ainsworth, 1983, *Ainsworth & Bisby's Dictionary of the Fungi*, 7th edition. Commonwealth Agricultural Bureaux, Slough.

Hennig, W., 1966, *Phylogenetic Systematics* [transl. D.D. Davis and R. Zangerl.] University of Illinois Press, Urbana.

Ochman, H., and A.C. Wilson, 1987, Evolution in bacteria: evidence for a universal substitution rate in cellular genomes, *Journal of Molecular Evolution* 26: 74-86.

Queiroz, K. de, and J. Gauthier, 1990, Phylogeny as a central principle in taxonomy: phylogenetic definitions of taxon names, *Systematic Zoology* 39: 307-322.

Sibley, C.G. and J.E. Ahlquist, 1990, *Phylogeny and Classification of Birds, a Study in Molecular Evolution*, Yale University Press, New Haven.

Simon, L., J. Bousquet, R.C. Lévesque, and M. Lalonde, 1993, Origin and diversification of endomycorrhizal fungi and coincidence with vascular land plants, *Nature* 363: 67-69.

Sogin, M.L., 1989, Evolution of eukaryotic microorganisms and their small subunit ribosomal RNAs, *American Zoologist* 29: 487-499.

Swofford, D.L., 1991, *PAUP: Phylogenetic Analysis Using Parsimony, Version 3.0.* Illinois Natural History Survey, Champaign.

Vilgalys, R., and D.S. Hibbett, 1993, Phylogenetic classification of fungi and our Linnean Heritage, *In: The Fungal Holomorph: Mitotic, Meiotic and Pleomorphic Speciation in Fungal Systematics* (D.R. Reynolds and J.W.Taylor, eds): 255-260. CAB International, Wallingford.

Wiley, E.O., 1981, *Phylogenetics, the theory and practice of phylogenetic systematics*, John Wiley and Sons, New York.

Wolfe, K.H., M. Gouy, Y.-W. Yang, P.M. Sharp, and W.-H. Li, 1989, Date of the monocot-dicot divergence estimated from chloroplast DNA sequence data, *Proceedings of the National Academy of Sciences, United States of America* 86: 6201-6205.

DISCUSSION

Malloch: As a plectomycete specialist, I am very concerned that you have selected only four taxa to defend your thesis on *Plectomycetes*, especially when three are well-known to be in the same family. A much wider range of very different cleistothecial fungi should have been considered, for example *Eremomyces* which has been said to have an ascostromatic type of development and the very different *Pseudeurotiaceae*. You also excluded *Pseudallescheria boydii* from the group although it is cleistothecial. I think the arguments are weak and that you have not successfully tested the hypothesis that the cleistothecium and perithecium are useful in defining higher taxonomic categories.

Berbee: Is a group of fungi defined by a character such as a closed fruiting body or a phylogenetic group ? My view is that the *Pyrenomycetes* and *Plectomycetes* are recognizable phylogenetic groups. The taxon sampling is not perfect but a more diverse group of fungi than might have been predicted, from ten orders recognized in the *Dictionary of the Fungi*, form very strongly supported monophyletic groups. I agree that there will be many fungi with cleistothecia that do not belong in the monophyletic group *Plectomycetes*. However, most of the taxa that Fennell would have placed in the subclass fall into this monophyletic group. The question is also, which taxa should be included here or in *Pyrenomycetes* to make them monophyletic groups.

Malloch: *Plectomycetes* is a very old term traditionally used for *all* cleistothecial fungi. It is therefore misleading to take so few taxa and then resurrect an old term used in a much wider sense than in the one you are applying it.

Berbee: An alternative might then be to introduce new names for these monophyletic groups. There is also structure within those groups. The levels to be assigned to those is not of primary concern to me. I use *Plectomycetes* and *Pyrenomycetes* because have a meaning for many mycologists which can be refined using molecular methods.

Taylor, J.: It is interesting that discussion in the Workshop has centred on names, whether morphological terms or higher taxonomic groups. As time goes on we may look back at this meeting as one of the last where the names were so prominent; I hope that we will increasingly focus on relationships. The beauty of molecular data and cladistic analysis is that relationships are emphasized. As datasets and phylogenetic analyses become more accessible through computer networks, relationships may be communicated better by trees than by names.

Malloch: My main concern is that the name *Plectomycetes* has been applied in a sense that it has never been applied before.

Taylor, J.: The question is whether to refine an old name or introduce one and at what level. I prefer to use *Plectomycetes* because of the image it has in many textbooks of closed ascomata with passively discharging asci.

Blackwell: Fennell describes the *Plectomycetes* as anything with a cleistothecium. The term carries much baggage which is why it has been dropped. There is also a group including genera such as *Nigrosabulum* which are distinct morphologically and it would have been good to consider some of those.

Eriksson: In principle I think we should avoid formal supraordinal taxa for the next ten years.

Kurtzman: Could differences in rates of evolutionary change impact on what is being seen ?

Berbee: The philosophical difference is that I feel it is important to define a group as monophyletic based on common ancestry, not characters all members possess. An organism with shared common ancestry in a monophyletic group could lack all the characters considered typical for the group.

Hawksworth: Could not your concept of *Plectomycetes* be treated as a single order ?

Berbee: Sure. That would be fine.

Cannon: *Ascosphaera, Eremascus*, *Monascus* and the *Trichocomaceae* are linked by the ecological niche to which they are adapted, one of high osmotic potential. While this does not exclude the possibility of monophyly, it must be borne in mind that extreme selection pressure is more likely to result in convergent evolution.

Taylor, J.: Some members of the *Trichocomaceae* live in wet environments also.

Berbee: It is curious that *Ascosphaera apis* and *Eremascus albus* group strongly together; the first is a weak parasite of bees and the second associated with pollen, a rather dry habitat.

Lodha: At one time perithecia and cleistothecia were considered very distinct, but this is not generally accepted today. Molecular biology can tell us whether such differences are real. I would also like to see more molecular work on cases where single teleomorph genera appear to have different anamorphs.

Rossman: We are all testing hypotheses, as in our work on *Hypocreales*. We refine them and then test them again. My problem in this case is that in order to test the hypothesis whether the cleistothecial fungi were monophyletic a group was selected which was not really representative of the group. There are several cleistothecial fungi that molecular studies show belong in *Hypocreales*, including *Mycoarachis* and *Roumegueriella*. As cleistothecial fungi were thought of in such a wide sense, if a large sample was taken, they would not come out as monophyletic.

Berbee: I think cleistothecia can arise as reversals, but I am not sure that applies to perithecia.

Verkley: We have been discussing mainly higher systematic levels but those concerned with the most practical aspects of systematics are interested in learning how molecular methods can assist at the generic level and below. What RNA genes are particularly useful at lower levels ?

Eriksson: I am not sure but it has been claimed the ITS1 region between 18S and 5S should be good, and ITS2 between 5.8S and 26S. However, recent work indicates that there are also conserved regions in these ITS units.

Spatafora: Genera and species are not necessarily equivalent which complicates this issue, but it is relatively easy to amplify the large subunits and apply four-base cutter restriction enzymes to provide a preliminary screen.

Kimbrough: Momal, working with several families of *Pezizales* was unable to resolve certain taxa of the *Ascodesmidaceae* and *Otideaceae* using the 5.8S region, but did succeed using the ITS2 sequence.

Taylor, J.: LoBuglio *et al.* (*Mycologia* 85, 592-604, 1993) have studied species of *Talaromyces* using the two segments of ITS plus the 5.8S as well as the mitochondrial small subunit RNA gene. Trees based on these nuclear and mitochondrial sequences were not incongruent, but neither were they identical. The mitochondrial sequences were easier to align because they were more conserved. One potential problem to keep in mind when examining closely related species is that polymorphic genes present in ancestral populations may still be found in derived species, and these species may still be capable of exchanging genes. In such cases, a population genetic analysis involving several different loci is more appropriate than a phylogenetic analysis of one locus. Homothallic species such as those of *Talaromyces* may have clonal relationships even when they are extremely closely related.

RELATIONSHIPS OF *TUBER*, *ELAPHOMYCES*, AND *CYTTARIA* (*ASCOMYCOTINA*), INFERRED FROM 18S rDNA STUDIES

S. Landvik and O.E. Eriksson

Department of Ecological Botany
University of Umeå
S-90187 Umeå, Sweden

SUMMARY

18S rDNA sequences from members of the genera *Elaphomyces* and *Tuber* (hypogeous ascomycetes) and from *Cyttaria* were compared with homologous sequences from members of the *Leotiales*, *Neolectales*, and *Pezizales*. Preliminary results from cladistic parsimony analyses indicated that *Elaphomyces* (*Elaphomycetales*) and *Tuber* (*Tuberales*) may be nested within *Pezizales*, and that *Cyttaria* (*Cyttariales*) may be an ingroup in the *Leotiales*. The isolated position of *Neolecta* was confirmed.

INTRODUCTION

The position of the hypogeous ascomycetes has been uncertain. They have been referred to one or two separate orders, or they have been included in *Pezizales*. There have also been diverging opinions on the position of the genus *Cyttaria*.

A taxon named *Tuberae*, corresponding to the order *Tuberales*, was erected by Vittadini (1831: 12; Saccardo 1889: 863, changed to *Tuberoideae*). The order has accomodated all hypogeous ascomycetes and been generally accepted by mycologists. Trappe (1979) saw *Tuberales* as an "artificial, anachronistic order", retained some of the families with emendation and transferred them to *Pezizales*. The family *Elaphomycetaceae*, however, he placed in a separate order, *Elaphomycetales* (Trappe, 1979: 330). This classification of the hypogeous ascomycetes was accepted by Eriksson (1981) and in the following *Outlines* of the ascomycetes (e.g. Eriksson and Hawksworth, 1991). However, this group of ascomycetes differs in morphology and ontogeny so much from other ascomycetes (Parguey-Leduc et al., 1991) that molecular data have been needed for a safer classification.

The genus *Cyttaria* is confined to *Nothofagus* and has coevolved with the host genus. It is restricted to the Gondwana regions in the Southern Hemisphere, and has attracted the interest of several mycologists and evolutionary biologists. It has been referred to three different orders: *Leotiales* (e.g. Santesson, 1945), *Pezizales* (e.g. Korf, 1983), and *Cyttariales* (e.g. Gamundí, 1971). Gamundí (1991: 69) summarized recent studies on the genus and concluded that the order "occupies a somewhat isolated position, as evidenced by the chemical data particularly, perhaps closer to *Leotiales* than *Pezizales*". No one previously has made any DNA studies of this genus.

In order to determine the relationships of these aberrant ascomycetes, we have compared the phylogenetically informative 18S rDNA sequences from representatives of the type genera of the three orders mentioned above: *Tuber* (*Tuberales*), *Elaphomyces* (*Elaphomycetales*), and *Cyttaria* (*Cyttariales*). The sequences have been compared with homologoues sequences from discomycetes in the orders *Leotiales*, *Pezizales*, and *Neo-*

lectales, that were treated in a previous study focused on clarifying the position of the genus *Neolecta*, *Neolectaceae* (Landvik et al., 1993).

MATERIALS AND METHODS

The new 18S rDNA sequences reported in this paper were amplified from DNA isolated from the following material: *Tuber* cf. *rapaeodorum* (Sweden, Västerbotten, Sävar par., Bådelögern, in a potato field, 26 September 1992, *G. Hållstam*, det. L.E. Kers, **UME** 29400); *Elaphomyces* cf. *granulatus* (Sweden, Västerbotten, Bureå par., near Harrsjöbäcken, pine forest, July 1991, *K. Forssell*, **UME** 29399); *Cyttaria darwinii* (Argentina, Tierra del Fuego, Parque Nacional Bahia Ensenada, on *Nothofagus betuloides*, 5 February 1991, *S. Sivertsen* 91-29, **UME** 29224).

The new sequences were compared with the homologous sequences from the following species (all from Landvik et al., 1993): *Peziza badia* (*Pezizales*), *Gyromitra esculenta* (d.o.), *Inermisia aggregata* (d.o.), *Plectania nigrella* (d.o.), *Leotia lubrica* (*Leotiales*), *Cudonia confusa* (d.o.), *Spathularia flavida* (d.o.), and *Neolecta vitellina* (*Neolectales*). The sequences of *Neurospora crassa* (*Sordariales*) and *Saccharomyces cerevisiae* (*Saccharomycetales*) were obtained from the EMBL Data Library.

The 18S rDNA sequences from *Tuber*, *Elaphomyces*, and *Cyttaria* have been obtained by cyclic sequencing (Murray, 1989) and by the same primers and procedures as described in Landvik et al. (1993). To increase the readability of bases per sequencing gel, we labelled the primers with P^{33} instead of P^{32}. P^{33} gives better resolution in the parts of the gel where the bases are densest, due to lower energy, and therefore sharper bands. We also ran the fragments on a 5% Long Ranger gel (AT Biochem, Malvern, PA, US), which has a wedge-like spacing effect, compacting the fragments in the lower part of the gel. It is possible to read 500 bases from each primer in this way, when running the sample through two different time intervals.

We highly recommend the sequencing of both strands of the DNA. Most parts of the *Tuber* and *Elaphomyces* genes were double-stranded sequenced. A large part of the *Cyttaria* gene is only single-strand sequenced. The sequences were manually aligned to those in Landvik et al. (1993). Double stranded sequences of the species added in the present study are underlined. Gaps are marked "-" and unresolved nucleotides or unknown sequences are indicated by "?". Informative sites are printed in boldface and those used as characters in the cladistic analyses are marked by "!". Bases that could not be unambiguously aligned within variable regions (not printed in boldface) were excluded from the analyses (Appendix 1).

Cladistical analyses were performed using the PAUP 3.0s package (Swofford, 1991). *Neolecta vitellina* and *Saccharomyces cerevisiae* were tested as outgroup (cf. Landvik et al., 1993).

RESULTS AND DISCUSSION

In a branch-and-bound analysis of the 13 genera (1746 characters, 137 of which are informative, gaps not considered), *Tuber* and *Elaphomyces* both nested in the *Pezizales* (Fig. 1, tree length = 341, consistency index (CI) = 0.58, retention index (RI) = 0.54). However, decay numbers, showing the support for each node (Donoghue et al., 1992) indicate that *Tuber* is only weakly associated to *Gyromitra*, and even to the *Pezizales* (2 steps). The grouping of *Elaphomyces* and *Inermisia* is better supported (3 steps), and their association with *Plectania* has a high decay number (7 steps). *Cyttaria* appeared among the *Leotiales* with the relatively strong decay number of 5 steps. However, parts of the *Cyttaria* gene have been sequenced in only one direction, but we do not expect any surprises from further sequencing of the gene.

When including the gaps in the analyses (7 additional characters), two most parsimonious trees were obtained (tree length = 364, CI = 0.58, RI = 0.54, trees not shown). One was identical to the tree in Fig. 1, whereas in the other, *Tuber* became a sister group to the pezizalean species, and *Gyromitra* branched off near the base of the

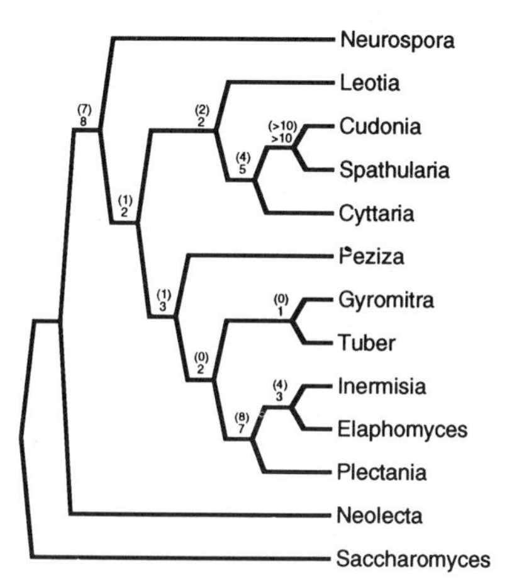

Figure 1. The most parsimonious tree after a branch-and-bound analysis based on 137 informative characters from 13 genera. Decay numbers (Donoghue et al., 1992) are shown. An analysis also with gaps from the alignment included (7 additional characters) gives somewhat different decay numbers; these are placed in parantheses.

order. There were no changes in the placement of either *Elaphomyces*, nor *Cyttaria*. The decay numbers for this data set is put in parentheses in Fig. 1.

It is notable that, when excluding *Saccharomyces* and using *Neolecta* as outgroup, the *Pezizales* became paraphyletic, with *Peziza* as a sister group to the rest of the taxa in the ingroup. The same results were obtained whether the gaps were considered or not. This could indicate that *Neolecta* is not always a good choice of outgroup as suggested in Landvik et al. (1993), or, as hinted in that paper, the *Pezizales* may actually be paraphyletic.

These preliminary results indicate that *Tuber* and *Elaphomyces* may be ingroups within the *Pezizales*, and that *Cyttaria* may belong to the *Leotiales*. However, more species of the genera should be studied, and the secondary structures may have to be examined in detail in order to find signature characters that might be informative as phylogenetic markers, before any new steps are taken in the classification of the orders.

ACKNOWLEDGEMENTS

We are obliged to the Swedish Natural Science Research Council for financial support. We are grateful to Dr S. Sivertsen (Trondheim) for material of *Cyttaria*, to Mrs K. Forssell for material of *Elaphomyces*, to Mr G. Hållstam for material of *Tuber*, and to Dr L.E. Kers for a preliminary identification of the *Tuber* material used.

REFERENCES

Donoghue, M.J., R.G. Olmstead, J.F. Smith, and J.D. Palmer, 1992, Phylogenetic relationships of *Dipsacales* based on *rbcL* sequences, *Annals of the Missouri Botanical Garden* 79: 333-345.

Eriksson, O.[E.], 1981, The families of bitunicate ascomycetes, *Opera Botanica* 60: 1-220.

Eriksson, O.E., and D.L. Hawksworth, 1991, Outline of the ascomycetes - 1990, *Systema Ascomycetum* 9: 39-271.

Gamundí, I., 1971, Las *Cyttariales* sudamericanas (Fungi-ascomycetes), *Darwiniana* 16: 461-510.

Gamundí, I., 1991, Review of recent advances in the knowledge of the *Cyttariales*, *Systema Ascomycetum* 10: 69-77.

Korf, R.P., 1983, *Cyttaria* (*Cyttariales*): coevolution with *Nothofagus* and evolutionary relationships to the *Boedijnopezizeae* (*Pezizales*, *Sarcoscyphaceae*), *Australian Journal of Botany, Supplement* 10: 77-87.

Landvik, S., O.E. Eriksson, A. Gargas, and P. Gustafsson, 1993, Relationships of the genus *Neolecta* (*Neolectales* ordo nov., *Ascomycotina*) inferred from 18S rDNA sequences, *Systema Ascomycetum* 11: 107-118.

Parguey-Leduc, A., M.-C. Janex-Favre, and C. Montant, 1991, L'ascocarp du *Tuber melanosporum* Vitt. (truffe noire du Périgord, discomycètes): structure de la glèbe. II. Les veines stériles, *Cryptogamie, Mycologie* 12: 165-182.

Murray, V., 1989, Improved double-stranded DNA sequencing using the linear polymerase chain reaction, *Nucleic Acids Research* 17: 8889.

Saccardo, P.A., 1889, *Sylloge Fungorum, omnium hucusque cognitorum*, Vol. 8, Patavii.

Santesson, R., 1945, *Cyttaria*, a genus of inoperculate discomycetes, *Svensk Botanisk Tidskrift* 39: 319-345.

Swofford, D.L., 1991, *PAUP: Phylogenetic Analysis Using Parsimony, Version 3.0s*, Illinois Natural History Survey, Champaign.

Trappe, J.M., 1979, The orders, families, and genera of hypogeous *Ascomycotina* (truffels and their relatives), *Mycotaxon* 11: 297-340.

Vittadini, C., 1831, *Monographia Tuberacearum*, Felices Rusconi, Milano.

DISCUSSION

Malloch: Why was *Glaziella* not included in the cladograms as you mentioned you had received material from Kimbrough ?

Eriksson: This has not yet been sequenced. That will be our next task.

Taylor, J.: I wondered what molecules you used?

Landvik: All the cladograms presented were from 18S rDNA sequences but not the flanking regions.

Hawksworth: I was interested in your results for *Cyttariales* as I had considered including these in *Leotiales* in the 1983 edition of the *Dictionary of the Fungi* after examining some recent collections from Chile. Would you now synonymize the two orders?

Landvik: My results are still preliminary. I would wish to have the results from a second gene sequence before proposing a formal change.

Appendix 1. 18S rDNA sequences: **Neu** = *Neurospora crassa*, **Leo** = *Leotia lubrica*, **Cud** = *Cudonia confusa*, **Spa** = *Spathularia flavida*, **Neo** = *Neolecta vitellina*, **Pez** = *Peziza badia*, **Gyr** = *Gyromitra esculenta*, ***Ine*** = *Inermisia aggregata*, **Ple** = *Plectania nigrella*, **Tub** = *Tuber* cf. *rapaeodorum*, **Ela** = *Elaphomyces granulatus*, **Cyt** = *Cyttaria darwinii*, **Sac** = *Saccharomyces cerevisiae.*

APPENDIX 1.

```
Neu  ATGCATGTCTAAGTTTAAGCAA--TTAAACCGCGAAACTGCGAATGGCTCATTAAATCAGTTATAGTTTATTTGATAGTACCTT-ACTACATGG-ATAAC
Leo  ??????????????????????????????????????????????????????????????AGTACGTTTATTTGATAGTACCTT-ACTACTTGG-ATAAC
Cud  ?????????TAAGTATAAGCAA-ACTATACCGTGAAACTGCGAATGGCTCATTAAATCAGTTATCGTTTATTTGATAGTACTTT-ACTACATGG-ATAAC
Spa  ?????GCTTAAGTATAAGCAA-ACTATACCGTGAAACTGCGAATGGCTCATTAAATCAGTTATCGTTTATTTGATAGTACTTT-ACTACATGG-ATAAC
Neo  ??GCATGTCTAAGTATAAGCAA-ATTGTACAGTGAAACTGCGAATGGCTCATTAAATCAGTTATAGTTTATTTGATAGTACCTT-ACTACTTGG-ATAAC
Pez  ???????????????????????????????????????????????CTCAT??AATCAG???TACGTTATT-GATAGTACCTT-ACTACTTG--ATAAC
Gyr  ATGCATGTCTAAGTATAAGCAA-AATATACAGTGAAACTGCGAATGGCTCATTAAATCAGTTATCGTTTATTTGATAGTACCTC-ATTACTTGG-ATAAC
Ine  ??????TCTAAGTATAAGCAA-TCTATACAGTGAAACTGCGAATGGCTCATTAAATCAGTTATCGTTTATTTGATAGTACCTT-ACTACTTGG-ATAAC
Ple  ATGCATGTCTAAGTATAAGCAA-TCTATACAGTGAAACTGCGAATGGCTCATTAAATCAGTTATCGTTTATTTGATAGTACCTT-ACTACTTGG-ATAAC
Tub  ATGCATGTCTAAGTATAAGCAA-TCTATACAGTGAAACTGCGAATGGCTCATTAAATCAGTTATCGTTTATTTGATAGTACCTTGCTACTTGG-ATAAC
Ela  ATGCATGTCTAAGTATAAGCAAATCTATACGGTGAAACTGCGAATGGCTCATTAAATCAGTTATCGTTTATTTGATAGTACCTC-ACTACTTGG-ATAAC
Cyt  ATGCATGTCTAAGTATAAGCAA-TCTATACGGTGAAACTGCGAATGGCTCATTAAATCAGTTATCGTTTATTTGATAGTACCCT-ACTACTTGG-ATAAC
Sac  ATGCATGTCTAAGTATAAGCAA-TTTATACAGTGAAACTGCGAATGGCTCATTAAATCAGTTATCGTTTATTTGATAGTTCCTTTACTACATGGTATAAC
          10        20        30        40        50        60        70        80        90

Neu  CGTGGTAATTCTAGAGCTAATACATGCTAAAAACCCCGACTTC--GGAAGGGGTGTATTTATTAGATTAAAAACCAATG-CCCTT-CGGGGCTA-ACTGG
Leo  CGTGGTAATTCTAGAGCTAATACATGCTAAAAACCCCGACTTC--GGAAGGGGTGTATTTATTAGATAAAAAACCAATG-CCCTT-CGGGGCTC-CTTGG
Cud  CGTGGTAATTCTAGAGCTAATACATGCTAAAAACCCCGACTTC--GGAAGGG-TGTATTTATTAGATAAAAAACCAATG-CCCTT-CGG??CTC-CCTGG
Spa  CGTGGTAATTCTAGAGCTAATACATGCTAAAAACCCCGACTTT--GGAAGGGGTGTATTTATTAGATAAAAAACCAATG-?CCTT-CGGGGCTC-CCTGG
Neo  CGTGGTAATTCTAGAGCTAATACATGCTAAAAATCCCGACTTTTTGGAAGGGATGTATTTATTAGATAAAAAACCAATGATCCTTACGGATCTC-CCTGG
Pez  CGTGGTAATTCTAGAGCTAATACATGCTAAAAACCCCGACTTCACGGA-GGGGTGTATTTATTAGATAAAAAACCAATG--CCTT-CGGGCCTC--CTGG
Gyr  CGTGGTAATTCTAGAGCTAATACATGCTAAAAATCCCGACCCCT-GGAAGGGATGTATTTATTAGATAAAAAACCAATG--CCTT-CGG-GCTCACTTGG
Ine  CGTGGTAATTCTAGAGCTAATACATGCTAAAAATCCCGACCTCT-GGAAGGGATGTATTTATTAGATAAAAAACCAATG--CCTT-CGG-GCTC-CTTGG
Ple  CGTGGTAATTCTAGAGCTAATACATGCTAAAAATCCCGACCCCT-GGAAGGGATGTATTTATTACATAAAAAACCAATG--CCTT-CGG-GCTC-CTTGG
Tub  CGTGGTAATTCTAGAGCTAATACATGCTAAAAATCCCGACCCCT-GGAAGGGATGTATTTATTAGATAAAAAACCAATG--GCTT-CGG-CCTC-CCTGG
Ela  CGTGGTAATTCTAGAGCTAATACATGCTTAAAATCCCGACCCCC-GGAAGGGATGTATTTATTAGATAAAAAACCAATG--TCTT-CGGA-CTC-TCTGG
Cyt  CGTGGTAATTCTAGAGCTAATACATGCTTAAAACCCCGAC?TC--GGAAGGGGTGTATTTATTAGATAAAAAACCAATG-CCCTC-CGGGGCTC-CTTGG
Sac  CGTGGTAATTCTAGAGCTAATACATGCTTAAAATCTCGACCCTTTGGAAGAGATGTATTTATTAGATAAAAAATCAATG--TCTT-CGG-ACTC-TTTGA
          110       120       130       140       150       160       170       180       190

                                                       SL12,21
Neu  TGATTCATAATAA-TTCTCGAATCGCATGGCCTTGCGCTGGCGATGGTTCATTCAAATTTCTGCCCTATCAACTTTCGACGGCTGGGTCTTGGCCAGCCA
Leo  TGATTCATAATAACTAACGAATCGCATGGCCTTGCGCCGGCGATGGTTCATTCAAATTTCTGCCCTATCAACTTTCGATGGTAGGATAGTGGCCTACCA
Cud  TGATTCATAATAACTTAACGAATCGCATGGCCTTGCGC-GGTGATGGTTCATTCAAATTTCTGCCCTATCAACTTTCGATGGTAGGATAGTGGCCTACCA
Spa  TGATTCATAATAACTTAACGAATCGCATGGCCTTGCGCCGGCGATGGTTCATTCAAATTTCTGCCCTATCAACTTTCGATGGTAGGATAGTGGCCTACCA
Neo  TGATTCATGATAACTTTTCGAATCGCATGGCCTTGCGCTGGCGATGGTTCATTCAAATTTCTGCCCTATCAACTTTCGATGGTAGGATAGAGGCCTACCA
Pez  TGATTCATGATAACTTAACGAATCGCATGGCCTTGTGCCGGCGATGGTTCATTCAAATTTCTGCCCTATCAACTTTCGATGGTAGGATAGTGGCCTACCA
Gyr  TGATTCATGATAACTTAACGAATCGCATGGCCTTGTGCCGGCGATGGTTCATTCAAATTTCTGCCCTATCAACTTTCGATGGTAGGATAGTGGCCTACCA
Ine  TGATTCATGATAACTTTACGAATCGCATGGCCTTGTGC-GGCGATGGTTCATTCAAATTTCTGCCCTATCAACTTTCGATGGTAGGATAGTGGCCTACCA
Ple  TGATTCATGATAACTTAACGAATCGCATGGCCTTGTGCCGGCGATGGTTCATTCAAATTTCTGCCCTATCAACTTTCGATGGTAGGATAGTGGCCTACCA
Tub  TGATTCATGATAACTTAACGAATCGCATGGCCTTGCGCCGGCGATGGTTCATTCAAATTTCTGCCCTATCAACTTTNNNNGGTAGGATAGTGGCCTACCA
Ela  TGATTCATGATAACTTTGCGAATCGCATGGCCTTGCGC-GGCGATGGTTCATTCAAATTTCTGCCCTATCAACTTTCGATGGTAGGATAGTGGCCTACCA
Cyt  TGATTCATAATAACTTCACGAATCGCATGGCCTTGTGC-GGCGATGGTTCATTCAAATTTCTGCCCTATCAACTTTCGATGGTAGGATAGTGGCCTACCA
Sac  TGATTCATAATAACTTTTCGAATCGCATGGCCTTGTGCTGGCGATGGTTCATTCAAATTTCTGCCCTATCAACTTTCGATGGTAGGATAGTGGCCTACCA
          210       220       230       240       250       260       270       280       290

Neu  TGGTGACAACGGGTAACGGAGGGTTAGGGCTCGACCCCGGAGAAGGAGCCTG-AGAAACGGCTACTACATCCAAGGAAGGCAGCAGGCGCGCAAATTACC
Leo  TGGTTTCAACGGGTAACGGGGAATTAGGGTTCTATTCCGGAGAGGGAGCCTG-AGAAACGGCTACCACATCCAAGGAAGGCAGCAGGCGCGCAAATTACC
Cud  TGGTTTCAACGGGTAACGGGGAATTAGGGTTCTATTCCGGAGAGGGAGCCTG-AGAAACGGCTACCACATCCAAGGAAGGCAGCAGGCGCGCAAATTACC
Spa  TGGTTTCAACGGGTAACGGGGAATTAGGGTTCTATTCCGGAGAGGGAGCCTG-AGAAACGGCTACCACATCCAAGGAAGGCAGCAGGCGCGCAAATTACC
Neo  TGGTTATAACGGGTAACGGGGAATTAGGGTTCGATTCCGGAGAGGGAGCCTGGAGAAACGGCTACCACATCCAAGGAAGGCAGCAGGCGCGCAAATTACC
Pez  TGGTTTCAACGGGTAACGGGGAATTAGGGTTCTATTCCGGAGAGGGAGCCTG-AGAAACGGCTACCACATCCAAGGAAGGCAGCAGGCGCGCAAATTACC
Gyr  TGGTTTCAACGGGTAACGGGGAATTAGGGTTCTATTCCGGAGAGGGAGCCTG-AGAAACGGCTACCACATCCAAGGAAGGCAGCAGGCGCGCAAATTACC
Ine  TGGTTTCAACGGGTAACGGGGAATTAGGGTTCTATTCCGGAGAGGGAGCCTG-AGAAACGGCTACCACATCCAAGGAAGGCAGCAGGCGCGCAAATTACC
Ple  TGGTTTCAACGGGTAACGGGGAATTAGGGTTCTATTCCGGAGAGGGAGCCTG-AGAAACGGCTACCACATCCAAGGAAGGCAGCAGGCGCGCAAATTACC
Tub  TGGTTTCAACGGGTAACGGGGAATTAGGGTTCTATTCCGGAGAGGGAGCCTG-AGAAACGGCTACCACATCCAAGGAAGGCAGCAGGCGCGCAAATTACC
Ela  TGGTTTCAACGGGTAACGGGGAATTAGGGTTCGATTCCGGAGAGGGAGCCTG-AGAAACGGCTACCACATCCAAGGAAGGCAGCAGGCGCGCAAATTACC
Cyt  TGGTTTCAACGGGTAACGGGGAATTAGGGTTCGACTCCGGAGAGGGAGCCTG-AGANACGGCTACCACATCCAAGGAAGGCAGCAGGCGCGCAAATTACC
Sac  TGGTTTCAACGGGTAACGGGGAATAAGGGTTCGATTCCGGAGAGGGAGCCTG-AGAAACGGCTACCACATCCAAGGAAGGCAGCAGGCGCGCAAATTACC
          310       320       330       340       350       360       370       380       390

Neu  CAATCCCGACACGGGGAGGTAGTGACAATAAATACTGATACAGGGCTCTTTTG-G-GTCTTGTAATTGGAATGAGTACAATTTAAATCCCTTAACGAGGA
Leo  CAATCCCGACACGGGGAGGTAGTGACAATAAATACTGATATTGGGGTCTTTAG-G-CTCTAATAATTGGAATGAGTACAATTTAAATCCCTTAACGAGGA
Cud  CAATCC?GACACGGGGAGGTAGTTACAATAAATACAGATACAGGGCTCTTTTG-G-GTCTTGTAATTGGAATGAGTACAATTTAAATCCCTTAACGAGGA
Spa  CAATCCCGACACGGGGAGGTAGTGACAATAAATACTGATACAGGGCTCTTTTG-G-GTCTTGTAATTGGAATGAGTACAATTTAAATCCCTTAACGAGGA
Neo  CAATCCCGACACGGGGAGGTAGTGACAATAAATAACAATACAGGGCTCTTTTG-G-GTCTTGTAATTGGAATGAGTACAATTTAAATCCCTTAACGAGGA
Pez  CAATCCCGACACGGG-AGGTAGTGACAATAAATACTAATACAGGTGG-TTTATGCCTCTTGTAATTGGAATGAGTACAATTTAAATCTCTTAACGAGGA
Gyr  CAATCC?GACACGGGGAGGTAGTGACAATAAATACTGATACAGGGCTCTTTTG-G-GTCTTGTAATTGGAATGAGTACAAT?????????????????A
Ine  CAATCCCGACACGGGGAGGTAGTGACAATAAATACTGATACAGGGCC-TTTCG-G-GTCTTGTAATTGGAATGAGTACAATTTAAATCCATTAACGAGGA
Ple  CAATCCCGACACGGGGAGGTAGTGACAATAAATACTGATACAGGGCC-TTTCG-G-GTCTTGTAATTGGAATGAGTACAAATTAAATCCCTTAACGAGGA
Tub  CAATCCCGACACGGGGAGGTAGTGACAATAAATACTGATACAGGGCTCTTTTG-G-GTCTTGTAATTGGAATGAGTACAATTTAAATCCCTTAACG?GGA
Ela  CAATCCCGACACGGGGAGGTAGTGACAATAAATACTGATGTAGGGCCATTTAT-G-GTCTTGCAATCGGAATGAGTACAATTTAAATCCCTTAACGA???
Cyt  CAATCCCGACACGGG?????????????????????????????????????????????????????????????????????????????????????
Sac  CAATCCTAATTCAGGGAGGTAGTGACAATAAATAACGATACAGGGCCCATTCG-G-GTCTTGTAATTGGAATGAGTACAATGTAAATACCTTAACGAGGA
          410       420       430       440       450       460       470       480       490

                 NS2,NS3
Neu  ACAATTGGAGGGCAAGTCTGGTGCCAGCAGCCGCGGTAATTCCAGCTCCAATAGCGTATATTAAAGTTGTTGAGGTTAAAAAGCTCGTAGTTGAACCTTG
Leo  ACAATTGGAGGGCAAGTCTGGTGCCAGCAGCCGCGGTAATTCCAGCTCCAATAGCGTATATTAAAGTTGTTGCAGTTAAAAAGCTCGTAGTTGAACCTCG
Cud  ACAATTGGAGGGCAAGTCTGGTGCCAGCAGCCGCGGTAATTCCAGCTCCAATAGCGTATATTAAAGTTGTTGCAGTTAAAAAGCTCGTAGTTGAACCTTG
Spa  ACAATTGGAGGGCAAGTCTGGTGCCAGCAGCCGCGGTAATTCCAGCTCCAATAGCGTATATTAAAGTTGTTGCAGTTAAAAAGCTCGTAGTTGAACCTTG
Neo  ACAATTGGAGGGCAAGTCTGGTGCCAGCAGCCGCGGTAACTCCAGCTCCAATAGCGTATATTAAAGTTGTTGCAGTTAAAAAGCTCGTAGTTGAACTTTG
Pez  ACAATTGGAGGGCAAGTCTGGTGCCAGCAGCCGCGGTAATTCCAGCTCCAATAGCGTATATTAAAGTTGTTGCAGTTAAAAAGCTCGTAGTTGAACCTTG
Gyr  ACAATTGGAGGGCAAGTCTGGTGCCAGCAGCCGCGGTAATTCCAGCTCCAATAGCGTATATTAAAGTTGTTGCAGTTAAAAAGCTCGTAGTTGAACCTTG
Ine  ACAATTGGAGGGCAAGTCTGGTGCCAGCAGCCGCGGTAATTCCAGCTCCAATAGCGTATATTAAAGTTGTTGCAGTTAAAAAGCTCGTAGTTGAACCTTG
Ple  ACAATTGGAGGGCAAGTCTGGTGCCAGCAGCCGCGGTAATTCCAGCTCCAATAGCGTATATTAAAGTTGTTGCAGTTAAAAAGCTCGTAGTTGAACCTTG
Tub  ACAATTGGAGGGCAAGTCTGGTGCCAGCAGCCGCGGTAATTCCAGCTCCAATAGCGTATATTAAAGTTGTTGCAGTTAAAAAGCTCGTAGTTGAACCTTG
Ela  ACAATTGGAGGGCAAGTCTGGTGCCAGCAGCCGCGGTAATTCCAGCTCCAATAGCGTATATTAAAGTTGTTGCAGTTAAAAAGCTCGTAGTTGAACCTTG
Cyt  ?????????????????????????????????????????????????????????TATTAAAGTTGTTGCAGTTAAAAAGCTCGTAGTTGAACCTTG
Sac  ACAATTGGAGGGCAAGTCTGGTGCCAGCAGCCGCGGTAATTCCAGCTCCAATAGCGTATATTAAAGTTGTTGCAGTTAAAAAGCTCGTAGTTGAACTTTG
          510       520       530       540       550       560       570       580       590
```

```
Neu  GGCTCGGCCGTC-GGTCCGCCTCACCGCGTGCACTGA---CTG-GGTCGG-GCCTTTTTTCCTGGAGAACC-GCATGCCCTTCACTGGGTGTGTCGGGGA
Leo  GGCCTGGCTGGCCGGTCCGCCTCACCGCGTGCACTGG---TCC-GGCCGG-GCCTTTCCTTCTGGGGACCC-GCATGCACTTCAGTGTGTGTGCTGGGGA
Cud  GGCCTGGTTGGC?GGTCCGCCTCACCGCGTGCACTGG---TCC-GACCGG-GCCTTTCCTTCTAGGGAGCC-GCATGCCCTTCATTGGGTGTGTTGGGGA
Spa  GGCCTGGTTGGCCGGTCCGCCTCACCGCGTGCACTGG---TCC-GACCGG-GCCTTTCCTTCTAGGGAGCC-GCATGCCCTTCATTGGGTGTGTCGGGGA
Neo  GGCCTGGTTGGCCGGTCCGCCTAACGGTGTGCACTGG---TCCCGACCGG-GCCTTTCCTTCTGACTAACCCGCATGCCCTTCATTGGGTGTGCGGGCGA
Pez  GACCTGCCCAACCGGTCTACCTCACCGTATGCACTGG---TTT-GGTCGG-GTCTTTCCTTCTGGCAAACT-GCATGCCCTTCACTGGGTGTGTTTGGGA
Gyr  GGTCTGGCTGGCCGGTCCGCCTCACCGCGTGCACTGG---TCC-GGCCGG-?TCTTTCCTTCTGGCTAGCC-GCATGCCCTTCGTTGGGTGTGTCGGGGA
Ine  GGTCTGGCTACC?GGTCCGCCGTAA-GCGTGCACTGG---AAA-CCCCGG-ATCTTTCCTTCTGGCTAACC-TCATGCCCTTTACTGGGTGTGCTGGGGA
Ple  GGCCTGGCTGGC?GGTCCACCTCACCGTGAGAACTGG---TCC-GGCCGG-GTCTTTCCTTCTGGCTAACC-TCATGCCCTTCACTGGGTGTGTTGGGGA
Tub  GGTCTGGCTGGCCGGTCCGCCTCACCGCGTGCACTGG---TCC-GGCCGG-ATCTTTCCTTCTGGCTAGCC-GCATGCCCTTCACTGGGTGTGTCGGGGA
Ela  GGCCTGGCTGACCGGTCCGCCTCACCGCGTGTACTGG---TTC-GGCCGG-GCCTTTCCTTCTGGCGAACC-GCATGCCCTTCACTGGGTGTGTTGGGGA
Cyt  GGCCTGGCTGGC-GGTCCGCCTCACCGCGTGCACTGG---TCC-GGCCGG-GCCTTTCCT?CTAGGGAGCC-GCATGCCCTTCACTGGGTGTGTCGGGGA
Sac  GGCCCGGTTGGCCGGTCCGATT-TTTTCGTGTACTGGATTTCC--AACGGGGCCTTTCCTTCTGGCTAACC-TTGAGTCCTT--GTGGCTCTTG--GCGA
              610       620       630       640       650       660       670       680       690

                                                                                     SL43
Neu  ACCAGGACTTTTACCGTGAACAAATCAGATCGCTCAAAGAAGGCCT-ATGCTCGAATGTACTAGCATGGAATAATAGAATAGGACGTGTGGTTCT-ATTT
Leo  ?CCAGGACTTTTACT?TGAAAAAATTAGAGTGTTCAAAGCAGGCCT-ATGCTCGAATACATTAGCATGGAATAATAGAATAGGACGTGTGGTTCT-ATTT
Cud  ACTAGGACTTTTACTTTGAAAAAATTAGAGTGTTCAAAGCAGGCCT-ATGCTCGAATACATTAGCATGGAATAATAGAATAGGACGTGTGGTTCT-ATTT
Spa  ACTAGGACTTTTACTTTGAAAAAATTAGAGTGTTCAAAGCAGGCCT-ATGCTCGAATACATTAGCATGGAATAATAGAATAGGACGTGTGGTTCT-ATTT
Neo  ATCAGGACTTTTACTTTGAAAAAATTAGAGTGTTCAAAGCAGGCCT-TTGCCCGAATACATTAGCATGGAATAATAGAATAGGACGTGTGGTTCT-ATTT
Pez  ACCAGGACTTTTACTTTGAAAAAATTAGAGTGTTCAAAGCAGGCAT-TAGCTCGAATACATTAGCATGGAATAATAGAATAGGACGTGTGGTTCT-ATTT
Gyr  ACCAGGACTTTTACTTTGAAAAAATTAGAGTGTTCAA?????????????????????CATTAGCATGGAATAATAGAATAGGACGTGCGGTTCT-ATTT
Ine  TCCAGGACTTTTACTTTGAAAAAATTAGAGTGTTCAAAGCAGGCAT-TTGCTCGAATACATTAGCATGGAATAATAGAATAGGACGTGCGGTTCT-ATTT
Ple  ACCAGGACTTTTACTTTGAAAAAATTAGAGTGTTCAAAGCAGGCAT-TTGCTCGAATACATTAGCATGGAATAATAGAATAGGACGTGCGGTTCT-ATTT
Tub  ACCAGGACTTTTACTTTGAAAAAATTAGAGTGTTCAAAGCAGGCAT-ATGCTCGAATACATTAGCATGGAATAATAGAATAGGACGTGCGGTTCT-ATTT
Ela  ACCAGGACTTTTACTTTGAAAAAATTAGAGTGTTCAAAGCAGGCAT-ATGCTCGAATACATTAGCATGGAATAATAGAATAGGACG-GCGGTTCT-ATTT
Cyt  ACTAGGACTTTTACTTTGAAAAAATTAGAGTGTTCAAAGCAGGCCT-AT-CTCGAATACATTAGCATGNAATAATAGAATAGGGACGCGCAGTCTTATTT
Sac  ACCAGGACTTTTACTTTGAAAAAATTAGAGTGTTCAAAGCAGGCGTATTGCTCGAATATATTAGCATGGAATAATAGAATAGGACGTTTGGTTCT-ATTT
              710       720       730       740       750       760       770       780       790

                                                                    SL34
Neu  TGTTGGTTTCTAGGACCGCCGTAATGATTAATAGGGACAGTCGGGGGCATCAGTATTCAATTGTCAGAGGTGAAATTCTTGGATTTATTGAAGACTAACT
Leo  TGTTGGTTTCTAGGACCGCCGTAATGATTAATAGGGATAGTCGGGGGCATCAGTATTCAATTGTCAGAGGTGAAATTCTTGGATTTATTGAAGACTAACT
Cud  TGTTGGTTTCTAGGACCGCCGTAATGATTAATAGGGATAGTCGGGGGTGTCAGTATTGCGTTGTCAGAGGTGAAATTCTTGGATTTACGCAAGACTAACT
Spa  TGTTGGTTTCTAGGACCGCCGTAATGATTAATAGGGATAGTCGGGGGTGTCAGTATTGCGTTGTCAGAGGTGAAATTCTTGGATTTACGCAAGACTAACT
Neo  TGTTGGTTTCTAGGACCGCCGTAATGATTAATAGGGATAGTCGGGGGCATTAGTATTTCGTTGTCAGAGGTGAAATTCTTGGATTTACGAAAGACTAACT
Pez  TGTTGGTTTCTAGGACCACCGTAATGATTAATAGGGATAGTCGGGGGCATCAGTATTCAATCGTCAGAGGTGAAATTCTTGGATCGATTGAAGACTAACT
Gyr  TGTTGGTTTCTAGGACCGCCGTAATGATTAATAGGGATAGTCGGGGGCATCAGTATTCAATTGTCAGAGGTGAAATTCTTGGATTTATTGAAGACTAACT
Ine  TGTTGGTTTCTAGGACCGCCGTAATGATTAATAGGGATAGTCGGGGGCATCCGTATTCAATTGTCAGAGGTGAAATTCTTGGATTTATTGAAGACGAACT
Ple  TGTTGGTTTCTAGGACCGCCGTAATGATTAATAGGGATAGTCGGGGGCATCCGTATTCAATTGTCAGAGGTGAAATTCTTGGATTTATTGAAGACGAACT
Tub  TGTTGGTTTCTAGGACCGCCGTAATGATTAATAGGGATAGTCGGGGGCATCAGTATTCAATTGTCAGAGGTGAAATTCTTGGATTTATTGAAGACTAACT
Ela  TGTTGGTTTCTAGGACCGCCGTAATGATTAATAGGGATAGTCGGGGGCATCCGTATTCAATTGTCAGAGGTGAAATTCTTGGATTTATTGAAGACGAACT
Cyt  TGTTGGTTTCTAAGACCGCCGTAATGATTAATAGGGATAGTCGGGGGCATCAGTATTCAATTGTCAGAGGTGAAATTCTTGGATTTACTGAAGACTAACT
Sac  TGTTGGTTTCTAGGACCATCGTAATGATTAATAGGGACGGTCGGGGGCATCGGTATTCAATTGTC-GAGGTGAAATTCTTGGATTTATTGAAGACTAACT
              810       820       830       840       850       860       870       880       890

Neu  ACTGCGAAAGCATTTGCCAAGGATGTTTTCATTAATCAG-GAACGAAAGTTAGGGGATCGAAGACGATCAGATACCGTCGTAGTCTTAACCATAAACTAT
Leo  ACTGCGAAAGCATTTGCCAAGGATGTTTTCATTAATCAGTGAACGAAAGTTAGGGGATCGAAGACGATCAGATACCGTCGTAGTCTTAACCATAAACTAT
Cud  ACTGCGAAAGCATTCACCAAGGATG?TTTCATTAATCAGTGAACGAAAGTTAGGGGATCGAAGACGATCAGATACCGTCGTAGTCTTAACCATAAACTAT
Spa  ACTGCGAA???ATTCACCAAGGATGTTTTCATTAATCAGTGAACGAAAGTTAGGGGATCGAAGACGATCAGATACCGTCGTAGTCTTAACCATAAACTAT
Neo  ACTGCGAAAGCATTTGCCAAGGATGTTTTCATTAATCAA-GAACGAAAGTTAGGGGATCAAAGACGATCAGATACCGTCGTAGTCTTAACCATAAACGAT
Pez  ACTGCGAAAGCATTTGCCAAGGATGTTTTCATTAATCAG?GAACGAAAGTTAGGGGATCGAAGACGATCAGATACCGTCGTAGTCTTAACCATAAACTAT
Gyr  ACTGCGAAAGCATTTGCCAAGGATGTTTTCATTAATCAGTGAACGAAAGTTAGGGGATCGAAGACGATCAGATACCGTCGTAGTCTTAACCATAAACTAT
Ine  ACTGCGAAAGCATTTGCCAAGGATGTTTTCATTAATCAG-GAACGAAAGTTGAGGGATCGAAGACGATCAGATACCGTCGTAGTCTCAACCATAAACTAT
Ple  ACTGCGAAAGCATTTGCCAAGGATGTTTTCATTAATCAG-GAACGAAAGTTGAGGGATCGAAGACGATCAGATACCGTCGTAGTCTCAACCATAAACTAT
Tub  ACTGCGAAAGCATTTGCCAAGGATGTTTTCATTAATCAGTGAACGAAAGTTAGGGGATCGAAGACGATCAGATACCGTCGTAGTCTTAACCATAAACTAT
Ela  ACTGCGAAAGCATTTGCCAAGGATGTTTTCATTAATCAG-GAACGAAAGTTGAGGGATCGAAGACGATCAGATACCGTCGTAGTCTCAACCATAAACTAT
Cyt  ACTGCGAAAG?ATTTGCCAAGGATGTTTTCATTAATCAGGGAACGAAAGTTAGGGGATCGAAGACGATCAGATACCGTCGTAGTCTTAACCATAAACTAT
Sac  ACTGCGAAAGCATTTGCCAAGGACGTTTTCATTAATCAA-GAACGAAAGTTAGGGGATCGAAGATGATCTGGTACCGTCGTAGTCTTAACCATAAACTAT
              910       920       930       940       950       960       970       980       990

Neu  GCCGATTAGGGATCGGACGGTGTTATTTTTT-GACCCGTTCGGCACCTTACGATAAATCAAAATGTTTGGGCTCCTGGGGGAGTATGGTCGCAAGGCTGA
Leo  GCCGACTAGGGATCGG?CGATGTTAT?TTTTTGACTCG?TCGGCACCTT?CGAGAAATCAAAGT?T?TGGGTTCTGGGGGGAGTATGGTCGCAAGGCTGA
Cud  GCCGACTAGGGATCAGGCGATGTTATCTTTTTGACTCGCTTGGCACCTTACGAGAAATCAAAGTCTTTGGGTTCTGGGGGGAGTATGGTCGCAAGGCTGA
Spa  GCCGACTAGGGATCAGGCGATGTTATCTTTTTGACTCGCTTGGCACCTTACGAGAAATCAAAGTCTTTGGGTTCTGGGGGGAGTATGGTCGCAAGGCTGA
Neo  GCCGACTAGGGATCGGGCGTGCTCTTTTCTT-GACTCGCTCGGCACCTTATGAGAAATCAAAGTCTTTGGGTTCTGGGGGGAGTATGGTCGCAAGGCTGA
Pez  GCCGACTAGGGATTGGGCGATGTTCTTTTTT-GACTCGCTCAGCACCTTACGAGAAATCAAAGTCTTTGGGTTCTGGGGGGAGTATGGTCGCAAGGCTGA
Gyr  GCCGACTAGGGATCGGGGGATGCTTACTAGATGGCTCCCTCGGCACCTTACGAGAAATCAAAGTCTTTGGGTTCTGGGGGGAGTATGGTCGCAAGGCTGA
Ine  GCCGACTAGGGATCGGGCGATGTTTTATTCT-GACTCGCTCGGCACCTTACGAGAAATCAAAGTCTTTGGGTTCTGGGGGGAGTATGGTCGCAAGGCTGA
Ple  GCCGACTAGGGATCGGGCGATGTACTTTTCGTGACTCGCTCGGCACCTTACGAGAAATCAAAGTCTTTGGGTTCTGGGGGGAGTATGGTCGCAAGGCTGA
Tub  GCCGACTAGGGATCGGGCGATGTTCTTTTTCTGACTCGCTCGGCACCTTACGAGAAATCAAAGTCTTTGGGTTCTGGGGGGAGTATGGTCGCAAGGCTGA
Ela  GCCGACTAGGGATCGGGCGATGTTTTTCTTTTGACTCGCTCGGAACCTTGCGAGAAATCAAAGTCTTTGGGTTCTGGGGGGAGTATGGTCGCAAGGCTGA
Cyt  GCCGACTAGGGATCGGGCGATGTTATCTTTTTGACTCGCTCGGCACCTTACGAGAAATCAAAGTCTTTGGGTTCTGGGGGGAGTATGGTCGCAAGGCTGA
Sac  GCCGACTAG--ATCGGGTGGTGTTTTTTTAATGACCCACTCGGTACCTTACGAGAAATCAAAGTCTTTGGGTTCTGGGGGGAGTATGGTCGCAAGGCTGA
              1010      1020      1030      1040      1050      1060      1070      1080      1090

      NS4,NS5
Neu  AACTTAAAGAAATTGACGGAAGGGCACCACCA-GGGGTGGAGCCTGCGGCTTAATTTGACTCAACACGGGGAAACTCACCAGGTCCAGACACGATGAGGA
Leo  AACTTAAAGAAATTGACGGAAGGGCACCACCA-GGAGTGGAGCCTGCGGCTTAATTTGACTCAACACGGGGAAACTCACCAGGTCCAGACACAAAAAGGA
Cud  AACTTAAAGAAATTGACGGAAGGGCACCACCA-GGAGTGGA??GTGCGGCTTAATTTGACTCAACACGGGGAAACTCACCAGGTCCAGACACAATAAGGA
Spa  AACTTAAAGAAATTGACGGAAGGGCACCACCA-GGAGTGGA???TGCGGCTTAATTTGACTCAACACGGGGAAACTCACCAGGTCCAGACACAATAAGGA
Neo  AACTTAAAGGAATTGACGGAAGGGCACCACCA-GGAGTGGAGCCTGCGGCTTAATTTGACTCAACACGGGGAAACTCACCAGGTCCAGACACAGTAAGGA
Pez  AACTTAAAGGAATTGACGGAAGGGCACCACCA-GGAGTGGAGC-TGCGGCTTAATTTGACTCAACACGGGGAAACTCACCAGGTCCAGACACATTAAGGA
Gyr  AACTTAAAGGAATTGACGGAAGGGCACCACCA-GGAGTGGAGCCTGCGGCTTAATTTGACTCAACACGGGGAAACTCACCAGGTCCAGACACATTAAGGA
Ine  AACTTAAAGGAATTGACGGAAGGGCACCACCA-GGAGTGGA???TGCGGCTTAATTTGACTCAACACGGGGAAACTCACCAGGTCCAGACACATTAAGGA
Ple  AACTTAAAGGAATTGACGGAAGGGCACCACCA-GGAGTGGAGCCTGCGGCTTAATTTGACTCAACACGGGGAAACTCACCAGGTCCAGACACATTAAGGA
Tub  AACTTAAAGGAATTGACGGAAGGGCACCACCA-GGAGTGG????TGCGGCTTAATTTGACTCAACACGGGGAAACTCACCAGGTCCAGACACATTAAGGA
Ela  AACTTAAAGGAATTGACGGAAGGGCACCACCA-GGAGTGGA??CTGCGGCTTAATTTGACTCAACACGGGGAAACTCACCAGGTCCAGACACATTAAGGA
Cyt  AACTTAAAGAAATTGACGGAAGGGCACCACCA-GGAGTGG?G--TGCGGCTTAATTTGACTCAACACGGGGAAACTCACCAGGTCCAGACACAATAAGGA
Sac  AACTTAAAGGAATTGACGGAAGGGCACCACTA-GGAGTGGAGCCTGCGGC-TAATTTGACTCAACACGGGGAAACTCACCAGGTCCAGACACAATAAGGA
              1110      1120      1130      1140      1150      1160      1170      1180      1190
```

```
                                                                                  SL56,65
Neu  TTGACAGATTGAGAGCTCTTTCTTGATTTCGTGGGTGGTGGTGCATGGCCGTT-CTTAGTTGGTGGAGTGATTTGTCTGCTTAATTGCGATAACGAACGA
Leo  TTGACAGATTGAGAGCTCTTTCTTGATCTTGTGGGTGGTGGTGCATGGCCGTT-CTTAGTTGGTGGAGTGATTTGTCTGCTTAATTGCGATAACGAACGA
Cud  TTGACAGATTGAG??CTCTTTCTTGATTTTGTG?GTGGTGGTGCATGGCCGTT-CTTAGTTGGTGGAGTGATTTGTCTGCTTAATTGCGATAACGAACGA
Spa  TTGACAGATTGAGAGCTCTTTCTTGATTTTGTGGGTGGTGGTGCATGGCCGTT-CTTAGTTGGTGGAGTGATTTGTCTGCTTAATTGCGATAACGAACGA
Neo  TTGACAGATTGAGAGCTCTTTCTTGATTCTGTGGGTGGTGGTGCATGGCCGTT-CTTAGTTGGTGGAGTGATTTGTCTGCTTAATTGCGATAACGAACGA
Pez  TTGACAGATTGAGAGCTCTTTCTTGATTATGTGGGTGGTGGTGCATGGCCGTT-CTTAGTTGGTGGAGTGATTTGTCTGCTTAATTGCGATAACGAACGA
Gyr  TTGACAGATTGAGAGCTCTTTCTTGATCATGTGGGTGGTGGTGCATGGCCGTT-CTTAGTTGGTGGAGTGATTTGTCTGCTTAATTGCGATAACGAACGA
Ine  TTGACAGATTGAGAGCTCTTTCTTGATCATGTGGGTGGTGGTGCATGGCCGTT-CTTAGTTGGTGGAGTGATTTGTCTGCTTAATTGCGATAACGAACGA
Ple  TTGACAGATTGAGAGCTCTTTCTTGATCATGTGGGTGGTGGTGCATGGCCGTT-CTTAGTTGGTGGAGTGATTTGTCTGCTTAATTGCGATAACGAACGA
Tub  TTGACAGATTGAGAGCTCTTTCTTGATCATGTGGGTGGTGGTGCATGGCCGTT-CTTAGTTGGTGGAGTGATTTGTCTGCTTAATTGCGATAACGAACGA
Ela  TTGACAGATTGAGAGCTCTTTCTTGATCATGTGGGTGGTGGTGCATGGCCGTT-CTTAGTTGGTGGAGTGATTTGTCTGCTTAATTGCGATAACGAACGA
Cyt  TTGACAGATTGAGAGCTCTTTCTTGATTTTGTGGGTGGTGGTGCATGGCCGTT-CTTAGTTGGTGGAGTGATTTGTCTGCTTAATTGCGATAACGAACGA
Sac  TTGACAGATTGAGAGCTCTTTCTTGATTTTGTGGGTGGTGGTGCATGGCCGTTTCTCAGTTGGTGGAGTGATTTGTCTGCTTAATTGCGATAACGAACGA
          1210      1220      1230      1240      1250      1260      1270      1280      1290

                                                                                             NS6,NS7
Neu  GACCTTAACCTGCTAAATAGCCCGTATTGCTTTGGCAGTACGCTGGCTTCTTAGAGGGACTATCGGCT---CAAGCCGATGGAAGTTTGAGGCAATAACA
Leo  GACCTTAACCTGCTAAATAGCCAGGCTAGCTTTGGCTGGTCGCCGGCTTCTTAGAGGGACTATCGGCT---CAAGCCGATGGAAGTTTGAGGCAATAACA
Cud  GACCTTAACCTGCTAAATAGCCAGGCTAGCTTTGGCTGGTCGCCGGCTTCTTAGAGGGACTA???GCT---CAAG??--TGGAAGTTTGAGGCAATAACA
Spa  GACCTTAACCTGCTAAATAGCCAGGCTAGCTTTGGCTGGTCGCCGGCTTCTTAGAGGGACTATCGGCT---CAAGCCGATGGAAGTTTGAGGCAATAACA
Neo  GACCTTAACCTGCTAAATAGCCCGGCCAGCTTTTGCTGGTCGCTGGCTTCTTAGAGGGACTATTGGCAT--AAAGCCAATGGAAGTTTGAGGCAATAACA
Pez  GACCTTAACCTGCTAAATAGTCAGGCCAGCTTCGGCTGGTTGCAGGCTTCTTAGAGGGACTATCGGCTT--CAAGCCGATGGAAGTTTGAGGCAATAACA
Gyr  GACCTTAACCTGCTAAATAGCCCGGCC?GCTTCTGCGGGTCGCTGGCTTCTTAGAGGGACTATCGGCT---CAAGCCGATGGAAGTTTGAGGCAATAACA
Ine  GACCTTAACCTGCTAAATAGCCAGGCC-GCTTTTGCGGGTCGCCGGCTTCTTAGAGGGACTATCGGATTT-CAAGTCGATGGAAGTTTGAGGCAATAACA
Ple  GACCTTAACCTGCTAAATAGCCCGGCCCGCTTTTGCGAGTCGCTGGCTTCTTAGAGGGACTATCGGATTT-CAAGACGATGGAAGTTTGAGGCAATAACA
Tub  GACCTTAACCTGCTAAATAGCCCGGCC?????????????????????TCTTAGAGGGACTATCGGCT---CAAGC?-ATGGAAGTTTGAGGCAATAACA
Ela  GACCTTAACCTGCTAAATAGCCAGGCTCGC--TTCGGGGTCGCCGGCTTCTTAGAGGGACTATCGGATTTACAAGACGATGGAAGTTTGAGGCAATAACA
Cyt  GACCTTAACCTGCTAAATAGCCCGGCTAGCCTCGGCTGGTCGCGGGCTTCTTAGAGGGACTATCGGCT---CAAGCCGATGGAAGTTT-AGGCAATAACA
Sac  GACCTTAACCTACTAAATAGTGGTGCTAGCATTTGCTGGTTATCCACTTCTTAGAGGGACTATCGGTT?--CAAGCCGATGGAAGTTTGAGGCAATAAGA
          1310      1320      1330      1340      1350      1360      1370      1380      1390

Neu  GGTCTGTGATGCCCTTAGA-TGTTCTGGGCCGCACGCGCGCTACACTGACACAGCCAGCGAGTAC-TC-CCTTGGCCGGAAGGTCCGGGTAATCTTGTTA
Leo  GGTCTGTGATGCCCTTAGA-TGTTCTGGGCCGCACGCGCGCTACACTGACAGAGCCAACGAGTTCATCACCTTGGCCGAAAGGCCTGGGTAATCTTGTTA
Cud  ?????????????????????????????????????????GTGACAGAGCCAACGAGTTCATCACCTTAGCCGAGAG?TTGGGTAATCTTGTTA
Spa  GGTCTGTGATGCCCTTAGA-TGTTCTGGGCCGCACGCGCGCTACACTGACAGAGCCAACGAGTTCATCACCTTAGCCGAAAGGTTTGGGTAATCTTGTTA
Neo  GGTCTGTGATGCCCTTCGA-TGTCCTGGGCCGCACGCGCGCTACACTGACGAAGCCAGCGAGTTAATCACCTTGGCCGGAAGGTCTGGGTAATCTTGTTA
Pez  GGTCTGTGATGCCCTTAGA-TGTTCTGGGCCGCACGCGCGCTACACTGACAGAGCCAACGAGTCTAT-ACCTTGGCCGAAACGTT-GGGTAATCTTGTGA
Gyr  GGTCTGTGATGCCCTTAGA-TGTTCTGGGCCGCACGCGCGCTACACTGACAGAGCCAACGAGTACATCACCTTGGCCGGAAGGTCTGGGTAATCTTGTTA
Ine  GGTCTGTGATGCCCTTAGA-TGTTCTGGGCCGCACGCGCGCTACACTGACAGAGCCAACGAGTACATCACCTTGGCCGGAAGGTCTGGGTAATCTTGTTA
Ple  GGTCTGTGATGCCCTTAGA-TGTTCTGGGCCGCACGCGCGCTACACTGACAGAGCCAACGAGTACATCACCTTGGCCGGAAGGTCTGGGTAATCTTGTTA
Tub  GGTCTGTGATGCCCTTAGA-TGTTCTGGGCCGCACGCGCGCTACACTGACAGAGCCAACGAGTTCATCACCTTTGCCGGAAGGTCTGGGTAATCTTGTTA
Ela  GGTCTGTGATGCCCTTAGA-TGTTCTGGGCCGCACGCGCGCTACACTGACAGAGCCAACGAGTTCTTCACCTTGGCCGGAAGGTCTGGGTAATCTTGTTA
Cyt  GGTCTGTGATGCCCTTAGA-TGTTCTGGGCCGCACGCGCGCTACACTGACAGAGCCAACGAGTTCATCCCCTTGGCCGAAAGGTCTGGGTAATCTTGTTA
Sac  GGTCTGTGATGCCCTTAGAACGTTCTGGGCCGCACGCGCGCTACACTGACGGAGCCAGCGAGTCTAA--CCTTGGCCGAGAGGTCTTGGTAATCTTGTGA
          1410      1420      1430      1440      1450      1460      1470      1480      1490

Neu  AACTGTGTCGTGCTGGGGATAGAGCATTGCAATTATTGCTCTTCAACGAGGAATCCCTAGTAAGCGCAAGTCATCAGCTTGCGTTGATTACGTCCCTGCC
Leo  AACTCTGTCGTGCTGGGGATAGAGCATTGCAATTATTGCTCTTCAACGAGGAATTCCTAGTAGGCGCAAGTCATCAGCTTGTGCCGACTACGTCCCTGCC
Cud  AACTCTGTCGTGCTGGGGATAGAGCATTGCAATTATTGCTCTTCAACGAGGAATTCCTAGTAAGCGCAAGTCATCAGCTTG???TGATTACGTCCCTGCC
Spa  AACTCTGTCGTGCTGGGGATAGAGCATTGCAATTATTGCTCTTCAACGAGGAATTCCTAGTAAGCGCAAGTCATCAGCTTGCGCTGATTACGTCCCTGCC
Neo  AACTTCGTCGTGCTGGGGATAGAGCATTGCAATTATTGCTCTTCAACGAGGAATTCCTAGTAAGCGCAAGTCATCAGCTTGCGTTGATTACGTCCCTGCC
Pez  AACTCTGTCGTGCTGGGGATAGAGCATTGCAATTATTGCTCTTCAACGAGGAATTCCTAGTAAGCGCAAGTCATCAGCTTGCGTTGACTACGTCCCTGCC
Gyr  AACTCTGTCGTGCTGGGGATAGAGCATTGCAATTATTGCTCTTCAACGAGGAATTCCTAGTAAGCGCAAGTCATCAGCTTGCGTTGATTACGTCCCTGCC
Ine  AACTCTGTCGTGCTGGGGATAGAGCATTGCAATTATTGCTCTTCAACGAGGAATTCCTAGTAAGCGCAAGTCATCAGCTTGCGTTGATTACGTCCCTGCC
Ple  AACTCTGTCGTGCTGGGGATAGAGCATTGCAATTATTGCTCTTCAACGAGGAATTCCTAGTAAGCGCAAGTCATCAGCTTGCGTTGATTACGTCCCTGCC
Tub  AACTCTGTCGTGCTGGGGATAGAGCATTGCAATTATTGCTCTTCAACGAGGAATTCCTAGTAAGCGCAAGTCATCAGCTTGCGTTGATTACGTCCCTGCC
Ela  AACTCTGTCGTGCTGGGGATAGAGCATTGCAATTATTGCTCTTCAACGAGGAATTCCTAGTAAGCGCAAGTCATCAGCTTGCGTTGATTACGTCCCTGCC
Cyt  AACTCTGTCGTGCTGGGGATAGAGCATTGCAATTATTGCTCTTCAACGAGGAATTCCTAGTAAGCGCAAGTCATCA??TTGCG??GATTACGTCCCTGCC
Sac  AACTGCGTCGTGCTGGGGATAGAGCATTGTAATTATTGCTCTTCAACGAGGAATTCCTAGTAAGCGCAAGTCATCAGCTTGCGTTGATTACGTCCCTGCC
          1510      1520      1530      1540      1550      1560      1570      1580      1590

Neu  CTTTGTACACACCGCCCGTCGCTACTACCGATTGAATGGCTCAGTGAGGCTTCCGGAC-TGGCCCAGGGAGGTCGGCAACGACCACCCAGGGCCGGAAAG
Leo  CTTTGTACACACCGCCCGTCGCTACTACCGATTGAATGGCTCAGTGAGGCTT-CGGACCTGGC?CAGGGAGGGCGGCAACAATACTACC???????????
Cud  CTTTGTACACACCGCCCGTCGCTACTACCGATTGAATGGCTAAGTGAGGCTTTCGGAC-TGGCTCAAGCAGATTGGCAACGATCAGCCCGAGCTGGAAAG
Spa  CTTTGTACACACCGCCCGTCGCTACTACCGATTGAATGGCTAAGTGAGGCTTTCAGAC-TGGCTTAAGCAGATTGGCAACGATCAGCCCGAGCTGGAAAG
Neo  CTTTGTACACACCGCCCGTCGCTACTACCGATTGAATGGCTTAGTGAGGCCTCAGGAC-TGGCTTTGGATGACTGGCAACGGTTGTCTGTCGCTGGAAAT
Pez  CTTTGTACACACCGCCCGTCGCTACTACCGATTGAATGGCTCAGTGAGGCCT-CGGAC-CGGTCCAGGAATGTCGGCAACGATC???????????????
Gyr  CTTTGTACACACCGCCCGTCGCTACTACCGATTGAATGGGTCAGTGAGGCCTTCGGAC-TGGCC??GG?AGATCGGCAACGATCACC?????????AAG
Ine  CTTTGTACACACCGCCCGTCGCTACTACCGATTGAATGGCTTAGTGAGGCTTCAGGAC-TGTCCAGTG?AGATCGGCAACGATCATCACCGGACGGAAAT
Ple  CTTTGTACACACCGCCCGTCGCTACTACCGATTGAATGGCTTAGTGAGGCCTTCGGAC-TGGCCCGAGGAGGTCGGCAACGACCACCTTGGCC?GGAAAG
Tub  CTTTGTACACACCGCCCGTCGCTACTACCGATTGAATGGCTCAGTGAGGCTT-CGGAG-TGCCCGATG-AGGTCGGCAACGACCACCGCCAGGTGCAAAG
Ela  CTTTGTACACACCGCC-GTCGCTACTACCGATTGAATGGCTTAGTGAGGCTT-GGGAC-TGATCTGCGGTCGTCGGCAACG--CAGCTGCGAGTCGGAAC
Cyt  CTTTGTACACACCGCCCGTCGCTACTACCGATTGAACGGCTCAGTGAGGCTTTCGGAC-TGGCCTAAGAAGAGTGCGAC?CTCGTCTAGGGCC-GGAAAG
Sac  CTTTGTACACACCGCCCGTCGCTAGTACCGATTGAATGGCTTAGTGAGGCCTCAGGATCTGCTTAGAGAAGGG-GGCAACTCCATCTCAGAGCGAGAA-T
          1610      1620      1630      1640      1650      1660      1670      1680      1690

Neu  CTATCCAAACTCGGTCATTTAGAGGAAGTAAAAGTCGTAACAAGGT
Leo  ??????????????????????????????????????????????
Cud  TTGTCCAAACTTGGTCATTTAGAGGAAGTAAAAGTCGTAACAAGGT
Spa  TTGTCCAAACTTGGTCATTTAGAGGAAGTAAAAGTCGTAACAAGGT
Neo  TTGGTCAAACTTGGTCATTTAGAGGAAGTAAAAGTCGTAACAAGG?
Pez  ??????????????????????????????????????????????
Gyr  T?GGTCAAACTTGGTCATTTAGAGGAA???????????????????
Ine  CTAGTCAAACTTGGTCATTTAGAGGAAGTAAAAGTCGTAACAAGG?
Ple  TTGGTCAAACTTGGTCATTTAGAGGAACTAAAAGTCGTAACAAGGT
Tub  TTGGTCAAACTTGGTC??????????????????????????????
Ela  CTAGTCAA-CTTGGTCATTTAGAGGAAGTAAAAGTCGTAACAAGGT
Cyt  TTGTTCAAACTTGGT??TTTA?????????????????????????
Sac  TTGGACAAACTTGGTCATTTAGAGGAACTAAAAGTCGTAACAAGGT
          1710      1720      1730      1740
```

CLADISTIC ANALYSIS OF PARTIAL ssrDNA SEQUENCES AMONG UNITUNICATE PERITHECIAL ASCOMYCETES AND ITS IMPLICATIONS ON THE EVOLUTION OF CENTRUM DEVELOPMENT

J.W. Spatafora[1] and M. Blackwell[2]

[1]Department of Botany, Duke University
Durham, North Carolina 27708, USA

[2]Department of Botany, Louisiana State University
Baton Rouge, Louisiana 70803, USA

SUMMARY

Cladistic analysis using maximum parsimony was performed on partial sequences of the nuclear-encoded small subunit ribosomal DNA among unitunicate perithecial ascomycetes, "pyrenomycetes". Two lines of evolution were inferred within the group. Centrum types as described by multiple investigators are mapped onto the gene tree, and the evolution of centrum development is discussed. The evanescent paraphyses of the *Clavicipitales* and *Hypocreales* are treated as homologous, whereas, the homology of the centrum pseudoparenchyma described in *Diaporthe* centrum type versus that described for the genera *Melanospora*, *Ceratocystis* and the order *Microascales* is questioned. The existence of two lineages within the pyrenomycetes argues that parallel evolution involving the formation of the central cavity is an important aspect of centrum development.

INTRODUCTION

Past classifications for ascomycetous fungi have been based upon one or few characters and have primarily reflected gross morphological similarities. The result has been artificial taxonomic assemblages that have proven unstable as our knowledge of these organisms has increased. A more logical approach is to construct a natural classification that conveys our understanding of the genealogical relationships, i.e. a phylogenetic classification. In addition to communicating evolutionary relationships, a phylogenetic classification provides information concerning character evolution, selection pressures and a host of other evolutionary factors.

The cornerstone of a stable phylogenetic classification is a stable phylogeny or, more accurately stated, a robust phylogenetic hypothesis. Such a hypothesis is inferred from a suite of characters and analyzed within the paradigm of a defined phylogenetic methodology. Characters may be morphological, ecological, behavioural, biochemical or molecular, and analyzed via distance and/or character-based algorithms (Swofford and Olsen, 1990). The resulting phylogenetic hypothesis is depicted as a series of bifurcating events that terminate in the taxa sampled, i.e. a dendrogram, cladogram, or tree. When we depict evolutionary relationships in the form of a tree, we acknowledge that genealogical relationships are hierarchical by nature. It is argued that such a hierarchy can be recovered by investigating those characters that are most likely to have recorded the speciation event(s) in question (Hennig, 1966). That is, two taxa can be united as sister taxa if one can identify characters or character states that are unique to them. These shared, derived characters (synapomorphies) are the discrete units of cladistic analysis, which allows for the detection of a hierarchical correlation or congruence among numerous putative synapomorphies. The literature is replete with arguments for and against cladistic theory, and a lengthy defence is not possible here (Swofford and Olsen, 1990). Suffice it to say that the data presented here were analyzed cladistically invoking maximum parsimony.

The focus of this study is to estimate phylogenetic relationships among those taxa that produce unitunicate asci within perithecial ascomata, i.e. pyrenomycetes. Phylogenetic hypotheses will be generated from the cladistic analysis of characters derived from the nuclear-encoded small subunit ribosomal DNA (ssrDNA). We have mapped characters derived from the pyrenomycete centrum as described by several investigators onto the ssrDNA gene tree. This approach allows for an independent assessment of homology that has been proposed for several characters, e.g. paraphyses and centrum pseudoparenchyma. Furthermore, new hypotheses, which can be tested by additional independent data sets, are proposed for these characters

This study represents the first broad scaled survey of these organisms in molecular systematics, however, the number of taxa sampled is inadequate with respect to the sheer numbers of these fungi. In addition, the phylogenetic hypotheses are drawn from a single gene, which in some studies have been shown to be inconsistent with the accepted organismal phylogeny. Therefore, the results inferred from these data should be interpreted within the strict boundaries of the taxa and characters sampled and treated as a working phylogenetic hypothesis and not a "phylogeny".

MATERIALS AND METHODS

The specifics of DNA isolation techniques, polymerase chain reactions (PCR) and DNA sequencing parameters, DNA sequence alignments and data analysis are addressed elsewhere (Spatafora and Blackwell, 1993a, b). Approximately, 900 base pairs (bp) were determined for the ssrDNA from 36 ascomycetes. The ingroup consists of 30 pyrenomycetes; the outgroup comprises five taxa from the *Endomycetales* and *Taphrina deformans*. Analyses were performed with and without the outgroup to detect if its inclusion had any topological effects within the ingroup. All analyses were performed using PAUP 3.0r (Swofford, 1990) on a MacIntosh IIfx. Only heuristic searches were possible due to the number of taxa. To ensure as thorough a search as possible, the input order of taxa were varied and multiple (ten) replications of branch swapping were performed.

Taxon sampling was designed to survey a wide range of morphological and ecological variation. Attention was paid to sampling those taxa that exhibit characters, or character states, which have historically played prominent roles in pyrenomycete classifications. Specifically, this sampling represents a survey of different modes of centrum development (centrum types) as defined in past and more contemporary studies. Designated ordinal rankings follow those of Hawksworth et al. (1983).

RESULTS

Cladistic analysis of the ingroup without the use of an outgroup resulted in three most parsimonious trees (mpts) of 326 steps with a CI, RI and RC of 0.564, 0.751 and 0.424, respectively. A strict consensus of these three mpts is 327 steps and has a CI, RI and RC of 0.563, 0.749 and 0.422, respectively (Fig. 1). The subsequent inclusion of the outgroup resulted in two mpts with identical ingroup topologies as the unrooted analysis. These two mpts have a length of 497 steps and CI, RI and RC of 0.549, 0.760 and 0.418, respectively. A strict consensus of these two mpts is 498 steps with a CI, RI and RC of 0.548, 0.759 and 0.416, respectively (Fig. 2). Bootstrap values were derived for both analyses from 100 bootstrap replications; those nodes that are supported by values greater than 75% are denoted by their respective bootstrap values (Figs 1-2).

The monophyly of the ingroup was maintained with respect to the outgroup. It received a bootstrap value of 100%. The ingroup consists of two subclades, which we have arbitrarily designated A and B (Fig. 2). Subclade A comprises taxa sampled from the orders *Clavicipitales*, *Hypocreales* and *Microascales* and the genera *Ceratocystis*, *Glom-*

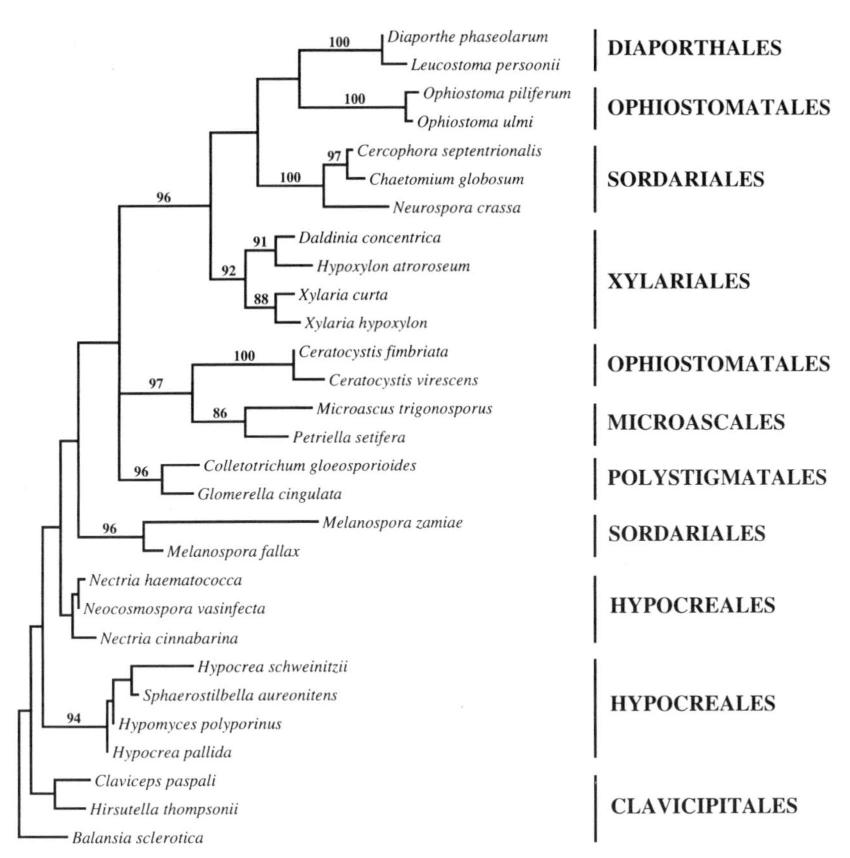

Figure 1. Strict consensus cladogram of three equally most parsimonious trees. This consensus tree is an unrooted network of the pyrenomycete ingroup. Numbers above the branches denote those branches that received >75% bootstrap values from 100 bootstrap replications.

erella and *Melanospora*. Subclade B consists of taxa sampled from the *Xylariales*, *Diaporthales* and *Sordariales* and the genus *Ophiostoma*. Subclade A is supported by a bootstrap value of 89%; subclade B did not receive a bootstrap value greater than 75%. Although the monophyly of the ingroup is strongly supported, taxon sampling is not sufficient to argue that all pyrenomycetes are part of this monophyletic group. Therefore, the term pyrenomycete will be used throughout this document as a matter of convenience; it is not meant to designate a formal taxonomic classification.

DISCUSSION

Maximum parsimony analysis of these data suggests that there exists two major lines of evolution (subclades A and B) within the pyrenomycetes sampled (Fig. 2). Two of the orders represented in this sampling have taxa in both subclades, thus arguing against their monophyly. These are the polyphyletic orders *Ophiostomatales* and *Sordariales*. Additionally, the *Hypocreales* was inferred to be a paraphyletic component of subclade A. The *Ophiostomatales* are addressed elsewhere (Hausner et al., 1992; Spatafora and Blackwell, 1993b; Blackwell and Spatafora, this volume) and will not receive extensive discussion here. Likewise, the non-monophyly of the *Hypocreales* has been addressed previously (Spatafora and Blackwell, 1993a). Taxa from both of these orders are discussed with respect to the evolution of centrum development.

The remaining non-monophyletic order is the *Sordariales*. There exists one clade that comprises *Neurospora crassa*, *Chaetomium globosum* and *Cercophora septentrionalis*. This group is a member of subclade B and is supported by a bootstrap value of

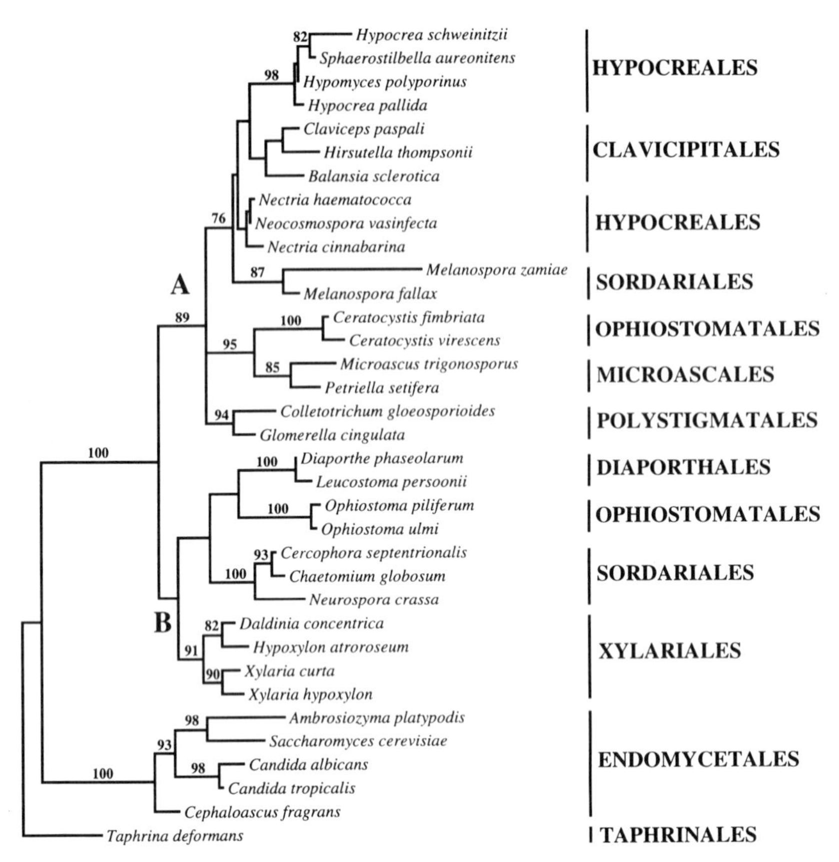

Figure 2. Strict consensus cladogram of three equally most parsimonious trees that were constructed using an outgroup. The outgroup comprises the *Endomycetales* and *T. deformans*. Numbers above the branches denote those branches that received >75% bootstrap values from 100 bootstrap replications.

100%. The second group consists of the genus *Melanospora* and is represented by the taxa *M. fallax* and *M. zamiae*. *Melanospora* has long been a problematic taxon. It has been placed in the *Diaporthales* (Luttrell, 1951), *Hypocreales* (Douget, 1955), *Sordariales* (Hawksworth et al., 1983), and *Xylariales* (Barr, 1990). The different placements of this genus emphasize different morphological characters. Luttrell's placement of *Melanospora* within the *Diaporthales* emphasized characters derived from centrum development. Its placement within the *Sordariales* and *Xylariales* are further variations of Luttrell's centrum concept and place an emphasis on ascospores. The placement of *Melanospora* within the *Hypocreales* emphasizes characters derived from asexual states (anamorphs). The anamorphs of *Melanospora* are enteroblastic phialides as are those of the *Hypocreales* (sensu Rogerson, 1970). The characters determined from the ssrDNA provide an independent test of these two sets of morphological characters, and the phylogenetic hypothesis proposed here is consistent with a hypocrealean treatment of the genus.

A similar situation is encountered with the taxa sampled from the *Clavicipitales*. Luttrell (1951) placed the *Clavicipitales* as a family within the *Xylariales* based upon centrum development. Barr (1990) proposed a similar treatment, although she retained the group as an order. Taxonomic treatments of the *Clavicipitales* previous to Luttrell emphasized characters derived from the stromata and anamorphs and placed the group as a sister taxon to the *Hypocreales* (Gäumann and Dodge, 1928; Gäumann, 1952). This latter treatment was supported by Rogerson (1970). Cladistic analysis of the ssrDNA suggests that the *Clavicipitales* is a sister taxon to the *Hypocreales* (Fig. 2; Spatafora and Blackwell, 1993a), thereby supporting the taxonomic treatments of Gäumann and Rogerson.

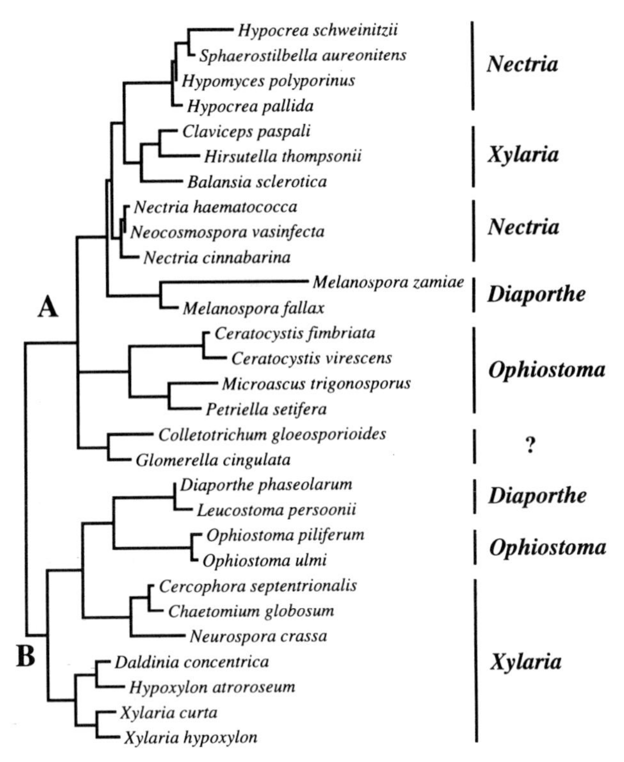

Figure 3. Enlargement of the pyrenomycete clade with centrum types as defined by Luttrell (1951) mapped onto the strict consensus tree. The *Diaporthe*, *Ophiostoma* and *Xylaria* centrum types are not congruent with the monophyletic groups as inferred from cladistic analysis of the ssrDNA.

The phylogenetic hypothesis proposed here suggests that similarities between the centra of the *Clavicipitales* and the *Xylaria* centrum types and *Melanospora* and the *Diaporthe* centrum types are cases of mistaken homology (Fig. 3). However, there are examples of traditional ideas on centrum development, which are consistent with the ssrDNA data. The groupings of the taxa sampled from the *Xylariales* and *Sordariales* are consistent with the *Xylaria* centrum type as defined by Luttrell (1951). Later studies by Huang (1976) and Uecker (1976) refined the centrum concept to delimit the centrum type found in the *Sordariales* as a variation of the *Xylaria* type. This development is referred to as the *Sordaria* centrum type. Uecker (1976) suggested that the *Sordaria* centrum type represents one extreme of a developmental continuum extending from the *Xylaria* type through the *Sordaria* type to the *Diaporthe* type. This hypothesis was later supported by Jensen (1983). This developmental continuum is consistent with the topology inferred from the ssrDNA (Fig. 4).

A brief review of salient centrum types is required here to assist in elucidating potential homologies among the pyrenomycetes. The *Xylaria* centrum type is the classic example of centrum development among pyrenomycetes (Luttrell, 1951; Alexopoulos and Mims, 1979). Typically, asci and paraphyses are produced from a hymenium, which occupies the basal to lateral inner perithecial walls. Luttrell included all taxa that possessed centra with true paraphyses in the order *Xylariales*. The *Clavicipitales* produces asci from an aparaphysate basal cluster but has lateral evanescent paraphyses. Luttrell treated the *Clavicipitales* centrum development as a variation of the pattern observed in *Chaetomium* (*Sordariales*), which possesses an aparaphysate basal cluster of asci and produce lateral persistent paraphyses (Corlett, 1966b). Additional studies within the *Sordariales* described the *Sordaria* centrum type as possessing centrum pseudoparenchyma and true paraphyses (Mirza and Khatoon, 1973; Huang, 1976). Uecker (1976) corroborated their findings and showed the asci are produced from a basal hymenium. Furthermore, the paraphyses are produced after the central cavity is developed, thus questioning

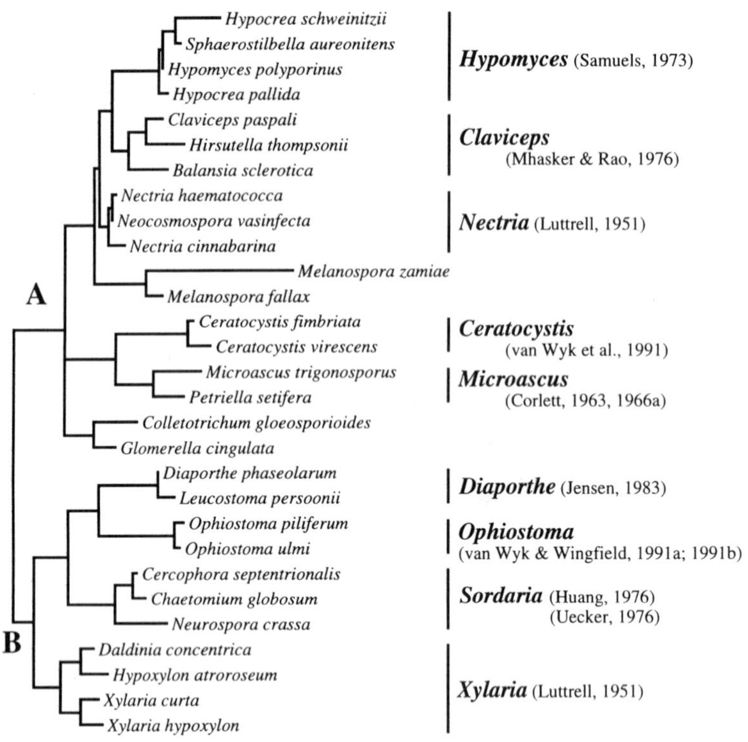

Figure 4. Englargement of the pyrenomycete clade with centrum types that are congruent with the ssrDNA gene tree mapped onto the strict consensus tree.

their role in formation of the central cavity as had been proposed previously (Luttrell, 1951).

The *Nectria* centrum type is considered diagnostic of the *Hypocreales*, and is characterized by apical, evanescent paraphyses (Luttrell, 1965). The central cavity is formed by the apical paraphyses that grow downward pushing the ascogenous system against a degenerative subhymenial layer of pseudoparenchyma (Doguet, 1956; Hanlin, 1971; Parguey-Leduc, 1967). The pattern of development in the genera *Hypomyces* and *Hypocrea* varies somewhat from the *Nectria* centrum type. Species of these genera produce centripetal paraphyses from the innermost layers of the perithecial walls (Hanlin, 1965; Canham, 1969; Samuels, 1973; Carey and Rogerson, 1981). The uppermost of these centripetal paraphyses elongate considerably relative to the more lateral ones and function in forming the central cavity. These elongated paraphyses are treated as homolgous to the apical paraphyses in *Nectria*. Samuels (1973) recognized the *Hypomyces* centrum type and suggested that centripetal paraphyses may be more widespread among the *Hypocreales*.

Mhasker and Rao (1976) recognized the *Claviceps* centrum type to distinguish it from the *Xylaria* centrum type. In *Epichloë typhina* and *E. cinera* (Mhasker and Rao, 1976) lateral, evanescent paraphyses are produced from the inner perithecial walls and are proposed to function in creating the central cavity. The similarities in function, evanescence and non-hymenial origin support the homology of paraphyses found in the *Clavicipitales* and *Hypocreales*. Therefore, the centrum development seen in *Chaetomium* may be viewed as a variation of the *Sordaria* centrum type, and similarities between the centra of *Chaetomium* and the *Clavicipitales* are the result of parallel evolution.

The *Diaporthe* centrum type is characterized by centrum pseudoparenchyma, and a basal hymenium that releases free, intact asci into the central cavity (Luttrell, 1951). The centrum cavity was originally described as lacking paraphyses, however, later they were demonstrated in *Gnomonia* (Morgan-Jones, 1959; Huang and Luttrell, 1982) and *Diaporthe phaseolarum* (Jensen, 1983; Uecker, 1988). These diaporthalean paraphyses are inconspicuous due to their short length and compression by the free asci filling the central cavity (Jensen, 1983). In the ssrDNA tree, the *Diaporthales* are part of a clade that is a sister group to the *Sordariales* (Fig. 2). The other members of this clade are the taxa sampled from the genus *Ophiostoma*. Curiously, the centrum development observed in *Ophiostoma* displays some similarities with that of the *Diaporthe* centrum type. The *Ophiostoma* centrum has been described as possessing pseudoparenchyma, paraphysis-like bands and a basal hymenium (Andrus, 1936; Hutchinson, 1950; Rosinski, 1961). Van Wyk and Wingfield (1991a, b) observed that the asci of *O. multiannulata* and *O. davidsonii* are produced from the base of the inner perithecium, then released into the central cavity. Once free within the central cavity the ascus walls evanesce and release free ascospores. This is not unlike the *Diaporthe* type centrum; asci are produced from a basal hymenium and then released into the central cavity. The production of evanescent asci should not be interpreted as inconsistent with the centrum development observed in *Ophiostoma* as representing a variation of the *Diaporthe* type. The production of evanescent asci has been shown to have evolved several times and is most likely the result of convergent evolution (Berbee and Taylor, 1992; Hausner et al., 1992; Spatafora and Blackwell, 1993b).

The remaining taxa that are reported to have a *Ophiostoma* centrum type are those sampled from the *Microascales* and *Ceratocystis*. Luttrell originally defined the *Ophiostoma* central cavity as being formed by the disintegration of pseudoparenchyma with asci produced from an ill-defined hymenium; paraphyses were lacking. In *Microascus* (Corlett, 1963) and *Petriella* (Corlett, 1966a) the ascogenous system is displaced to the apex of the central cavity by evanescent, sterile cells (hyphae) that grow inward from the base and sides. The ascogenous hyphae grow downward and produce terminal asci that become arranged irregularly throughout the centrum cavity. *Ceratocystis fimbriata* and *C. virescens* form the sister group to the *Microascales*. The centrum studies within *Ceratocystis* s.str. are in agreement with Luttrell's *Ophiostoma* centrum type (Elliott, 1925; Steirs, 1976; van Wyk et al., 1991). Centrum development appears to be quite variable within this clade and further studies are needed among the *Microascales* and *Ceratocystis* at both the developmental and molecular levels. Ironically, Luttrell's *Ophiostoma* type more accurately depicts the centrum development in the taxa currently circumscribed in *Ceratocystis* s.str. rather than *Ophiostoma*.

The centrum of *Melanospora* resembles the *Diaporthe* type in that it has been described as pseudoparenchymous. Asci are produced from a basal hymenium, however, free asci are not released into the central cavity. Rather, the ascus walls evanesce and release ascospores into the centrum, which is described as lacking paraphyses (Cannon and Hawksworth, 1982). The placement of *Melanospora* as a near relative to the *Hypocreales* does not support the homology assigned to the centrum pseudoparenchyma of *Melanospora* and the *Diaporthe* centrum type. The production of paraphyses in *Diaporthe* and not *Melanospora*, in conjuncton with the differences in the release of free asci into the central cavity mentioned above, may be indicative of non-homologous centrum pseudoparemchyma. Clearly, more studies are needed on the centrum development in *Melanospora*. The same can be said for *Glomerella*. Studies have begun (Uecker, unpubl.), and thus, discussion will be deferred until later.

The centrum as a whole is defined as the character and the variation observed in the components of the centrum constitute its character states. Taking this approach, the centrum types as described by various investigators are mapped onto the ssrDNA gene tree (Fig. 4). These data are congruent with the proposed developmental continuum extending from *Xylaria* through *Sordaria* to *Diaporthe* types of development. The use of an independent data set and broad taxon sampling suggests that this continuum may be extended to the genus *Ophiostoma*. Mapping of the characters states, given the proposed rooting of the ingroup, allows the centrum continuum of subclade B to be polarized. The *Xylaria* centrum type represents the more primitive character state of the continuum, and the *Diaporthe* type, including the genus *Ophiostoma*, is inferred as the more derived character state. The evolution of centrum development in subclade B involves a reduction in the prominence of paraphyses, a restriction of the production of asci to a more basal region of the inner perithecium, and ultimately the release of free asci into the central cavity. The evanescent asci of *Chaetomium* and *Ophiostoma* are derived independently.

The spatial relationships between paraphyses and asci, and qualitative aspects of paraphyses are important tendencies that partially define the character states of the centrum. The development of paraphyses and resulting terminology for them have long been a source of confusion (Luttrell, 1965). Following Jensen (1983), the apical paraphyses of the *Nectria* centrum are not treated as homologous to the pseudoparaphyses of loculoascomycetes (Strickman and Chadefaud, 1961), but the homology between the paraphyses of the *Nectria* and *Xylaria* centrum types is questioned. Also, the homology of centrum pseudoparenchyma across the taxa sampled is questioned.

The interpretation of the polarities and homologies among characters and character states discussed above is dependent on the placement of the root on the ingroup and taxon sampling within the ingroup in the molecular analysis. If the addition of taxa results in the survey of new character states not previously included in the analysis, the order and polarity of the characters may differ from previous analyses (Donoghue et al., 1989).

The use of independent data sets (molecular, morphological or otherwise) and explicit phylogenetic analyses allow one to test previous phylogenetic hypotheses based on intuitive investigations and generate new hypotheses. These new or refined hypotheses can be further tested by the inclusion of additional taxa and character sets. Congruence among multiple data sets will argue for strong phylogenetic hypotheses that more accurately reflect the genealogical relationships of the organisms in question.

ACKNOWLEDGMENTS

We thank Drs G.J. Samuels, F.A. Uecker, T. Harrington, D.W. Malloch, J.D. Rogers, and C.T. Rogerson for their assistance in providing cultures and Drs J.W. Taylor and M.L. Berbee for providing sequences prior to publication. Financial support from the National Science Foundation is greatfully acknowledged (NSF-BSR-9101088 to JWS; and NSF-BSR-8918157 and DEB-3209027 to MB).

REFERENCES

Alexopolous, C.J. and C.W. Mims, 1979, *Introductory Mycology*, 3rd edition. John Wiley, New York.

Andrus, C.F., 1936, Cell relations in the perithecium of *Ceratostomella multiannulata*, *Mycologia* 28: 133-153.

Barr, M.E., 1990, Prodromus to nonlichenized, pyrenomycetous members of class *Hymenoascomycetes*, *Mycotaxon* 39: 43-184.

Berbee, M.L. and J.W. Taylor, 1992, Convergence in ascospore discharge mechanism among pyrenomycete fungi based upon 18S ribosomal RNA gene sequence, *Molecular Phylogenetics and Evolution* 1: 59-71.

Cannon, P.F. and D.L. Hawksworth, 1982, A re-evaluation of *Melanospora* Corda and similar pyrenomycetes, with a revision of the British species, *Botanical Journal of the Linnean Society* 84: 115-160.

Canham, S.C., 1969, Taxonomy and morphology of *Hypocrea citrina*, *Mycologia* 61: 315-331.

Carey, S.T. and C.T. Rogerson, 1981, Morphology and cytology of *Hypomyces polyporinus* and its *Sympodiophora* anamorph, *Bulletin of the Torrey Botanical Club* 108: 13-24.

Corlett, M., 1963, The developmental morphology of two species of *Microascus*, *Canadian Journal of Botany* 41:253-266.

Corlett, M., 1966a, Developmental studies in the *Microascaceae*, *Canadian Journal of Botany* 44:79-88.

Corlett, M., 1966b, Perithecium development in *Chaetomium trigonosporum*, *Canadian Journal of Botany* 44: 155-162.

Donoghue, M.J., J.A. Doyle, J. Gauthier, A.G. Kluge and T. Rowe, 1989, The importance of fossils in phylogeny reconstruction, *Annual Review of Ecology and Systematics* 20: 431-460.

Douget, G., 1955, Le genre *Melanospora*, *Le Botaniste* 39:1-313.

Douget, G., 1956, Morphologie, organogénie du *Neocosmospora vasinfecta* E.F. Smith et du *Neocosmospora africana* von Arx., *Annales des Sciences Naturelle, Botanique, sér.* 11, 17: 353-370.

Douget, G., 1960, Morphologie, organogénie et évolution nucléaire de l'*Ephichloë typhina*. La place des *Clavicipitaceae* dans la classification, *Bulletin de la Société Mycologique de France* 76: 171-203.

Elliot, J.A., 1925, A cytological study of *Ceratostomella fimbriata* (E. & H.) Elliot, *Phytopathology* 5: 417-422.

Gäumann, E.A., 1952, *The Fungi: A Description of Their Morphological Features and Evolutionary Development*, Hafner Publishing, New York.

Gäumann, E.A. and C.W. Dodge, 1928, *Comparative Morphology of Fungi*, McGraw-Hill, New York.
Hanlin, R.T., 1965, Morphology of *Hypocrea schweinitzii*, *American Journal of Botany* 52: 570-579.
Hanlin, R.T., 1971, Morphology of *Nectria haematococca*, *American Journal of Botany* 58: 105-116.
Huang, L.H., 1976, Developmental morphology of *Triangularia backusii* (*Sordariaceae*), *Canadian Journal of Botany* 54:250-267.
Huang, L.H. and E.S. Luttrell, 1982, Development of the perithecium in *Gnomonia comari* (*Diaporthaceae*), *American Journal of Botany* 69: 421-443.
Hausner, G., J. Reid, and G.R. Klassen, 1992, Do galeate-ascospore members of the *Cephaloascaceae*, *Endomycetaceae*, and *Ophiostomataceae* share a common phylogeny?, *Mycologia* 84:870-881.
Hawksworth, D.L., B.C. Sutton and G.C. Ainsworth (eds.), 1983, *Ainsworth & Bisby's Dictionary of the Fungi*, 7th edition, Commonwealth Agricultural Bureaux, Slough.
Hennig, W., 1966, *Phylogenetic Systematics*, University of Illinios Press, Urbana.
Hutchinson, S.A., 1950, The perithecia of *Ophiostoma majus* (van Beyma) Goidànich, *Annals of Botany*, n.s. 14: 115-125.
Jensen, J.D., 1983, The development of *Diaporthe phaseolarum* variety *sojae* in culture, *Mycologia* 75: 1074-1091.
Luttrell, E.S., 1951, Taxonomy of the pyrenomycetes, *University of Missouri Studies* 3: 1-120.
Luttrell, E.S., 1965, Paraphysoids, pseudoparaphyses, and apical paraphyses, *Transactions of the British Mycological Society* 48: 135-144.
Mhasker, D.N. and V.G. Rao, 1976, Development of the ascocarp in *Epichloë cinerea* (*Clavicipitaceae*), *Mycologia* 68: 994-1001.
Mirza, J.H., and A. Khatoon, 1973, Studies on *Sordaria lumana* (Fuckel) Winter: the cytology of ascus development and developmental morphology of the perithecium, *Pakistan Journal of Botany* 5: 19-28.
Morgan-Jones, J.F., 1959, Morpho-cytological studies of the genus *Gnomonia*. III. Early stages of perithecial development, *Svensk Botanisk Tidskrift* 53: 81-101.
Parguey-Leduc, A., 1967, Recherches sur l'ontogénie et l'anatomie comparée des ascocarpes des Pyrénomycètes ascoloculaires. Seconde partie. Les ascocarpes des Pyrénomycètes ascoloculaires unituniques, *Annales des Sciences Naturelle, Botanique, sér.* 12, 8: 1-110.
Rogerson, C.T., 1970, The hypocrealean fungi (ascomycetes, *Hypocreales*), *Mycologia* 62: 865-910.
Rosinski, M.A., 1961, Development of the ascocarp of *Ceratocystis ulmi*, *American Journal of Botany* 48: 285-293.
Samuels, G.J., 1973, Perithecial development in *Hypomyces aurantius*, *American Journal of Botany* 60: 268-276.
Spatafora, J.W. and M. Blackwell, 1993a, Molecular systematics of unitunicate perithecial ascomycetes: the *Clavicipitales-Hypocreales* connection, *Mycologia*: in press.
Spatafora, J.W. and M. Blackwell, 1993b, The polyphyletic origins of the ophiostomatoid fungi, *Mycological Research*: in press.
Steirs, D.L., 1976, The fine structure of ascospore formation in *Ceratocystis fimbriata*, *Canadian Journal of Botany* 54: 1714-1723.
Strickman, E. and M. Chafaud, 1961, Recherches sur les asques et les périthèces des *Nectria* et réflexions sur l'evolution des ascomycetes, *Revue Génerale de Botanique* 68: 725-770.
Swofford, D.L., 1990, *PAUP: Phylogenetic analysis using parsimony, Version 3.0*, Illinios Natural History Survey, Champaign.
Swofford, D.L. and G.J. Olsen, 1990, Phylogeny reconstruction, *In*: *Molecular Systematics* (D.M. Hillis and C. Moritz, eds): 411-515. Sinaur Associates, Sunderland, Mass.
Uecker, F. A., 1976, Development and cytology of *Sordaria humana*, *Mycologia* 68: 30-46.
Uecker, F. A., 1988, A timed sequence of development of *Diaporthe phaseolarum* (*Diaporthaceae*) from *Stokesia laevis*, *Memoirs of the New York Botanic Garden* 49: 38-50.
van Wyk, P.W.J., M.J. Wingfield, and P.S. van Wyk, 1991, Ascospore development on *Ceratocystis moniliformis*, *Mycological Research* 95: 96-103.
van Wyk, P.W.J. and M.J. Wingfield, 1991a, Ultrastructure of ascosporogenesis in *Ophiostoma davidsonii*. *Mycological Research* 95: 725-730.
van Wyk, P.W.J. and M.J. Wingfield, 1991b, Ascospore ultrastructure and development in *Ophiostoma cucullatum*, *Mycologia* 83: 698-707.

DISCUSSION

Wingfield: It is clear that many characters morphologists have placed weight upon are not holding up to molecular examination. This is particularly evident in the ophiostomoid fungi. With biodiversity being so important we have to have names and I am concerned that we look at relationships based on molecular data and fewer people are studying characters. There is now a desperate need to find characters based on what we learn from molecular studies. Much more energy needs to be directed towards such activities. I am also unsure how many of the fungi we know can be examined from a molecular angle - many are known from fragmentary specimens.

Blackwell: What much of the molecular work shows is that in many cases we have been right with morphology by testing our hypotheses with an independent data set, as indicated in our paper. Spatafora has refined what Luttrell proposed and has been discredited so that we can return to morphological characters more easily.

Spatafora: I do not see a separation between searching for a common ancestor and looking for more characters. You cannot do one without the other, and you also need to define and analyze morphological characters more clearly. The problem with the majority of morphological classifications is that they are intuitive and not as objective as a defined and explicit methodology.

Kurtzman: Ascospore shape, as a taxonomic character in the yeasts, is not only weak from molecular comparisons but also from genetic studies (e.g. variations in ascospore types in the progeny from crossing different parent species).

Eriksson: It is best to have research groups with both traditional and molecular biologists. Applications for grants to undertake molecular work should include necessary morphological studies. Molecular biologists alone tend not to ask the correct questions, and morphologists alone cannot undertake the molecular studies.

Landvik: Now the PCR method is established we can start to think more carefully about what sequences tell us.

Hennebert: What is important at this meeting is the interactive stimulation between morphologists and molecular biologists. This process must be developed more. In my collaborative work with molecular biologists, I have come to appreciate that the dendrogram from different parts of the RNA molecule can be different. Whether the number of species considered is low or high can also make significant differences, as can the choice of an outgroup. The largest number of species should always be included, especially before any formal changes in nomenclature are made.

Blackwell: There is also a problem in how cladograms are presented, for example they can be rotated on the branch axis. The topology from different fungal studies has, however, been in good agreement. We must also be wary in that most are based only on 18S ribosomal genes. Spatafora tried vacuolar ATPase on a subset of the same taxa as a test whether that was correct. It was gratifying how close they were. I concur that we need to remain cautious about introducing nomenclatural changes.

Kurtzman: If we are a little conservative and do not move too quickly that should be fine.

Taylor, J.: If conflicts between trees arise, the first step is to apply statistical tests to ascertain if the apparent difference is significant, and the second is to reexamine the data. There may be no real conflict, or one data set may be inappropriate for the question being addressed.

Tehler: The differences emerging between molecular and morphological workers arise from differences in phylogenetic methods. Molecular biologists have to use cladistic methods to organize their data while many morphologists are not using such approaches. Morphologists are being challenged to organize their data in a way that the phylogenies based on the two types of data can be properly compared. In time all types of data will be used as total evidence.

MOLECULAR DATA SETS AND BROAD TAXON SAMPLING IN DETECTING MORPHOLOGICAL CONVERGENCE

M. Blackwell[1] and J.W. Spatafora[2]

[1]Department of Botany
Louisiana State University
Baton Rouge, LA 70803, USA

[2]Department of Botany
Duke University
Durham, NC 27708, USA

SUMMARY

Previous hypotheses of relationships between *Pyxidiophora* and other perithecial ascomycetes with morphological features facilitating arthropod dispersal were examined. Partial sequences of the nuclear encoded small subunit rDNA for about fifty taxa provided a database for phylogenetic analysis. The resulting analysis using parsimony criteria failed to support previous hypotheses of relationships involving taxa with arthropod associations. The following hypotheses are supported: (1) taxa with evansecent asci are in six independent lineages; (2) *Pyxidiophora*, *Subbaromyces*, and *Kathistes* are basal to the larger clade of derived perithecial species; (3) *Pyxidiophora* does not share a common ancestory with the *Hypocreales*, *Ophiostoma*, or *Ceratocystis*; and (4) *Pyxidiophora* is not allied with any of the yeasts, including *Cephaloascus* and *Ambrosiozyma*. More taxa remain to be sampled since *Pyxidiophora* is in a region of the tree that is still somewhat unstable.

INTRODUCTION

The attempt to bring order to the classification of ascomycetes has generally progressed slowly, with significant progress occurring when new taxonomic characters and new taxa have been recognized and applied to the solution of the problem. Periodic broad reviews of the voluminous literature are additionally helpful. However, a newer innovation to mycology, the use of explicit methods of data analysis, are essential to the mission of achieving a phylogenetic system of classification. Here we discuss our efforts to employ both new characters and cladistic analysis to obtain insight into an old question, the degree of convergent evolution among arthropod dispersed perithecial ascomycetes. Our discussion will centre around efforts to determine the ancestry of *Pyxidiophora*.

Morphological characters provide the basis of our current fungal systematics. When used carefully they have provided a wealth of information to distinguish taxa. Sometimes they serve to group similar taxa; less often they allow higher level taxa to be hierarchically arranged, and this is reflected in most recent systems of ascomycete classifications that do not attempt to group orders within higher taxa (Hawksworth, 1985). However, Barr (1987, 1990) provides a notable exception in her placing of orders within subclasses and classes, thereby providing explicit testable hypotheses.

Nevertheless, morphological characters present several problems. In ascomycetes such characters are relatively few in number, and their use is complicated by the difficulty of recognizing the effects of parallel and convergent evolution or, alternatively, rapid morphological divergence. Rapid divergence likely results in the many unique, derived (autapomorphic) characters that usually delimit orders well, but may fail to be predictive of higher level relationships. In addition character polarity is often problematical. However, greater problems with our characters have been the failure to use empirical methods of analysis and the inability to test the hypotheses with independent data sets (Blackwell, 1993).

PLACEMENTS PROPOSED FOR *PYXIDIOPHORA*

Pyxidiophora is an interesting candidate for discussion because its life history involves a complex dispersal system and its morphology has features that may be strongly selected to enhance arthropod dispersal. Lundqvist (1980) made an important contribution when he brought together species from eight genera and clarified the generic limits. He recognized that changes observed during ascospore maturation were the basis of the taxonomic confusion, and, additionally, pointed out an association of mature ascospores of several species with mites. *Pyxidiophora* is now a well-defined genus characterized by a basal cluster of early evanescent asci; long, usually one-septate, ascospores with a darkened apical region developing in age; and several distinct types of mitospores. A *Chalara* stage (Lundqvist, 1980) and an unnamed holoblastically formed conidium with percurrent or sympodial proliferation (Blackwell and Malloch, 1989b; Blackwell et al., 1993) have played a role in ordinal classification. Subsequent studies have focused on life history studies involving dispersal by mites that are in turn dispersed by beetles. The phoretic mode of dispersal is well established for a number of species (Blackwell et al., 1986a, b, 1989; Blackwell and Malloch, 1989a, b, 1990; Malloch and Blackwell, 1992).

Pyxidiophora was originally placed in the *Hypocreales* (Tulasne and Tulasne, 1865). Rogerson (1970), Arnold (1971), and Lundqvist (1980) retained the genus in the *Hypocreales*. In addition, Arnold (1971) erected the family *Pyxidiophoraceae* in recognition of the distinctive nature of the genus. The family was expanded by Lundqvist (1980) to include the cleistothecial genus *Mycorhynchidium*. Parguey-Leduc and Janex-Favre (1981) suggested a relationship between *Lulworthia* and *Ceratocystis*, *Pyxidiophora*, and *Thielavia* on the basis of ascospore and perithecium morphology and *Chalara*-type anamorphs, but made no nomenclatorial changes. There recently has been greater interest in the classification of *Pyxidiophora*, beginning with the classification of von Arx and van der Walt (1987), who placed the *Pyxidiophoraceae* in the *Ophiostomatales*. *Ceratocystis* and *Cryptendoxyla* also were moved to the family, primarily on the basis of reported *Chalara* conidial stages; they further suggested a derived relationship with the *Pyxidiophoraceae* and the yeast family *Metschnikowiaceae*.

Ontogenic studies of *Pyxidiophora* led Blackwell and Malloch (1989a) to propose a new model of evolution for the *Laboulbeniales* with *Pyxidiophora* as a sister group. While a strong case for a close relationship between *Pyxidiophora* and the *Laboulbeniales* was presented, the more difficult problem, assessing the relationships of *Pyxidiophora* among other mycelial perithecial ascomycetes, was not resolved. Based upon morphology of the holomorph and, to some extent, habitat and substrate, several orders - the *Sordariales*, *Ophiostomatales*, and *Hypocreales* - were considered and disregarded as possible relatives. *Pyxidiophora* species and other arthropod-dispersed perithecial ascomycetes were suggested to be convergent for characters that promote arthropod dispersal (long perithecial necks, evanescent asci, and yeast stages), the very characters used for taxonomy! This is certainly not a new idea for arthropod-associated species (Cain and Weresub, 1957). There is a large body of morphological, physiological, and life history data to separate *Ceratocystis* and *Ophiostoma* (de Hoog and Scheffer, 1984). Barr (1990) retained the *Pyxidiophoraceae* in the *Hypocreales*. But based upon the arguments of Blackwell and Malloch (1989a), Eriksson and Hawksworth (1991) suggested transferring *Pyxidiophora* to the *Laboulbeniales*.

The reiteration of the history of the classification of *Pyxidiophora* serves to emphasize the need for a data set independent of morphology to provide support for one of the competing hypotheses, or to reject them all in favour of a new hypothesis. It is also important to note that all of these hypotheses were arrived at intuitively. Without the use of emperical methods of data analysis an independent worker could fail at their reconstruction.

RELATIONSHIPS AMONG OPHIOSTOMATOID FUNGI

Malloch and Blackwell (1990, 1993) discussed eleven genera of ophiostomatoid fungi, including *Pyxidiophora*. Of this group, species of *Ceratocystis*, *Kathistes*, *Ophiostoma*, *Sphaeronaemella*, *Subbaromyces*, and *Pyxidiophora*, were included in the sequencing study. Cultures or specimens of the other ophiostomatoid genera (*Klasterskya*, *Rhynchonectria*, *Spumatoria*, and *Treleasia*) were not available and *Ceratocystiopsis* was not included. However, Hausner et al. (1982) included *Ceratocystiopsis* within *Ophiostoma*.

Because convergent evolution was central to the study it also was necessary to include representatives of a variety of perithecial ascomycete orders (*Hypocreales*, *Clavicipitales*, *Sordariales*, *Microascales*, *Polystigmatales*, *Diaporthales*, and *Xylariales*). *Chaetomium globosum* and *Melanospora* species of this group have evanescent asci as well. Members of other ascomycete groups (*Eurotiales*, *Dothideales*, *Ostropales*, *Endomycetales*, and *Taphrinales*) and three basidiomycete species (for outgroup rooting) also were included. However, outside the perithecial ascomycetes, taxon sampling was inadequate. This factor will be discussed again below. Whenever possible type species of taxa were used to facilitate potential nomenclatorial changes. Techniques are identical to those discussed in Spatafora and Blackwell (this volume).

The results of the phylogenetic analysis (Fig. 1) provide support for six independent lineages of ascomycetes with evanescent asci: (1) *Melanospora* species; (2) *Ceratocystis* species, *Sphaeronaemella fimicola*, *Halosphaeriopsis mediosetigera*, *Petriella setifera*, and *Microascus trigonosporus*; (3) *Ophiostoma* species; (4) *Chaetomium globosum*; (5) *Kathistes* species; and (6) *Pyxidiophora* species and *Subbaromyces splendens*. In most instances arthropod associations are known or strongly suspected in the species. These results are discussed in more detail elsewhere (Spatafora and Blackwell, 1993). Berbee and Taylor (1992) used complete small subunit rDNA sequences to provide evidence of convergence for evanescent asci in *Chaetomium*, *Ophiostoma*, and *Microascus* among the taxa they studied; they also showed a clade comprised of *Leucostoma persoonii*, *Ophiostoma ulmi*, *O. stenoceras*, and *Sporothrix schenkeii*. Hausner et al. (1992) used partial sequences of small subunit rDNA and large subunit rDNA to hypothesize a distant relationship for *Ceratocystis* s.str. with species of *Ophiostoma* and *Europhium*. Additionally, Hausner (1993) suggested inclusion of *Ceratocystis* in the *Microascales*.

Our results support those of Berbee and Taylor (1992), Hausner et al. (1992), and Hausner (1993), and expand the data set by adding additional controversial taxa. The inclusion of more non-ophiostomatoid taxa is expecially useful in providing evolutionary perspective.

Hypotheses including *Pyxidiophora* in the *Hypocreales* (Rogerson, 1970; Arnold, 1971; Lundqvist, 1980; Barr, 1990) and in the *Ophiostomatales* (von Arx and van der Walt, 1987) are not supported. Also a derived yeast connection (Redhead and Malloch, 1977; von Arx and van der Walt, 1987) with any perithical ascomycete is not supported, although a number of the arthropod-associated taxa are dimorphic. Hausner et al. (1992) also tested this hypothesis and our results support their conclusions. The laboulbenialean hypothesis has not yet been tested (Blackwell and Malloch, 1989a)*.

CONCLUSIONS

We now return to our original purpose, that of determining the relationships of *Pyxidiophora*. The use of an independent data set of molecular characters and an empirical method of analysis has only partially answered the questions posed concerning the relatives of *Pyxidiophora*. Our evidence does not support any previously hypothesized

---

**Note added in proof:* Analysis of sequence data recently acquired for *Rickia* sp. (*Laboulbeniales*) places it as a sister taxon to the clade of the two species of *Pyxidiophora* (Fig. 1) with no other changes in the tree topology (*Mycologia* 86: 1).

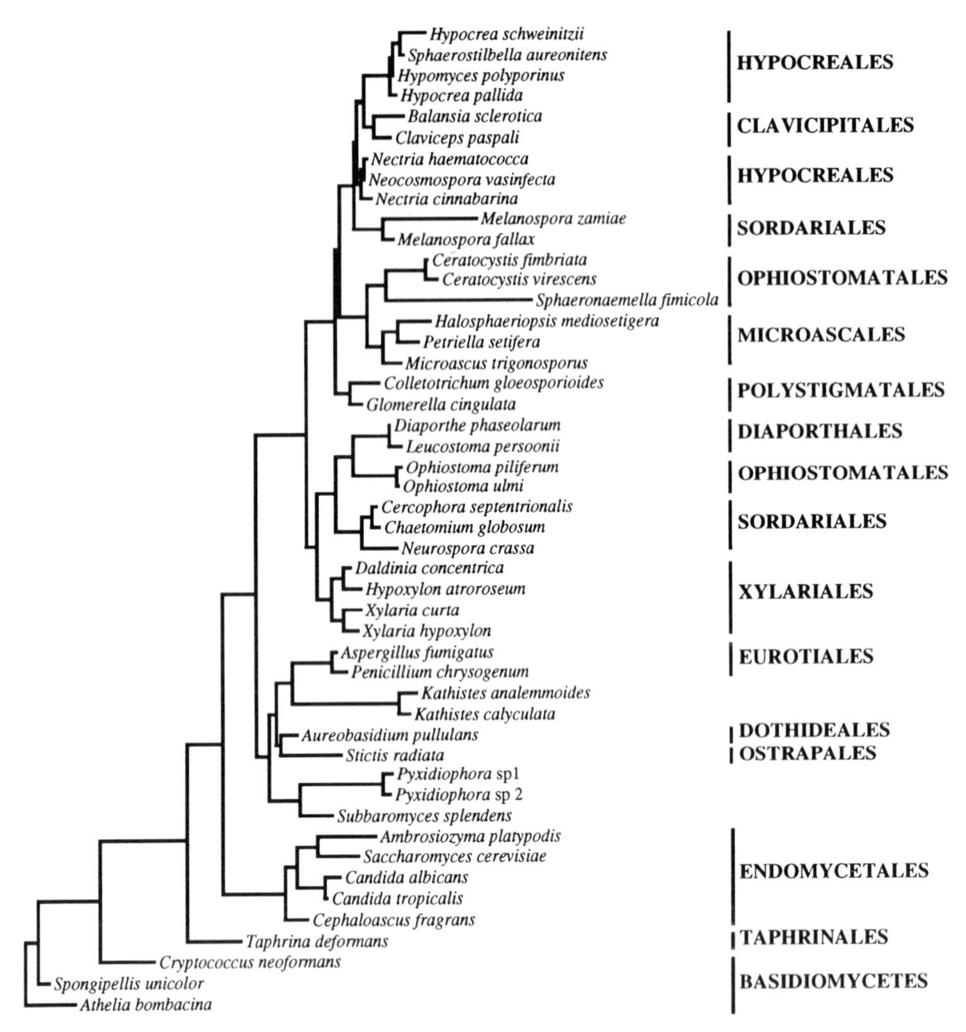

Figure 1. A single most parsimonious tree derived using the computer software package PAUP (Swofford, 1990). The tree is 924 steps long and has a consistency index of 0.439, a retension index of 0.695, and a rescaled consistency index of 0.305. Orders and included taxa (Hawksworth et al., 1983) are indicated. Several observations can be made from the phylogram: (1) six independent lineages contain species with evanescent asci; (2) as delimited the *Ophiostomatales* are polyphyletic; and (3) *Kathistes*, *Pyxidiophora*, and *Subbaromyces* are basal to and excluded from the clade of derived perithecial ascomycetes. Source of specimens and sequences is given in Spatafora (1992).

relationships. However, *Pyxidiophora*, *Subbaromyces*, and *Kathistes* are basal to and excluded from the larger group of perithecial ascomycetes. This was not predicted on the basis of previous morphologically-based hypotheses. Nevertheless, another molecular data set (the B subunit of the vacuolar ATPase gene) supports the placement (Spatafora and Blackwell, unpubl.). However, because there has not been a broad sampling of taxa in this region of the tree, it is relatively unstable and only can provide evidence of an excluded basal position for the three genera. While we do not know exactly what *Pyxidiophora* is in relation to a higher taxon, we certainly know what it is not and have a good basis for eventually completing the task.

While morphological criteria may fail to resolve a particular question, they need not necessarily lead us astray. *Pyxidiophora*, *Subbaromyces*, and *Kathistes* previously have presented particular problems in ordinal placement. When Hesseltine (1953) described

Subbaromyces he did not place the genus in a higher level taxon. Similarly, Malloch and Blackwell (1990) described *Kathistes* in the *Kathistaceae*, but neglected to place the family in an order. Again, they were equivocal about *Pyxidiophora* when after careful consideration they were unable to settle on a likely order for the genus (Blackwell and Malloch, 1989a). Perhaps these instances of improper taxonomic practice are vindicated on the basis of the hypothesis presented here.

ACKNOWLEDGEMENTS

We thank Drs M.L. Berbee and J.W. Taylor, who kindly provided two sequences before publication. Dr K. Jones made helpful comments on an earlier version of the manuscript. Financial support from the National Science Foundation and the Louisiana Board of Reagents is greatfully acknowledged (NSF-BSR-8918157 and NSF-DEB-9208027 to MB, and NSF-BSR-9101088 to MB and JWS; and NSF-LASER-EPSCoR 92-96-ADP-02 to the Louisiana State University Molecular Systematics Cluster).

REFERENCES

Arnold, G.W.R., 1971, Über einige neue taxa und Kombinationen der *Sphaeriales*, *Zeitschrift für Pilzkunde* 37: 187-198.

von Arx, J.A., and J.P. van der Walt, 1987, *Ophiostomatales* and *Endomycetales*, *Studies in Mycology, Baarn* 30: 167-176.

Barr, M.E., 1987, Prodromus to class *Loculoascomycetes*. M.E. Barr, Amherst.

Barr, M.E., 1990, Prodromus to nonlichenized pyrenomycetous members of class *Hymenoascomycetes*, *Mycotaxon* 39: 43-184.

Berbee, M.L., and J.W. Taylor, 1992, Convergence in ascospore discharge mechanism among pyrenomycete fungi based on 18S ribosomal RNA gene sequence, *Molecular Phylogenetics and Evolution* 1: 59-71.

Blackwell, M., 1993, Phylogenetic systematics and ascomycetes, *In*: *The Fungal Holomorph: Mitotic, Meiotic and Pleomorphic Speciation in Fungal Systematics* (D.R. Reynolds and J.W. Taylor, eds): 93-103. CAB INTERNATIONAL, Wallingford.

Blackwell, M., and D. Malloch, 1989a, *Pyxidiophora*: a link between the *Laboulbeniales* and hyphal ascomycetes, *Memoirs of the New York Botanical Garden* 49: 23-32.

Blackwell, M., and D. Malloch, 1989b, *Pyxidiophora*: life histories and arthropod associations of two species, *Canadian Journal of Botany* 67: 2552-2562.

Blackwell, M., and D. Malloch, 1990, Discovery of a *Pyxidiophora* with *Acariniola*-type ascospores, *Mycological Research* 94: 415-417.

Blackwell, M., J.R. Bridges, J.C. Moser, and T.J. Perry, 1986a, Hyperphoretic dispersal of a *Pyxidiophora* anamorph, *Science* 232: 993-995.

Blackwell, M., T.J. Perry, J.R. Bridges, and J.C. Moser, 1986b, A new species of *Pyxidiophora* and its *Thaxteriola* anamorph, *Mycologia* 78: 605-612.

Blackwell, M., J.C. Moser, and J. Wisniewski, 1989, Ascospores of *Pyxidiophora* on mites associated with beetles in trees and wood, *Mycological Research* 92: 397-403.

Blackwell, M., J.W. Spatafora, D. Malloch, and J. W. Taylor, 1993, Consideration of higher taxonomic relationships involving *Pyxidiophora, In*: *Ceratocystis and Ophiostoma: Taxonomy, Ecology and Pathology* (M.J. Wingfield, K.A. Seifert and J.A. Webber, eds): 105-108. American Phytopathological Society Press, St Paul.

Cain, R.F., and L.K. Weresub, 1957. Studies of coprophilous ascomycetes. V. *Sphaeronaemella fimicola*, *Canadian Journal of Botany* 35: 119-131.

Eriksson, O.E., and D.L. Hawksworth, 1991, Notes on ascomycete systematics - Nos 969-1127, *Systema Ascomycetum* 9: 1-38.

Hausner, G., 1993, *Molecular Taxonomy of Ceratocystis sensu lato*. PhD thesis, University of Manitoba, Winnipeg.

Hausner, G., J. Reid, and G.R. Klassen, 1992, Do galeate ascospore members of the *Cephaloascaceae*, *Endomycetaceae* and *Ophiostomataceae* share a common phylogeny?, *Mycologia* 84: 870-881.

Hawksworth, D.L., 1985. Problems and prospects in the systematics of the *Ascomycotina*, *Proceedings of the Indian Academy of Science, Plant Science* 94: 319-339.

Hawksworth, D.L., B.C. Sutton, and G.C. Ainsworth, 1983, *Ainsworth & Bisby's Dictionary of the Fungi*, 7th edition, Commonwealth Agricultural Bureaux, Slough.

Hesseltine, C.W., 1953. Study of trickling filter fungi, *Bulletin of the Torrey Botanical Club* 80: 507-514.
de Hoog, G. S., and R.J. Scheffer, 1984, *Ceratocystis* versus *Ophiostoma*: a reappraisal, *Mycologia* 76: 292-299.
Lundqvist, N., 1980, On the genus *Pyxidiophora sensu lato* (pyrenomycetes), *Botaniska Notiser* 133: 121-144.
Malloch, D., and M. Blackwell, 1990, *Kathistes*, a new genus of pleomorphic ascomycetes, *Canadian Journal of Botany* 68: 1712-1721.
Malloch, D., and M. Blackwell, 1992, Dispersal of fungal diaspores, *In*: *The Fungal Community: Its Organization and Role in the Ecosystem* (G.C. Carroll and D.T. Wicklow, eds): 147-171. 2nd edition. Marcel Dekker, New York.
Malloch, D., and M. Blackwell, Dispersal biology of ophiostomatoid fungi, *In*: *Ceratocystis and Ophiostoma: Taxonomy, Ecology and Pathology* (M.J. Wingfield, K.A. Seifert and J.F. Webber, eds): 195-206. American Phytopathological Society, St. Paul.
Parguey-Leduc, A., and M.-C. Janex-Favre, 1981, The ascocarps of ascohymenial pyrenomycetes, *In*: *Ascomycete Systematics: The Luttrellian Concept* (D.R. Reynolds, ed.): 102-123. Springer-Verlag, New York.
Redhead, S.A., and D. Malloch, 1977, The *Endomycetaceae*: new concepts, new taxa, *Canadian Journal of Botany* 55: 1701-1711.
Rogerson, C.T., 1970, The hypocrealean fungi (ascomycetes, *Hypocreales*), *Mycologia* 62: 865-910.
Spatafora, J.W., 1992, *The Molecular Systematics of Unitunicate, Perithecial Ascomycetes*. PhD dissertation, Louisiana State University, Baton Rouge.
Spatafora, J.W., and M. Blackwell, 1993, The polyphyletic origins of ophiostomatoid fungi, *Mycological Research* 98: 1-9.
Swofford, D.L., 1989, *PAUP -- phylogenetic analysis using parsimony, version 3.0 (User's manual and program)*. Illinois Natural History Survey, Champaign.
Tulasne, L.R., and C. Tulasne, 1865. *Selecta Fungorum Carpologia*. Vol. 3. Typographie Imperiale, Paris.

ORDERS AND FAMILIES OF ASCOSPOROGENOUS YEASTS AND YEAST-LIKE TAXA COMPARED FROM RIBOSOMAL RNA SEQUENCE SIMILARITIES

C.P. Kurtzman and C.J. Robnett

Microbial Properties Research
National Center for Agricultural Utilization Research
Agricultural Research Service
US Department of Agriculture
1815 N. University Street
Peoria, Illinois 61604, USA

SUMMARY

Extent of divergence in partial nucleotide sequences from large and small subunit ribosomal RNAs was used to assess the placement of genera among families and orders of ascosporogenous yeasts and yeast-like fungi. These data indicate the taxa comprise two orders: the *Schizosaccharomycetales* (genus *Schizosaccharomyces)* and the *Saccharomycetales* (*Endomycetales*; all genera of yeasts and yeast-like fungi except the fission yeasts). The data also suggest that certain currently accepted families are artificial. Furthermore, the rRNA sequence comparisons indicate that the ascosporogenous yeasts are not reduced forms of extant filamentous fungi.

INTRODUCTION

Since the time of Guilliermond (1912) and before, the phylogeny of the ascosporogenous yeasts has been vigorously debated. Some have viewed the yeasts as primitive fungi while others perceived them to be reduced forms of more evolved taxa. Cain (1972) has been a proponent of this latter idea, arguing that hat(galeate)-spored genera such as *Pichia* and *Cephaloascus* are likely to be reduced forms of the genus *Ceratocystis*. Redhead and Malloch (1977) and von Arx and van der Walt (1987) accepted this argument and comingled yeasts and mycelial taxa in their treatments of the *Saccharomycetales* (*Endomycetales*) and *Ophiostomatales*.

Measurements of nucleotide sequence divergence in ribosomal DNA (rDNA) or its transcript ribosomal RNA (rRNA) have provided a genetic means for estimating the phylogeny of the fungi. Comparisons of a limited number of taxa have indicated the ascosporogenous yeasts, with the exception of *Schizosaccharomyces*, to form a monophyletic group distinct from the filamentous fungi (Barns et al., 1991; Bruns et al., 1992; Hausner et al., 1992; Hendriks et al., 1992; Kurtzman, 1993; Nishida and Sugiyama, 1993; Walker, 1985). In the present study, we compared partial sequences of small and large subunit rRNAs from the type species of all known cultivatable ascosporogenous yeasts and yeast-like genera and demonstrated that all of these taxa are members of a monophyletic group separate from all the filamentous ascomycetes examined to date. These data provide a broad outline of the phylogeny of the ascomycetous yeasts and allow comment on the various proposals for their classification.

Table 1. Strain designations and selected phenotypic characters of the species compared

Species	Strain designations[1] NRRL	CBS	Co-Q[2]	Type of budding[3]	Hyphal septum[4]	Ascospore shape[5]
Saccharomyces cerevisiae	Y-12632T	1171	6	M		Sph
Zygosaccharomyces rouxii	Y-229T	732	6	M		Sph
Torulaspora delbrueckii	Y-866T	1146	6	M		Sph
Kluyveromyces wickerhamii	Y-8286T	2745	6	M		Ren
Pachytichospora transvaalensis	Y-17245T	2186	6	M		Sph/Elp
Arxiozyma telluris	YB-4302T	2685	6	M		Sph
Issatchenkia orientalis	Y-5396T	5147	7	M		Sph
Pichia membranaefaciens	Y-2026T	107	7	M		Sph/Hat
Saturnospora dispora	Y-1447T	794	7	M		Sat
Dekkera bruxellensis	Y-12961T	74	9	M		Sat
Cyniclomyces guttulatus	Y-17561A		6	M		Elg/Sph
Saccharomycodes ludwigii	Y-12793T	821	6	B		Sph/Ldg
Hanseniaspora valbyensis	Y-1626T	6622	6	B		Hat
Ashbya gossypii	Y-1056A	109.51	6	-		Spd
Eremothecium ashbyi	Y-1363A		7	-		Spd
Nematospora coryli	Y-12970T	2608	5,6	M		Spd
Holleya sinecauda	Y-17231T	8199	9	M		Spd
Pachysolen tannophilus	Y-2460T	4044	8	M		Hat
Citeromyces matritensis	Y-2407T	2764	8	M		Sph
Nadsonia fulvescens	Y-12810T	2596	6	B		Sph
Wickerhamia fluorescens	YB-4819T	6778	9	B		Elg/Ldg
Williopsis saturnus	Y-1304T	5761	7	M		Sat
Metschnikowia bicuspidata	YB-4993T	5575	9	M		Drt
Clavispora lusitaniae	Y-11827T	6936	8	M		Clv
Debaryomyces hansenii	Y-7426T	767	9	M		Sph
Wingea robertsii	Y-6670T	2934	9	M		Len
Cephaloascus fragrans	Y-6742T	121.29	9	M	CP	Hat
Ascoidea africana	Y-6762T	377.68	8	-	PL	Hat
Hormoascus platypodis	Y-6732T	4111	7	M	PP	Hat

Ambrosiozyma monospora	Y-1484T	2554	7	M	PP	Hat
Saccharomycopsis fibuligera	Y-2388T	2521	8	M	PL	Hat
Arthroascus javanensis	Y-1483T	2555	8	M/F	MP	Sat
Guilliermondella selenospora	Y-1357T	2562	8	M	PL	Sph/Ren
Botryoascus synnaedendrus	Y-7466T	6161	8	M	PL	Hat
Lodderomyces elongisporus	YB-4239T	2605	9	M		Elg
Dipodascus albidus	Y-12859T	766.85	9	F		Elp
Galactomyces geotrichum	Y-7366T	772.71	9	F	MP	Elp
Sporopachydermia lactativora	Y-11591T	6192	9	M		Sph
Stephanoascus ciferrii	Y-10943A	5295	9	M	MP	Hem/Ldg
Yarrowia lipolytica	YB-423T	6124	9	M	MP	Sph/Ldg
Wickerhamiella domercqiae	Y-6692T	4351	9	M		Elg
Zygoascus hellenicus	Y-6591T	4099	9	M	MP	Hat
Dipodascopsis uninucleata	Y-1268A	740.74	9	M	CP	Elp/Ren
Lipomyces starkeyi	Y-11557T	1807	9	M		Elp
Zygozyma oliphaga	Y-17247T	7107	8	M		Elp
Eremascus fertilis	Y-1463A	209.39		-	CP	Sph
Emericella nidulans	22233A			-	CP	Sph/Ldg
Ceratocystis fimbriata	13496A	146.53		-	CP	Hat
Protomyces inundatus	Y-6349A		10?	M		Elp
Taphrina deformans	T-857A		10	M		Elp
Schizosaccharomyces pombe	Y-12796T	356	10	F		Sph/Elp
Filobasidiella neoformans	Y-170A	882	10	M	DO	-

[1] T = ex-type strain, A = authentic strain. *S. capsularis*, the type species of *Saccharomycopsis*, was not included in this study.

[2] Co-Q = number of isoprene units on the side chain of coenzyme Q or ubiquinone. Data from Barnett et al. (1990), von Arx and van der Walt (1987).

[3] M = multilateral budding, B = bipolar budding, F = fission.

[4] DO = dolipore, CP = central pore, MP = micropore, PL = plasmodesmata, PP = plugged pore

[5] Clv = clavate, Drt = dart-like, Elg = elongate, Elp = ellipsoidal, Hat = hat-shaped (galeate), Hem = hemispheroidal, Ldg = with subequatorial ledge, Len = lenticular, Sat = saturn-shaped, Spd = spindle-like, Sph = spheroidal, Ren = reniform, Var = various shapes.

MATERIALS AND METHODS

Source of Strains

The strains used in this study are listed in Table 1, and they are maintained in the ARS Culture Collection (NRRL), National Center for Agricultural Utilization Research.

Isolation and Purification of rRNA

Cells were grown at 25°C in 100ml of YM liquid medium (Wickerham, 1951) on a rotary shaker at 200 rpm for *c.* 16 hr and harvested by centrifugation. Isolation and purification of rRNA was as described by Kurtzman and Liu (1990).

Sequencing Reactions and Sequence Comparisons

Sequencing of rRNA was accomplished with specific oligonucleotide primers and the dideoxynucleotide chain termination method (Lane et al., 1985; Sanger et al., 1977). Two regions were sequenced from the large (25S) subunit and one region from the small (18S) subunit (Peterson and Kurtzman, 1991). The large subunit primers and first bases of the rRNA sequences copied are:

5'-GGTCCGTGTTTCAAGACGG(635) and
5'-TTGGAGACCTGCTGCGG(1841).

The small subunit primer and first base of the rRNA sequence copied is: 5'-ACGGGCGGTGTGTAC(1627). Nucleotide numerical designations are referenced to the primary structure of *Saccharomyces cerevisiae* (Georgiev et al., 1981; Mankin et al., 1986; Rubtsov et al., 1980). For reference in the text, the regions sequenced are referred to as 25S-635, 25S-1841, and 18S-1627, and the number of nucleotides examined per region were 371, 330, and 323, respectively. Primers were synthesized with an Applied Biosystems Model 381A DNA synthesizer. Nucleotide fragments generated in the chain termination reactions were separated on 8% acrylamide-8M urea gels and visualized by autoradiography. The sequencing of region 25S-635 for the species *Dekkera bruxellensis, Hanseniaspora valbyensis,* and *Sporopachydermia lactativora* was an exception to the foregoing protocol. For these species, sequences were obtained from polymerase chain reaction amplified rDNA by standard methods as described by O'Donnell (1992). Sequences were manually aligned, and the sequence data were analyzed with the program PAUP, Version 3.0s (Swofford, 1991). Bootstrap analysis of the data (100 replications) was performed using PAUP. Sequences have been deposited with GenBank.

RESULTS AND DISCUSSION

Results of our analysis of phylogenetic relationships among ascomycetous yeast genera are shown in Fig. 1. Before these data are discussed, we need to consider the nucleotide variability found among strains of a species and among the species assigned to various representative genera. Comparisons of divergence in the highly variable large subunit region 25S-635 (domain D_2 of Guadet et al., 1989) for various heterothallic and homothallic ascomycetous yeasts have shown that strains of genetically defined species generally have less than 1 percent nucleotide divergence and that closely related species differ by 1-5 percent (Kurtzman and Liu, 1990; Kurtzman and Robnett, 1991; Liu and Kurtzman, 1991; Mendonça-Hagler et al., 1993; Peterson and Kurtzman, 1990, 1991). Consequently, sequence differences in region 25S-635 offer reliable separation of nearly all yeast species. Extent of divergence among all known species from each of eight genera was also measured during the preceding studies. For region 25S-635, species regarded as congeneric showed 3-23 percent nucleotide divergence (Table 2). In part, this range of variation among species in different genera is a reflection of time elapsed since the species diverged but unequal rates of substitution may also account for some of the divergence because the percent substitutions between 25S-635 and the other two rRNA regions sequenced are not proportionally the same for all genera (Table 2).

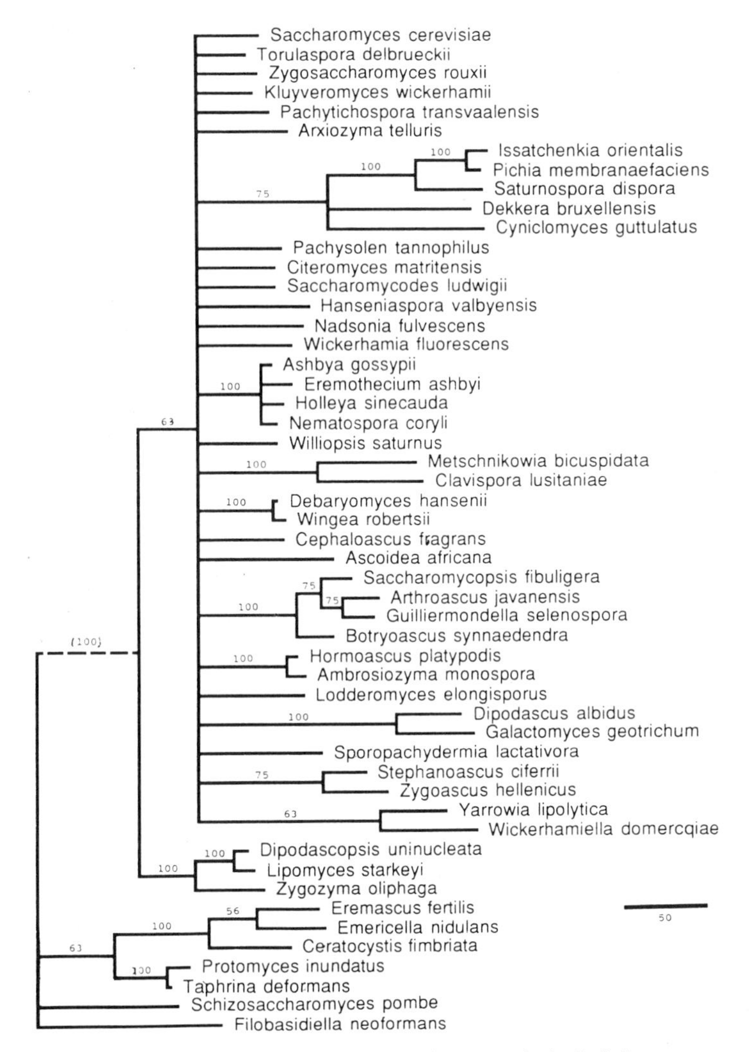

Figure 1. A phylogenetic tree derived from maximum parsimony analysis depicting ascomycetous yeasts, yeast-like fungi, and various reference species. The phylogram was calculated with the program PAUP, version 3.0s, using the combined small and large subunit sequences described in the text. Branch lengths are proportional to nucleotide differences and the marker bar represents a length of 50 nucleotide substitutions. Numbers given on branches are the percentages of frequencies with which a given branch appeared in 100 bootstrap replications. Branches without numbers had frequencies of less than 50%. The branch supporting the yeast clade is shown as a broken line to denote the uncertainty of whether the yeasts or the other taxa are ancestral (see text for discussion).

Can genera be defined from rRNA/rDNA nucleotide similarity? When sequences from the preceding eight genera are aligned, species of each genus exhibit a pattern of common nucleotides that appear genus-specific. It is only in the genera *Williopsis* and *Metschnikowia* that several of the species are sufficiently divergent to partially mask the genus-specific patterns shown by the remainder of the species. Consequently, the range of sequence divergence (*c.* 3-20%) seen among species in the apparently well-defined genera discussed above should apply to other ascomycetous yeasts and can provide a prediction of whether the numerous monotypic genera included in this study are genetically separate or merely phenotypically distinct species of previously described genera. Such an assessment is necessary if families are to be defined.

Table 2. Greatest number of rRNA nucleotide differences between species in selected yeast genera[1].

Genus (Number of species)	Percent divergence per rRNA region[2]		
	25S-635	25S-1841	18S-1627
Debaryomyces (12)	10	2	2
Lipomyces (5)	9	8	4
Metschnikowia (7)	20(40)[3]	21(28)[3]	15(17)[3]
Saccharomyces (10)	10	7	4
Saturnospora (4)	8	1	1
Schizosaccharomyces (3)	18	13	4
Torulaspora (3)	3	1	0.3
Williopsis (5)	23	13	14

[1]Data from Kurtzman (unpubl.), Kurtzman and Liu (1990), Kurtzman and Robnett (1991), Liu and Kurtzman (1991), Mendonça-Hagler et al. (1993).
[2]*C.* 300 nucleotides per region.
[3]Numbers in parentheses represent percentages with inclusion of the outlying species *M. hawaiiensis*.

In Fig. 1, the ascosporogenous yeasts form a clade that is distinct from those comprising the filamentous ascomycetes, *Protomyces/Taphrina*, and *Schizosaccharomyces* (and synonyms *Octosporomyces* and *Hasegawaea*). *Dipodascopsis, Lipomyces* (and synonym *Waltomyces*) and *Zygozyma* form a divergent subgroup of the yeast clade. Consequently, these data suggest that the yeasts are monophyletic and that they are not derived forms from a variety of extant filamentous species.

The phylogram in Fig. 1 excludes depiction of branching among taxa when bootstrap values for those branches are less than 50%. Phylogenetic signals would be strengthened with the analysis of longer sequences, but another factor impacting on resolution is the diversity of taxa in the present comparison. This diversity renders many phylogenetically informative nucleotides unrecognizable, but the eventual inclusion of all known species from each genus in the analysis should help resolve genus boundaries.

The nucleotide differences between closely related genera are given in Table 3. On the basis of the comparisons presented in Table 2, these data demonstrate the monotypic genus *Wingea* to be a synonym of the genus *Debaryomyces*. The major phenotypic difference between the two genera is found in the glabrous lenticular shape of ascospores produced by *W. robertsii* which contrasts with those of *Debaryomyces* species which are spheroidal to ellipsoidal and roughened by circular protuberances and/or elongated ridges. Additional close relationships are shown between *Issatchenkia/ Pichia, Hormoascus/ Ambrosiozyma,* and *Dipodascopsis/ Lipomyces.* The taxonomic status of these latter genera will be reconsidered once sequences are available from all known species. *Ashbya, Eremothecium, Holleya* and *Nematospora* comprise a small clade whose intergeneric distances are no greater than the distances found between species assigned to *Saccharomyces* and *Lipomyces.* All are plant pathogens, have similar elongated ascospores, and may represent members of a single genus.

Aside from the broad phylogenetic relationships discussed earlier and the closely allied genera noted in Table 3, the present data are insufficiently robust to define families with certainty. The analysis, however, does allow comment on previous taxonomic proposals, which are primarily based on differences in ascospore shape and ornamentation, presence or absence of hyphae and, for a few taxa, the occurrence of unique biochemical properties. The proposals of Kreger-van Rij (1984) and von Arx and van der Walt (1987) are given in Table 4. Traditionally, the fission yeast *Schizosaccharomyces* has been assigned to either a separate family or subfamily of the *Endomycetales*. The great phylogenetic distance of this genus from the budding yeasts (Kurtzman and Robnett, 1991) merits its reassignment to a separate order, the *Schizosaccharomycetales*, as discussed by Kurtzman (1993). Our analysis supports the idea that the remaining ascomycetous yeasts and yeast-like species are members of a single order, the *Saccharomycetales* (*Endomyce-*

Table 3. Number of rRNA nucleotide differences between the closely associated genera shown in Fig. 1 and more distantly related reference pairs

Genus pair	Percent divergence per rRNA region[1]		
	25S-635	25S-1841	18S-1627
Issatchenkia/Pichia	6	2	2
Hanseniaspora/Saccharomycodes	10	8	8
Ashbya/Eremothecium	3	2	3
Ashbya/Nematospora	3	2	1
Ashbya/Holleya	4	1	3
Eremothecium/Nematospora	5	2	3
Eremothecium/Holleya	6	2	4
Holleya/Nematospora	5	1	3
Clavispora/Metschnikowia	19	21	11
Debaryomyces/Wingea	1	0	0
Ambrosiozyma/Hormoascus	5	3	4
Arthroascus/Guilliermondella	10	2	4
Dipodascus/Galactomyces	13	10	8
Dipodascopsis/Lipomyces	5	2	2
Saccharomyces/Debaryomyces	17	6	4
Saccharomyces/Lipomyces	18	11	7

[1]Number of nucleotides per region: 25S-635=371; 25S-1841=330; 18S-1627=323.

tales). A possible exception is the outlying family *Lipomycetaceae* (*Lipomyces, Dipodascopsis* and *Zygozyma*), which may require ordinal status once phylogenetic relationships are better understood.

Of the phenotypic criteria used for definition of families and genera, ascospore shape is, in most cases, the least reliable. This has been demonstrated here for *Wingea*, for species of *Debaryomyces* (Kurtzman and Robnett, 1991) and for *Pichia ohmeri* (Wickerham and Burton, 1954). Hat-shaped ascospores are produced by many yeast genera and by *Ceratocystis* (Fig. 1). Most systems of classification include *Metschnikowia* in the same family as *Ashbya* because the dart-like ascospores of *Metschnikowia* were perceived as phylogenetically similar to the somewhat spindle-shaped ascospores formed by genera of the *Ashbya* clade. Our data show no close association between *Metschnikowia* and *Ashbya*. By contrast, the type of hyphal septum formed by mycelial species does seem to fall along phylogenetic lines (Fig. 1). Some congruence is also noted between placement of genera in the phylogram and their major type of coenzyme Q. A significant exception in correlation is seen for the clade comprised of *Ashbya* (Q-6), *Eremothecium* (Q-7), *Holleya* (Q-9), and *Nematospora* (Q-5, Q-6).

The present study gives a general overview of generic relationships within the ascosporogenous yeasts. It is clear that additional less homoplasic data are needed if family assignments are to be resolved. In order for families to represent the natural history of the yeasts, such phylogenetically based taxa are likely to be variable in size. We might also expect that once the genetic basis of phenotypic characters is understood, they may better correlate with a molecular phylogeny than is now evident.

One additional issue that invites comment is whether or not the budding, ascosporogenous yeasts are ancestral to the filamentous ascomycetes and the basidiomycetes. The work of Bruns et al. (1992) and of Nishida and Sugiyama (1993) shows the basidiomycetes to be ancestral to the ascomycetes. Both groups of authors note, however, that this placement is statistically weak. The branching order of *Filobasidiella neoformans* and *Schizosaccharomyces pombe* is unresolved in the present study but both appear ancestral to the budding ascosporogenous yeasts when *F. neoformans* is designated as the outgroup in the phylogenetic analysis. However, designation of *Saccharomyces cerevisiae* as outgroup results in the yeast clade becoming ancestral. Consequently, the branch supporting the yeast clade is represented by a broken line in the phylogram. In all analyses of our data, the yeast clade showed a bootstrap value of 100%.

Table 4. Recent proposals for classification of the ascosporogenous yeasts and yeast-like fungi.

Kreger-van Rij (1984)	von Arx and van der Walt (1987)
Order: **Endomycetales**	Order: **Endomycetales**
Families, subfamilies, and genera	Families and representative genera
Spermophthoraceae	**Saccharomycetaceae**
Coccidiascus	*Saccharomyces*
Metschnikowia	*Issatchenkia*
Nematospora	**Endomycetaceae**
Saccharomycetaceae	*Endomyces*
Schizosaccharomycetoideae	*Pichia*
Schizosaccharomyces	**Saccharomycopsidaceae**
Nadsonioideae	*Saccharomycopsis*
Hanseniaspora	*Williopsis*
Nadsonia	**Lipomycetaceae**
Saccharomycodes	*Lipomyces*
Wickerhamia	*Zygozyma*
Lipomycetoideae	**Dipodascaceae**
Lipomyces	*Dipodascus*
Saccharomycetoideae	*Schwanniomyces*
Ambrosiozyma	**Metschnikowiaceae**
Arthroascus	*Metschnikowia*
Citeromyces	*Nematospora*
Clavispora	**Saccharomycodaceae**
Cyniclomyces	*Saccharomycodes*
Debaryomyces	*Nadsonia*
Dekkera	**Schizosaccharomycetaceae**
Guilliermondella	*Schizosaccharomyces*
Hansenula	**Unassigned Genera**
Issatchenkia	*Kluyveromyces*
Kluyveromyces	*Guilliermondella*
Lodderomyces	*Yarrowia*
Pachysolen	*Stephanoascus*
Pachytichospora	*Zygoascus*
Pichia	*Debaryomyces*
Saccharomyces	
Saccharomycopsis	Order: **Ophiostomatales**
Schwanniomyces	Families and representative genera
Sporopachydermia	**Ophiostomataceae**
Stephanoascus	*Ceratocystiopsis*
Torulaspora	*Ophiostoma*
Wickerhamiella	**Cephaloascaceae**
Wingea	*Cephaloascus*
Zygosaccharomyces	*Ambrosiozyma*
	Hormoascus
	Pyxidiophoraceae
	Ceratocystis
	Pyxidiophora
	Pseudeurotiaceae
	Emericellopsis
	Pseudeurotium

REFERENCES

von Arx, J.A., and J.P. van der Walt, 1987, *Ophiostomatales* and *Endomycetales*, *Studies in Mycology, Baarn* 30: 167-176.

Barnett, J.A., R.W. Payne, and D. Yarrow, 1990, *Yeasts: Characteristics and Identification*. 2nd edition. Cambridge University Press, Cambridge.

Barns, S.M., D.J. Lane, M.L. Sogin, C. Bibeau, and W.G. Weisburg, 1991, Evolutionary relationships among pathogenic *Candida* species and relatives, *Journal of Bacteriology* 173: 2250-2255.

Bruns, T.D., R. Vilgalys, S.M. Barns, D. Gonzalez, D.S. Hibbett, D.J. Lane, L. Simon, S. Stickel, T.M. Szaro, W.G. Weisburg, and M.L. Sogin, 1992, Evolutionary relationships within the fungi: analyses of nuclear small subunit rRNA sequences, *Molecular Phylogenetics and Evolution* 1: 231-241.

Cain, R.F., 1972, Evolution of the fungi, *Mycologia* 64: 1-14.

Georgiev, O.I., N. Nikolaev, A.A. Hadjiolov, K.G. Skryabin, V.M. Zakharyev, and A.A. Bayev, 1981, The structure of the yeast ribosomal RNA genes. 4. Complete sequence of the 25S rRNA gene from *Saccharomyces cerevisiae, Nucleic Acids Research* 9: 6953-6958.

Gaudet, J., J. Julien, J.F. Lafey, and Y. Brygoo, 1989, Phylogeny of some *Fusarium* species, as determined by large subunit rRNA sequence comparison. *Molecular Biology and Evolution* 6: 227-242.

Guilliermond, A., 1912, *Les Levures*. O. Doin et Fils, Paris.

Hausner, G., J. Reid, and G.R. Klassen, 1992, Do galeate-ascospore members of the *Cephaloascaceae, Endomycetaceae* and *Ophiostomataceae* share a common phylogeny? *Mycologia* 84: 870-881.

Hendriks, L., A. Goris, Y. van de Peer, J.-M. Neefs, M. Vancanneyt, K. Kersters, J -F. Berny, G.L. Hennebert, and R. De Wachter, 1992, Phylogenetic relationships among ascomycetes and ascomycete-like yeasts as deduced from small ribosomal subunit RNA sequences, *Systematic and Applied Microbiology* 15: 98-104.

Kreger-van Rij, N.J.W., 1984, Systems of classification of the yeasts. *In*: *The Yeasts - A Taxonomic Study* (N.J.W. Kreger-van Rij, ed.): 2-13. 3rd edn. Elsevier Science Publishers, Amsterdam.

Kurtzman, C.P., 1993, Systematics of the ascomycetous yeasts assessed from ribosomal RNA sequence divergence, *Antonie van Leeuwenhoek* 63: 165-174.

Kurtzman, C.P., and Z. Liu, 1990, Evolutionary affinities of species assigned to *Lipomyces* and *Myxozyma* estimated from ribosomal RNA sequence divergence, *Current Microbiology* 21: 387-393.

Kurtzman, C.P., and C.J. Robnett, 1991, Phylogenetic relationships among species of *Saccharomyces, Schizosaccharomyces, Debaryomyces* and *Schwanniomyces* determined from partial ribosomal RNA sequences, *Yeast* 7: 61-72.

Lane, D.J., B. Pace, G.J. Olsen, D.A. Stahl, M.L. Sogin, and N.R. Pace, 1985, Rapid determination of 16S ribosomal RNA sequences for phylogenetic analyses, *Proceedings of the National Academy of Sciences, United States of America* 82: 6955-6959.

Liu, Z., and C.P. Kurtzman, 1991, Phylogenetic relationships among species of *Williopsis* and *Saturnospora* gen. nov. as determined from partial rRNA sequences, *Antonie van Leeuwenhoek* 60: 21-30.

Mankin, A.S., K.G. Skryabin, and P.M. Rubtsov, 1986, Identification of ten additional nucleotides in the primary structure of yeast 18S rRNA, *Gene* 44: 143-145.

Mendonça-Hagler, L.C., A.N. Hagler, and C.P. Kurtzman, 1993, Phylogeny of *Metschnikowia* species estimated from partial rRNA sequences, *International Journal of Systematic Bacteriology* 43: 368-373.

Nishida, H., and J. Sugiyama, 1993, Phylogenetic relationships among *Taphrina, Saitoella,* and other higher fungi, *Molecular Biology and Evolution* 10: 431-436.

O'Donnell, K., 1992, Ribosomal DNA internal transcribed spacers are highly divergent in the phytopathogenic ascomycete *Fusarium sambucinum* (*Gibberella pulicaris*). *Current Genetics* 22: 213-220.

Peterson, S.W., and C.P. Kurtzman, 1990, Phylogenetic relationships among species of the genus *Issatchenkia* Kudriavzev, *Antonie van Leeuwenhoek* 58: 235-240.

Peterson, S.W., and C.P. Kurtzman, 1991, Ribosomal RNA sequence divergence among sibling species of yeasts. *Systematic and Applied Microbiology* 14: 124-129.

Redhead, S.A., and D.W. Malloch, 1977, The *Endomycetaceae*: new concepts, new taxa, *Canadian Journal of Botany* 55: 1701-1711.

Rubtsov, P.M., M.M. Musakhanov, V.M. Zakharyev, A.S. Krayev, K.G. Skryabin, and A.A. Bayev, 1980, The complete structure of yeast ribosomal RNA genes. I. The complete nucleotide sequence of the 18S ribosomal RNA gene from *Saccharomyces cerevisiae, Nucleic Acids Research* 8: 5779-5794.

Sanger, F., S. Nicklen, and A.R. Coulson, 1977, DNA sequencing with chain-terminating inhibitors, *Proceedings of the National Academy of Sciences, United States of America* 74: 5463-5467.
Swofford, D.L., 1991, *PAUP: Phylogenetic analysis using parsimony, 3.0s*. Illinois Natural History Survey, Champaign, Illinois.
Walker, W.F., 1985, 5S ribosomal RNA sequences from ascomycetes and evolutionary implications, *Systematic and Applied Microbiology* 6: 48-53.
Wickerham, L.J., 1951, Taxonomy of yeasts, *United States Department of Agriculture Technical Bulletin* 1029: 1-56.
Wickerham, L.J., and K.A. Burton, 1954, A clarification of the relationships of *Candida guilliermondii* to other yeasts by a study of their mating types, *Journal of Bacteriology* 68: 594-597.

Biology and Species Concepts

EVOLUTIONARY PROCESSES AFFECTING ADAPTATION TO SAPROTROPHIC LIFE STYLES IN ASCOMYCETE POPULATIONS

A.D.M. Rayner

School of Biology and Biochemistry
University of Bath, Claverton Down
Bath BA2 7AY, UK

SUMMARY

An understanding of sources of phenotypic variation in natural ascomycete populations is critical to the evaluation of criteria for taxonomic delimitation. Such variation may reflect genetic differences between individuals due to mutational and recombinatorial processes, or result from epigenetic changes in developmental patterns within heterogeneous habitats.

Amongst saprotrophic fungi, the degree to which an organism is adapted to colonization of disturbed or undisturbed habitats may influence its breeding biology and somatic development. Disturbance imposes R-selection, favouring organisms that are quick to arrive at newly available domains and to exploit readily assimilable resources, hence reducing genetic and epigenetic variation. Organisms that colonize undisturbed habitats are S- or C-selected, being able respectively to survive selectively stressful conditions or a potentially high incidence of competitors; they tend to exhibit both genetic and epigenetic variation.

Patterns of variation amongst ascomycete populations and individuals are discussed in relation to environmental heterogeneity and compared with those exhibited by basidiomycetes. Possible mechanisms resulting in phenotypic instability are introduced, and the basis for general differences between ascomycetes and basidiomycetes.

INTRODUCTION

Aims

When evaluating the biological and taxonomic significance of observed character differences between individuals, it is important to understand how phenotypic variation is generated and regulated in natural populations of the organisms concerned. There are two main components to such understanding. The first involves ascertaining the types of selection pressures operating under particular environmental circumstances, and identifying the adaptive responses that will maximize evolutionary fitness in those circumstances (Andrews, 1992; Anderson et al., 1992). The second involves knowledge of the generative processes, and associated mechanisms, that determine the *scope* for phenotypic diversification given particular organizational attributes (Andrews, 1992; Rayner et al., 1993; *cf.* Gould and Lewontin, 1979). In other words, it is important to clarify the relationship between what is *compelled* by the environment and what is *impelled* by the self-organizing properties of the organism. It may then be possible to relate the modes of reproduction and development of organisms to their evolutionary niche, and hence to their taxonomic affinities, in a fully accountable manner. This is the long-term aim of the present contribution, with respect to understanding how phenotypic variation in mycelial ascomycetes is related to saprotrophic life-styles. In the short-term, however, it will be evident that much additional information is needed from population and developmental biological studies, before a legitimate synthesis can be made. The attempted generalizations that follow are therefore intended to highlight the fundamental issues that need to be resolved rather than to establish definitive conclusions.

Saprotrophy and Systematics

In considering the environmental context within which mycelial ascomycetes operate, it immediately becomes important to distinguish between somatic and reproductive life-cycle phases. By their very nature, these phases occupy distinctive microenvironments, have contrasting functions, and so may be thought to be acted upon independently by natural selection. This raises the question of the relative taxonomic importance of these phases.

As with vascular plants, it is the characteristics of reproductive phases which have usually been held to be of most value in ascomycete taxonomy, because these phases exhibit (apparently) greater diversity in form but less phenotypic plasticity than somatic phases. Correspondingly, environmental factors affecting propagule formation, dispersal, and survival may be expected to be more relevant taxonomically than factors, such as nutritional mode, affecting patterns of energy assimilation and allocation.

Why, then, should adaptation to saprotrophy, a somatic function, concern the systematist? There are several reasons. First, it is important, when placing an organism within a classification hierarchy, that *all* available information about the *whole* organism should, without prejudice, be taken into account. The general culturability of saprotrophic fungi means that a wealth of information can be gleaned concerning their mycelial characteristics, but some rationale is needed to enable this information to be interpreted in a taxonomically meaningful way (*cf.* Jahns and Ott, this volume). Second, habitat selectivity, which is presumably largely due to somatic attributes, has been used widely as a diagnostic character in ascomycetes. It is important to know how soundly-based such usage is, and in particular to establish how adaptation to habitat may affect gene flow within and between populations (Anderson et al., 1992). Third, the form and function of reproductive phases are not so independent from the properties of mycelial phases as they may at first seem. The scale and indeed the existence of reproductive structures is dependent on resources supplied from the mycelium. Moreover, since these structures are themselves derived from hyphal components, the pattern-generating processes involved in their formation will inevitably be subject to the opportunities and constraints implicit in mycelial organization. Finally, certain somatic attributes depend on reproductive processes, in particular whether the means of propagation is clonal or recombinatorial.

The interrelationships between environmental circumstances and mycelial and reproductive form and function are therefore of key importance to the systematist, and provide the basis for this discussion.

SELECTION PRESSURES IN HETEROGENEOUS AND HOMOGENEOUS ENVIRONMENTS

In the quest for consistency and ease of interpretation, experimental studies with fungal mycelia have usually emphasized the need for constancy and uniformity of growth conditions. Consciously or unconsciously, this has encouraged a view of the mycelium as a *fundamentally* homogeneous, purely assimilative structure, which at least to a good approximation can be treated dynamically as an additive assemblage of discrete hyphal growth units (e.g. Prosser, 1991). Observed departures from homogeneity have correspondingly been regarded as the result of imprecision, failure to maintain exact conditions, rather than a manifestation of the mycelium as a complex dynamic system able, by feedback, to respond sensitively to changes in its circumstances. However, the notion of heterogeneity as a "fault" rather than an organizational property can only impede progress in understanding the significance of mycelial characters and using them for taxonomic purposes.

That the mycelium should fundamentally be regarded as structurally, functionally, and evolutionarily heterogeneous is, on the other hand, evident from a variety of geometric and ecological considerations. Firstly, from the moment when it first branches, the mycelium has a measurable fractal dimension, filling space non-uniformly and indeterminately (Ritz and Crawford, 1990; Crawford and Ritz, 1993). As it expands, this structure maintains a dynamic balance between explorative, exploitative, conservative, and redistributive processes that enable it to operate as an energy efficient structure (see below). Such efficiency is fundamental to evolutionary fitness in heterogeneous natural environments where resources may vary in quality, quantity, and spatiotemporal location, and be subject to the competing demands of neighbouring organisms. Here, the mycelium may be required, at locally unpredictable times and places, to locate sources of water and nutrients, assimilate simple substrates, digest refractory substrates, mate, compete and do battle with neighbours, and organize itself into reproductive and migratory structures. To do all this as a homogeneous system would be akin to playing chess equipped only with pawns.

The primary question to be addressed about the adaptational context in which saprotrophy occurs therefore concerns the processes by which environmental heterogeneity arises in space and time and how these impose selection pressure on patterns of genetic and developmental variation.

Temporal Pressures

Temporal heterogeneity, changes in conditions over time, has long been regarded as the principal cause of successional change within fungal communities. Correspondingly, the stage in community development at which a fungus predominates may be thought to reflect its phenotypic attributes and to be of taxonomic value. One approach to understanding how phenotype and community dynamics are connected is through ecological strategies theory (Cooke and Rayner, 1984; Andrews, 1992), which relates the individual life-span of an organism to the incidence of disturbance, competitors or stress within its natural habitat. A simplified discussion of this theory, as it may be applied to saprotrophic ascomycetes, follows.

Disturbance may be defined as any environmental event or process, which, by enrichment of the living space or by destruction of residents makes available new resources for exploitation. Immediately following disturbance, conditions are liable to be at their most homogeneously unrestrictive. Disturbance therefore imposes R-selection, favouring opportunist or "ruderal" organisms, equipped for rapid arrival and exploitation of readily assimilable resources. However, as conditions aggravate or competitors establish, pioneers unable to adapt to these changes will come under increasing pressure to disseminate their genes.

R-selected fungi are therefore likely to have short individual life-spans and correspondingly to reproduce rapidly, without genetic diversification and with minimal resources, and to have a limited developmental repertoire. Just such fungi are the most readily cultured in the laboratory, and it may be this which has encouraged the simplistic view of mycelial dynamics already referred to. However, in the absence of human intervention, disturbance and R-selection are by no means ubiquitous features of natural fungal communities.

Stress may be defined as any more or less continuously imposed environmental feature, other than competition, which limits the productivity of the majority of organisms under consideration. The minority of "S-selected" organisms that are able, by means of specialized attributes, to develop effectively under stressful conditions may therefore do so in the relative absence of competitors. They may generally be expected to possess long individual life spans, either because of the maintenance of stress conditions over extended time periods or because of an ability to persist once these conditions are alleviated. Commitment to reproduction is therefore liable to be slow, though the final investment of resources, both assimilable and refractory, may be considerable. Since conditions are liable to vary with time, a versatile developmental repertoire that enables suitable adjustment to changed circumstances would be advantageous. Organisms which colonize living plants in latent form, prior to becoming saprotrophically active when their host is stressed, e.g. probably many host-selective species of *Hypoxylon* provide a good example (Rayner and Boddy, 1988; Boddy and Griffith, 1989; see below).

Whether reproduction should be genetically diversifying may depend on whether the dominant sources of stress operate locally, i.e within a set of individually unique habitats such as a variable population of plant hosts, or widely, as with temperature stress on the edge of an organism's geographical range. In the latter case, recombinatorial processes may disrupt a favoured genotype within a widespread niche, and so clonal propagation would be advantageous (e.g. Rayner, 1992). On the other hand, combining the possession of a long life-span with clonal population structure may enhance the risk to evolutionary fitness posed by transmissible infection.

The occurrence of a widespread but nonetheless specialized niche may also be a general feature of radical, irregular shifts in environmental conditions, like those which could potentially be induced by global warming. Such shifts have been said by Brasier (1987) to cause "episodic selection", as distinct from the "routine" selection pressures identified in ecological strategies theory. Episodic selection has probably played an important role in the recent evolution of the Dutch elm disease pathogen, *Ophiostoma novo-ulmi* (Brasier, 1987, 1991). It is of interest that populations of this fungus are clonal at epidemic fronts but genetically variable elsewhere (Brasier, 1988), where the spread of a transmissible double-stranded RNA disease factor (d-factor) is impeded by somatic rejection between different genotypes. Such rejection, or somatic incompatibility, mechanisms probably serve generally to demarcate genetically different, but indeterminate individuals within natural populations of both ascomycetes and basidiomycetes (e.g. Rayner, 1991; Anderson et al., 1992; see also below). Using a term originating from plant population biology, these individuals may usefully be referred to as "genets" (Brasier and Rayner, 1987).

In the relative absence of disturbance and selective stress, the potential incidence of competitors, both of the same and different species, within a habitat may be considerable and "C-selected" or combative organisms will be favoured. These may either retain resources captured during earlier stages of colonization under stressful or disturbed conditions, or replace former inhabitants. A useful indication of the combative ability of fungi can sometimes be obtained from mycelial interaction experiments in laboratory culture, which can also provide helpful diagnostic data (Rayner and Webber, 1984).

Combative organisms are prone to have long individual life-spans during which they encounter considerable biotic and abiotic heterogeneity. The pressure on them will therefore be to diversify both genetically and developmentally. In many habitats, the most combative fungi in terms of ability to replace other species, are often basidiomycetes (Cooke and Rayner, 1984). However, many ascomycetes, for example members of the *Xylariaceae*, may be effective in retaining resources and it is sometimes possible to rank organisms according to their relative combative ability (Coates and Rayner, 1985; Chapela et al., 1988). For example, *Daldinia concentrica* isolated from *Fraxinus excelsior* is generally more combative under atmospheric conditions than *Hypoxylon rubiginosum*, but less combative than the basidiomycete *Trametes versicolor* (Boddy et al., 1985).

Spatial Ordering

The spatial distribution patterns of mycelial genets are important both with respect to adaptation to local variations in microenvironmental conditions and the actual size of the physiological domain occupied. The latter can affect resource allocation to and hence the scale of reproductive structures.

Where nutritional resources are packaged into discrete units, such as individual leaves, fruits, twigs, etc., the physiological domain of a fungus will depend on whether it can produce migratory structures capable of connecting between these units. Fungi with this ability may be described as "non-unit-restricted" (Cooke and Rayner, 1984; Rayner et al., 1985) and on occasion may occupy genetic territories measurable in hectares (Smith et al., 1992). Good examples are provided by rhizomorphic and mycelial cord-forming basidiomycetes. These fungi locate resource units efficiently by means of versatile "foraging strategies" akin to those of army ants and stoloniferous plants (e.g. Dowson et al., 1988; Rayner and Franks, 1987), and are often highly combative, being able to replace former residents. Such behaviour does not seem to occur widely in saprotrophic ascomycetes, however, and unless they occupy diffuse habitats, such as comminuted plant residues in soil, these organisms are for the most part "resource-unit-restricted".

The physiological domains of unit-restricted fungi can be no greater than the physical boundaries of the resource units they occupy, and the volumes enclosed by these boundaries will therefore also limit reproductive output. For example, the conidial and perithecial stromata of *Xylaria carpophila* inhabiting *Fagus sylvatica* cupules are necessarily smaller than those of *X. hypoxylon* and *X. polymorpha* inhabiting twigs, logs, or tree stumps.

The other factor dictating the domain of unit-restricted fungi is micro-environmental heterogeneity due either to inherent properties of the resource itself (e.g. partitioning into distinctive regions such as bark, sapwood and heartwood), or the presence of other inhabitants, including competitors. The incidence of competitors will depend on the circumstances under which colonization occurs, and may be understood, as already described, in terms of ecological strategy theory. Where the incidence is high, due to relatively non-stressful, undisturbed conditions, physiological domain and consequently reproductive output per individual mycelium will be restricted. Where the incidence is low, extensive mycelial individuals will be capable of allocating resources to relatively large reproductive structures.

Host-selective (or apparently host-selective) species of the *Xylariaceae* illustrate this point. It is probable that many of these fungi establish in trees latently, only establishing active mycelial systems when the woody tissues become aerated following withdrawal of or restriction in water supply (Rayner and Boddy, 1988; Boddy and Griffith, 1989). Whereas some of these fungi, e.g. *Hypoxylon fuscum* on *Corylus avellana* and *H. fragiforme* on *Fagus sylvatica*, form swarms of separate, individually small perithecial stromata, others such as *Daldinia concentrica* on *Fraxinus excelsior* and *Biscogniauxia nummularia* on *Fagus sylvatica* form relatively voluminous or extensively effuse stromata. This distinction is associated with marked differences in the size of individual genets typically produced by the fungi in the hosts mentioned. Whereas *H. fragiforme* and *H. fuscum* produce numerous, mutually exclusive, small genets, originating in localized decay pockets that develop as the wood is dried, *D. concentrica* and *B. nummularia* produce few, axially extensive genets (Sharland, 1987; Chapela and Boddy, 1988; Boddy et al., 1985; A. Inman and A.D.M. Rayner, unpubl.; S.J. Hendry, pers. comm.). In fact, both *D. concentrica* and *B. nummularia* can often be isolated from trees other than those which are regarded as their typical hosts (e.g. Sharland, 1987; Boddy and Griffith, 1989; R. Whiteside and A.D.M. Rayner, unpubl.), but fail to form large genets therein. The apparent host-selectivity may therefore reflect constraints arising from the demands on resources involved in a genetic requirement to produce large teleomorphs, which can only be met when colonization occurs in particular types of trees, rather than limitations on ability to colonize *per se*.

GENERATIVE PROCESSES

Now that the distinctive demands of the environmental circumstances under which saprotrophic, mycelial ascomycetes operate have been outlined, the organizational attributes that constrain and enable responses to these demands can be addressed. Here, it is necessary to understand how diversity may be lessened in homogeneous, short-lived environments and enhanced in heterogeneous, longer lasting, locally unpredictable ones. It is also important to resolve between two kinds of diversity-generating processes (cf. Andrews, 1992; Anderson et al., 1992). Genetic processes operate predominantly at the population level and involve actual changes in genetic information content. Epigenetic processes operate predominantly at the individual level and involve variation in the way that genotype is converted into phenotype. This distinction is particularly interesting in organisms, such as mycelial ascomycetes, which combine developmentally determinate and indeterminate modes within their life-cycle. Whilst determinate development is apposite when future demands on the current generation are certain, both in character and timing, indeterminate development, in which growth potential is maintained indefinitely, is suited to continued exploration and exploitation of unpredictable environments. Broadly, determinate development and genetic diversification of subsequent generations are attributes of reproductive phases, whereas indeterminate development and epigenetic diversification within current generations occur in somatic phases. However, since these phases are not entirely independent, complexities may arise where their developmental paths and diversification processes overlap.

Reproductive Pathways

Commitment to reproduction is inevitably made at a cost to indeterminate development, and so may only be appropriate when the future of the latter is in doubt, for example following attainment of a resource boundary. However, it is an important property of mycelial systems that this cost may be varied according to whether the mode of reproduction is clonal or recombinatorial.

Conidiation allows an individual to distribute all, rather than on average only half, its genes to the next generation, to produce potentially cooperative rather than competitive progeny, and to circumvent the requirements for mating and (often) investment in substantive sporophores. Ostensibly, this process, which is much more widespread in saprotrophic ascomycetes than saprotrophic basidiomycetes, might therefore seem to be less costly, both genetically and energetically, than production of a teleomorph. However, the genetic saving may depend on the conditions determining the fitness of offspring being the same as those operating on their progenitor. In heterogeneous environments this is liable to apply only when dispersal occurs over a short range, as is the likelihood that spores from a common source will arrive at the same destination (Rayner, 1992). Equally, the energetic cost of producing teleomorphs with a large biomass may be offset by the capacity of these phases to disseminate spores over a long range, and by their better protection against dissipation of stored energy to the environment.

Such considerations provide a rationale for understanding the deep relationship between reproductive mode, stage in community development, and mycelial properties. Conidiation does not usually necessitate much long-range translocation of resources, and is an asset at stages of colonization when mycelia are, or have recently been, predominantly in an assimilative and therefore dissipative mode (*cf.* Fig. 1; see also below). The process enables local proliferation of a successfully established genet, either directly or *via* a role in sexual fertilization. Less usually it would be responsible for dissemination over a long-range. By contrast, teleomorph production is, as the term implies, a late developmental phase, predominantly serving in long-range dissemination to, and establishment, in new sites. This phase is beneficial when assimilable resources (though not necessarily refractory ones) have long since been depleted and mycelia are in an energy-conserving, distributive mode (*cf.* Fig. 2; see also below), capable of long range translocation to sporophore initials.

Whereas conidiation may therefore be interpretable largely as an R-selected attribute, production of a teleomorph implies C- or S-selection. However, with regard to genetic diversification, it is not just the question of anamorph *versus* teleomorph that matters, but also whether the teleomorph and its ascospores are produced by a non-outcrossing or outcrossing mechanism (Anderson et al., 1992). The latter terms are distinct from and preferred to "homothallism" and "heterothallism" for reasons discussed by Rayner (1990).

Whereas non-outcrossing yields clonal offspring, outcrossing, *via* recombination, yields variable offspring. Which of these breeding mechanisms is operative can generally be discerned simply by examining patterns of variation and mycelial interaction between single ascospore isolates derived from the same sporophore. When, for example, this was done for various species of xylariaceous fungi isolated from wood samples from European woodlands, it was found that outcrossing operated generally within populations of *Biscogniauxia nummularia*, *Daldinia concentrica*, *Hypoxylon fragiforme*, *H. fuscum*, *H. mammatum*, *H. rubiginosum*, *H. serpens* and *Rosellinia mammiformis*, whereas non-outcrossing was detected in *H. multiforme*, the undescribed "*H. purpureum*", and *Rosellinia desmazieresii* (Sharland, 1987; Sharland and Rayner, 1986, 1989a, b; Sharland et al., 1988).

The ecological significance of this distinction is not clear, though the possibility that non-outcrossing may be associated with reduced spatial heterogeneity has been raised (see above). However, the question is complicated as the population structure of both *H. multiforme* and *R. desmazieresii* is genetically diverse in spite of non-outcrossing, and, in *H. multiforme* at least, there is evidence for a facultative outcrossing mechanism. Clearly, more population studies are needed in xylariaceous and other ascomycetes to provide a better understanding of the occurrence and biological and taxonomic significance of these alternative breeding mechanisms (Anderson et al., 1992). The occurrence of outcrossing and non-outcrossing populations has been detected in some saprotrophic basidiomycete species, and here there is more indication of a relationship with geographical distribution (e.g. Ainsworth et al., 1990a).

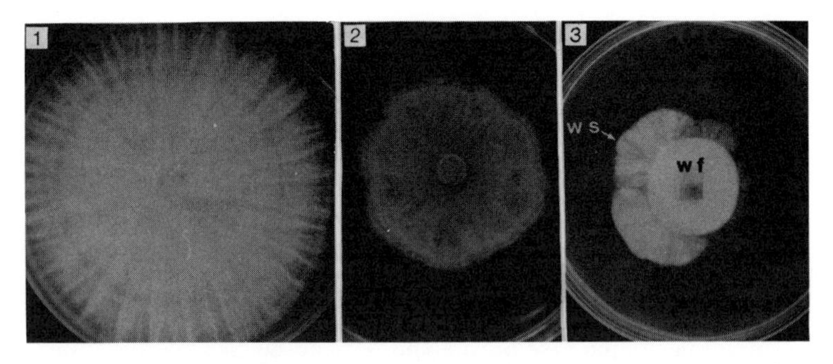

Figures 1-3. Mycelial polymorphism in *Hypoxylon serpens* (aggr.) on 2% malt agar. **Fig. 1**. Single ascospore-derived isolate exhibiting "white, silky" morphology; few, if any conidia are produced, and the colony develops pseudosclerotia as it ages. **Fig. 2**. Single ascospore isolate of the same age, showing "grey conidial" morphology; the colony has a powdery covering of conidiophores and does not produce pseudosclerotial plates. **Fig. 3**. A "white felty" conidiogenous colony form derived by subculture from an aged white silky colony. The "slow-dense" white felty form (wf) is producing "fast-effuse" white silky sectors (ws). Preliminary indications are that white silky forms are characteristic of wood undergoing active decay following relatively recent colonization, whereas grey conidial forms are characteristic of older, more degraded wood (from Sharland, 1987).

Developmental Options

As has been implied, the indeterminacy of mycelial systems brings with it the opportunity for a fungal individual to explore alternative developmental options, allowing it to respond sensitively and appropriately to changing circumstances in spatiotemporally heterogeneous environments. One aspect of this versatility, the capacity to give rise to different reproductive modes at distinctive stages in the colonization process has already been mentioned. More generally, the ability to operate in distinctive functional modes (Gregory, 1984), sequentially or in combination, is evident in the diverse organizational patterns adopted by mycelial systems, and by the sometimes abrupt transitions that can occur from one pattern to another (e.g. Rayner and Coates, 1987; Stenlid and Rayner, 1989; Rayner et al., 1993). The resultant changeability in mycelial morphology, even within the same strain, may at first sight seem to obviate the use of mycelial characters, or at least greatly reduce their value, for systematic purposes. However, if consistent themes, or organizational principles underlying the changes can be identified, then the task of making systematic sense of mycelial variation may greatly be eased. The following summary, based on accounts given by Rayner and Coates (1987), Rayner (1991), and Rayner et al. (1993), attempts such an identification, based on a consideration of the relationship between the functional and dynamic properties of mycelia.

The process of mycelium formation begins when a spore germinates. This usually involves a phase of isotropic expansion prior to emergence of one or more polarized germ tubes. However, even at this early stage, options between determinate and indeterminate development can occur, resulting in mycelial-yeast dimorphism and secondary spore formation. Production of a unicellular phase aids dispersion through liquid environments and is an important property of fungi that establish within the circulatory systems of plant and animal hosts. It represents a mechanism whereby an individual genet could become extensively established prior to producing active mycelium. There is recent evidence that it occurs in response to interaction with *Fagus sylvatica* callus in *Biscogniauxia nummularia* and "*Hypoxylon purpureum*", both of which form axially extensive individuals in the sapwood of this tree (Hendry et al., 1993; see above).

Once germ tubes have emerged, the hyphae they give rise to become partitioned internally by septa into uninucleate, binucleate or multinucleate compartments. The septa act as valves, closeable by blockage with Woronin bodies, and their formation is closely correlated with the initiation of lateral branches (Trinci, 1978). Branching results in a radiating structure whose fractal dimension (space-filling capacity) progressively increases (Ritz and Crawford, 1990; Crawford and Ritz, 1993), prior to the formation of anastomoses. The latter have the fundamental dynamic consequence of establishing a communications network behind the expansive colony margin (e.g. Rayner, 1991).

As branches emerge, they may do so with variable frequency and at variable angles, signifying differential allocation of effort to radial (explorative) or tangential (exploitative/consolidative) vectors. Where such changes in branching pattern are coordinated, possibly as a result of anastomosis, they give rise to shifts in the organization and extension rate of the mycelial margin. Where these shifts occur abruptly, they result in what have been termed slow-dense/fast-effuse dimorphisms (Rayner and Coates, 1987), and are often manifest as sectoring phenomena (Fig. 3).

Another important aspect of mycelial coordination depends on the capacity of hyphae to develop either diffusely or in compact associations. The latter may have migratory or connective roles in the case of rhizomorphs and mycelial cords, protective roles in sclerotia and pseudosclerotia, as well as reproductive roles in stromata and ascomata.

Such roles also depend on another fundamental property of hyphal systems, the capacity to shift between assimilative and non-assimilative modes. Whereas, in relative terms, assimilative modes are necessarily open to the environment and therefore dissipative, non-assimilative modes are sealed off and therefore distributive or conservative. The striking dimorphism between grey, conidiogenous and white, silky, pseudosclerotial mycelial forms of *Hypoxylon serpens* (Figs 1-2) illustrate the different morphogenetic patterns that may result from variable operation of these modes. Extraordinarily similar dimorphisms occur in certain basidiomycetes, sometimes correlated with homokaryon-heterokaryon transitions (Stenlid and Rayner, 1989). Overall, it is tempting to suggest that the predominantly heterokaryotic mycelial systems of basidiomycetes (see also below) are more prone to exhibit non-assimilative properties than the predominantly homokaryotic systems of ascomycetes. This might be linked in turn to the general differences, already mentioned, between these organisms with respect to their reproductive development, foraging patterns and combative ability.

As a mycelium expands, there is a probability that parts of it will become functionally redundant and yet still act as resource sinks. This is circumvented by autodegenerative processes that allow redistribution to metabolically active phases, and may also provide a means of isolating the latter from potential pathogens.

The above outline may have served to demonstrate that, fundamentally, the mycelium is a creatively unstable system. Its apparently haphazard behaviour in spite of stringent efforts to constrain it within predictable, homogeneous environments is an expression of its developmental repertoire, but nonetheless challenges the systematist anxious to find consistency. Is there any way in which the richness of this repertoire can be explored systematically without generating a confusion of inconsistent data?

One answer may be to reverse the long-standing tradition of culturing mycelia in homogeneous conditions and to simulate in some way the heterogeneity that characterizes their natural environments. That way it may be possible not only to induce them to display their versatility to the full, but actually to specify where and when particular functional modes will be produced. A promising approach, under current investigation at Bath, involves growing mycelia in interconnected matrices of sites containing media designed to select for distinctive mycelial attributes, e.g. varied nutrient content, C:N ratio.

The above approach is ecological, manipulating the circumstances which in natural environments are thought to influence mycelial functioning. It may also be possible to actuate the mechanisms of mycelial instability by more direct intervention, providing these mechanisms are known.

INSTABILITY MECHANISMS

Genetic *versus* Environmental Regulation

Determinate development involves a sequence of ontogenetic events, any significant departure from which is generally catastrophic, ordered by a strict genetic programme executed by regulatory genes (e.g. Pritchard, 1986). Such development is anticipative rather than responsive, and adaptation to unpredictable circumstances is only possible through natural selection operating on genetic variation produced between generations.

Indeterminate development is, on the other hand, potentially more responsive to circumstance and hence more liable to be environmentally regulated. Nonetheless, the mechanism for an appropriate response to environmental change has to be in place before the change occurs, and in predictable environments there will be benefit in anticipating events endogenously. Often it is the timing rather than the nature of the change that is unpredictable. Moreover, as specialization for some niche component occurs, some developmental options will become increasingly redundant and therefore may be lost from the repertoire, as may occur in R-selected fungi. Alternatively a particular developmental option may be reinforced to the point when it is genetically imperative. In many ways, mycelial development may therefore be finely balanced at the interface between exogenous and endogenous controls (Stenlid and Rayner, 1989). This raises the question of the relationship between these controls, and how they may feed back on one another.

Metabolic Feedback

According to a recent hypothesis (Rayner et al., 1993), the mycelium is a non-linear system in which expansive processes resulting from input of energy are counteracted by constraints that prevent absolute dispersion. When it experiences a drop in internal energy charge, the system activates metabolic pathways that lead either to the rigidification and sealing off ("insulation") of hyphal boundaries or to the onset of degeneration. These pathways involve the production of hydrophobic aromatic and terpenoid compounds, certain polypeptides known as "hydrophobins" (Wessels, 1991, 1992) and the operation of free radical-generating phenol-oxidizing enzymes. By such means, mycelia can vary the deformability and penetrability of their external boundaries, and the partitioning of their interior, to accord with fluctuations in assimilable energy supplies. In so doing, they generate variable amounts of hydraulic thrust, and, thereby, diverse developmental patterns. The exact pattern depends on the relative rates of uptake of water and nutrients from the environment, and throughput of energy following such uptake to sites of deformation on the system's boundary. These pattern-generating processes are fundamentally the same in all systems, but the parameters governing their operation in individual cases are specified genetically. For example, ascomycetes may generally be less prone to insulate their boundaries than basidiomycetes, accounting for their generally more R-selected attributes. Adaptation therefore lies not so much in the pattern-generating processes themselves, which are organizationally impelled, as in their pre-set or variable tuning to meet the predictable and unpredictable demands of diverse niches, definable in terms of ecological strategies.

Genetic Disharmony

From the argument so far, the developmental instability of mycelial systems has more to do with epigenetic than genetic heterogeneity. This accords with the fact that ascomycetes probably do not in general produce vigorous, independently growing mycelia of heterokaryotic origin in nature (e.g. Anderson et al., 1992). The formation of heterokaryons is potentiated by hyphal fusions between genetically distinct (non-self) mycelia of the same or closely related species. Temporarily at least, these fusions bring both nuclei and mitochondria from different origins within the same protoplasm. This raises the question of the effects, in particular the possible induction of mitochondrial disfunction (Rayner and Ross, 1991), of introducing potentially disparate genetic information into a feedback system which is already delicately counterpoised on the brink of degeneracy.

That stable associations are possible following non-self fusion is evident from the fact that heterokaryotic mycelia are produced as a consequence of mating in many saprotrophic basidiomycetes. However, these mycelia do not appear to be capable of maintaining stable associations of more than two genetically disparate types of nuclei and more than one mitochondrial type (e.g. Rayner, 1991; Rayner and Ross, 1991). Moreover, when strains from different geographical regions or ecological niches are mated, the outcome may be extensive degeneration, genomic takeover or complex patterns of phenotypic expression (Ainsworth et al., 1992; Ainsworth et al., 1990a, b).

Instability is also a feature of non-self fusions between ascomycete homokaryons. Often, this has the effect of causing degeneration of fused segments in a mutual exclusion (rejection) zone between adjacent genets, just as occurs between heterokaryotic genets of basidiomycetes. Pigmentation, pseudosclerotial plate formation and conidiation are often induced in or abutting this zone. However, there is also evidence, for example in various xylariaceous fungi (Sharland, 1987; Sharland and Rayner, 1986, 1989a, 1989b) and the diatrypaceous *Eutypa spinosa* (Sharland, 1987; S.J. Hendry, pers. comm.), for temporary heterokaryosis. This leads to the formation of hydrophobic aerial mycelium, prone to break down into one or other or both progenitors or producing a recombinant on subculture, and sometimes leading to genomic takeovers in the interaction itself. Interestingly, *E. spinosa* is often isolated in heterokaryotic form from wet parts of the wood in beech trees, but as mutually exclusive homokaryons from drier parts, indicating the probable importance of aeration in determining the onset of degenerative processes (S.J. Hendry, unpubl.).

Non-self fusions therefore provide a potential frontier where reproductive and somatic, genetic and epigenetic processes come into direct interplay in the generation of evolutionary complexity within saprotrophic fungi. Their study may therefore provide key insights for the systematist.

REFERENCES

Ainsworth, A.M., J.R. Beeching, S.J. Broxholme, B.A. Hunt, A.D.M. Rayner, and P.T. Scard, 1992, Complex outcome of reciprocal exchange of nuclear DNA between two members of the basidiomycete genus *Stereum*, *Journal of General Microbiology*, 138: 1147-1157.

Ainsworth, A.M., A.D.M. Rayner, S.J. Broxholme, and J.R. Beeching, 1990a, Occurrence of unilateral genetic transfer and genomic replacement between strains of *Stereum hirsutum* from non-outcrossing and outcrossing populations, *New Phytologist* 115: 119-128.

Ainsworth, A.M., A.D.M. Rayner, S.J. Broxholme, J.R. Beeching, J.A. Pryke, P.T. Scard, J. Berriman, K.A. Powell, A.J. Floyd, and S.K. Branch, 1990b, Production and properties of the sesquiterpene, (+)-torreyol, in degenerative mycelial interactions between strains of *Stereum*, *Mycological Research* 94: 799-809.

Anderson, J.B., L.M. Kohn, and J.F. Leslie, 1992, Genetic mechanisms in fungal adaptation, *In*: *The Fungal Community* (G.C. Carroll and D.T. Wicklow, eds): 73-98. Marcel Dekker, New York.

Andrews, J.H., 1992, Fungal life-history strategies, *In*: *The Fungal Community* (G.C. Carroll and D.T. Wicklow, eds): 119-145. Marcel Dekker, New York.

Boddy, L., O.M. Gibbon, and M.A. Grundy, 1985, Ecology of *Daldinia concentrica*: effect of abiotic variables on mycelial extension and interspecific interactions, *Transactions of the British Mycological Society* 85: 201-211.

Boddy, L. and G.S. Griffith, 1989, Role of endophytes and latent invasion in the development of decay communities in sapwood of angiospermous trees, *Sydowia* 41: 41-73.

Brasier, C.M., 1987, The dynamics of fungal speciation, *In*: *Evolutionary Biology of the Fungi* (A.D.M. Rayner, C.M. Brasier and D. Moore, eds): 231-260. Cambridge University Press, Cambridge.

Brasier, C.M., 1988, Rapid changes in genetic structure of epidemic populations of *Ophiostoma ulmi*, *Nature* 332: 538-541.

Brasier, C.M., 1991, *Ophiostoma novo-ulmi* sp. nov., causative agent of current Dutch Elm disease pandemics, *Mycopathologia* 115: 151-161.

Brasier, C.M. and A.D.M. Rayner, 1987, Whither terminology below the species level in fungi?, *In*: *Evolutionary Biology of the Fungi* (A.D.M. Rayner, C.M. Brasier and D. Moore, eds): 379-388. Cambridge University Press, Cambridge.

Chapela, I. and L. Boddy, 1988, Fungal colonization of attached beech branches. II. Spatial and temporal organization of communities arising from latent invaders in bark and functional sapwood, under different moisture regimes, *New Phytologist* 110: 47-57.

Chapela, I., L. Boddy, and A.D.M. Rayner, 1988, Structure and development of fungal communities in beech logs four and a half years after felling, *FEMS Microbiology Ecology* 53: 59-70.

Coates, D. and A.D.M. Rayner, 1985, Fungal population and community development in beech logs. III. Spatial dynamics, interactions and strategies, *New Phytologist* 101: 183-198.

Cooke, R.C. and A.D.M. Rayner, 1984, *Ecology of Saprotrophic Fungi*, Longman, London.

Crawford, J. and K. Ritz, 1993, Origin and consequence of colony form in fungi: a reaction-diffusion mechanism for morphogenesis, *In*: *Shape and Form in Plants and Fungi* (D.S. Ingram, ed.): in press. Academic Press, London.

Dowson, C.G., A.D.M. Rayner, and L. Boddy, 1988, Foraging patterns of *Phallus impudicus*, *Phanerochaete laevis* and *Steccherinum fimbriatum* between discontinuous resource units in soil, *FEMS Microbiology Ecology* 53: 291-298.

Gould, S.J. and R.C. Lewontin, 1979, The spandrels of San Marco and the Panglossian paradigm: a critique of the adaptionist programme, *Proceedings of the Royal Society of London, B* 205: 581-598.

Gregory, P.H., 1984, The fungal mycelium: an historical perspective, *Transactions of the British Mycological Society* 82: 1-11.

Hendry, S.J., L. Boddy, and D. Lonsdale, 1993, Interactions between callus cultures of European beech, indigenous ascomycetes and derived fungal extracts, *New Phytologist* 123: 421-428.

Pritchard, D.J., 1986, *Foundations of Developmental Genetics*, Taylor and Francis, London.

Prosser, J.I., 1991, Mathematical modelling of vegetative growth of filamentous fungi, *In*: *Handbook of Applied Biology*, Vol. 1 (D.H. Arora, B. Rai, K.J. Mukerji and G.R. Knudsen, eds): 591-623. Marcel Dekker, New York.

Rayner, A.D.M., 1990, Natural genetic transfer systems in higher fungi, *Transactions of the Mycological Society of Japan* 31: 75-87.

Rayner, A.D.M., 1991, The challenge of the individualistic mycelium, *Mycologia* 83: 48-71.

Rayner, A.D.M., 1992, Monitoring genetic interactions between fungi in terrestrial habitats, *In*: *Genetic Interactions among Microorganisms in the Natural Environment* (E.M.H. Wellington and J.D. van Elsas, eds): 267-285. Pergamon Press, Oxford.

Rayner, A.D.M. and L. Boddy, 1988, *Fungal Decomposition of Wood*, John Wiley, Chichester.

Rayner, A.D.M. and D. Coates, 1987, Regulation of mycelial organisation and responses, *In*: *Evolutionary Biology of the Fungi* (A.D.M., Brasier, C.M. and D. Moore, eds): 115-136. Cambridge University Press, Cambridge.

Rayner, A.D.M. and N.R. Franks, 1987, Evolutionary and ecological parallels between ants and fungi, *Trends in Ecology and Evolution* 2: 127-133.

Rayner, A.D.M., G.S. Griffith, and H.G. Wildman, 1993, Differential insulation and the generation of mycelial patterns, *In*: *Shape and Form in Plants and Fungi* (D.S. Ingram, ed.): in press. Academic Press, London.

Rayner, A.D.M. and I.K. Ross, 1991, Sexual politics in the cell, *New Scientist* 129: 30-33.

Rayner, A.D.M., R. Watling, and J.C. Frankland, 1985, Resource relationships - an overview, *In*: *Developmental Biology of Higher Fungi* (D. Moore, L.A. Casselton, D.A. Wood and J.C. Frankland, eds): 1-40. Cambridge University Press, Cambridge.

Rayner, A.D.M. and J.F. Webber, 1984, Interspecific mycelial interactions - an overview, *In*: *The Ecology and Physiology of the Fungal Mycelium* (D.H. Jennings and A.D.M. Rayner, eds): 383-417. Cambridge University Press, Cambridge.

Ritz, K. and J. Crawford, 1990, Quantification of the fractal nature of colonies of *Trichoderma viride*, *Mycological Research* 94: 1138-1141.

Sharland, P.R., 1987, *Mycelial Biology of Xylariaceous Fungi*, PhD thesis, University of Bath.

Sharland, P.R. and A.D.M. Rayner, 1986, Mycelial interactions in *Daldinia concentrica*, *Transactions of the British Mycological Society* 86: 643-650.

Sharland, P.R. and A.D.M. Rayner, 1989a, Mycelial interactions in outcrossing populations of *Hypoxylon*, *Mycological Research* 93: 187-198.

Sharland, P.R. and A.D.M. Rayner, 1989b, Mycelial ontogeny and interactions in non-outcrossing populations of *Hypoxylon*, *Mycological Research* 93: 273-281.

Sharland, P.R., A.D.M. Rayner, A.U. Ofong, and D.K. Barrett, 1988, Population structure of *Rosellinia desmazieresii* causing ring dying of *Salix repens*, *Transactions of the British Mycological Society* 90: 654-656.

Smith, M.L., J.N. Bruhn, and J.B. Anderson, 1992, The fungus *Armillaria bulbosa* is among the largest and oldest living organisms, *Nature* 356: 428-431.

Stenlid, J. and A.D.M. Rayner, 1989, Environmental and endogenous controls of developmental pathways: variation and its significance in the forest pathogen, *Heterobasidion annosum*, *New Phytologist* 113: 245-258.

Trinci, A.P.J., 1978, The duplication cycle and vegetative development in moulds, *In*: *The Filamentous Fungi*, Vol. 3 (J.E. Smith and D.R. Berry, eds): 132-163. Edward Arnold, London.

Wessels, J.G.H., 1991, Fungal growth and development: a molecular perspective, *In*: *Frontiers in Mycology* (D.L. Hawksworth, ed.): 27-48. CAB International, Wallingford.

Wessels, J.G.H., 1992, Gene expression during fruiting in *Schizophyllum commune*, *Mycological Research* 96: 609-620.

DIFFERENT SPECIES TYPES IN LICHENIZED ASCOMYCETES

J. Poelt

Institut für Botanik
Holteigasse 6
A-8010 Graz, Austria

SUMMARY

Species in lichenized ascomycetes can, according to their biology, be arranged roughly into three different categories. Many belong to a group of very stable sexual species, which are homogeneous over large areas. A second group, mostly silicicolous lichens, conists of genetically very variable taxa, where thalli with varied combinations of small, but distinct differences grow side by side; examples can be found within *Bellemerea, Fuscidea, Graphis, Lecidea* and *Sporastatia*; taxonomy in this group is difficult. The third group consists of taxa with solely or predominantly asexual reproduction. Such taxa are apomicts, genetically isolated from their sexual relatives. Contrary to a modern trend, but in agreement with taxonomic practice in other groups of plants and fungi, those taxa should be retained as species. Synonymizing would result in the loss of most important information about their evolution and history.

INTRODUCTION

Lichenization occurred in different groups of fungi, probably at very different times. Therefore, lichens are not a natural, but a polyphyletic group. In spite of this, most lichens are distinguished from most non-lichenized ascomycetes by the longevity of their thalli as well as of their ascocarps.

This important difference has strong significance for the understanding of the systematics of both groups. Lichenized ascomycetes are often exposed for a very long time to harsh ecological conditions. Therefore, they show a very high amount of modificability. Extremely modified thalli of one species often look more different than different species do. Lichenologists often described mainly phenotypes in the past. A large number of names of taxa: species, varieties, and forms, was introduced to describe nothing but ecologically controlled variability. All old catalogues are full of such names. Zahlbruckner (1927) cited no less than 59 varieties and forms for *Cladonia squamosa*, and 23 infraspecific taxa for *C. uncialis*. Even the comparatively rare *C. strepsilis* has eight infraspecific names. In some cases, biological units are really hidden under such names, in rare cases they may belong to "chimeras", as shown by Jahns (1987), but the greatest amount of all these varieties and forms are nothing but modifications, developmental stages, or even damaged forms, all connected by intermediates. For some decades there has been a strong tendency to refuse or even forget all those names. It is the aim of taxonomy to define and describe biological units, stirpes, not modifications.

Unfortunately, this tendency in lichenology has led to a disreputation of observations on variability. At this time, only the species as the basic unit is of interest, and commonly thought to be a homogeneous type with distinct borderlines against other such units. The question arises, if such a view is correct.

If many different lichen species are compared, it is possible to come to the conclusion that there are at least three different types of species in lichenized ascomycetes.

STABLE SEXUAL SPECIES

Stable sexual species, or more exactly, species with regular reproduction through ascospores, which need to combine with a corresponding alga. These species may show a high plasticity due to the influence of ecological factors, but they are distinctly separated from allied species and also rather monomorphic. Very many lichens from very different families and genera belong to this group, which is not discussed further here. There may be a distinct genetic variation of such lichens in their often very large areas, for example between Europe and North America, but that has not been studied in detail.

GENETICALLY VARIABLE SEXUAL SPECIES

In several genera of principally crustose mostly silicicolous lichens, we can observe another type of species. This may be first demonstrated by some members of the genus *Fuscidea*. *F. cyathoides* (Fig. 1C) is widely distributed in western, more rarely in central, Europe. It grows on steep shaded side surfaces of very hard and mostly nutrient-deficient siliceous rocks. The crustaceous thalli develop abundant apothecia and spermogonia. Quite often, the lichen forms large populations of very many thalli, with only a few other lichens intermixed. But if such populations, growing under identical ecological condition, are observed, it can be demonstrated that all thalli are indeed similar in their principal characters, but that they differ markedly in characters of a second or third value: thickness and areolation of thalli, colour of thalli (more whitish or more greenish/brownish), size of apothecia, shape of apothecia (roundish or very early undulate in outline), and position of apothecia (sessile or somewhat immersed). All these characters can occur in different combinations. Hardly two thalli are identical. The same observations can be made in different localities. The characters described are presumably not subject to strong selective pressure, and might be interpreted as neutral in evolution, at least in optimum habitats. However, the corticolous populations of *F. cyathoides*, often separated as var. *corticola*, seem to be genetically homogeneous.

Fuscidea kochiana, frequent and often abundant on steep sides of exposed hard siliceous rocks, for example in the Alps, shows similar phemomena; in this species the pruinosity of the discs can also vary strongly from individual to individual.

Attempts to define varieties and forms in both these *Fuscidea* species have been made (Magnusson, 1925), but as all combinations of characters occur with no clear cut differences, such taxa cannot be well-circumscribed.

A similar situation exists in the genus *Bellemerea*. *B. alpina* (Fig. 1A) shows, on the same rock, under the same conditions, and in varying thalli, small but distinct differences in form, size and thickness of areoles, in colour of the surface (whitish, whitish grey, more or less bluish grey), and in position, size and pruinosity of apothecia. Here also, variable combinations of such characters are common and prohibit any distinction of infraspecific taxa. The same holds true for *B. cinereorufescens* (Fig. 1B). Even the well-known *B. diamarta*, characterized by the ochraceous to rusty red thallus, is far from homogeneous; some thalli are more intensely rusty in colour than others, which are rather pale ochraceous. It may be questioned, whether the species is well-defined at all. No doubt, *Bellemerea*, except for the sorediate *B. subsorediza*, is composed of genetically variable taxa, probably with gene flow between the main types, and thus connected by intermediate forms. It was shown by Culberson et al. (1988), in a pioneering study, that gene flow between different taxa of lichens can occur.

Some further examples from other lichenized groups may be cited. The arctic-oreophytic *Sporastatia testudinea* (Magnusson, 1936; Grube and Poelt, 1993) is represented in its wide geographic distribution by many different phenotypes with varying combinations of morphological characters of low value, growing not rarely side by side and under the same conditions. The width and form of areoles, elongation of marginal areoles, colour of surfaces, development of prothallus, presence of thallospores, size and form of ascocarps, and probably also the form of the ascospores, vary strongly. However, as these characters occur in many different combinations, they do not permit any distinction of clear infraspecific taxa. In addition, these characters are often concealed by modifications in the most extreme biotopes. Even influences from the substratum may contribute to the modificability: on rocks containing some calcium, the thalli are not seldom pruinose.

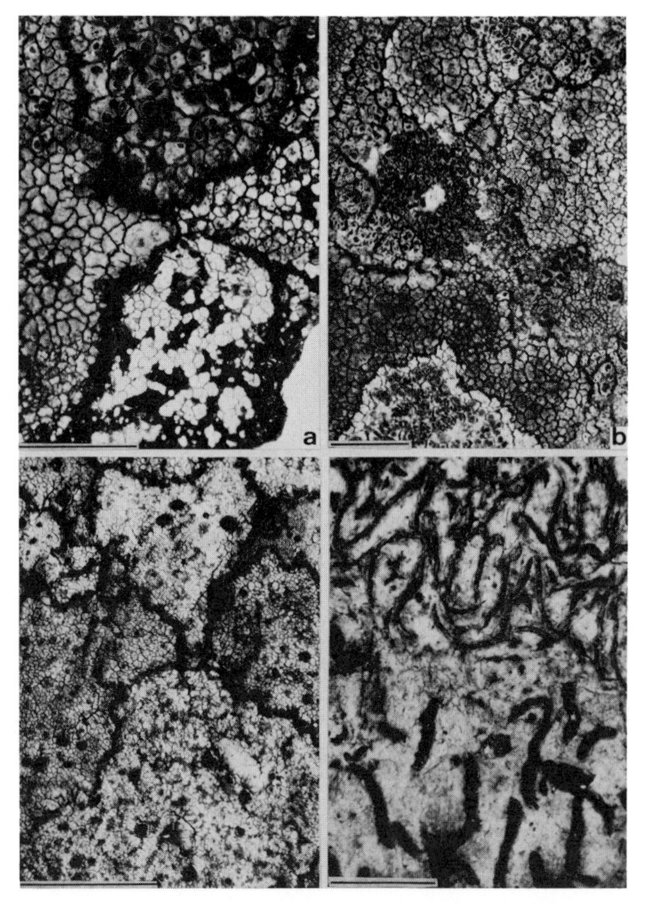

Figure 1. A, *Bellemerea alpina*; B, *Bellemerea cinereorufescens*; C, *Fuscidea cyathoides*; D, *Graphis scripta*. Several, or at least two (*Graphis*), phenotypes grown side by side. Scale = 1 cm.

Another example is the well-known *Graphis scripta* (Fig. 1D). On the same tree, under identical conditions, sharply separated thalli with differently formed and situated ascocarps can occur. In some thalli they are elongate and not or sparsely divided, in others they are short and strongly divided, they may be more protruding or more immersed and their discs can be pruinose or not. In addition, the shape and orientation of ascocarps can also be influenced by ecological conditions, especially by the structure of the host bark. But, without doubt, there is a distinct genetical variability. Also in *G. scripta*, attempts have been made to distinguish many different infraspecific taxa. Zahlbruckner (1923) enumerates no less than 72 varieties and forms, but as the characters can occur in variable combinations, no clear-cut borderlines between infraspecific taxa can be drawn.

Similar cases can be found in the notoriously difficult genus *Lecidea*. The type species, *L. fuscoatra*, seems to be represented by many races with varying combinations of characters of low value; sometimes such different forms grow side by side. Many published infraspecific names provide evidence for this variability. Even more difficult is the case of *L. lapicida*. The variability of this complex involves also chemistry and the structure of ascocarps. Quite often, different forms can be seen growing side by side; Schwab (1986) was unable to recognize distinct taxa inside this complex, but probably there are some infraspecific units, which could be defined as subspecies or varieties.

Most probably, the enormous taxonomic difficulties in the complex *Rhizocarpon geographicum* have a similar basis. A corticolous counterpart of this type of species may be seen in some groups of *Arthonia*.

In foliose lichens, comparable cases seem to be rare. However, the high variability of *Umbilicaria cylindrica* may, at least partly, be seen in this way. In *Parmelia omphalodes*, in the same locality, and under identical conditions, thalli are found with distinct differences in the number and breadth of lobes, presence of secondary lobes, development of pseudocyphellae, and colour (whitish grey, grey, brownish, and also nearly black). But the "cornerstone" types are connected by intermediates defined by varying combinations of characters. This lichen quite rarely produces ascocarps, and it is presumed that some races became somewhat independent by solely vegetative reproduction, at least for a while. According to this, the taxonomy of the species taxonomically complex must be difficult and it was treated in different ways by Skult (1984, 1987) and Hale (1987). The somewhat similar complex of *Parmelia somloensis* (syn. *Xanthoparmelia somloensis*; Hale, 1990) is even more complicated and not discussed here.

Only relatively few lichens are regularly, and by distinct means, able to propagate at the same time by ascospores and in addition by some type of asexual diaspore. One example is the well-known *Baeomyces rufus*, a successful colonizer of eroded soil, because it simultaneously develops apothecia, schizidia, and soredia. In several short-lived pioneer lichens, true conidia and apothecia are sometimes developed at the same time.

APOMICTIC SPECIES

As a rule, lichens with a distinct ability for asexual reproduction by means of soredia, isidia, etc., only rarely or never develop ascocarps, or, if ascocarps are produced, they often contain no well-developed asci and(or) ascospores.

A well-known example for this third, very important group of species is *Hypocenomyce scalaris*, a very common corticolous lichen in large parts of Europe. Apothecia are rare, but not uncommon, if the lichen grows on burnt timber. Asci are rare, and, as a rule not well-developed (Timdal, 1984; Poelt, unpubl.). *H. scalaris* is the type species of its genus, which could not be correctly classified, as far as concerns ascus type, because well-developed asci were not found by Hafellner (1984). *H. scalaris* is, through its large geographic area, a monomorphous apomictic species.

In many lichen species with asexual reproduction, the situation seems to be similar. Ascocarps are produced only rarely or very rarely and often only under optimal ecological conditions; mature ascospores are rare or not produced at all. But such cases are not well-studied, partly to save the rare fruits. At least, sexual reproduction by means of ascospores plays no, or at least a very reduced, roles, in such species.

In discussing asexual reproduction, a short general review of the main types of dispersal bodies in lichens is necessary. For discussion, the well-studied foliose lichens, which can, besides rarer types not discussed here, produce soredia, isidia or phyllidia, may be used. Phyllidia are the most elaborate structures; they resemble lobules, with a dorsiventral structure and are like-wise corticate. Isidia are also corticate, but show no dorsiventrality. Soredia are very simple non-corticate bodies, consisting mostly of a few, at least one, algal cell(s) and some hyphae between and around these. If these different types of diaspores germinate, they develop firstly, as far as we know, some undifferentiated clumps of tissue (Ott, 1986), which later produce a thallus primordium or may be transformed into a primordium. It was shown quite recently (Stocker-Wörgötter and Türk, 1993), that the regeneration of whole thalli of *Cladonia furcata* takes place through soredia-like dedifferentiated bodies. In any case, phyllidia and isidia become dedifferentiated to be able to produce a new thallus. In this view, the histologically most differentiated phyllidia are the most primitive asexual diaspores, while soredia are the most derived (and successful); isidia stand inbetween.

The outcome of these three types, can be tested on European foliose lichens containing green phycobionts. Clearly phyllidiate taxa are very rare. *Parmelia horrescens* and *Heterodermia propagulifera* are extremely rare western species. *Physconia grisea* subsp. *lilacina* is probably the rarest *Physconia* in Europe, with a submediterranean distribution. *Umbilicaria deusta* is not rare, but ecologically very specialized; *U. freyi* is rare and mostly mediterranean. In other parts of the world, in temperate and tropical zones, phyllidiate taxa are much more common. The reason for this difference seems to be their distinct histories (see below). At least, the sorediate species are the most common amongst the asexual foliose lichens.

For many years, taxa with regular asexual reproduction have been treated, following Du Rietz (1924), as independent species. Recently, there has been a tendency to reduce these taxa to forms (Tehler, 1983), or to treat them as taxonomically identical with non-sorediate related taxa (Sipman, 1983; Niebel-Lohmann and Feuerer, 1993). This may be correct in cases where thalli are able to simultaneously propagate sexually and asexually. However, as discussed above, such cases are rare. The problems connected with this trend will be discussed shortly.

Dirina ceratoniae, including *D. massiliensis*, is a widespread, probably genetically not homogeneous, lichen of the southern and western coasts of Europe (Tehler, 1983). It has never been found inland in central Europe, but both in the mediterranean region as well in central Europe related sorediate types do exist, probably representing several taxa. They have scarcely ever been found with ascocarps and well-developed ascospores. So, practically, they represent fully apomictic lichens as they are obviously genetically separated from *D. ceratoniae*. The genetical difference may be low, but as apomicts the sorediate taxa show no gene flow to the sexual types. Therefore, the treatment of one of these types, *D. stenhammari*, as a form of *D. ceratoniae* seems untenable and contradicts the treatment of apomicts in other groups of plants and fungi.

The shortcoming of synonymous treatments of sexual and asexual taxa can be demonstrated with quite a new example. In Europe several sorediate-blastidiate taxa of broad-lobed *Xanthoria*'s occur, which have been badly studied for a long time. It has been shown that there are at least five different species (Poelt and Petutschnig, 1992a, b). Some of these, *X. fallax* and *X. ulophyllodes,* are distinguished from the *X. parietina* s.str. group, with which they were thought to be related, by their well-developed true rhizines. There are no sexual counterparts of these *Xanthoria*'s in Europe, but from north-east Asia a new *Xanthoria* species, *X. oxneri* is being described (Kondratyuk and Poelt, 1994) which is obviously the sexual counterpart of the holarctic blastidiate *X. ulophyllodes*; it differs in the lack of blastidia and the common occurrence of apothecia. For *X. fulva*, a sexual counterpart most probably exists in North America. It is necessary to be cautious in considering the European blastidiate species as of Asiatic or American origin respectively. Most probably, during the ice-age when the European lichen biota became impoverished, the more vulnerable sexual taxa became extinct in Europe in the same way as many trees, shrubs, and other flowering plants.

This hypothesis is corroborated by the well-known genus *Hypogymnia*. In Europe, the genus is represented only by secondary species, with soredia or soredia-like diaspores. Sexual taxa without soredia have been known in the holarctic for a long time, from eastern Asia, and eastern and western North America. For many years, they remained united as *H. enteromorpha*. During the last decades this "species" has been separated into a number of different taxa with mostly small distribution areas. Some may be the primary non-sorediate counterparts of the widespread taxa occurring also in Europe. Anyhow, their distributional areas match quite well those of many relict phanerogamic plants,

Information in taxonomy is linked with names. If we cancel the names, we cancel the wealth of information combined with those names. If we synonymize sexual and asexual, primary and secondary taxa, we renounce the information they can yield about their evolution and vegetational history. So, from a biological view and for their historical reasons, we should retain secondary, mainly asexual taxa as independent species, except in the few cases where we have obvious reasons for other treatments.

ACKNOWLEDGEMENTS

I thank Mr M. Grube for discussions, M. Grube, Professor D.L. Hawksworth and Ch. Scheuer for linguistic help, and M. Suanjak for work on the manuscript.

REFERENCES

Culberson, C.F., W.L. Culberson and A. Johnson, 1988, Gene flow in lichens, *American Journal of Botany* 75: 1135-1139.

Du Rietz, G.E., 1924, Die Soredien und Isidien der Flechten, *Svensk Botanisk Tidskrift* 18: 371-396.

Grube, M. and J. Poelt, 1993, Beiträge zur Kenntnis der Flechtenflora des Himalya X. *Sporastatia testudinea,* ihre Variabilitat, ihre Ökologie und ihre Parasiten in Hochasien, *Fragmenta floristica et geobotanica, Suppl.* 2 (1): 113-122.

Hafellner, J., 1984, Studien in Richtung einer natürlichen Gliederung der Sammelfamilien *Lecanoraceae* und *Lecideaceae, Beiheft zur Nova Hedwigia* 79: 241-371,

Hale, M.E., 1987, A monograph of the lichen genus *Parmelia* Acharius sensu stricto, *Smithsonian Contributions to Botany* 66: 1-55.

Hale, M.E., 1990, A synopsis of the lichen genus *Xanthoparmelia* (Vainio) Hale, *Smithsonian Contributions to Botany* 74: 1-250.

Jahns, H.M., 1987, New trends in developmental morphology of the thallus, *Bibliotheca Lichenologica* 25:17-33.

Kondratyuk, S. and J. Poelt, 1994, On two new Asian *Xanthoria* species (*Teloschistaceae*): in preparation.

Magnusson, A.H., 1925, Studies on the *Rivulosa*-group of the genus *Lecidea, Göteborgs Kungl. Vetenskaps och Vitterhets-Samhälles Handlingar* 29(4): 1-40.

Magnusson, A.H., 1936, *Acarosporaceae* und *Thelocarpaceae, Rabenhorst's Kryptogamen-Flora von Deutschland, Oesterreich und der Schweiz* 9, 5(1): 1-318.

Niebel-Lohman, A. and T. Feuerer, 1992, Die Gatting *Pertusaria* (*Lichenes*) in Schleswig-Holstein. Morphologie, Taxonomie und Verbreitung, *Mitteilungen aus dem Institut für Allgemeine Botanik in Hamburg* 24: 199-252.

Ott, S., 1986, The juvenile development of lichen thalli from vegetative diaspores, *Symbiosis* 3: 57-74.

Poelt, J. and W. Petutschnig, 1992a, Beiträge zur Kenntnis der Flechtenflora des Himalayas IV. Die Gattungen *Xanthoria* und *Teloschistes* zugleich Versuch einer Revision der *Xanthoria candelaria*-Gruppe, *Nova Hedwigia* 54: 1-36.

Poelt, J. and W. Petutschnig, 1992b, *Xanthoria candelaria* und ähnliche Arten in Europa. *Herzogia* 9: 103-114.

Schwab, A., 1986, Rostfarbene Arten der Sammelgattung *Lecidea* (*Lecanorales*) Revision der Arten Mittel- und Südeuropas, *Mitteilungen aus der botanischen Staatssammlung München* 22: 221-476.

Sipman, H.J.M., 1983, A monograph of the lichen family *Megalosporaceae, Bibliotheca Lichenologica* 18: 1-241.

Skult, H., 1984, The *Parmelia omphalodes* (ascomycetes) complex in eastern Fennoscandia. Chemical and morphological variation, *Annales Botanici Fennici* 21: 117-142.

Skult, H., 1987, The *Parmelia omphalodes* complex in the Northern Hemisphere. Chemical and morphologhical aspects, *Annales Botanici Fennici* 24: 371-381.

Stocker-Wörgötter, E. and R. Türk, 1993, Redifferentiation of the lichen *Cladonia furcata* ssp. *furcata* from cultivated lichen tissue, *Cryptogamic Botany* 3: 283-289.

Tehler, A., 1983, The genera *Dirina* and *Roccellina* (*Roccellaceae*), *Opera Botanica* 70: 1-86.

Timdal, E., 1984, The genus *Hypocenomyce* (*Lecanorales, Lecideaceae*) with special emphasis on the Norwegian and Swedish species, *Nordic Journal of Botany* 4: 83-108.

Zahlbruckner, A., 1922-24, *Catalogus Lichenum Universalis.* Vol. 2. Bornträger, Leipzig.

Zahlbruckner, A., 1926-27, *Catalogus Lichenum Universalis.* Vol. 4. Bornträger, Leipzig.

Problems and Perspectives

PERIDIAL MORPHOLOGY AND EVOLUTION IN THE PROTOTUNICATE ASCOMYCETES

R.S. Currah

Department of Botany
University of Alberta
Edmonton, Alberta T6G 2E9, Canada

SUMMARY

The prototunicate ascomycetes include fungi with small, unicellular, hyaline or pale-coloured ascospores that develop within globose, thin-walled, evanescent asci. Ascospore release is passive and mediated by the activities of micro- and macrofauna. A diverse range of peridial types has evolved within the group in response to selective pressures affecting spore dispersal from enclosed areas such as animal burrows and small protected spaces in decaying plant and animal matter and in soil. Passive release of hyaline spores has lead to a simplification and modification of peridial structure so that apparently unrelated groups share similar peridial types. Anamorph characteristics and substrate preferences support the supposition that at least two distinct evolutionary lines, representing descendants related to the operculate and inoperculate discomycetes and the *Hypocreales*, share characteristics encompassed by the prototunicate ascomycetes. With this in mind, the genera of the prototunicate ascomycetes in the orders *Eurotiales* and *Onygenales* are reviewed and some are realigned to provide a basis for analysis by molecular techniques.

The new combination *Auxarthon llanense* (Varsavsky & Orr) Currah is made.

INTRODUCTION

The prototunicate ascomycetes represent an assemblage of fungi united by four main characteristics: (1) ascospores tiny, single-celled, hyaline to light-coloured, conglobate in more or less globose ascal clusters; (2) asci evanescent and borne in irregular clusters within ascomata; (3) peridium mesh-like, membranous or lacking; ascomata discrete or confluent in clusters or in loose or cottony or sclerotioid stromata which may be stalked or sessile; and (4) anamorphs of aleurio-, arthro-, or phialoconidia. There are two main groups or orders within this assemblage: the *Eurotiales* and the *Onygenales*. Kendrick (1992) includes also the *Ophiostomatales* and the *Laboulbeniales* because of their deliquescent asci, but these orders are not dealt with further in the following discussion.

The recent set of guns added to our taxonomic armamentarium are analyses of relationships among taxa using nucleic acid data and a few studies exemplify how useful these data will be in resolving problems concerning the natural affinities of taxa of reduced or structurally simplified ascomycetes (Berbee and Taylor, 1992a, 1992b; Bowman et al., 1992; van de Peer et al., 1992). The expedient use of these techniques, and our ability to draw significant conclusions with the results are dependent on the availability of sound hypotheses regarding relationships among the taxa in question.

The objectives of this paper are to review the taxonomy of the genera and families in the *Eurotiales* and *Onygenales*, to provide some discussion of the range of morphological and ecological characteristics associated with them, and to suggest some realignment

Ascomycete Systematics: Problems and Perspectives in the Nineties
Edited by D.L. Hawksworth, Plenum Press, New York, 1994

of taxa within the orders to provide some sound hypotheses concerning the evolutionary process in the group, for testing with molecular methods.

MORPHOLOGICAL AND ECOLOGICAL DATA

Gametangia

Gametangia are morphologically undifferentiated structures in the *Eurotiales* and *Onygenales* and no clear distinction can be made between antheridial and ascogonial cells (Figs 1-4). These usually consist of a single crooked or coiled hypha, or a pair of hyphae that coil about each other forming a double helix. Neither trichogynes nor spermatia are involved in the process of nuclear transfer. In heterothallic forms wall dissolution presumably occurs between the filaments and one (or several) nuclei pass from one of the

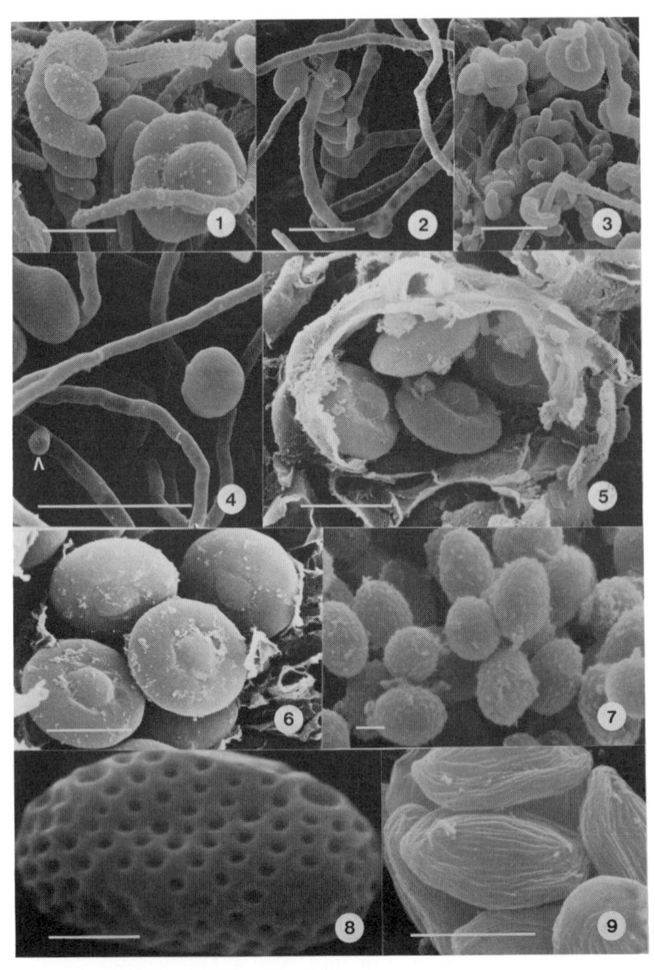

Figures 1-9. Fig. 1, Gametangia of *Gymnascella udagawae* (UAMH 5726), scale bar = 4 μm. **Fig. 2.** Gametangia of *G. littoralis* (UAMH 3865), scale bar = 10 μm. **Fig. 3.** Gametangia of *G. dankaliensis* (UAMH 3552), scale bar = 10 μm. **Fig. 4.** Gametangia of *G. punctata* (UAMH 3164), scale bar = 10 μm. **Fig. 5.** Ascus and ascospores of *G. citrina* (UAMH 3073), scale bar = 4 μm. **Fig. 6.** Ascospores of *Gymnascella aurantiaca* (UAMH 3814), scale bar = 4 μm. **Fig. 7.** Ascospores of *Talaromyces emersonii* (UAMH 6427), scale bar = 1 μm. **Fig. 8.** Ascospore of *Renispora flavissima* (UAMH 4205), scale bar = 1 μm. **Fig. 9.** Ascospores of *Myxotrichum stipitatum* (UAMH 7291), scale bar = 3 μm.

gametangia to the other to form the dikaryon. Elaboration of the asci in chains or clusters, with or without crozier formation, is presumably concomittant with karyogamy and meiosis.

Gametangia are often mentioned in descriptions of prototunicate taxa but, because of their simplicity and common morphology, they are either not very valuable in inferring relationships among groups, or, our groups are too heterogeneous for us to detect significant patterns. Malloch (1986), after sorting the genera of the *Trichocomaceae* into two subfamilies on the basis of ascospore, ascoma, and conidial characters, noted a general trend in the morphology of the gametangial initials between the two groups in that the *Trichocomoideae* had variable initials with irregular or nonexistent coiling while the *Dichlaenoideae* had regular to irregular helical coils.

It is difficult to imagine the environmental or evolutionary pressures that would maintain morphological distinctions in gametangial initials within a taxon. The structures do not require consistency to encourage recognition between compatible strains. Presumably all that is necessary to effect nuclear transfer is hyphal contact and wall dissolution.

Ascospores

Morphological characteristics of the mature ascospores (size, shape, colour, surface features) are considered to be extremely useful in all taxonomic systems (Benny and Kimbrough, 1980; Currah, 1985; Horie, 1980; Malloch, 1986; Udagawa and Horie, 1973; von Arx, 1987). Although the ascospores of the prototunicates are small and single-celled, their characteristics are distinctive, stable and easily observed (Figs 5-9). Scanning electron microscopy is now almost routinely used to gather information to define taxa.

Von Arx (1987), on the basis of spore shape, redistributed many of the prototunicates among four families: *Eurotiaceae*, with saturniform or equally bivalvate spores; *Gymnoascaceae*, with discoid, lenticular or unequally bivalvate spores; *Onygenaceae*, with ellipsoidal, fusiform or cylindrical spores; and *Amauroascaceae*, with spherical spores (Table 1). With the possible exception of the *Eurotiaceae*, these groups are rather artificial, are not congruent with other character sets (anamorph morphology, ecology) and have minimal predictive value. The group of taxa included in the *Eurotiaceae*, parallels somewhat the taxa included by Malloch (1986) in the *Dichlaenoideae* (Table 2).

Currah (1985, 1988), in resorting taxa traditionally distributed among the *Gymnoascaceae*, placed more emphasis on spore sculpturing characteristics and their degree of congruence with other characteristics, namely anamorph morphology and enzymatic abilities, and defined four groups of genera among the families: *Arthrodermataceae* (spores smooth and lenticular, tendency toward multiseptate, rhexolytically dehiscing conidia, keratinolytic enzymes); *Gymnoascaceae* (spores smooth or lumpy, oblate, often with equatorial and polar thickenings; Figs 5-6); *Myxotrichaceae* (elongate striate spores, Fig. 9; *Oidiodendron* anamorphs, and cellulolytic enzymes); and *Onygenaceae* (spores pitted, Fig. 8; conidia aleurio- or arthroconidia, keratinolytic or keratinophilic tendencies; Table 3). The *Eurotiales* were not considered in this latter treatment although some of the *Gymnoascaceae* (*sensu* Currah) clearly have strong resemblances to this order which has oblate, discoid, bivalvate or ellipsoidal spores which are smooth, or have ridges, broad or narrow spines. The ascospores of some *Eurotiales* are reticulate (e.g. *Eupenicillium sinaicum*; Udagawa and Ueda, 1982) and resemble the spores of the *Onygenaceae* (e.g. *Amauroascus*) with the exception that they are also strongly bivalvate.

Asci

As a common denominator in the prototunicates, the evanescent ascus (Fig. 5) is not particularly useful for defining taxa within the group although the disposition of the ascospores within the ascus is useful in defining some taxa (e.g. *Gymnoascoideus petalosporus*). The membranous structure itself is too short-lived and simple to be useful in identifying mature specimens. Ultrastructural characteristics might be of some use and some comparative work using TEM is required to check on this, although, as Malloch (1986) points out, when active ascospore discharge was lost, all associated structures, i.e. opercula, pores and thickenings, probably disappeared.

Table 1. Families and genera of the *Eurotiales* based primarily on ascospore shape (von Arx, 1987)

Eurotiaceae	Gymnoascaceae	Onygenaceae	Amauroascaceae
ascospores bivalvate or saturniform; often with equatorial frills	ascospores dorsiventrally flattened, lenticular, discoid or unequally bivalvate and usually smooth	ascospores elongate, ellipsoidal, cylindrical or fusiform (occasionally nearly sperhical), smooth, striate, punctulate or spinulose	ascospores globose to oblate, alveolate to punctate
Chaetosartorya	*Arthroderma*	*Ascocalvatia*	*Ajellomyces*
Cristaspora	*Ctenomyces*	*Byssoascus*	*Amauroascus*
Dichlaena	*Gymnoascus*	*Byssochlamys*	*Aphanoascus*
Dichotomomyces	*Leucothecium*	*Cephalotheca*	*Apinisia*
Emericella	*Nannizzia*	*Eremascus*	*Arachnomyces*
Eupenicillium	*Narasimhella*	*Hamigera*	*Arachnotheca*
Eurotium	*Uncinocarpus*	*Monascella*	*Emmonsiella*
Hemicarpenteles	*Xynophila*	*Monascus*	*Leiothecium*
Mallochia		*Myxotrichum*	*Pleuroascus*
Neosartorya		*Onygena*	*Xanthothecium*
Saitoa		*Penicilliopsis*	*Xylogone*
		Pseudogymnoascus	
		Renispora	
		Talaromyces	
		Thermoascus	
		Trichocoma	

Anamorph

Most of the prototunicates do form an asexual state and anamorph taxa in the prototunicates are highly indicative of evolutionary lineages. The *Onygenales* and the *Eurotiales* have distinctive patterns of ontogeny and release of the asexual spores. Conidia in the *Onygenales* develop as slight intercalary and terminal modifications of vegetative hyphae producing arthroconidia and aleurioconidia respectively (Sigler, 1989), while those of the *Eurotiales* are produced repetitively from a stable conidiogenous locus, usually in the form of a phialide.

Arthroconidia develop from vegetative hyphae and may look almost incidental, in that the conidia are irregular in length and variable in number, or they may develop in distinctive and repetitive patterns along the length of a conidiogenous hypha. Terminal equivalents of intercalary arthroconidia are often single celled and bulbous terminal regions of conidiogenous hyphae. Terminal and intercalary conidia often occur together in the same formations and appear distinctive enough to be recognized as individual taxa in genera such as *Chrysosporium* (predominantly terminal conidia) or *Malbranchea* (predominantly intercalary conidia). Irregular, sparse arthroconidia by themselves should be regarded with suspicion by taxonomists because they represent a simple and non-specific method of forming vegetative propagules. We should expect this type of conidiogenesis to have arisen independently several times. In fact, arthroconidia are found rather sporadically throughout both the ascomycetes (*Onygenales*, *Pezizales*; Paden, 1975) and basidiomycetes (Tsuneda et al., 1993).

Morphological similarities among or between anamorphs in *Onygenales* can be misleading in the absence of solid supporting data. For example, although the anamorph of *Renispora* resembles the anamorph of *Ajellomyces capsulatus*, Bowman and Taylor (1993) did not find that sequence data placed it particularly close to *Ajellomyces*. Characteristics of the ascomata of these two taxa do not indicate a particularly close association within the *Onygenaceae* and the large, tuberculate spiny conidia they form is a good example of convergence in anamorphic features. In contrast, *Uncinocarpus reesii* was found by Bowman and Taylor (1993) to be closest in their analysis to the pathogen

Table 2. Classification of the *Trichocomaceae* (*Eurotiales*) by Malloch (1986); von Arx (1987) family assignments in parentheses

Trichocomoideae	**Dichlaenoideae**
ascospores spherical to ellipsoidal, roughened, lacking furrows	ascospores oblate to ellipsoidal, usually with a furrow
stromata lacking	stromata with pseudoparenchymatrous walls or Hulle cells
associated with wood, cellulose	associated with starchy or oily seeds and fruits; often proteolytic
anamorphs *Penicillium* *Paecilomyces*	anamorphs *Aspergillus* *Penicillium* *Merimbla* *Paecilomyces* *Polypaecilium*
Byssochlamys *Dendrosphaera* *Sagenoma* *Talaromyces (Onygenaceae)* *Trichocoma (Onygenaceae)*	*Chaetosartorya (Eurotiaceae)* *Cristaspora (Eurotiaceae)* *Dichlaena (Eurotiaceae)* *Dichotomomyces (Eurotiaceae)* *Edyuillia* *Emericella (Eurotiaceae)* *Eupenicillium (Eurotiaceae)* *Eurotium (Eurotiaceae)* *Fennellia* *Hamigera (Onygenaceae)* *Hemicarpenteles (Eurotiaceae)* *Hemisartorya* *Neosartorya (Eurotiaceae)* *Penicilliopsis* *Petromyces* *Sclerocleista* *Thermoascus* *Warcupiella*

Coccidioides immitis, a taxon for which a teleomorph is unknown. In this case, the *Malbranchea*-like anamorphs of the two species, which are morphologically similar, are indicative of a close relationship.

Phialoconidia are much more limited and restricted in their distribution among the prototunicate taxa and among the fungi in general. The presence of a phialidic anamorph in a series of ascomycete taxa carries high predictive value concerning their affinity. In the prototunicates, the presence of a phialidic anamorph (*Penicillium*, *Aspergillus*) in the life-cycle automatically indicates eurotialean affinities.

The presence of this type of anamorph in both the *Hypocreales* and *Eurotiales* suggests that the two orders may be related by descent, with the *Hypocreales* ancestral and the *Eurotiales* derived (Malloch, 1981, 1986). One possible problem in this proposed lineage is the potential difference between true chain phialoconidiogenesis as observed in the *Eurotiales* and false or no chain conidiogenesis in hypocrealean anamorphs (Minter et al., 1982). Nevertheless, the idea deserves testing with molecular data.

Von Arx (1987) presents a very different approach to assessing the significance of the presence of phialoconidia and arthro- or aleurioconidia in the prototunicate taxa. He interpreted both types of conidiogenesis as being a very similar process in that both phialoconidia (exemplified by *Aspergillus*, *Paecilomyces*, and related genera) and arthroconidia (e.g. *Malbranchea*) develop from a similar "meristematic filament". Conidia at maturity are separated by disjunctives which are short in the phialoconidial species and longer in the arthroconidial species.

Table 3. Families and genera of the *Onygenales* (Currah, 1985)

Arthrodermataceae	Gymnoascaceae	Myxotrichaceae	Onygenaceae
ascomata smooth, discoid	ascospores oblate with thickenings, never punctate	ascospores fusiform, smooth or striate	ascospores variable in shape, always punctate
Arthroderma	*Acitheca*	*Byssoascus*	*Ajellomyces*
Ctenomyces	*Arachniotus*	*Myxotrichum*	*Amauroascus*
Nannizzia	*Gymnascella*	*Pseudogymnoascus*	*Aphanoascus*
	Gymnoascoideus		*Apinisia*
	Gymnoascus		*Ascocalvatia*
			Auxarthron
			Keratinophyton
			Kuehniella
			Nannizziopsis
			Neogymnomyces
			Neoxenophila
			Onygena
			Pectinotrichum
			Renispora
			Shanorella
			Spiromastix
			Uncinocarpus
			Xynophila

Peridia and other ascospore enclosures

The peridium (or the structures associated with the developing asci and ascospores; Figs 10-23) has been emphasized as a source of characteristics to indicate relationships among groups of prototunicates.

Characteristics of the structures enclosing the spores in both orders exhibit an interesting range of morphological types, from the ancestral closed peridium of closely packed hyphal cells (e.g. *Eurotium*, *Dichotomomyces*, *Aphanoascus*), to cottony enclosures (e.g. *Byssochlamys*, *Arachniotus*, *Talaromyces*, Figs 13, 15-17; and some species of *Gymnascella*), cage-like enclosures that resemble miniature tumble weeds (Figs 10-12; e.g. *Gymnoascus*, *Auxarthron*, *Myxotrichum*) and no enclosures (naked asci, Figs 14, 19, 20; e.g. *Edyuillia*, *Byssoascus*, *Gymnascella*). These "peridial" types can be found singly, gregariously or in stalked (e.g. *Onygena*, *Penicilliopsis*, Fig. 21) or sessile (e.g. *Eupenicillium*, *Shanorella*, Fig. 23) stromatic conglomerations of varying complexity. Superficial similarities among enclosure devices have lead to a number of unusual classifications of the taxa of prototunicates (e.g. *Gymnoascaceae* s. lat. of Apinis, 1964, and Benjamin 1956; see also Benny and Kimbrough, 1980) that often have very little congruence with groupings made on the basis of other characteristics.

The mesh-like peridium with passively liberated, hyaline ascospores, which typifies the old idea of the *Gymnoascaceae*, is almost certainly a structure which has evolved independently in different groups. We routinely expect and look for this type of convergence in animals and plants. A similar dispersal method in the angiosperms can be found in the tumble weed mechanisms at work in the *Chenopodiaceae* (*Kochia*, *Salsola*) and *Fabaceae* (*Psoralea*). They use the same mechanism to disperse the seeds from mature plants but they are distantly related taxa. It is the mechanism that fosters the development of similar superficial similarities in morphology. The tumbleweed ascoma in the prototunicates is an indicator of animal involvement in spore dispersal. These ascomata are also often provided with hooks (e.g. *Gymnoascus reessii*), crooks (e.g. *Myxotrichum chartarum*), barbs (e.g. *Myxotrichum deflexum*; Fig. 10) and spring-loaded leghold devices (e.g. *Arthroderma*). Still others function like grenades which explode when touched. An example of this type of ascoma is found in *Shanorella* (Figs 16, 22-24), an ascomycete in which the peridium itself disarticulates into a series of thick-walled, curved, or Y-shaped hyphal fragments (Fig. 22) that resemble irregular Hülle cells. Morphologically similar elements are found in ascomata of *Fennellia* and in some species of *Gymnascella*.

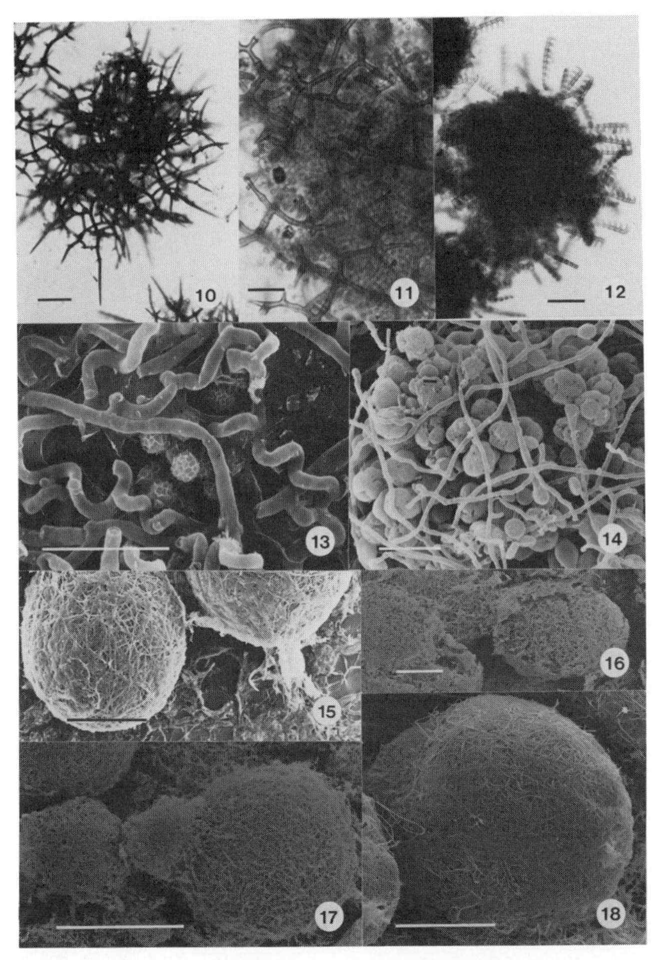

Figures 10-18. Fig. 10. Ascoma of *Myxotrichum stipitatum* (RSA 1556), scale bar = 40 μm. **Fig. 11.** Ascoma of *Aurxarthron reticulatum* (RSA 1528), scale bar = 25 μm. **Fig. 12.** Ascoma of *Ctenomyces serratus*, scale bar = 40 μm. **Fig. 13.** Peridial elements on outer surface of ascoma of *Amauroascus aureus* (UAMH 7290), scale bar = 15 μm. **Fig. 14.** Ascoma of *Gymnascella udagawae* (UAMH 5726), scale bar = 10 μm. **Fig. 15.** Ascomata of *Neosartorya fischeri* (UAMH 6422), scale bar = 100 μm. **Fig. 16.** Ascomata of *Talaromyces flavus* (UAMH 6424), scale bar = 100 μm. **Fig. 17.** Ascomata of *Talaromyces emersonii* (UAMH 6427), scale bar = 200 μm. **Fig. 18.** *Shanorella spirotricha* (UAMH 3062), scale bar = 200 μm.

With these potential convergences in mind, we should be cautious when tempted to group together superficially similar taxa. Unrelated taxa among the prototunicates may only look the same because they share a common dispersal mechanism.

Ecological Characteristics

The *Eurotiales* and *Onygenales* together occupy a wide range of niches in soil, composting vegetation, sugary, starchy, oily, salty substrates, and substrates rich in keratin or cellulose. They are all decomposers or in some cases necrotrophs. Several *Myxotrichaceae* are involved in endomycorrhizal symbioses with *Ericaceae* (Dalpé, 1989) but other biotrophic associations involving *Eurotiales* and *Onygenales* and plants are unknown. None apparently degrades lignin.

An interesting dichotomy is apparent between some *Onygenales* and the *Eurotiales* in the ability to break down keratin-containing structures. These keratin degrading enzymes are found in the species I have grouped in the *Arthrodermataceae* and in the *Onygenaceae* (Currah, 1985). In nature, keratinous structures, such as feathers, hoof, horn, hair and hide which is on or in soil and dung are frequently colonized by species in the *Onygenaceae* and *Arthrodermataceae*. Currah (1985) assigned considerable taxonomic weight to the presence of keratin degrading enzymes and the degree of correlation with morphological characteristics of ascospores and anamorphs. The justification for this was that the ability to enzymatically rupture the disulphide bridges between peptide chains of keratin molecules is not widespread among microorganisms and the likelihood of this ability arising several times in the *Ascomycotina* was extremely low. Therefore, it was considered a good marker indicating monophyly in ascomycetes which possessed the proven ability to degrade native keratin.

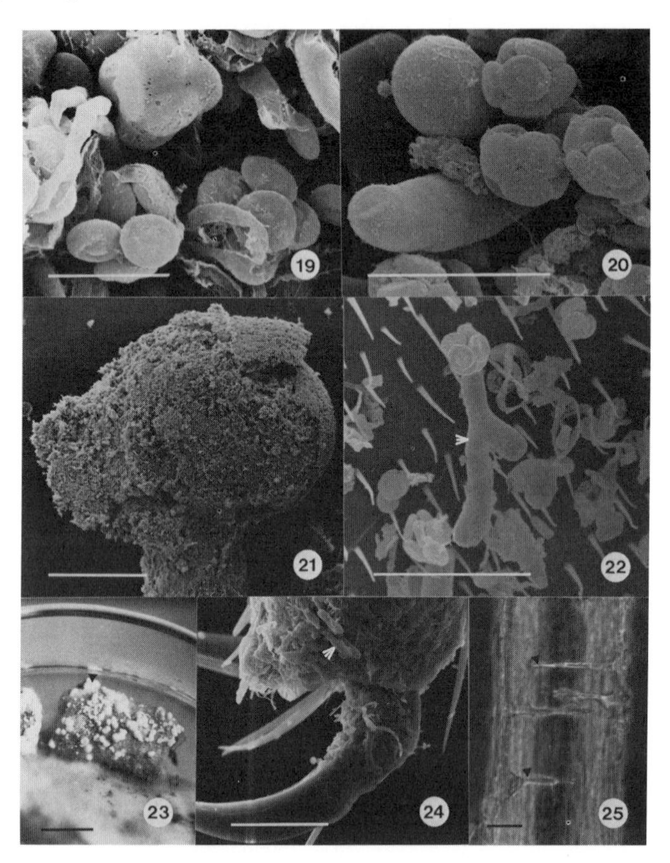

Figures 19-25. Fig. 19. Asci of *Gymnascella aurantiaca*, scale bar = 10 μm. **Fig. 20.** Asci of *Shanorella spirotricha*, scale bar = 10 μm. **Fig. 21.** Stromatic head of *Onygena equina*, scale bar = 400 μm. **Fig. 22.** Peridial element and ascospores on dermis of *Dermestes* beetle, scale bar = 20 μm. **Fig. 23.** Sessile sclerotial masses of ascomata of *Shanorella spirotricha* on natural substrate, scale bar = 20 μm. **Fig. 24.** Toe of *Dermestes* beetle with adhering ascospores and peridial elements from a crushed ascoma of *S. spirotricha*, scale bar = 50 μm. **Fig. 25.** Hair shaft with pentrating bodies (arrows) produced by *Chrysosporium keratinophilum*, scale bar = 10 μm.

Species which actively degrade keratin often cause striking morphological alterations in the hairs used in the standard keratin assay which involves inoculating the fungus in question on hair on a starvation medium. Strongly keratinolytic organisms (e.g. *Onygena*) cause the development of a frayed appearance of the hair shafts as the linear protein molecules are broken apart. Others form cone-shaped "penetrating bodies" as the shaft is degraded (e.g. *Aphanoascus*; Fig. 25). Fungi which have limited keratinolytic abilities often cause only minimal fraying of the shafts after extended periods of incubation and in these cases the results may be difficult to interpret. A better assay is required in these cases.

Other prototunicates in these orders are cellulolytic. Cellulolytic abilites are particularly well-developed in the *Myxotrichaceae* and possibly in some *Trichocomoideae*.

The remaining prototunicates are often found associated with substrates of high osmolarity and in environments subjected to extremes in temperature where simple polysaccharides are the substrate. The *Trichocomaceae* encompasses most species in these nutritional categories.

GENERA OF *EUROTIALES* AND *ONYGENALES*

A summary of the following comments concerning the taxonomic status or redisposition of some genera in the *Onygenales* and *Eurotiales* is presented in Table 4.

Table 4. Realignment of genera and families in the prototunicate orders *Eurotiales* and *Onygenales*

EUROTIALES -	ONYGENALES -
Trichocomaceae	**Onygenaceae**
(***Trichocomoideae***)	*Ajellomyces*
Byssochlamys	? *Amauroascopsis*
Dendrosphaera	*Amauroascus*
Gymnoascus	*Aphanoascus*
Gymnascella	*Ascocalvatia*
Talaromyces	*Auxarthron*
Trichocoma	*Brunneospora*
	Byssoonygena
(***Dichlaenoideae***)	? *Nannizziopsis*
Arachniotus	*Onygena*
Chaetosartorya	*Renispora*
Cristaspora	*Shanorella*
Dichlaena	*Spiromastix*
Dichotomomyces	*Uncinocarpus*
Edyuillia	
Emericella	
Eupenicillium	**Arthrodermataceae**
Eurotium	*Arthroderma*
Fennellia	*Ctenomyces*
Gymnascella	*Gymnoascoideus*
Hamigera	*Nannizzia*
Hemicarpenteles	
Hemisartorya	
Neosartorya	**Myxotrichaceae**
Penicilliopsis	*Byssoascus*
Petromyces	*Gymnostellatospora*
Sclerocleista	*Myxotrichum*
Thermoascus	*Pseudogymnoascus*
Warcupiella	

The *Onygenales* as defined by Currah (1985, 1987) included genera distributed among four families: *Arthrodermataceae*, *Onygenaceae*, *Myxotrichaceae* and *Gymnoascaceae*. The *Onygenaceae* and *Arthrodermataceae* are closely related and relatively homogeneous taxa based on their shared ability to degrade keratin (a few exceptions exist), ascospore morphology, and the production of rhexolytically dehiscing conidia. A close relationship betwen the onygenaceous taxa *Ajellomyces dermatitidis* and *A. capsulata* and *Arthroderma* (*Trichophyton rubrum*) has been demonstrated by Bowman et al. (1992) using sequence data from small subunit ribosomal RNA. Further analysis (Bowman and

Taylor, 1993) supported a close evolutionary relationship among a selection of the pathogenic and saprophytic species of *Onygenales* and indicated that the pathogenic species do not constitute a single lineage among the other species in the order. A wider study involving more taxa and more than one strain of each taxon would be of considerable interest. In order to interpret the results of this type of study some working hypotheses concerning relationship must be suggested on the basis of morphology and ecology of the organisms on trial.

Since the publication of my monograph (Currah, 1985), a number of new taxa have been described and others have been reevaluated on the basis of new taxonomic data.

I continue to maintain *Ajellomyces* and *Emmonsiella* in the same genus, although von Arx (1987) considered each of them unique within the *Amauroascaceae* (defined as a eurotialean family with spherical ascospores) on the basis of differences in surface features of the ascospores. As far as I am aware the ascospores of both are punctate.

The genus *Amauroascus* needs some revision as it currently melds with *Auxarthron* (see below) on the basis of ascospore and peridal characteristics. The genus may also have some significance evolutionarily. Kimbrough (1989) considered it, along with *Eleutherascus* and *Ascodesmis*, in his family *Ascodesmidaceae* (*Pezizales*; Kimbrough, 1989). These three genera grow well in culture and should be examined, along with other simple, putatively reduced discomycete taxa (e.g. *Monascella*; Guarro and von Arx, 1986) using molecular techniques to test their degree of relatedness. If the idea has some substance, the *Ascodesmidaceae*, in having a keratinophilic genus with rhexolytic conidia would strengthen the speculation that the *Onygenales* share common antecedents with the *Pezizales*. Guarro et al. (1992) point out that there are no ascus ultrastructure data supporting the placement of *Amauroascus* along with operculate taxa. Von Arx (1987) referred *Eleutherascus* to the *Tuberales*, which also lack opercula, reiterating his earlier suggestion (von Arx, 1971) that *Eleutherasucs* had strong pezizalean affinites. The genus *Mallochia* (von Arx and Samson, 1986) seems to fit here also.

The genus *Aphanoascus* has been revised recently by Cano and Guarro (1990) who broadened the generic concept to include species formerly considered in *Keratinophyton*, and *Xynophila mephitale*. They admit the existence of two distinct clades within the genus that correspond to the dichotomy based on ascospore morphology and anamorph characteristics corresponding with the definitions of *Aphanoascus* and *Keratinophyton* (Cano et al., 1990).

A similar situation exists with *Arthroderma* and *Nannizzia*. *Arthroderma* is the teleomorph genus in which some sexual phases of the dermatophyte genera are classified. Weitzman et al. (1986) transferred here all species of the similar teleomorph genus *Nannizzia*, but it is probably more useful to maintain the two genera as distinct taxa. The significance or depth of the dichotomies in these two examples would be an interesting question to examine once molecular data become available.

Ctenomyces is one of the most distinctive genera in the *Onygenales*. The species *C. serratus* has a close relationship with *T. rubrum* (*Arthroderma*?) according to Bowman and Taylor (1993).

Auxarthron remains a relatively large genus within the family. On the basis of similarities in peridial morphology, anamorph characteristics and enzymatic ability, *Pectinotrichum llanense* should be tranferred here[1].

I have not observed material of *Amauroascopsis*, but the original description (Guarro et al., 1992) indicates that it is almost definitely an onygenaceous genus. Based on the sperhical ascospores, the authors placed the new species in the *Amauroascaceae*. It bears a strong resemblance to other species in *Amauroascus* and may be disposed here eventually. This species, along with *Byssoonygena reticulata* (Guarro et al., 1987a) and *Byssoonygena ceratinophila* (Guarro et al., 1987b), which have similar (but ellipsoidal) ascospores, may be treated best in a single genus.

Nannizziopsis has been reviewed by Guarro et al. (1991). They considered white (or light-coloured) ascomata and globose punctate-reticulate ascospores characteristics of the genus. They described a new species in this genus (*N. hispanica*) and transferred *Kuehniella racovitzae* and *Amauroascus albicans* here. This is now a heterogenous genus and needs revision.

---

[1]**Auxarthron llanense** (Varsavsky & Orr) Currah **comb. nov.**, basionym: *Pectinotrichum llanense* Varsavsky & Orr, *Mycopathologia et Mycologia Applicata* 43: 229 (1971) Holotype: NY 228.

Onygena is an example of a stromatic ascoma and is possibly one of the most ancestral forms among the species in this order. There is an interesting parallel in *Penicilliopsis* of the *Eurotiales*. Any evaluation involving the origin and evolutionary patterns among the prototunicates and based on molecular data should involve these taxa.

Shanorella spirotricha remains a monotypic genus in which the peridial elements thicken and disarticulate at maturity. In this aspect, ascomata resemble those of *Gymnascella nodulosa*, which has swollen, irregular thick-walled peridal elements or Hülle cells like those found in *Fennellia*.

Spiromastix is an enigmatic genus among the *Onygenales*. Species in the genus lack anamorphs and ascospores of the teleomorph are only slightly pitted. Keratinolytic abilities are barely detectable using the hair plate assay.

The *Myxotrichaceae* seems to comprise a monophyletic group based on ascospore morphology, peridial characteristics and in having arthroconidial anamorphic states in the genus *Oidiodendron*. The family encompasses four genera: *Byssoascus*, *Myxotrichum*, *Pseudogymnoascus*, and *Gymnostellatospora* (Udagawa et al., 1993). The latter genus resembles *Emericellopsis* (*Pseudoeurotiaceae* fide von Arx, 1987) in having ascospores that have longitudinal ridges, flanges or wings.

There are strong indications that the group plays a specific ecological role because some species (in *Byssoascus* and *Myxotrichum*) have been shown to form mycorrhizas with ericaceous host plants (Dalpé, 1989). Many *Oidiodendron* species are also shown to form ericoid mycorrhizas (Dalpé, 1991), an ability they share with the inoperculate discomycete *Hymenoscyphus ericae* which, coincidentally, also has an arthroconidial anamorph (Egger and Sigler, 1993). The colour and thick-walled hyphae comprising the conidiophores of *Oidiodendron* species are reminiscent of the sterile elements of the ascomata of *Myxotrichum*. A particularly good example of this resemblance is *Oidiodendron setiferum* (Udagawa and Toyazaki, 1987) which has conidiophores with sterile elements resembling the peridial hyphae of *Myxotrichum*. The relationship of the exclusively anamorphic *Oidiodendron* species with the *Myxotrichaceae*, and with the inoperculate *H. ericae* should be examined using molecular methods. It is possible that the *Myxotrichaceae* represents a distinct evolutionary line derived from the inoperculate discomycetes and merits placement in its own order.

The *Gymnoascaceae*, as outlined by Currah (1985), is a heterogeneous group. Some genera are obviously closely related to the *Eurotiales* and should in fact be placed in that order.

Arachniotus ruber has bivalvate ascospores with a double equatorial flange, a characteristic of *Trichocomaceae* (e.g. *Eurotium*). The peridium is scant and weft-like, resembling *Edyuillia* or *Talaromyces* (*Eurotiales*), in overall morphology. The arthroconidial anamorph prevents making a transfer to the *Eurotiales* without some reservations, although the arthroconidia are elongate, irregular, and sparsely produced and atypical of the regular, alternate arthroconidia found in most *Onygenales*.

Gymnascella is a large genus of species with similar hyphal peridia, and oblate to discoid ascospores with equatorial rims or grooves and with or without polar thickenings. Many lack asexual states. The ascospores with pronounced equatorial modifications (i.e. broad bands or grooves; e.g. *G. confluens*, *G. marginospora*, *G. punctata*) would be accommodated easily in the *Eurotiales*.

Gymnoascoideus petalosporus, with smooth discoid spores, and a distinctive arthroconidial anamorph, is better disposed in the *Arthrodermataceae*.

The *Eurotiales* includes a single family (*Trichocomaceae*) with about 20 genera (Table 4). The *Trichocomaceae* includes two subfamilies: *Trichocomoideae* and *Dichlaenoideae* (Malloch, 1986). These subfamilies are differentiated on the basis of the morphology of the hyphal structures enclosing the asci, ascospore shape, related anamorph taxa and substrate preferences. Because of the phialidic anamorphs, it has been suggested that members of the order share a common ancestor with the *Hypocreales* or perhaps with the *Leotiales* (Malloch, 1986). Cleistothecial taxa, with phialidic anamorphs in *Acremonium* such as *Albertiniella*, *Emericellopsis* and possibly *Nigrosabulum*, considered under the *Pseudoeurotiaceae* (fide von Arx, 1987), may represent intermediates in this pattern.

Table 4 presents a realignment of the genera and families of the prototunicate orders *Eurotiales* and *Onygenales*. The *Eurotiales* is expanded slightly to include some genera of the *Gymnoascaceae* among the two subfamilies of the *Trichocomaceae*. The *Onyge-*

nales is now restricted to keratinolytic genera with distinctive anamorphs. I have left the *Myxotrichaceae* here but suggest it merits formal placement in its own order.

CONCLUSION

The prototunicates are a group defined on the basis of common morphological characteristics associated with both teleomorph and anamorph phases among several presumably distinct evolutionary lineages. Among the prototunicates with reduced or mesh-like containments around the ascospores, there may be several distinct lineages from both operculate and inoperculate discomycete ancestors and the *Hypocreales*. Convergences are difficult to decipher using morphological data, but more progress is possible using classical sources of data if the degree of congruence among character sets is taken into account and if the resulting classification has some predictive value. This approach will be essential for developing sound hypotheses for testing with molecular techniques and for assigning some significance to the results.

ACKNOWLEDGEMENTS

This work was undertaken with support from the University of Alberta Central Research Fund and the Natural Sciences and Engineering Research Council of Canada. The laboratory assistance of Sarah Hambleton and Russell Baron is gratefully acknowledged.

REFERENCES

Apinis, E.A., 1964, Revision of the British *Gymnoascaceae*, *Mycological Papers* 96: 1-56.

von Arx, J.A., 1971, On *Arachniotis* and related genera of *Gymnoascaceae*, *Persoonia* 6: 371-380.

von Arx, J.A., 1987, A re-evaluation of the *Eurotiales*, *Persoonia* 13: 273-300.

von Arx, J.A., and R.A. Samson, 1986, *Mallochia*, a new genus of the *Eurotiales*, *Persoonia* 13: 185-188.

Benjamin, R.K., 1956, A new genus of the *Gymnoascaceae* with a review of the other genera, *Aliso* 3: 301-328.

Benny, G.L., and J.W. Kimbrough, 1980, A synopsis of the orders and families of plectomycetes with keys to genera, *Mycotaxon* 12: 1-91.

Berbee, M.L., and J.W. Taylor, 1992a, Detecting morphological convergence in true fungi, using 18S rRNA gene sequence data, *BioSystems* 28: 117-125.

Berbee, M.L., and J.W. Taylor, 1992b, Two ascomycetes classes based on fruiting body characters and ribosomal DNA sequence, *Molecular Biology and Evolution* 92: 278-284.

Bowman, B.H., J.W. Taylor, and T.J. White, 1992, Molecular evolution of the fungi: human pathogens, *Molecular Biology and Evolution* 9: 893-904.

Bowman, B.H., and J.W. Taylor, 1992, Molecular phylogeny of pathogenic and non-pathogenic *Onygenales*, *In*: *The Fungal Holomorph: mitotic, meiotic and pleomorphic speciation in fungal systematics* (D.R. Reynolds and J.W. Taylor, eds): 169-178. CAB International, Wallingford.

Cano, J., and J. Guarro, 1990, The genus *Aphanoascus*, *Mycological Research* 94: 355-377.

Cano, J., J. Guarro, and L. Zaror, 1990, Two new species of *Aphanoascus* (*Ascomycotina*), *Mycotaxon* 38: 161-166.

Currah, R.S., 1985, Taxonomy of the *Onygenales*: *Arthrodermataceae*, *Gymnoascaceae*, *Myxotrichaceae* and *Onygenaceae*, *Mycotaxon* 24: 1-216.

Currah, R.S., 1988, An annotated key to the genera of the *Onygenales*, *Systema Ascomycetum* 7: 1-12.

Dalpé, Y., 1989, Ericoid mycorrhizal fungi in the *Myxotrichaceae* and *Gymnoascaceae*, *New Phytologist* 113: 523-527.

Dalpé, Y., 1991, Statut endomycorrhizien du genre *Oidiodendron*, *Canadian Journal of Botany* 69: 2206-2212.

Egger, K.N., and L. Sigler, 1993, Relatedness of the ericoid endophytes *Scytalidium vaccinii* and *Hymenoscyphus ericae* inferred from analysis of ribosomal DNA, *Mycologia* 85: 219-230.

Guarro, J., and J.A. von Arx, 1986, *Monascella*, a new genus of the *Ascomycota*, *Mycologia* 78: 869-871.

Guarro, J., J. Cano, and C.H. de Vroey, 1991, *Nannizziopsis* (*Ascomycotina*) and related genera, *Mycotaxon* 42: 193-200.

Guarro, J., J. Gene, and C.H. de Vroey, 1992, *Amauroascopsis*, a new genus of *Eurotiales*, *Mycotaxon* 45: 171-178.

Guarro, J., L. Punsola, and J. Cano, 1987a, *Byssoonygena ceratinophila*, gen. et sp. nov. a new keratinophilic fungus from Spain, *Mycopathologia* 100: 159-191.

Guarro, J., L. Punsola, and M.J. Figueras, 1987b, *Brunneospora reticulata* gen. et spec. nov. a keratinophilic ascomycete from Spain, *Persoonia* 13: 387-390.

Horie, Y., 1980, Ascospore ornamentation and its applications to the taxonomic re-evaluation in *Emericella*, *Transactions of the Mycological Society of Japan* 21: 483-493.

Kendrick, B., 1992, *The Fifth Kingdom*. 2nd edn. Mycologue Publications, Waterloo.

Kimbrough, J.W., 1989, Arguments towards restricting the limits of the *Pyronemataceae* (ascomycetes, *Pezizales*), *Memoirs of the New York Botanical Garden* 49: 326-335.

Malloch, D., 1981, The plectomycete centrum, *In*: *Ascomycete Systematics: The Luttrellian Concept* (D.R. Reynolds, ed.): 73-91. Springer, New York.

Malloch, D., 1986 ["1985"], The *Trichocomaceae*: relationships with other ascomycetes, *In*: *Advances in Penicillium and Aspergillus Systematics* (R.A. Samson and J.I. Pitt, eds): 365-378. Plenum Press, New York.

Minter, D.W., P.M. Kirk,, and B.C. Sutton, 1982, Holoblastic phialides, *Transactions of the British Mycological Society* 79: 75-93.

Paden, J.W., 1975, Ascospore germination, growth in culture, and imperfect spore formation in *Cookeina sulcipes* and *Phillipsia crispata*, *Canadian Journal of Botany* 53: 56-61.

van de Peer, Y., L. Hendricks, A. Goris, J.-M. Neefs, M. Vancanneyt, K. Kersters, J.-F. Berny, G.L. Hennebert, and R. de Wachter, 1992, Evolution of basidiomycetous yeasts as deduced from small ribosomal subunit RNA sequences, *Systemematic and Applied Microbiology* 15: 250-258.

Sigler, L., 1989, Problems in application of the terms "blastic" and "thallic" to modes of conidiogenesis in some onygenalean fungi, *Mycopathologia* 106: 155-161.

Tsuneda, A., S. Murakami, L. Sigler, and Y. Hiratsuka, 1993, Schizolysis of dolipore parenthesome septa in an arthroconidial fungus associated with *Dendroctonus ponderosae* and in similar anamorphic fungi, *Canadian Journal of Botany*: in press.

Udagawa, S., and Y. Horie, 1973, Surface ornamentation of ascospores in *Eupenicillium* species, *Antonie van Leeuwenhoek* 39: 313-319.

Udagawa, S., and N. Toyazaki, 1987, A new species of *Oidiodendron*, *Mycotaxon* 28: 233-240.

Udagawa, S., S. Uchiyama, and S. Kamiya, 1993, *Gymnostellatospora*, a new genus of the *Myxotrichaceae*, *Mycotaxon*: in press.

Udagawa, S., and S. Ueda, 1982, A new *Eupenicillium* species with reticulately ornamented ascospores, *Mycotaxon* 14: 266-272.

Weitzman, I., M.R. McGinnis, A.A. Padhye, and L. Ajello, 1986, The genus *Arthroderma* and its later synonym *Nannizzia*, *Mycotaxon* 25: 505-518.

TRADITIONAL DISCOMYCETE TAXONOMY: SHOULD WE ALSO SHIFT TO A SECOND GEAR?

S. Huhtinen

Herbarium, University of Turku
FIN - 20500 Turku, Finland

SUMMARY

Nine points, concerning the study of inoperculate discomycetes, are presented. The ideology, on which present day taxonomic work is based, is shown to have major deficiencies. We gather and distribute only part of the valuable information content of fungi. The imbalance between the effort needed before the specimens are available for study and the short time spent at the microscope, is often prominent. Much of the information extracted never reaches the scientific community. Along with the traditional documentation in hard copy, we need a system to gather and distribute basic data of studied populations. Also, vital taxonomy and cultural studies are too easy and too informative to be neglected, as is frequently done, if we are to work efficiently. Documentation of ascal development, and of the mountants used, are recommended as standard procedures.

INTRODUCTION

The purpose of this contribution is to promote methodological awareness in the study of discomycetous fungi. The nine guidelines are mainly directed to a junior mycologist starting a study of inoperculate discomycetes. It is hoped that by doing this, taxonomic revisions of the future, will be produced based on a different ideology. The paper tries to focus on the inefficiency of our present day ideology. It is not the traditionalism which is criticized, the huge percentage of not-yet-described taxa (Hawksworth, 1991; Korf, 1990; Zhuang and Korf, 1991) demands pioneering descriptive taxonomy, but the following paragraphs will show major pitfalls in extracting and distributing our raw material, i.e. data from fungal populations. A change in methodological attitude is needed, with little or no loss of time, to clearly increase the output: shared knowledge on existing taxa.

Recent developments in the systematics of fungi, highlighted for example by Hawksworth and Bridge (1988), are not treated here. The focus is on traditional, descriptive taxonomy, but reflects no sentiment on the value of old *versus* new techniques. Surprisingly, although we have had over a century to refine the traditional methods, young scientists are still insufficiently trained in the profession. After a hundred years, many mycologists still use the conventional light microscope as they wish and store information only to their minds. As mycologists work usually as isolated individuals, only a small fragment of the information gained is distributed to the scientific community. With limited manpower and a huge task ahead of us, can we afford to lose information? Can we afford methodological inaccuracy, unsatisfactory illustrations, or excessively individualistic approaches? Or should we rather shift into a second gear?

Ascomycete Systematics: Problems and Perspectives in the Nineties
Edited by D.L. Hawksworth, Plenum Press, New York, 1994

Is it superfluous to demand more accurate microscopical observations? At first sight it may seem so. It can, of course, be claimed that the study of inoperculate discomycetes will be occupied with tiny undescribed, characteristic species for decades. Almost any new area visited reveals numerous novelties. A relatively gloomy view of the task ahead was given by Korf (1990), who estimated that two thirds of the small discomycetes remain undescribed. A justifiable claim would seem to be, that it is most important to get them described, even using mediocre accuracy. On the whole this ideology does not save our resources to proceed faster. Seldom we seem to have time to work thoroughly, but always to have the time to do things over and over again! Data on variability, data from distant populations, awareness of methodological errors, and discussions before publishing, would all shorten the future lists of synonyms.

Many present day descriptions and herbarium packet annotations are similar in scope to those dating from the last century. The tool, the conventional light microscope, has experienced a drastic development, but not much has happened to the output. Too often only the ink has changed.

As consumers of fungal taxonomy, we have become content with the product. It is easy to be happy about reportedly clear taxonomic units when one is dealing with a group that is not too difficult. As long as we can forget naming populations of, for example, *Mollisia*, *Pezizella*, or *Cistella*, we can say that casual light microscopy gives satisfactory results. But isn't this short-sighted? Sooner or later we will face the truly difficult groups. Sooner or later we will become more interested in variability. Why not do our microscopy well enough now to face the future challenges?

RAW MATERIAL: MINUTES VERSUS MILLIONS?

Collecting raw material, data on fungal populations, is time-consuming. Exploration demands much time spent in monetary arrangements, report writing, travelling, etc. Likewise, curatorial procedures in herbaria are costly, both in time and money. The effort in getting the specimens under a cover glass and on a microscope stage, measured in the expert's time available for science, may be huge. The second part is extracting the information from the mounts. The expert's skills are used to do the actual job, but the information available under the cover glass too often remains unseen, or does not reach the scientific community.

Here minutes turn into money. We report the hairs as aseptate due to low contrast in Melzer's reagent (MLZ). We report only smooth hairs as we have dissolved the granulation with the mountant. We do not see the pigments inside paraphyses, as we have killed the material. Paraphyses are reported as smooth because we do not open the bottle of Congo red (CR) but instead dissolve the encrustation in cotton blue (CB). We do not observe ascal bases, as the mount is imperfectly stained or squashed. And, worst of all, we do not make drawings nor annotate the collection. After all the money spent up to this point, and the time lost in curatorial herbarium bureaucracy, we are careless! The minutes we are saving may become excessively costly. Imperfect or erroneous observations will haunt the future literature, collections, and taxonomists.

Hyaloscypha aureliella will serve as an example of such a "ghost". The species has long been known as *Hyaloscypha stevensonii*, a trivial species occurring on softwood from Svalbard to the Philippines. The history of the "ghost" is marked both with accurate and inaccurate microscopical studies: Nylander (1869), Berkeley and Broome (1875), Starbäck (1895), Höhnel (1903), Nannfeldt (1932, 1936), Velenovsky (1934), Dennis (1949), Graddon (1972), Svrček (1978, 1983), Breitenbach and Kränzlin (1981), Spooner and Dennis (1985), and Huhtinen (1990). The "ghost" survived in literature, waisting time and money, for more than a hundred years: numerous printed pages, many illustrations and, in 1981, a coloured portrait under an alias. All this just because sometimes the resin on hairs, the amyloid nodules, the presence of croziers, or the strict ecology were observed and sometimes not. This "ghost" was a very trivial and characteristic one. Nothing to compare with the laborious and demanding species delimitations we are facing in numerous genera.

Do your microscopy with time and caution, to avoid future taxonomists turning into ghostbusters.

ANNOTATIONS: FOR YOU OR FOR US ALL?

The mere presence of microscopically studied, but unannotated, specimens in world herbaria is of mediocre value. In inoperculate discomycetes we seldom benefit from macroscopical appearance. A specimen is most valuable when its microscopical details are available without a painstaking restudy. Accordingly, the demand for standardized microscopy is especially acute in this group.

It is an astounding feeling to receive an illustrated herbarium annotation sheet from a colleague. From a set of detailed camera lucida drawings one learns more than from a published paper. The effort made in drawing results in a transfer of information on the raw material, the populations. Certainly, preparing a camera lucida drawing (or carefully prepared freehand drawing) takes more time than a study with sketchy notes only one person can utilize. But, already with two investigators, the time spent is payed back. Likewise, making a camera lucida drawing forces the investigator to spend more time observing and to use various mountants. The result is more detailed observations.

It is clear that our efficiency in the future will depend on how effectively we can distribute the characteristics of populations. A somewhat naïve question is whether the objective of discomycetologists is to produce papers or to add to the shared knowledge on these fungi. Let's be more naïve and idealistic: we would study population after population and the output (when we decide to share) would be generally available as soon as ready. It would have to be in the same format to be most usable. The camera lucida would be the key item of apparatus, more accessible and usable than, for example, photographic equipment or image analysis systems. The annotated drawings would create a general pool of information on populations which could be made available on, for example, CD-ROM, all over the world. The same information, which at present is inaccessible, in our minds or in our drawers, is already in the graveyard.

Of course such annotations would not remove the need for microscopical restudies, but they would direct that need. Nor is it necessary to annotate every collection. But the specimens we decide to study more closely should be annotated to the point where their information content can be transferred. A camera lucida drawing beats individual memory when considering populations and variability.

Think ahead and store the information extracted from specimens into well-annotated drawings.

THE MONOGRAPHER AND THE JIGSAW PUZZLE

The pieces of the generic jigsaw puzzle are scattered all over the world. From world herbaria and reference collections they are gathered to the desk of a monographer. Solving the puzzle puts a monographer out of circulation for years. With limited human resources, all this is a high price to pay.

The monographer more or less solves the puzzle. He or she is the only living person who has an idea of the jigsaw puzzle as a whole. The place and characteristics of each piece has laboriously been solved. The amount of basic information gathered at the one desk is huge. If all these pieces are returned to the world herbaria unannotated (as is often done), we have worked inefficiently. The information content in returned, unannotated specimens is: "these were included in the monographers concept of this species". But where do they stand in the variational scale? Are they the marginal populations where the limit between two taxa was drawn? Which characteristics were considered to result from high altitude or geographic remoteness? Is the spore size more variable on this substratum? All these are questions once judged and answered by the monographer, but somewhere along the road the answer has been lost.

Even when the specimens are annotated and camera lucida drawings are made, it is not sufficient just to return the specimens. These annotations should be gathered taxon by taxon and filed in a recognized collection. In the future such records will prevent the answers from getting lost. Anyone who wants to check the correctness of delimitations, will find such information valuable. The pieces are all there, so when microscopic restudies are necessary, they can be directed reasonably.

The days needed to file one set of annotations are negligible in comparison to the years spent working with the revision.

THE "HIDE AND SEEK" OF VARIABILITY

A revision is a filtered product based on the pieces in the jigsaw puzzle. Consequently, it can only transfer part of the information. Fortunately so, as the total information content would confuse the reader (and can be distributed by other means). But what is the part that is traditionally transferred?

In a revision we simplify the presentation to enable the user to identify easily specimens to species. We concentrate the populations into one diagnosis, and too seldom comment on the causes and patterns of variability. In a variable species we know the whole puzzle. We know where the critical delimitations lie. We know the shortcomings. We know that variability between the populations cannot be expressed in only a formal diagnosis, as the characteristics are linked in various combinations. We simplify to the point where our concern for a reader's convenience turns against us. By focusing solely on the stability seen in typical populations, we hinder the recognition of divergent populations.

As pointed out by Sigler and Hawksworth (1987), there is a tendency to rely on published descriptions when new taxa are described. We will never rid taxonomy from this abuse. Accordingly, new taxa based on insufficient knowledge of the full range of variation will be erected to burden future work.

Why should we keep on illustrating "unvariability"? Why does a revision include only one, "central" illustrated population per species? Why are, for example, the critical, divergent populations not illustrated? Limited space in journals might be proposed as a potential answer, but this is not the case. We seem to have the resources to print overwhelming collection data, repetitions of descriptions in Latin, gigantic illustrations to fill the page, or worst of all, the blank space following almost every article. Hence, there already is space to devote to the causes and expression of variability. To some extent this has been done in words, but not in illustrations.

Serve the reader, give some idea of overall variation between populations in a species, preferably with illustrations.

REVIEW INFORMATION TRANSFER, NOT STYLE

Professional editors and reviewers often concentrate on stylistic matters. As a result, we can enjoy neatly printed studies, which, however, may be devoid of basic data such as the mountants which have been used. Often there is a striking contrast between the correctly positioned colons and the hastily compiled illustrations. The ideology behind making a line drawing is curious. If we are not able to write Latin descriptions or use statistical methods we ask for help. But if we are not able to make a good illustration, we just do not care. On this point a reviewer should be strict. There should be a strong demand for line drawings to be informative. These illustrations are essential in distributing the characteristics of a taxon to the readership.

The need for vital taxonomy is discussed below. But even when working with dry material, different mountants are an important source of variability. The differences in ascal and spore measurements in *Hyaloscypha* were summarized by Huhtinen (1990). A gross difference in ascal width between CB and MLZ mounts was found to be around 10 %. Using ammoniacal Congo red, another 10 % can easily be added, as the asci in many genera seem to be widest in CR-mounts.

The appearance of apothecia can be drastically changed by different mountants. In the two forms of *Protounguicularia barbata*, the apothecia are different almost beyond recognition in KOH, CR, and MLZ. But in heated lactic acid, dissolving the crust-like resin from the hairs, the conspecifity of the two forms can easily be seen (Huhtinen, 1987). Conscious observations in different mountants also adds to our knowledge on variability of hair exudates, as seen in, for example, *Cistella tenuicula* (Huhtinen, 1993). Hein (1981) introduced the ornamentation of paraphysis tips as a taxonomic character in some mollisioid genera and reported on mountant-specific changes. An example of one such species is given in Fig. 1.

When acting as a reviewer, demand that the measurements and the informative plates include data on the mountants. If not, call the paper a potential boomerang.

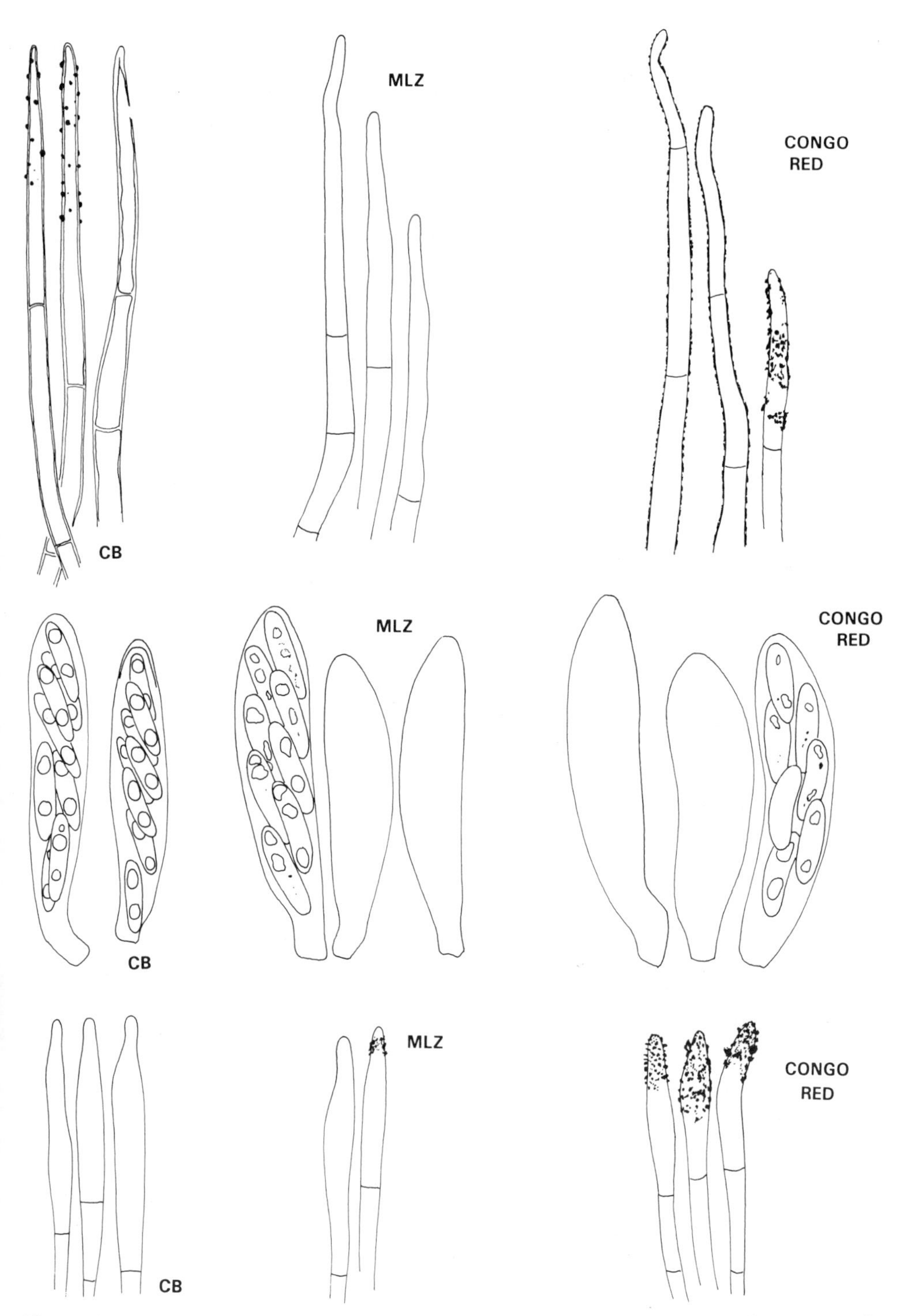

Figure 1. Mountant-induced changes in the hairs and hymenial characters of *Coronellaria caricinella* (*Fungi Fenniae exsiccati* no. 645, H).

LIVING VERSUS DEAD

Baral (1992) summarized the importance of vital taxonomy, i.e. taxonomy working with living fungal material. Another equally innovative paper was recently presented by Common (1991). Such novelties are the stimulants of traditional taxonomy.

Working on a local basis, we can easily put novelties into use. We can make a small number of collections to be studied without delay. But a conflicting situation arises when we have the rare change to collect in a remote area. Should we then collect only the amount we can study in a fresh condition and make cultures of, i.e. study these organisms as a whole? Or should we make a greater number of collections to be dried, at the same time losing a great part of their information content? There is no definite answer.

Whenever possible, a taxonomic study should deal with both living and dead material. We can never totally be free from the study of dead herbarium material, but that is no excuse for not studying small discomycetes fresh whenever we have a chance to do so. As the monographer should illustrate much of the total variability, the populations selected inevitably originate from world herbaria, but at least the minimum of one population should be illustrated in a vital condition. Measurements should be given separately to both living and dead material.

Before starting your next discomycete study, read Baral (1992).

THE SIMPLICITY OF CULTURAL STUDIES

Cultures are relatively easy to obtain. They can be made in the field. A petri dish can be carried to the arctic tundra or to a remote mountain-top. There are no excuses for not obtaining cultures, except lack of the will to do so. Both the anamorph-teleomorph connections and general cultural characters are valuable tools in traditional discomycete taxonomy.

Why should we travel to the other side of the globe, to a mycological "terra incognita", just to fetch 50% of the information content of the organisms growing there? Can we claim to work efficiently if the other 50% must be fetched by a scientist interested in anamorphs? We only have to take petri dishes or small test tubes along when we go for an expedition. If not for all the material, but at least for those which seem to be the interesting ones, and definitely those that are under closer study. These strains can then be compared with the more controlled single-spore isolates.

What is your excuse for not undertaking cultural work?

ASCI FROM TOP TO BOTTOM

Ascal development has in general been neglected as a taxonomic character. The emphasis has been on ascal plugs and one could even speak of "top priority". Numerous experts of many ascomycete genera have never even thought of checking the ascal bases. The idea has scarcely penetrated into lichenology, except for the studies of Schmidt (1970) and Tibell (1978, 1987). This neglect was briefly summarized by Huhtinen (1990). The presence of croziers *versus* simple septa offers a rapid and valuable additional character for classification. More observations are, however, needed to give a general idea of the taxonomic implications of this feature in various genera. In Table 1, some preliminary results on the distribution of the two basic ascus base types are given.

As erroneous observations on ascal bases have also been reported, the importance of correct methodology must be stressed. In vital taxonomy, the hymenium is easily squashed and the living asci easy to check for this feature. In herbarium material, mountants giving a low contrast (such as MLZ, KOH) are a source of error. Ammoniacal Congo red is recommendated for those not familiar with the feature. The asci are more readily separated and are stained to give contrast in this mountant. Also, a well-prepared and heated CB-mount, with all the blue in the background removed by lactic acid and with thorough staining, is usable. One should seek juvenile asci, where the character is most easily seen.

Make checking ascal bases a routine procedure.

Table 1. Preliminary numbers of taxa (published and tentative) with asci arising from croziers or simple septa in some leotiaceous genera. Data mainly from H.O. Baral (pers. comm.), and Huhtinen (unpubl.).

Genera	Croziers	Simple
Allophylaria	9	2
Bisporella	8	2
Ciboria	12	4
Cistella	22	5
Dasyscyphella	7	2
Hyaloscypha	12	15
Incrucipulum	0	4
Lachnum	19	28
Lasiobelonium	7	0
Pezicula	13	0
Phialina	8	0

BE AWARE OF THE WORD "TYPE"

Imagine yourself making a type catalogue from the works of a productive nineteenth century mycologist, who desribed over 2000 fungi from myxomycetes to coelomycetes. Art. 8.3 of the Code, often turning the mere citation of a specimen as "type" to an unintentional lectotypification, would drive you into despair. Every study where original material was studied during the last decades, needs to be checked for the word "type" or equivalent. The unintentional typifications may refer to syntypes of poor quality. Their expression may be vague and cause various interpretations. In fact, Art. 8.3. makes it possible to be vague in tens of different ways! Anyone making such a type catalogue will soon have an overwhelmingly abundant collection of such expressions. The problems caused by Art. 8.3. were discussed by Barrie et al. (1992a, 1992b).

Be careful what you cite as type. When making formal lectotypifications, hunt down all the earlier citations in legends, lists of studied material, etc.

ACKNOWLEDGEMENTS

I express my appreciation to my colleagues in mycology (including lichenologists) for valuable discussions. Special thanks are due to my friend Mr H.O. Baral for his contributions to my manuscript.

REFERENCES

Baral, H.O., 1992, Vital versus herbarium taxonomy: morphological differences between living and dead cells of ascomycetes, and their taxonomic implications, *Mycotaxon* 44: 333-390.

Barrie, F.R., C.E. Jarvis, and J.L. Reveal, 1992a, The need to change Article 8.3. of the Code, *Taxon* 41: 508-512.

Barrie, F.R., C.E. Jarvis, and J.L. Reveal, 1992b, Two proposals to amend Article 8 of the Code, *Taxon* 41: 600-601.

Berkeley, M.J. and C.E. Broome, 1875, Notices of British fungi, *Annals and Magazine of Natural History, ser.* 4, 15: 28-41.

Breitenbach, J. and F. Kränzlin, 1981, *Pilze der Schweiz 1. Ascomyceten*, Verlag Mykologia, Luzern.

Common, R.S., 1991, The distribution and taxonomic significance of lichenan and isolichenan in the *Parmeliaceae* (lichenized *Ascomycotina*), as determined by iodine reactions. 1. Introduction and methods. 2. The genus *Alectoria* and associated taxa, *Mycotaxon* 41: 67-112.

Dennis, R.W.G., 1949, A revision of the British *Hyaloscyphaceae*, with notes on related European species, *Mycological Papers* 32: 1-97.

Graddon, W.D., 1972, Some new discomycete species 2, *Transactions of the British Mycological Society* 58: 147-159.

Hawksworth, D.L., 1991, The fungal dimension of biodiversity: magnitude, significance, and conservation, *Mycological Research* 95: 641-655.

Hawksworth, D.L. and P.D. Bridge, 1988, Recent and future developments in techniques of value in the systematics of fungi, *Mycosystema* 1: 5-19.

Hein, B., 1981, Zum Wert von Paraphysenauflagerungen für die Taxonomie des *Hysteropezizella*-Komplexes (*Dermateaceae, Mollisioideae*), *Nova Hedwigia* 34: 449-466.

Höhnel, F., 1903, Mykologische Fragmente, *Annales Mycologici* 1: 391-414.

Huhtinen, S., 1987, The genus *Protounguicularia* in Europe, *Beiträge zur Kenntnis Pilze Mitteleuropas* 3: 457-463.

Huhtinen, S., 1990, A monograph of *Hyaloscypha* and allied genera, *Karstenia* 29: 45-252.

Huhtinen, S., 1993, Some hyaloscyphaceous fungi from tundra and taiga, *Sydowia* 45: 188-198.

Korf, R.P., 1990, Discomycete systematics today: a look at some unanswered questions in a group of unitunicate ascomycetes, *Mycosystema* 3: 19-27.

Nannfeldt, J.A., 1932, Studien über die Morphologie und Systematik der nicht-lichenisierten inoperculaten Discomyceten, *Nova Acta Regiae Societatis Scientiarum Upsaliensis, ser.* iv, 8 (2): 1-368.

Nannfeldt, J.A., 1936, Notes on type specimens of British inoperculate discomycetes (First part, notes 1-50), *Transactions of the British Mycologial Society* 20: 191-206.

Nylander, W., 1869, Observationes circa Pezizas Fenniae, *Notiser ur Sällskapets pro Fauna et Flora Fennica Förhandlingar* 10: 1-97.

Schmidt, A., 1970, Ascustypen in der Familie *Caliciaceae* (Ordnung *Caliciales*), *Vorträge aus dem Gesamtgebiet der Botanik, n.f.* 4: 127-137.

Sigler, L. and D.L. Hawksworth, 1987, International Commision on the Taxonomy of Fungi (ICTF): Code of practice for systematic mycologists, *The Mycologist* 1: 101-105.

Spooner, B.M. and R.W.G. Dennis, 1985, New or interesting ascomycetes from the Highlands and Islands, *Sydowia* 38: 294-316.

Starbäck, K., 1895, Discomyceten-Studien, *Bihang till Konglikje Svenska Vetenskaps-Akademiens Handlinger* 21: 1-42.

Svrček, M., 1978, New or less known *Discomycetes*. 9, *Česká Mykologie* 32: 202-204.

Svrček, M., 1983, New or less known *Discomycetes*. 12, *Česká Mykologie* 37: 65-71.

Tibell, L., 1978, The genus *Microcalicium*, *Botaniska Notiser* 131: 229-246.

Tibell, L., 1987, Australasian *Caliciales*, *Symbolae Botanicae Upsalienses* 27(1): 1-280.

Velenovsky, J., 1934, *Monographia Discomycetum Bohemiae*, 2 vols. Prague.

Zhuang, W.-Y. and R.P. Korf, 1991, Our current knowledge of the tropical non-lichenized inoperculate discomycetes, *Mycosystema* 4: 129-139.

PROBLEMS IN THE SYSTEMATICS OF *PEZIZALES*

J. van Brummelen

Rijksherbarium/Hortus Botanicus
P.O. Box 9514, 2300 RA Leiden
The Netherlands

SUMMARY

The most important recent views concerning the relationships between families and genera within the *Pezizales* are summarized. Attention is also paid to the delimitation of this order from the *Tuberales* and the *Plectomycetes*. Many ideas of the great pioneers of the systematics of discomycetes at the turn of the last century are still of importance in present classifications. Many macroscopic and microscopic morphological characters they used proved to be very practical for identification, though some of them appear to be of value only at a low taxonomic level. In recent years characters obtained from cytology, histochemistry, ultrastructure, possible anamorph connections, and molecular structure of proteins and DNA became available for classification. The correct evaluation of these new characters is one of the main problems of the present systematics of the *Pezizales*.

HISTORY

When in 1857 the Crouan brothers (Crouan and Crouan, 1857), two naturalists living in the neighbourhood of Brest, discovered the operculum at the top of the ascus in species of *Ascobolus*, they thought that the operculum was a unique feature of this genus. In the next few years many mycologists started to study *Ascobolus*. Among them were Karsten (1861, 1870), Coemans (1862), Cooke (1864), de Notaris (1864), Berkeley and Broome (1865), Fuckel (1866), and Boudier (1869). Boudier especially studied the genus in detail. He started his mycological studies with the help of his famous neighbour at Montmorency, who was one of the discoverers of the basidium, J.H. Léveillé.

Boudier distinguished several genera within the "Ascobolés" and discovered the operculum at the top of the ascus in many genera of discomycetes, while he also found many other genera without an operculum. Boudier (1879, 1885) published his ideas on a natural classification of the discomycetes, based on ascus dehiscence, blue staining of the ascus wall with iodine, gross morphology of the fruit body, presence or absence of an apothecial margin, and microscopic characters. He clearly defined the operculate and inoperculate discomycetes as taxa of high rank.

Other important classifications were proposed by Fuckel (1870) and Karsten (1869, 1871, 1885), who both followed the systematic arrangement of discomycetes proposed by Fries (1822, 1849) and raised many of his "tribes" of *Peziza* to generic rank, but they also distinguished many new genera using microscopic features.

Saccardo (1884, 1889) abandoned the Friesian system completely by introducing one which put much weight upon ascospore characters. He distinguished large genera

Ascomycete Systematics: Problems and Perspectives in the Nineties
Edited by D.L. Hawksworth, Plenum Press, New York, 1994

with a great number of subgenera. His system was of great importance and of lasting influence.

Cooke (1875-79) and Rehm (1887-96) contributed much to the knowledge of species by their descriptions based on original observations of authentic specimens sent to them by other mycologists.

In 1907 Boudier completed his earlier classification of the European discomycetes and recognized seven families, 12 tribes, and 72 genera within the operculates (Boudier, 1907). But, unfortunately many of his taxonomic decisions in this work were based on descriptions from literature rather than on the study of material.

In his survey of the North American cup-fungi, Seaver (1928) used a practical system based on easily observed, but completely unnatural characters.

Corner (1929a, 1929b, 1930, 1931) laid a sound basis for the study of the morphology and ontogeny of ascomata in discomycetes.

Chadefaud started a series of studies on the apical apparatus of asci. He also studied several operculates (Chadefaud, 1942, 1946). Both he and Le Gal (1946) focused their attention on the ascus structure in *Sarcoscypha* and some related genera of the *Sarcoscyphaceae* and described in this group a structure they called "para-operculate" and "suboperculate" respectively. It was considered a transition between the ascus of the "Inoperculati" (*Leotiales*) and the "true Operculati". The fungi concerned were believed to be the most primitive among the operculates and placed in a separate high ranking taxon.

In her thesis, Le Gal (1947) studied the development of ascospore ornamentation in the operculate discomycetes, and believed that these spore characters should take precedence over all others. She modified Boudier's system accordingly, brought in her experience with tropical discomycetes (Le Gal, 1953, 1959), and included (Le Gal, 1963) the results of cytological studies by Berthet (1961, 1963, 1964). Her classification was followed by many mycologists (e.g. Nannfeldt, 1949; Dennis, 1960, 1968, 1978; Gäumann, 1964; Rifai, 1968; Korf, 1954, 1970, 1972, 1970, 1973). Korf extended Le Gal's classification in several ways and contributed much to the clarification of the nomenclature of discomycetes.

A re-evaluation of the taxonomy of the genera of the operculate discomycetes by Eckblad (1968) considered all generic names in this group, and put much emphasis on morphological and anatomical characters of the excipulum, much in the good Nordic tradition of Karsten (1861, 1869, 1871, 1885), Nylander (1868), Starbäck (1895), and Nannfeldt (1932). Like Gäumann (1926, 1949, 1964) and Nannfeldt (1932, 1937), Eckblad does not consider the operculates to be derived from the inoperculates, and throws doubt upon the extreme importance attached to the "suboperculate ascus" by Le Gal. He considered the *Thelebolaceae* the most primitive family within the *Pezizales* and the common origin of both main groups of discomycetes.

During the last two decades, taxonomic surveys of the whole of the *Pezizales* have been rare. Only recently have Korf and Zhuang (1991a, 1991b, 1991c, 1991d, 1991e, 1991f), in their preliminary discomycete flora of Macaronesia, adopted new findings in the arrangement of families and genera.

Since 1982, Eriksson (1982) and Eriksson and Hawksworth (1991) have published annually revised *Outlines of the ascomycetes*. These try to keep the classification up to date by bringing together the often rather scattered contributions of potential taxonomic relevance. However, as far as the *Pezizales* are concerned, there is a need for specialist evaluation of the entries.

Some important taxonomic revisions of genera published recently include: *Aleurina* (syn. *Jafneadelphus*; Zhuang, 1986), *Arpinia* (Hohmeyer, 1988), *Ascodesmis* (van Brummelen, 1981a), *Boudiera* (Dissing and Schumacher, 1979), *Cheilymenia* (Moravec, 1988, 1989a, 1989b, 1990), *Helvella* (Abbott and Currah, 1988), *Inermisia* (syn. *Byssonectria*; Dennis and Itzerott, 1973), *Lasiobolus* (Bezerra and Kimbrough, 1975), *Marcelleina* (syn. *Pulparia*; Donadini, 1976; Moravec, 1987), *Octospora* (incl. *Lamprospora*; Dennis and Itzerot, 1973; Khare and Tewari, 1978; Caillet and Moyne, 1980, 1987a, 1987b; Benkert, 1987; Wang and Kimbrough, 1992), *Pachyella* (Pfister, 1973), *Pseudoplectania* (Donadini, 1987), *Sarcoscypha* (Harrington, 1990), *Scutellinia* (Kullman, 1982; Schumacher, 1990), *Sowerbyella* (Moravec, 1985), *Tricharina* (Yang and Korf, 1985), and *Wynnea* (Pfister, 1979).

DELIMITATION OF *PEZIZALES* FROM *TUBERALES*

With the exception of *Elaphomyces*, which was assigned to the monotypic *Elaphomycetales*, Trappe (1979) transferred all genera of the polyphyletic order *Tuberales* to nine families of the *Pezizales*. Although no one has ever demonstrated a case of opercula permanently losing their function, since Trappe the inclusion of several hypogeous genera in families of operculate discomycetes has been suggested.

The transfer of *Balsamia*, *Hydnotrya*, *Gymnohydnotrya*, and *Choiromyces* to the *Helvellaceae* was made plausible (e.g. Donadini, 1986; Zhang, 1992a, 1992b) because the ascospore nuclear number was constantly four. This is now considered to be one of the main characters of that family. Pfister (1984) saw great similarities in peridial anatomy and ascospore morphology between species of *Genea* and *Aleurina* (syn. *Jafneadelphus*). Since the ascospores are uninucleate in both genera Zhang (1992b) suggested merging the *Geneaceae* with the *Pyronemataceae*.

A rather homogeneous family *Terfeziaceae* (including *Delastria*, *Hydnobolites*, *Pachyphloeus*, and *Terfezia*) was recently redefined by Zhang (1992a, 1992b) to include only fungi with uninucleate ascospores. Because of the blue staining of ascus walls with iodine, the presence of typical electron-dense biconvex bands in the septal pores of asci, and lamellate structures in the pores of excipular cells in *Hydnobolites*, Kimbrough et al. (1991) expect that most members of this family will turn out to be members of the *Pezizaceae*.

In a series of extensive and detailed studies Parguey-Leduc and Janex-Favre with co-workers (Parguey-Leduc and Janex-Favre, 1977, 1981, 1987; Parguey-Leduc et al., 1987, 1988, 1989, 1990, 1991; Janex-Favre and Parguey-Leduc, 1980, 1983, 1985, 1988) demonstrated that at least 10 species of *Tuber*, studied so far, show a development and a structure of ascoma and ascus in many respects very different from those in the operculate discomycetes. The most unique features of *Tuber* are in the organization of the ascus. The young asci are globular and show successively a longitudinal and a concentric polarity on further development. The initiation of the ascospores proceeds independently for each spore and starts from membrane complexes in the ascus top; an ascus vesicle of the *Euascomycetes* being absent. Usually there are less then eight spores per ascus, surrounded by a unique "post-sporal" membrane. The structure of the ascospore wall is complex, with a characteristic pattern of electron-dense fibrils in the ornamentation. Moreover, there is a distinct structure and development of the ascoma, while the septal pores in vegetative hyphae are of the *Sordaria*-type (Parguey-Leduc et al., 1991). This is considered of sufficient importance not to include *Tuber* in the *Pezizales* and maintain it in the probably monotypic order of the *Tuberales*.

DELIMITATION OF *PEZIZALES* FROM *PLECTOMYCETES*

Cain (1959) and Malloch (1979, 1981) share the view that the genera of cleistothecial ascomycetes, currently arranged either in the *Plectomycetes* (Malloch, 1979; Benny and Kimbrough, 1980) or in the *Eurotiales* (von Arx, 1967, 1987a, 1987b; Fennell, 1973), should be distributed among nine families of *Loculoascomycetes*, *Pyrenomycetes*, and *Discomycetes*. They believe that those fungi represent the evolutionary termination of a large number of unrelated and highly evolved taxa. Based on this presumption, with only rather fragmented and superficial evidence, Malloch (1979, 1981) placed *Guillermondia*, *Monascus*, the whole family *Onygenaceae*, *Warcupia*, *Xenomyces*, *Cleistothelebolus*, and *Orbicula* in the *Pezizales*. In their critical review of the *Plectomycetes*, Benny and Kimbrough (1980) argued against all these transfers and proposed to place the last four genera, with some others, in the *Eoterfeziaceae*, a rather vague family of doubtful position, until more data becomes available on their cytology, ontogeny, and ultrastructure.

Guarro and von Arx (1986) considered the families *Monascaceae* and *Onygenaceae* as indistinguishable, with a taxonomic position intermediate between the *Pezizales* and the *Eurotiales* (incl. *Gymnoascales*).

The ultrastructure of ascospore development in species of *Eleutherascus* was shown by van Brummelen (1989a, 1989b) to be identical with the distinctive type of development found in species of *Ascodesmis*. Both genera were placed in the *Ascodesmidaceae*.

CHARACTERS IN THE MODERN CLASSIFICATION OF THE *PEZIZALES*

For a sound classification of the *Pezizales* with well-defined families and genera, we need sets of characters that are clear and independent, with distinct variation at each of these taxonomic levels. Although in taxonomy single unique characters are sometimes used to discriminate between taxa at higher levels, a stable classification should be based on multiple independent characters. If we compare the different classifications so far proposed for the *Pezizales*, we see a gradual shift in importance from mainly morphological, microscopical, cytological, and developmental characters, towards ultrastructural, karyological, chemical, and molecular features.

In reviewing some of the most important modern characters, a few examples are given and possible problems indicated. This does not mean that some of the older characters have lost their importance in classification. But characters such as pigment composition, anamorph connections, microscopical, anatomical, ontogenetical, and histochemical features have not been the subject of a general study in the *Pezizales* during the last two decades. A re-evaluation of characters in these fields is badly needed.

Ultrastructure of the Ascus Tip

One of the main features used to define families within the *Pezizales* is the ascus, its structure, opening mechanisms, and histochemistry. The ultrastructure of asci has been studied and compared throughout the *Pezizales* (van Brummelen, 1974, 1975, 1977, 1978, 1981b, 1986, 1987, 1989a, 1989b, 1993; Samuelson, 1975, 1978a, 1987b, 1987c, 1987d; Hung, 1977; Kimbrough and Benny, 1978; Samuelson and Kimbrough, 1978; Samuelson et al., 1980; Bellemère et al., 1990).

In my opinion each family of the *Pezizales* can be recognized by a characteristic major type of ascus apical structure. Comparative analysis of the different ascus structures are very helpful in determining affinities at familial and sometimes generic level. Elements that form the features of the apical structures are:

(1) Blue staining of the periascus (*Ascobolaceae*, *Pezizaceae*).
(2) Presence of an internal circular indentation (*Ascobolaceae*, *Pezizaceae*).
(3) Presence of a strongly thickened inner wall layer in the tip, often with sublayering (*Sarcoscyphaceae*, *Thelebolaceae*).
(4) Differences in reactivity to tests on polysaccharides, e.g. Thiéry test (e.g. *Sarcoscyphaceae vs. Sarcosomataceae*).
(5) Presence of a more or less evident subapical ring in the wall at a place where the relative thickness of wall layers changes abruptly (*Thelebolaceae*, *Pyronemataceae*).
(6) Moment of first differentiation of the wall in the apex (early in *Ascobolaceae*, *Pezizaceae*, and *Sarcoscyphaceae*; late in *Thelebolaceae*, *Ascodesmidaceae*, *Pyronemataceae*, *Helvellaceae*, *Morchellaceae*, and *Sarcosomataceae*).
(7) Width of the zone of dehiscence (very narrow in *Ascobolaceae*, *Pezizaceae*, *Helvellaceae*, *Morchellaceae*, and *Sarcoscyphaceae*, resulting in a smooth ascostome; rather broad in *Thelebolaceae*, *Pyronemataceae*, and *Sarcosomataceae*, resulting in irregular rupturing or a rough ascostome).
(8) Presence of a tractus and a funiculus in the acroplasm (all operculate families, except *Ascodesmidaceae*).

Chadefaud (1946) and Le Gal (1946) believed that the operculate ascus was derived from the pored ascus, and that the *Sarcoscyphaceae* represented an intermediate between inoperculates and operculates. Eckblad (1968), Samuelson (1975, 1978d), and van Brummelen (1975, 1978), however, clearly demonstrated that the *Sarcoscyphaceae* are truly operculate and do not form a bridge between the *Leotiales* and *Pezizales*. There is therefore no evidence for the hypothesis that the operculate ascus was derived from the inoperculate ascus. On the contrary, there is a lot of evidence to indicate the operculate ascus is an independent line of development with asci such as those found in the *Thelebolaceae* to be the most primitive. Especially in *Thelebolus*, we see a wide variety of dehiscence mechanisms, from pseudo-bitunicate with irregular rupturing in the multispored asci to more or less regularly operculate in the 8-spored species (van Brummelen, 1978;

Samuelson, 1978d). Genera like *Ascozonus*, *Coprotus* (syn. *Leporina*), *Coprotiella*, *Dennisiopsis*, *Lasiobolus*, *Leptokalpion*, *Mycoarctium*, *Ochotrichobolus*, *Pseudascozonus*, and *Trichobolus* would fit very well with the *Thelebolaceae*.

The ascus type in the *Pyronemataceae* could be considered a direct continuation of that found in the *Thelebolaceae*, with the series ending in that found in of the *Helvellaceae* and *Morchellaceae*.

In a recent ultrastructural study, mainly based on revived herbarium specimens, Bellemère et al. (1990) reinstated the concept of the "suboperculate ascus". Their new concept of the term "suboperculum" differs from earlier concepts of Le Gal (1946) and Samuelson (1975) and is based on a persistent differentiation with the Thiéry reaction of two sublayers of the inner wall layer ("*d* layer") in the top of the ascus. They attached far-reaching consequences for taxonomy to this feature. The definition of the *Sarcoscyphaceae* and the *Sarcosomataceae* were emended in such a way that representatives of the *Sarcoscyphaceae* (e.g. *Cookeina*, *Pithya*, *Pseudopithyella*, *Sarcoscypha*, *Wynnea*) should have two sublayers in the "*d* layer" and those of the *Sarcosomataceae* (e.g. *Plectania*, *Sarcosoma*, and *Urnula*) a "*d* layer" that is not subdivided. The genus *Galiella*, without a "*d* layer", should be placed apart from both families.

Ultrastructure of the Septa

Our knowledge of the septal ultrastructure of the *Pezizales* is mainly based on studies by Curry, Kimbrough, and Gibson. It seems that the structure of the septal pore plugs at the base of the asci in particular, and to a somewhat lesser degree, that in the ascogenous hyphae, can be used in the classification of *Pezizales*.

Most families show considerable uniformity in these septal structures. In the *Ascodesmidaceae* (*Ascodesmis*, *Eleutherascus*) the ascal plugs are highly differentiated unilateral domes at the ascal side with radiating tubular elements (Carroll, 1967; van Brummelen, 1989a). In the *Ascobolaceae* (*Ascobolus*, *Saccobolus*, *Thecotheus*) the ascal pore plugs are also rather elaborate with stratified, electron-dense, hemispherical structures (Kimbrough and Curry, 1985). In the *Pezizaceae* (*Peziza*, *Plicaria*, *Iodophanus*) the ascal plugs are simple with thin convex or biconvex, electron-dense bands in the pores (Curry and Kimbrough, 1983). The ascal pore in the *Helvellaceae* (*Helvella*, *Gyromitra*) is characterized by electron-dense, usually laminate, hemispherical or dumbbell-shaped structures with narrow electron-transparent bands just at the edge of the pore (Kimbrough and Gibson, 1989; Kimbrough, 1991). In *Geopyxis* the ascal plugs are very similar to those in the *Helvellaceae* (Kimbrough and Gibson, 1990). The *Pyronemataceae* show a wide variation in the structure of the ascal plugs, probably indicating that this family is heterogeneous (Kimbrough and Curry, 1986a, 1986b; Kimbrough, 1989; Kimbrough and Gibson, 1990). From these studies it could be concluded that *Thecotheus* belongs to the *Ascobolaceae*, *Iodophanus* to the *Pezizaceae*, *Gyromitra* to the *Helvellaceae*, and that *Geopyxis*, while in many respects close to the *Helvellaceae*, does not belong to that family, because it has no tetra-nucleate ascospores.

Ultrastructure of Ascospores

The initial ascospore wall formation in the *Pezizales* is a general process almost identical in all representatives of the *Euascomycetes* (e.g. Reeves, 1967; Merkus, 1973; Beckett, 1981; Czymmek and Klomparens, 1992). A different process is found in the *Hemiascomycetes* (*Endomycetales*, *Taphrinales*) and in the genus *Tuber* (Janex-Favre and Parguey-Leduc, 1980, 1983; Parguey-Leduc and Janex- Favre, 1981, 1987).

The ascospore wall develops between two closely spaced double membranes, at first as a homogeneous, electron-transparent primary wall. The inner double membrane becomes the sporoplasmalemma and the outer one the spore investing membrane. Secondary wall material is laid down between the primary wall and the investing membrane. All species of the *Pezizales* have a secondary wall outside the primary wall, and its further differentiation determines the appearance of the mature ascospore. During further ripening, usually after differentiation of the secondary wall, the primary wall differentiates internally through a redistribution of material into an inner electron-transparent endospore and an outer, often laminated, epispore (e.g. Merkus, 1973; van Brummelen, 1989a, 1993). At maturity the epispore becomes a very resistant layer and forms the delimitation of the spore proper. The secondary wall should be regarded as an extra wall layer, that sometimes even separates from the primary wall (e.g. *Ascobolaceae*, *Cheilymenia*, *Scutellinia*).

An extensive study of the development of the ascospore wall in the *Pezizales* was carried out by Merkus (1973, 1974, 1975, 1976). She distinguished seven groups of secondary wall formation. In later studies in this field (van Brummelen, 1986, 1987, 1989a; Duby and Kimbrough, 1987; Gibson and Kimbrough, 1988a, 1988b; Kimbrough et al., 1990; Wu and Kimbrough, 1991, 1992a, 1992b, 1993) these views were mainly confirmed and extended. Ultrastructural studies explained, for instance, that smooth ascospores may have a different structure and development.

The features of ascospore development are relevant at different taxonomic levels. The process of initial ascospore delimitation is considered of importance above the rank of order. *Ascodesmidaceae* and *Ascoboloideae* are each characterized by a unique type of development of the secondary wall. Several representatives of the *Thelebolaceae* show a simpler epispore, consisting of only a single electron-dense lamina, while the secondary wall is always very uniform it is without ornamentation and does not disappear. The greatest diversity in spore wall development occurs in the *Pyronemataceae*.

Often the development of the ascospore wall becomes important as a generic or subgeneric character if the structure of the ascus epiplasm and the sporoplasm are also considered. The importance of oil drops in ascospores is sometimes underestimated. No ultrastructural basis could be found for the occurrence of "de Bary Bubbles". This phenomenon may reveal more about the methods used by the observer than about establishment of a new character (*cf.* Barral, 1992).

Ascospore Nuclear Number

Of great influence on the creation of more stable families in the *Pezizales* are the comparative studies on the numbers of nuclei in ascospores, hyphae, and paraphyses by Berthet (1961, 1963, 1964). The nuclear number in ascospores has proved to be very constant for families. Ascospores were found to be multinucleate in *Morchellaceae*, *Sarcoscyphaceae*, and *Sarcosomataceae*; tetranucleate in *Helvellaceae*; and uninucleate in other families studied (*Pezizaceae*, *Ascobolaceae*, *Pyronemataceae*). As noted above, some genera of the former *Tuberales* can be included in the *Helvellaceae* because of their tetranucleate ascospores.

Ploidy of Nuclei

In a special study, Weber (1992) established the relative DNA content of 84 species of *Pezizales* by means of fluorometry. She found that in the haplophase only 1% of the species of the *Pezizales* are monoploid; 99% showed higher ploidy levels. The highest levels ever recorded in fungi were found in the genera *Neottiella* and *Octospora* (respectively 50x and 18x). In general, within otherwise closely related taxonomic groups more primitive and older species have lower ploidy levels than derived and younger ones.

Molecular Characters

Macromolecules can be used to derive evolutionary relationships. This discovery led to the development of a new discipline in biology, molecular evolution. Of the methods used to obtain molecular data for phylogenetic studies, the polymerase chain reaction (PCR) has proved to be one of the most efficient and reliable. This technique involves the amplification of specific segments of DNA or ribosomal RNA. The fragments of amplified DNA from different fungi can be compared by using restriction fragment length polymorphisms (RFLPs; Vilgalys and Hester, 1990) or direct sequencing (Bowman et al., 1992). It proved to be possible to amplify DNA fragments from very small pieces of mycelium or herbarium material. Analyses of restriction patterns by Wingfield and Wingfield (1993) showed the DNA to be identical in eleven year old dried material and living material of the same origin. Most fungi studied so far by these methods are yeasts or filamentous fungi.

In recent studies of ribosomal RNA with these techniques in *Wilcoxina* and *Tricharina*, Egger (1992) reported sequences that were highly divergent between both genera, but more conserved within each genus. On the other hand, three varieties of *Tricharina praecox* described by Yang and Korf (1985) had nearly identical sequences that could not be distinguished, suggesting that the varietal status is unwarranted in this case.

The PCR technique can be used to address problems at any taxonomic level. One of the greatest problems with applying PCR is the risk of contamination with foreign DNA.

CONCLUSIONS

I expect that molecular characters will sustain and supplement classifications based mainly on morphological and chemical characters. I also hope that DNA characters in particular will confirm the ideas behind our taxonomic systems and phylogenies. But in cases where DNA characters contradict a relationship based on reliable morphological characters, we should keep in mind that only certain fragments of DNA are tested, that still very little is known of how sequences in the tested DNA fragments have developed, and that DNA characters and morphological characters have probably not always evolved in a parallel way.

ACKNOWLEDGEMENTS

The author wishes to thank Dr S.M. Francis for her efforts to improve the linguistics of this paper.

REFERENCES

Abbott, S.P. and R.S. Currah, 1988, The genus *Helvella* in Alberta, *Mycotaxon* 33: 229-250.

Arx, J.A. von, 1967, *Pilzkunde*, J. Cramer, Lehre.

Arx, J.A. von, 1987a, A re-evaluation of the *Eurotiales*, *Persoonia* 13: 273-300.

Arx, J.A. von, 1987b, Plant pathogenic fungi, *Beiheft zur Nova Hedwigia* 87: 1-288.

Baral, H.O., 1992, Vital versus herbarium taxonomy: morphological differences between living and dead cells of ascomycetes, and their taxonomic implications, *Mycotaxon* 44: 333-390.

Beckett, A., 1981, Ascospore formation. *In*: *The Fungal Spore: Morphogenetic Controls* (G. Turian and H.R. Hohl, eds): 107-129. Academic Press, London.

Bellemère, A., M.C. Malherbe, H. Chacun, and L.M. Meléndez-Howell, 1990, L'étude ultrastructurale des asques et des ascospores de l'*Urnula helvelloides* Donadini, Berthet et Astier et les concepts d'asque suboperculé et de *Sarcosomataceae*, *Cryptogamie, Mycologie* 11: 203-238.

Benkert, D., 1987, Beiträge zur Taxonomie der Gattung *Lamprospora* (*Pezizales*), *Zeitschrift für Mykologie* 53: 195-271.

Benny, G.L. and J.W. Kimbrough, 1980, A synopsis of the orders and families of *Plectomycetes* with keys to genera, *Mycotaxon* 12: 1-91.

Berkeley, M.J. and C.E. Broome, 1865, Notices of British fungi, *Annals and Magazine of Natural History, ser.* III, 15: 444-452.

Berthet, P., 1961, Variation du nombre des noyaux dans les articles mycéliens des Discomycètes, en rapport avec la systématique et la phylogénie, *Compte Rendu Hebdomadaire des Séances de l'Académie des Sciences, Paris* 252: 3855-3857.

Berthet, P., 1963, Le nombre des noyaux dans la spore et son intérêt pour la systématique des Discomycètes operculés, *Compte Rendu Hebdomadaire des Séances de l'Academie des Sciences, Paris* 256: 5185-5186.

Berthet, P., 1964, *Essai biotaxinomique sur les Discomycètes*, PhD thesis, University of Lyon.

Bezerra, J.L. and J.W. Kimbrough, 1975, The genus *Lasiobolus* (*Pezizales*, *Ascomycetes*), *Canadian Journal of Botany* 53: 1206-1229.

Boudier, J.L.É., 1869, Mémoire sur les *Ascobolés*, *Annales des Sciences Naturelles, Botanique, sér.* V, 10: 191-268.

Boudier, J.L.É., 1879, On the importance that should be attached to the dehiscence of asci in the classification of the *Discomycetes*, *Grevillea* 8: 45-49.

Boudier, J.L.É., 1885, Nouvelle classification naturelle des Discomycètes charnus connus généralement sous le nom de Pezizes, *Bulletin de la Société Mycologique de France* 1: 91-120.

Boudier, J.L.É., 1907, *Histoire et classification des Discomycètes d'Europe*, P. Klincksieck, Paris.

Bowman, B.H., J.W. Taylor, A.G. Brownlee, J. Lee, S.-D. Lu, and T.J. White, 1992, Molecular evolution of the higher fungi: Relationship of basidiomycetes, ascomycetes and chytridiomycetes, *Molecular Biology and Evolution* 9: 285-296.

Brummelen, J. van, 1974, Light and electron microscopic studies of the ascus top in *Ascozonus woolhopensis*, *Persoonia* 8: 23-32.

Brummelen, J. van, 1975, Light and electron microscopic studies of the ascus top in *Sarcoscypha coccinea*, *Persoonia* 8: 259-271.

Brummelen, J. van, 1977, The operculate ascus and allied forms, *In*: *Second International Mycological Congress, Tampa, Florida*: 2. International Mycological Congress, Tampa, Florida.

Brummelen, J. van, 1978, The operculate ascus and allied forms, *Persoonia* 10: 113-128.

Brummelen, J. van, 1981a, The genus *Ascodesmis* (*Pezizales*, *Ascomycetes*), *Persoonia* 11: 333-358.

Brummelen, J. van, 1981b, The operculate ascus and allied forms. *In*: *Ascomycete Systematics. The Luttrellean Concept* (D, R. Reynolds, ed.): 27-48. Springer-Verlag, New York, Heidelberg, Berlin.

Brummelen, J. van, 1986, Ultrastructure of the ascus top and the ascospore wall in *Fimaria* and *Pseudombrophila* (*Pezizales*, *Ascomycotina*), *Persoonia* 13: 213-230.

Brummelen, J. van, 1987, Ultrastructure of the ascus and the ascospores in *Pseudascozonus* (*Pezizales*, *Ascomycotina*), *Persoonia* 13: 369-377.

Brummelen, J. van, 1989a, Ultrastructure of the ascospore wall in *Eleutherascus* and *Ascodesmis* (*Ascomycotina*), *Persoonia* 14: 1-17.

Brummelen, J. van, 1989b, Ultrastructural comparison of different types of ascospore ornamentation in *Eleutherascus tuberculatus* (*Pezizales*, *Ascomycotina*), *Studies in Mycology, Baarn* 31: 41-48.

Brummelen, J. van, 1993, Ultrastructure of the ascus and the ascospore wall in *Scutellinia* (*Pezizales*, *Ascomycotina*), *Persoonia* 15: 129-148.

Caillet, M. and G. Moyne, 1980, Contribution à l'étude du genre *Octospora* Hedw. ex S.F. Gray emend. Le Gal. Espèces à spores ornementées, globuleuses ou subglobuleuses, *Bulletin trimestriel de la Société Mycologique de France* 96: 175-211.

Caillet, M. and G. Moyne, 1987a, Contribution à l'étude du genre *Octospora* Hedw. ex S.F. Gray (*Pezizales*), *Bulletin trimestriel de la Société Mycologique de France* 103: 179-226.

Caillet, M. and G. Moyne, 1987b, Contribution à l'étude du genre *Octospora* (Hedw. ex S.F. Gray) (*Pezizales*). Écologie et morphologie, *Bulletin trimestriel de la Société Mycologique de France* 103: 277-304.

Cain, R.F., 1959, The *Plectascales* and *Perisporiales* in relation to the evolution of the *Ascomycetes*. *In*: *Ninth International Botanical Congress, Abstracts* 2: 56.

Carroll, G. C., 1967, The fine structure of ascus septum in *Ascodesmis sphaerospora* and *Saccobolus kerverni*, *Mycologia* 59: 527-532.

Chadefaud, M., 1942, Études d'asques, II: Structure et anatomie comparée de l'appareil apical des asques chez divers Discomycètes et Pyrénomycètes, *Revue de Mycologie* 7: 57-88.

Chadefaud, M., 1946, Les asques para-operculés et la position systématique de la Pézize *Sarcoscypha coccinea* Fries ex Jaquin, *Compte Rendu Hebdomadaire des Séances de l'Academie des Sciences, Paris* 222: 753-755.

Coemans, E., 1862, Spicilège mycologique. I.- Notice sur les *Ascobolus* de la flore belge, *Bulletin de la Société Royale de Botanique de Belgique* 1: 76-91.

Cooke, M.C., 1864, The genus *Ascobolus*, with descriptions of the British species, *Journal of Botany, London* 2: 147-154.

Cooke, M.C., 1875-1879, *Mycographia seu icones fungorum*. Vol. I. *Discomycetes*, Williams and Norgate, London.

Corner, E.J.H., 1929a, Studies in the morphology of *Discomycetes*. I. The marginal growth of apothecia, *Transactions of the British Mycological Society* 14: 263-274.

Corner, E.J.H., 1929b, Studies in the morphology of *Discomycetes*. II. The structure and development of the ascocarp, *Transactions of the British Mycological Society* 14: 275-291.

Corner, E.J.H., 1930, Studies in the morphology of *Discomycetes*. IV. The evolution of the ascocarp, *Transactions of the British Mycological Society* 15: 121-134.

Corner, E.J.H., 1931, Studies in the morphology of Discomycetes. V. The evolution of the ascocarp (continued), *Transactions of the British Mycological Society* 15: 332-350.

Crouan, P.L. and H.M. Crouan, 1857, Note sur quelques *Ascobolus* nouveaux et sur une espèce nouvelle de *Vibrissea*, *Annales des Sciences Naturelles, Botanique, sér.* IV, 7: 173-178.

Curry, K.J. and J.W. Kimbrough, 1983, Septal structures in apothecial tissues of the *Pezizaceae* (*Pezizales*, *Ascomycetes*), *Mycologia* 75: 781-794.

Czymmek, K.J. and K.L. Klomparens, 1992, The ultrastructure of ascosporogenesis in freeze-substituted *Thelebolus crustaceus*: enveloping membrane system and ascospore initial development, *Canadian Journal of Botany* 70: 1669-1683.

De Notaris, G., 1864, Proposte di alcune rettificazione al profilo dei Discomyceti, *Commentario Società Crittogamologica Italiana* 1: 357-388.
Dennis, R.W.G., 1960, *British Cup Fungi and their Allies*, Ray Society, London.
Dennis, R.W.G., 1968, *British Ascomycetes*, Cramer, Lehre.
Dennis, R.W.G., 1978, *British Ascomycetes.* Revised and enlarged edn, Cramer, Vaduz.
Dennis, R.W.G. and H. Itzerott, 1973, *Octospora* and *Inermisia* in Western Europe, *Kew Bulletin* 28: 5-23.
Dissing, H. and T. Schumacher, 1979, Preliminary studies in the genus *Boudiera*, taxonomy and ecology, *Norwegian Journal of Botany* 26: 99-109.
Donadini, J.-C., 1976, Le genre *Pulparia* Karsten en France, *Revue de Mycologie* 40: 255-272.
Donadini, J.-C., 1986, Les Balsamiacées sont des Helvellacées: Cytologie et scanning de *Balsamia vulgaris* Vitt. et de *Balsamia platyspora* Berk. et Br., *Bulletin trimestriel de la Sociéte Mycologique de France* 102: 373-387.
Donadini, J.-C., 1987, Étude des *Sarcoscyphaceae* ss. Le Gal (1) *Sarcosomataceae* et *Sarcoscyphaceae* ss. Korf. Le genre *Pseudoplectania* emend. nov., *P. ericae* sp. nov. (*Pezizales*), *Mycologia Helvetica* 2: 217-246.
Duby, S.D. and J.W. Kimbrough, 1987, A comparitive ultrastructural study of ascospore ontogeny in selected species of *Peziza* (*Pezizales*, *Ascomycetes*), *Botanical Gazette* 148: 284-296.
Eckblad, F.-E., 1968, The genera of the operculate *Discomycetes.* A re-evaluation of their taxonomy, phylogeny and nomenclature, *Norwegian Journal of Botany* 15: 1-191.
Egger, K.N., 1992, Systematics of *Wilcoxina* and *Tricharina* (*Pezizales*) inferred from ribosomal DNA sequences, *Newsletter, Mycological Society of America* 43: 30.
Eriksson, O.[E.], 1982, Outline of the ascomycetes - 1982, *Mycotaxon* 15: 203-248.
Eriksson, O.E. and D.L. Hawksworth, 1991, Outline of the ascomycetes - 1990, *Systema Ascomycetum* 9: 39-271.
Fennell, D.I., 1973, *Plectomycetes, Eurotiales. In*: *The Fungi. An Advanced Treatise* (G.C. Ainsworth, F.K. Sparrow, and A.S. Sussman, eds) 4A: 45-68. Academic Press, New York.
Fries, E.M., 1822, *Systema Mycologicum* 2: 1-274, Berling, Lund.
Fries, E.M., 1849, *Summa vegetabilium Scandinaviae. Sectio posterior*, Typographia Academica, Uppsala.
Fuckel, K.W.G.L., 1866, Ueber rheinische Ascobolus-Arten, *Hedwigia* 5: 1-5.
Fuckel, K.W.G.L., 1870, Symbolae mycologicae. Beiträge zur Kenntniss der rheinischen Pilze, *Jahrbuch des Nassauischen Vereins für Naturkunde* 23-24: 1-459.
Gäumann, E., 1926, *Vergleichende Morphologie der Pilze*, G. Fischer, Jena.
Gäumann, E., 1949, *Die Pilze, Grundzüge ihrer Entwicklungsgeschichte und Morphologie*, Birkhäuser, Basel.
Gäumann, E., 1964, *Die Pilze, Grundzüge ihrer Entwicklungsgeschichte und Morphologie*, 2nd edn. Birkhäuser, Basel.
Gibson, J.L. and J.W. Kimbrough, 1988a, Ultrastructural observations on *Helvellaceae* (*Pezizales*). Ascosporogenesis of selected species of *Helvella, Canadian Journal of Botany* 66: 771-783.
Gibson, J.L. and J.W. Kimbrough, 1988b, Ultrastructural observations on *Helvellaceae* (*Pezizales*). II. Ascosporogenesis of *Gyromitra esculenta, Canadian Journal of Botany* 66: 1743-1749.
Guarro, J. and J.A. Arx, 1986, *Monascella*, a new genus of *Ascomycotina*, *Mycologia* 78: 869-871.
Harrington, F.A., 1990, *Sarcoscypha* in North America (*Pezizales*, *Sarcoscyphaceae*), *Mycotaxon* 38: 417-458.
Hohmeyer, H.H., 1988, The genus *Arpinia* (*Pyronemataceae*, *Pezizales*), *Mycologia Helvetica* 3: 221-232.
Hung, C.-Y., 1977, Ultrastructural studies of ascospore liberation in *Pyronema domesticum*, *Canadian Journal of Botany* 55: 2544-2549.
Janex-Favre, M.C. and A. Parguey-Leduc, 1980, Formation et évolution des ascospores du *Tuber mesentericum* Vitt, *Bulletin trimestriel de la Sociéte Mycologique de France* 96: 225-237.
Janex-Favre, M.C. and A. Parguey-Leduc, 1983, Étude ultrastructurale des asques et des ascospores de Truffes du genre *Tuber*. II. - Les ascospores, *Cryptogamie, Mycologie* 4: 353-373.
Janex-Favre, M.C. and A. Parguey-Leduc, 1985, Les asques et les ascospores du *Terfezia claveryi* Ch. (Tubérales), *Cryptogamie, Mycologie* 6: 87-99.
Janex-Favre, M.C. and A. Parguey-Leduc, 1988, Les asques des *Tuber* (Discomycètes, Tubérales): particularités morphologiques et structurales, *Atti II Congresso Internationale Tartufo* 1988: 111-120.
Karsten, P.A., 1861, *Synopsis Pezizarum et Ascobolorum Fenniae. Öfversigt af i Finland funna arter af svampslägtena Peziza och Ascobolus. I-III*, J.C. Frenckell & Son, Helsingfors.
Karsten, P.A., 1869, Monographia Pezizarum fennicarum, *Notiser ur Sällskapets pro Fauna et Flora Fennica Förhandlingar* 10: 99-206.

Karsten, P.A., 1870, Monographia Ascobolorum Fenniae, *Notiser ur Sällskapets pro Fauna et Flora Fennica Förhandlingar* 11: 197-210.
Karsten, P.A., 1871, Mycologia fennica. Pars prima: Discomycetes, *Bidrag till Kännedom af Finlands Natur och Folk* 19: 1-264.
Karsten, P.A., 1885, Revisio monographica atque synopsis Ascomycetum in Fennia hucusque detectorum, *Acta Societatis pro Fauna et Flora Fennica, ser.* II, 6: 1-174.
Khare, K.B. and V.P. Tewari, 1978, Taxonomy and relationship within the genus *Octospora*, *Canadian Journal of Botany* 56: 2114-2118.
Kimbrough, J.W., 1989, Arguments towards restricting the limits of the *Pyronemataceae* (*Ascomycetes, Pezizales*), *Memoirs of the New York Botanic Garden* 49: 326-335.
Kimbrough, J.W., 1991, Ultrastructural observations on *Helvellaceae* (*Pezizales, Ascomycetes*). V. Septal structures in *Gyromitra*, *Mycological Research* 95: 421-426.
Kimbrough, J.W. and G.L. Benny, 1978, The fine structure of ascus development in *Lasiobolus monascus* (*Pezizales*), *Canadian Journal of Botany* 56: 862-872.
Kimbrough, J.W. and K.J. Curry, 1986a, Septal structures in apothecial tissues of the tribe *Aleurieae* in the *Pyronemataceae* (*Pezizales, Ascomycetes*), *Mycologia* 78: 407-417.
Kimbrough, J.W. and K.J. Curry, 1986b, Septal structure in apothecial tissues of taxa in the tribes *Scutellinieae* and *Sowerbyelleae* (*Pyronemataceae, Pezizales, Ascomycetes*), *Mycologia* 78: 735-743.
Kimbrough, J.W. and J.L. Gibson, 1989, Ultrastructural observations on *Helvellaceae* (*Pezizales; Ascomycetes*). III. Septal structures in *Helvella*, *Mycologia* 81: 914-920.
Kimbrough, J.W. and J.L. Gibson, 1990, Ultrastructural and cytological observations of apothecial tissues of *Geopyxis carbonaria* (*Pezizales, Ascomycetes*), *Canadian Journal of Botany* 68: 243-257.
Kimbrough, J.W., C.-G. Wu, and J.L. Gibson, 1990, Ultrastructural observations on *Helvellaceae* (*Pezizales, Ascomycetes*). IV. Ascospore ontogeny in selected species of *Gyromitra* subgenus *Discina*, *Canadian Journal of Botany* 68: 317-328.
Kimbrough, J.W., C.-G. Wu, and J.L. Gibson, 1991, Ultrastructural evidence for a phylogenetic linkage of the truffle genus *Hydnobolites* to the *Pezizaceae* (*Pezizales, Ascomycetes*), *Botanical Gazette* 152: 408-420.
Korf, R.P., 1954, A revision of the classification of operculate *Discomycetes* (*Pezizales*), *Rapport du Committee de Sectione 18-20, VIIIe Congress Internationale de Botanique, Paris*: 80.
Korf, R.P., 1970, Nomenclatural notes. VII. Family and tribe names in the *Sarcoscyphineae* (Discomycetes) and a new taxonomic disposition of the genera, *Taxon* 19: 782-786.
Korf, R.P., 1972, Synoptic key to the genera of the *Pezizales*, *Mycologia* 64: 937-994.
Korf, R.P., 1973, Discomycetes and *Tuberales*. *In*: *The Fungi. An Advanced Treatise* (G.C. Ainsworth, F.K. Sparrow, and A.S. Sussman, eds): 249-319. Academic Press, New York, London.
Korf, R.P. and W.-Y. Zhuang, 1991a, A preliminary discomycete flora of *Macaronesia*: Part 11, *Sarcoscyphineae*, *Mycotaxon* 40: 1-11.
Korf, R.P. and W.-Y. Zhuang, 1991b, A preliminary discomycete flora of Macaronesia: Part 12, *Pyronematineae* and *Pezizineae*, *Ascobolaceae*, *Mycotaxon* 40: 307-308.
Korf, R.P. and W.-Y. Zhuang, 1991c, A preliminary discomycete flora of Macaronesia: Part 13, *Morchellaceae*, *Helvellaceae*, *Mycotaxon* 40: 287-294.
Korf, R.P. and W.-Y. Zhuang, 1991d, A preliminary discomycete flora of Macaronesia: Part 14, *Pezizaceae*, *Mycotaxon* 40: 395-411.
Korf, R.P. and W.-Y. Zhuang, 1991e, A preliminary discomycete flora of Macaronesia: Part 15, *Terfeziaceae*, and *Otideaceae*, *Otideoideae*, *Mycotaxon* 40: 413-433.
Korf, R.P. and W.-Y. Zhuang, 1991f, A preliminary discomycete flora of Macaronesia: Part 16, *Otideaceae*, *Scutellinioideae*, *Mycotaxon* 40: 79-106.
Kullman, B.B., 1982, *A revision of the genus Scutellinia (Pezizales) in the Soviet Union.* [in Russian], Nauka, Tallin.
Le Gal, M., 1946, Les Discomycètes suboperculés, *Bulletin trimestriel de la Sociéte Mycologique de France* 62: 218-240.
Le Gal, M., 1947, Recherches sur les ornementations sporales des Discomycètes operculés, *Annales des Sciences Naturelles, Botanique, sér.* XI, 8: 73-297.
Le Gal, M., 1953, *Les Discomycètes de Madagascar*, Laboratoire de Cryptogamie, Paris.
Le Gal, M., 1959, Discomycètes du Congo belge d'après récoltes de Madame Goossens-Fontana, *Bulletin du Jardin Botanique de l'État à Bruxelles* 29: 73-132.
Le Gal, M., 1963, Valeur taxinomique particulière de certains caractères chez les Discomycètes supérieurs, *Bulletin trimestriel de la Sociéte Mycologique de France* 79: 456-470.
Malloch, D., 1979, Plectomycetes and their anamorphs. *In*: *The Whole Fungus* ([W.] B. Kendrick, ed.)1: 153-165. National Museum of Natural Sciences, Ottawa.

Malloch, D., 1981, The plectomycete centrum. *In*: *Ascomycete Systematics: The Luttrellian Concept* (D.R. Reynolds, ed.): 73-91. Springer Verlag, New York.
Merkus, E., 1973, Ultrastructure of the ascospore wall in *Pezizales* (*Ascomycetes*) I. *Ascodesmis microscopica* and *A. nigricans*, *Persoonia* 7: 351-366.
Merkus, E., 1974, Ultrastructure of the ascospore wall in *Pezizales* (*Ascomycetes*) II. *Pyronemataceae* sensu Eckblad, *Persoonia* 8: 1-22.
Merkus, E., 1975, Ultrastructure of the ascospore wall in *Pezizales* (*Ascomycetes*) III. *Otideaceae* and *Pezizaceae*, *Persoonia* 8: 227-247.
Merkus, E., 1976, Ultrastructure of the ascospore wall in *Pezizales* (*Ascomycetes*) IV. *Morchellaceae*, *Helvellaceae*, and *Sarcoscyhaceae*. General discussion, *Persoonia* 9: 1-38.
Moravec, J., 1985, A taxonomic revision of the genus *Sowerbyella* Nannfeldt (*Discomycetes*, *Pezizales*), *Mycotaxon* 23: 483-496.
Moravec, J., 1987, A taxonomic revision of the genus *Marcelleina*, *Mycotaxon* 30: 473-499.
Moravec, J., 1988, *Cheilymenia fraudans* and remarks on the genera *Cheilymenia* and *Coprobia*, *Mycotaxon* 31: 483-489.
Moravec, J., 1989a, *Cheilymenia megaspora* comb. nov. A new combination in the genus *Cheilymenia* (*Discomycetes*, *Pezizales*, *Pyronemataceae*), *Mycotaxon* 35: 65-69.
Moravec, J., 1989b, A taxonomic revision of the genus *Cheilymenia* - I. Species close to *Cheilymenia rubra*, *Mycotaxon* 36: 169-186.
Moravec, J., 1990, Taxonomic revision of the genus *Cheilymenia* in a new emendation, *Mycotaxon* 38: 459-484.
Nannfeldt, J.A., 1932, Studien über die Morphologie und Systematik der nicht-lichenisierten inoperculaten Discomyceten, *Nova Acta Regiae Societatis Scientarum Upsaliensis, ser.* IV, 8: 1-368.
Nannfeldt, J.A., 1937, Contributions to the mycoflora of Sweden. 4. On some species of *Helvella*, together with a discussion of the natural affinities within *Helvellaceae* and *Pezizaceae* trib. *Acetabuleae*, *Svensk Botanisk Tidskrift* 31: 47-66.
Nannfeldt, J.A., 1949, Contributions to the mycoflora of Sweden. 7. A new winter discomycete, *Urnula hiemalis* Nannf. n. sp., and a short account of the Swedish species of *Sarcoscyphaceae*, *Svensk Botanisk Tidskrift* 43: 468-484.
Nylander, W., 1868, Observationes circa Pezizas Fenniae, *Notiser ur Sällskapets pro Fauna et Flora Fennica Förhandlingar* 10: 1-97.
Parguey-Leduc, A. and M.C. Janex-Favre, 1977, L'organisation des asques de deux truffes: *Tuber rufum* Pico et *Tuber aestivum* Vitt., *Revue de Mycologie* 41: 1-32.
Parguey-Leduc, A. and M.C. Janex-Favre, 1981, Étude ultrastructurale des asques et des ascospores de truffes du genre *Tuber*. I. Les asques, *Cryptogamie, Mycologie* 2: 37-53.
Parguey-Leduc, A. and M.C. Janex-Favre, 1987, Formation et évolution des ascospores de *Tuber melanosporum* (truffe noire du Périgord, Discomycètes), *Canadian Journal of Botany* 65: 1491-1503.
Parguey-Leduc, A., M.C. Janex-Favre, and C. Montant, 1988, L'ascocarpe de *Tuber melanosporum* Vitt. (Truffe noire du Périgord, Discomycètes): Structure de la glèbe. I. Les veines fertiles, *Atti II Congresso Internationali Tartufo* 1988: 101-109.
Parguey-Leduc, A., M.C. Janex-Favre, C. Montant, and M. Kulifaj, 1989, Ontogénie et structure de l'ascocarpe du *Tuber melanosporum* Vitt. (Truffe noire du Périgord, Discomycètes), *Bulletin trimestriel de la Sociéte Mycologique de France* 105: 227-246.
Parguey-Leduc, A., M.C. Janex-Favre, and C. Montant, 1990, L'appareil sporophytique et les asques du *Tuber melanosporum* Vitt. (Truffe noire du Périgord, Discomycètes), *Cryptogamie, Mycologie* 11: 47-68.
Parguey-Leduc, A., M.C. Janex-Favre, and C. Montant, 1991, L'ascocarpe du *Tuber melanosporum* Vitt. (Truffe noire du Périgord, Discomycètes): structure de la glèbe. II. Les veines stériles, *Cryptogamie, Mycologie* 12: 165-182.
Parguey-Leduc, A., C. Montant, and M. Kulifaj, 1987, Morphologie et structure de l'ascocarpe adulte du *Tuber melanosporum* Vitt. (Truffe noire du Périgord, Discomycètes), *Cryptogamie, Mycologie* 8: 173-202.
Pfister, D.H., 1973, The psilopezioid fungi. IV. The genus *Pachyella* (*Pezizales*), *Canadian Journal of Botany* 51: 2009-20032.
Pfister, D.H., 1979, A monograph of the genus *Wynnea* (*Pezizales*, *Sarcoscyphaceae*), *Mycologia* 71: 144-159.
Pfister, D.H., 1984, *Genea-Jafneadelphus* - a tuberalean-pezizalean connection, *Mycologia* 76: 170-172.
Reeves Jr., F., 1967, The fine structure of ascospore formation in *Pyronema domesticum*, *Mycologia* 59: 1018-1033.

Rehm, H., 1887-1896, Ascomyceten: Hysteriaceen und Discomyceten, *Rabenhorst's Kryptogamen-Flora van Deutschlands, Östereich und der Schweiz* 1: 1-1275.
Rifai, M.A., 1968, The Australasian *Pezizales* in the herbarium of the Royal Botanic Garden Kew, *Verhandelingen der Koninklijke Nederlandsche Akademie van Wetenshappen, Natuurkunde* 57 (3): 1-295.
Saccardo, P.A., 1884, Conspectus generum Discomycetum hucusque cognitorum, *Botanische Zentralblatt* 18: 213-220, 247-256.
Saccardo, P.A., 1889, *Sylloge fungorum omnium hucusque cognitarum.* Vol. 8., Patavii.
Samuelson, D.A., 1975, The apical apparatus of the suboperculate ascus, *Canadian Journal of Botany* 53: 2660-2679.
Samuelson, D.A., 1978a, Asci of the *Pezizales*. I. The apical apparatus of iodine-positive species, *Canadian Journal of Botany* 56: 1860-1875.
Samuelson, D.A., 1978b, Asci of the *Pezizales*. II. The apical apparatus of representatives in the *Otidea-Aleuria* complex, *Canadian Journal of Botany* 56: 1876-1904.
Samuelson, D.A., 1978c, Asci of the *Pezizales*. III. The apical apparatus of eugymnohymenial representatives, *American Journal of Botany* 65: 748-758.
Samuelson, D.A., 1978d, Asci of the *Pezizales*. VI. The apical apparatus of *Morchella esculenta*, *Helvella crispa*, and *Rhizina undulata*. General discussion, *Canadian Journal of Botany* 56: 3069-3082.
Samuelson, D.A., G.L. Benny, and J.W. Kimbrough, 1980, Asci of the *Pezizales*. VII. The apical apparatus of *Galiella rufa* and *Sarcosoma globosum*: reevaluation of the suboperculate ascus, *Canadian Journal of Botany* 58: 1235-1243.
Samuelson, D.A. and J.W. Kimbrough, 1978, Asci of the *Pezizales* V: The apical apparatus of *Trichobolus zukalii*, *Mycologia* 70: 1191-1200.
Schumacher, T., 1990, The genus *Scutellinia* (*Pyronemataceae*), *Opera Botanica* 101: 1-107.
Seaver, F.J., 1928, *The North American Cup-fungi (Operculates)*, F.J. Seaver, New York.
Starbäck, K., 1895, Discomyceten-Studien, *Bihang till Konglikje Svenska Vetenskapsakademiens Handlingar, ser.* III, 21 (5): 1-42.
Trappe, J.M., 1979, The orders, families, and genera of hypogeous *Ascomycotina* (truffles and their relatives), *Mycotaxon* 9: 297-340.
Vilgalys, R. and M. Hester, 1993, Rapid genetic identification and mapping of enzymatically amplified ribosomal DNA from several *Cryptococcus* species, *Journal of Bacteriology* 172: 4238-4246.
Wang, Y.-Z. and J.W. Kimbrough, 1992, Monographic studies of North American species of *Octospora* previously ascribed to *Lamprospora* (*Pezizales*, *Ascomycetes*), *Special Publications of the National Museum of Natural History, Taiwan* 4: i-vii, 1-68.
Weber, E., 1992, Untersuchungen zu Fortpflanzung und Ploidie verschiedener Ascomyceten, *Bibliotheca Mycologica* 140: 1-186.
Wingfield, B.D. and M.J. Wingfield, 1993, The value of dried fungal cultures for taxonomic comparison using PCR and RFLP analysis, *Mycotaxon* 46: 429-436.
Wu, C.-G. and J.W. Kimbrough, 1991, Ultrastructural investigation of *Humariaceae* (*Pezizales*, *Ascomycetes*). II. Ascosporogenesis in selected genera of the *Ciliarieae*, *Botanical Gazette* 152: 421-438.
Wu, C.-G. and J.W. Kimbrough, 1992a, Ultrastructural studies of ascosporogenesis in *Ascobolus immersus*, *Mycologia* 84: 459-466.
Wu, C.-G. and J.W. Kimbrough, 1992b, Ultrastructural investigation of *Humariaceae* (*Pezizales*, *Ascomycetes*). III. Ascosporogenesis of *Mycolachnea hemisphaerica* (tribe *Lachneae*), *International Journal of Plant Science* 153: 128-135.
Wu, C.-G. and J.W. Kimbrough, 1993, Ultrastructure of ascospore ontogeny in *Aleuria*, *Octospora*, and *Pulvinula* (*Otidiaceae*, *Pezizales*), *International Journal of Plant Science* 154: 334-349.
Yang, C.S. and R.P. Korf, 1985, A monograph of the genus *Tricharina* and of a new segregate genus, *Wilcoxina* (*Pezizales*), *Mycotaxon* 24: 467-531.
Zhang, B.-C., 1992a, Ascospore nuclear number and taxonomy of truffles, *Micologia Vegetale Mediterranea* 7: 47-53.
Zhang, B.-C., 1992b, Nuclear numbers in *Geneaceae* and *Terfeziaceae* ascospores and their taxonomic value, *Systema Ascomycetum* 11: 31-37.
Zhuang, W.-Y., 1986, A monograph of the genus *Aleurina* Massee (= *Jafneadelphus* Rifai), *Mycotaxon* 26: 361-400.

PROBLEMS IN *LECANORALES* SYSTEMATICS

J. Hafellner

Institut für Botanik, Karl-Franzens-Universität
Holteigasse 6, A-8010 Graz, Austria

SUMMARY

A further attempt to circumscribe the order *Lecanorales* is made. So far the only distinguishing characters for lecanoralean fungi are connected with the ascus-type, the peculiarities of which are pointed out. Some other features which might influence details in ascus structure are discussed as well as some problems of character selection and their weighting, depending on the group studied. It is recommended to reinclude *Peltigerales*, *Pertusariales* and *Teloschistales* as suborders as was originally proposed.

INTRODUCTION

Even with a narrow concept which excludes *Peltigerales*, *Pertusariales* and *Teloschistales*, the order *Lecanorales* is one of the largest among the *Ascomycotina*. The latest edition of the *Dictionary of the Fungi* (Hawksworth et al., 1983) mentions a figure of about 5650 species. If the numbers of the three other mentioned orders are included, taken out from the same source, the total number of species would be more than 7500, about half the number of all known species of lichenized fungi. So, this is a really large order, having about the same size as the *Dothideales* in which the bulk of fissitunicate ascomycetes is classified. Thirty-nine families have been accepted in the most recent outline of *Lecanorales* (Eriksson and Hawksworth, 1991), and the number recognized is still rapidly increasing.

A DEFINITION OF LECANORALEAN FUNGI

For several reasons (*viz.* very diverse thallus organization, diverse secondary chemistry, many structural peculiarities, high numbers of species, and several, often parallel, evolutionary lines) the first important problem is encountered when an attempt is made to give a clear circumscription of *Lecanorales*. Reading earlier diagnoses of the order (e.g. Henssen and Jahns 1973; Poelt 1974; Hawksworth et al., 1983; Hafellner, 1988), and after screening hundreds of representative species, the following characters must be regarded as the most significant:

- • ascomycetes living together with algae (e.g. *Graphidales*).
- • ascomycetes producing apothecioid ascomata (e.g. *Leotiales*).
- • development of ascomata ascohymenial (e.g. *Leotiales*).
- • ascomata long-lived (e.g. *Graphidales*).
- ♦ ascus lecanoralean and dehiscence of either the rostrum (incl. variants) or the chimney-type.
- ♦ DNA-data - (still lacking?)
- • not a distinguishing character (i.e. this character is known in other orders such as those in parenthesis); ♦ distinguishing character.

Ascomycete Systematics: Problems and Perspectives in the Nineties
Edited by D.L. Hawksworth, Plenum Press, New York, 1994

There is thus only one anatomical, directly observable distinguishing group of characters, those of the ascus. Which characters are suitable for a definition of the lecanoralean ascus (the "archaeascé-type" sensu Chadefaud)?

For many lecanoralean mycobionts, the asci have already been investigated and the interpretations of their fine structural peculiarities have either been published as line drawings (e.g. Chadefaud, 1969, 1973; Chadefaud et al., 1963, 1967; Hafellner, 1984; Letrouit-Galinou, 1973) or as drawings accompanied by photographs (e.g. Bellemère, this volume; Bellemère and Hafellner, 1983; Bellemère and Letrouit-Galinou, 1981, 1987; Hafellner and Bellemère, 1982a, 1982b, 1982c; Honegger, 1978, 1980, 1982b). It should be stressed that the ascal wall is not just a single character but provides many characters. Several of them must be regarded as independent and not necessarily linked one to each other. The main elements which build up the upper part of the ascal wall of lecanoralean fungi have been summarized in a drawing by Hafellner (1984: 255). It is obvious that most of the characters are not suitable for an ordinal circumscription as they are present only in a more or less limited number of species within the order. Looking at numerous species of *Lecanorales* it becomes evident that the following features must be regarded as distinguishing lecanoralean asci from other ascus types (terminology according to Bellemère, 1971, this volume):

(1) Outermost wall layer (layer "*a*") amyloid or hemi-amyloid (Baral, 1987), secreting a layer of ascal gel which either completely or at least partly surrounds the ascal wall. This wall layer is very thin and may by discernible only in TEM. The presence of the ascal gel is easily demonstrated by staining with Lugol's solution. External gelatinous caps may play an important role in dehiscence.

(2) Layer "*c*" is more or less of equal thickness and generally non-amyloid. In TEM it may be possible to demonstrate that layer "*c*" itself is stratified, but this is usually invisible with the light microscope. The "*c*" layer determines the external shape of the ascus and provides a certain rigidity. It is rarely involved in dehiscence.

(3) Layer "*d*" is commonly of unequal thickness, laterally thin or even absent, apically mostly enlarged and forming the apical dome or tholus. In fresh living material, it may also be rather thin. Layer "*d*" is often stainable with Lugol's solution. Staining may be of different quality (amyloid or more rarely hemi-amyloid) and intense in either the whole apical dome or parts of it. It is in layer "*d*" where further structural features, such as the axial body, tube structures, or internal caps may occur. Layer "*d*" is essential in the dehiscence process as it carries the ascospores up to the hymenial surface.

The modes of dehiscence of the different main types of lecanoralean asci were drawn schematically by Honegger (1982a), Hafellner (1984), and Bellemère and Letrouit-Galinou (1987). Two types are extremely common and can be found in very different groups: the rostrate-type and the chimney- type. The main difference is that in the first case layer "*c*" does not contribute to a tube, which is built only by layer "*d*" and which brings the ascospores up to the hymenial surface. In the second case, all wall layers take part in forming such a pipe. The partly fissitunicate asci as observed in *Rhizocarpon* (Honegger, 1980) are explainable also as dehiscence variants of the rostrate-type, and this may be possible even for the *Peltigera*-type (Magne, 1946; Honegger, 1978; Bellemère and Letrouit-Galinou, 1981, 1987) if it can be proved that the protruding part, which functions as an endoascus, is only (part of) the expanded layer "*d*". Some drawings (Chadefaud et al., 1967: 94; Honegger, 1978: 61) point in this direction, and, in consequence, this would mean that the fissitunicate ascus of a dothidealean ascomycete and that of a lecanoralean ascomycete are functional convergences but structurally not the same.

EVOLUTION AND CROSSWISE INFLUENCE OF CHARACTERS IN *LECANORALES*

This is still poorly understood and only a few cases have yet been studied. Some aspects will be discussed briefly.

As far as the ascus is concerned, structure and function may be influenced either by the surrounding tissues (shape of ascoma, presence of an epithecium, density and viscosity of the hymenial gel), or by characters of the enclosed ascospores (shape, size, polyspory, formation of ascal conidia), or by autecology. The case study of *Thelocarpon* (incl. *Ahlesia*) revealed a rather dramatic correlation of ascoma shape with the shape and structural details of the ascus wall as well as with the paraphyses (Poelt and Hafellner, 1975), which is best explained as a series of reductions (Poelt, 1987). Whether such reductions can also be observed in *Pertusaria* is not yet known. In very preliminary investigations comparing *P. bryontha* with apothecioid ascomata and *P. pertusa* with perithecioid ascomata, surprisingly, no significant differences could be observed (Hafellner, unpubl.), but this needs further studies.

Recalling hymenial characters of members of *Lecanorineae*, that is to say *Lecanorales* having asci with an apical dome provided with an axial body, it is obvious that one-celled ascospores are often correlated with *Lecanora*-type asci, while species producing septate and often more elongate ascospores mostly have *Bacidia*-type asci. One might think that the shape of the axial body has been influenced by the shape of the ascospores. However, there are some genera which do not show this correlation. In *Bacidina phacodes* and *Scoliciosporum umbrinum*, the ascospores of which are very narrow and needle-shaped, the asci have very broad axial bodies, whereas the asci of *Tephromela atra*, with broadly ellipsoid ascospores, show conical axial bodies that vary very little (*cf.* Hertel and Rambold, 1985). The shape of the axial body must therefore be an independent character useful in the classification of the *Lecanorineae*.

If we consider *Lecanorales* producing muriform ascospores, especially those with one single, large ascospore per ascus, it cannot be overlooked that these fungi, although some of them are certainly not closely related (e.g. *Sporopodium*, *Lopadium*), do have ascus apices which look very similar, the main characteristic being a cap-shaped apical dome reacting rather uniformly with Lugol's solution. In this case we might have parallel coevolutionary lines, and asci functioning in a manner we do not yet understand. While no close relatives of *Lopadium* are known, we can make some statements as to the systematic position of *Sporopodium*. *Sporopodium*, and other genera of *Ectolechiaceae*, build campylidia, a peculiar type of conidioma (Sérusiaux, 1986) which might be used as a key character for genera of *Ectolechiaceae* (Vezda, 1986). Such campylidia are also produced by *Badimia dimidiata*. However, the asci of that species are 8-spored and their tholi are provided with a tube structure. So, the correct position of the family *Ectolechiaceae* seems to be in suborder *Cladoniineae*.

Another phenomenon which we have just started to understand is, that polyspory may have an influence on ascus characters. Within the suborder *Lecanorineae*, polyspory may correlate with the width of the axial body. The ascal axial bodies of polyspored species, such as *Pleopsidium chlorophanum* and *Strangospora pinicola*, is broader than those of species regarded as their closest 8-spored relatives, *Lecanora sommervellii* and *Scoliciosporum umbrinum* respectively. However, such a correlation is not always detectable: *Maronina australiensis* has rather typical *Lecanora*-type asci, and in *Fuscideaceae* (*Teloschistineae*) polyspory in *Maronea* does not significantly change ascal characters.

The prototunicate asci of *Lichinaceae* have been interpreted as an adaptation to special ecological conditions (Poelt, 1987). The asci of *Peltula* species (*Peltulaceae*), which often grow side by side with species of *Lichinaceae*, still release their spores in an active process and the construction, as well as mode of dehiscence of the ascal wall, resembles asci of the *Teloschistes*-type (Büdel, 1987: Figs 39-73). On the other hand, the families *Lichinaceae* and *Heppiaceae*, both having the same *Lichina*-type ascus, do not fit in the circumscription of *Lecanorales* given above, and the order *Lichinales* seems acceptable.

CHARACTER WEIGHTING IN *LECANORALES*

Further serious problems in the systematics of this order arise from character weighting dependent on the investigator, as well as the investigated group, which is partly linked with the unequal stage of knowledge in lecanoralean macrolichens and microlichens. During comparative studies in many lecanoralean genera, a high degree of congruence regarding hymenial characters between *Euopsis* and *Harpidium* has been noted (Hafellner, 1984). Henssen et al. (1987), however, included both these genera in *Lichinaceae*, pointing to similarities in thallus structure and ascocarp development between some species of *Pyrenopsis* and the mentioned genera (i.e. *Euopsis* and *Harpidium*), thus allowing a great variety of ascus types in one family. With today's knowledge, there can be no doubt that the *Harpidiaceae* belong to the suborder *Lecanorineae*, and that similarities in thallus structure probably result from the similar habitats these species live in.

Judging which value a character has is indeed a problem and will be a matter of ongoing dispute. One reproach that is often raised is that taxonomic proposals based mainly on a few hymenial characters will not necessarily elucidate the interrelationships between natural groups, especially as we know little about the evolution of the characters themselves. However, the rest of the fungal system is also based on characters of the reproductive organs. As non-lichenized groups of fungi generally do not show many thallus characters, in those groups the hymenial characters are not questioned. And competing proposals for a systematic arrangement of lichenized fungi, which are based on characters of supporting tissues and the thallus, and which more or less neglected the hymenial characters, did not yield more convincing results. It must be admitted that observable hymenial characters, such as ascus-types, may also be the result of convergent evolutionary processes. Such processes must, however, be very slow, as large numbers of species of genera with a world-wide distribution, living under a great variety of ecological conditions, may have hymenia with identical essential features (e.g. *Caloplaca*). Indeed, if convergences in ascus types were ever proven (e.g. asci with tube structures), the logical consequence would be a further splitting and an even more complex system!

In groups where sufficient thallus characters are available for a classification of species and genera, hymenial characters have been investigated only exceptionally, often because there was not much difference expected. In studying macrolichens, very sophisticated chemical methods or scanning electron microscopy are often applied rather than looking in an ordinary light microscope and trying to analyze critically the hymenial characters. For instance, in cetrarioid lichens this was not done until very recently (Kärnefelt et al., 1992). Therefore, the systematics of macrolichens and microlichens is based on only partly overlapping character sets, and the characters used for a definition of a genus may be almost completely different depending on the degree of thallus evolution. One example will serve to demonstrate this.

In the generic description of *Parmelia* s.str. (Hale, 1987: 18), less than two out of the 13 lines of text are devoted to hymenial characters, and, although highly significant, the layered ascospore wall is mentioned without any further comment. In fact, this may be one of the very few characters in *Parmelia* s.lat. which is directly comparable to a character set used in the taxonomy of microlichens. Incidently, the layered ascospore wall in species of *Parmelia* s.str. is the main reason for me (besides the occurrence of some specific lichenicolous fungi on *Parmelia* s.str., such as *Homostegia piggotii* and *Lichenopuccinia poeltii*) to accept seggregates of *Parmelia* s.lat. The difficulty is not to distinguish between *Parmelia* s.str. and one of the seggregates, but to find sufficient distinguishing characters between some of those seggregates.

A VOTE FOR THE USE OF SUBORDERS IN *LECANORALES*

For the future ordinal arrangement of the *Ascomycotina* as a whole it must be strongly recommended to use the subordinal level, the only additional rank available between order and family, for adjusting suprafamiliar entities which are doubtless related to the *Lecanoraceae* (*Lecanorineae*). Thus, *Pertusariales*, *Teloschistales*, and *Peltigerales* should be classified as suborders within *Lecanorales*, as proposed by Rambold and Triebel (1992), and which was also the original view of Henssen and Jahns (1973) and Poelt (1974). This point of view can be justified by recalling the characters of selected species. There is no doubt that *Lecanora allophana* is more closely related to *Psora decipiens* than to either *Graphis scripta* (*Graphidales*) or *Baeomyces rufus* (*Leotiales*). This statement is also true for other species which have recently been classified in separate orders. Also, *Letrouitia vulpina*, *Maronea constans*, *Ochrolechia upsaliensis*, *Varicellaria rhodocarpa*, *Placynthium nigrum* and *Solorina crocea* are more closely related to *Lecanora allophana* than to *Graphis scripta*. This opinion is supported by the fact that all the above mentioned species, other than *Graphis scripta* and *Baeomyces rufus*, share very similar ascus wall characters and conform exactly to the description of *Lecanorales* given above. Raising the rank of the mentioned entities causes a loss of information rather than a gain, because in an alphabetical arrangement of orders (e.g. Eriksson and Hawksworth, 1991) the taxa, which we know to be closely related, look clearly separated. Even more importantly, I consider that it is almost impossible to give good descriptions of these groups so as to distinguish them from the remaining *Lecanorales*. I am convinced that DNA-data will support, when available, this view.

ACKNOWLEDGEMENTS

I want to express my sincere thanks to Dr B.J. Coppins (Royal Botanic Garden Edinburgh) for his critical reading of the text.

REFERENCES

Baral, H.O., 1987, Lugol's solution/IKI versus Melzer's reagent: hemiamyloidity, a universal feature of the ascal wall, *Mycotaxon* 29: 399-450.

Bellemère, A., 1971, Les asques et les apothécies des discomycètes bituniqués, *Annales der Sciences Naturelle Botanique Biologie Vegetale* 12: 429-464.

Bellemère, A., and J. Hafellner, 1983, L'appareil apical des asques et la paroi des ascospores du *Catolechia wahlenbergii* (Ach.) Flotow ex Körber et de l' *Epilichen scabrosus* (Ach.) Clem. ex Haf. (lichens, *Lecanorales*): étude ultrastructurale, *Cryptogamie, Bryologie et Lichénologie* 4: 1-36.

Bellemère, A., and M.A. Letrouit-Galinou, 1981, The lecanoralean ascus: an ultrastructural preliminary study, *In*: *Ascomycete Systematics. The Lutrellian Concept* (D.R. Reynolds, ed.): 54-70, Springer, New York, Heidelberg, Berlin.

Bellemère, A., and M.-A. Letrouit-Galinou, 1987, Differentiation of lichen asci including dehiscence and sporogenesis: an ultrastructural survey, *Bibliotheca Lichenologica* 25: 137-161.

Büdel, B., 1987, Zur Biologie und Systematik der Flechtengattungen *Heppia* und *Peltula* im südlichen Afrika, *Bibliotheca Lichenologica* 23: 1-105.

Chadefaud, M., 1969, Remarques sur les parois, l'appareil apical et les réserves nutritives des asques, *Österriche Botanische Zeitschrift* 116: 181-202.

Chadefaud, M., 1973, Les asques et la systématique des ascomycètes, *Bulletin de la Société Mycologique de France* 89: 127-170.

Chadefaud, M., M.-A. Letrouit-Galinou, and M.-C. Favre, 1963, Sur l'évolution des asques et du type archaeascé chez les discomycètes de l'ordre des *Lécanorales*, *Compte Rendu Hebdomadeine des Séances de l'Academie des Sciences, Paris* 257: 4003-4005.

Chadefaud, M., M.-A. Letrouit-Galinou, and M.-C. Janex-Favre, 1967, Sur l' origine phylogénetique et l'évolution des ascomycètes des lichens, *Mémoires de la Société Botanique de France, Colloque sur les Lichens*: 79-111.

Eriksson, O.E., and D.L. Hawksworth, 1991, Outline of the ascomycetes - 1990, *Systema Ascomycetum* 9: 39-271.

Hafellner, J., 1984, Studien in Richtung einer natürlicheren Gliederung der Sammelfamilien *Lecanoraceae* und *Lecideaceae*, *Beiheft zur Nova Hedwigia* 79: 241-371.

Hafellner, J., 1988, Principles of classification and main taxonomic groups, *In*: *CRC Handbook of Lichenology* (M. Galun, ed.) 3: 41-52, CRC Press, Boca Raton.

Hafellner, J., and A. Bellemère, 1982a, Elektronenoptische Untersuchungen an Arten der Flechtengattung *Bombyliospora* und die taxonomischen Konsequenzen, *Nova Hedwigia* 35: 207-235.

Hafellner, J., and A. Bellemère, 1982b, Elektronenoptische Untersuchungen an Arten der Flechtengattung *Brigantiaea*, *Nova Hedwigia* 35: 237-261.

Hafellner, J., and A. Bellemère, 1982c, Elektronenoptische Untersuchungen an Arten der Flechtengattung *Letrouitia* gen. nov., *Nova Hedwigia* 35: 263-312.

Hale, M.E., 1987, A monograph of the lichen genus *Parmelia* Acharius sensu stricto (*Ascomycotina*: *Parmeliaceae*), *Smithsonian Contributions to Botany* 66: 1-55.

Hawksworth, D.L., B.C. Sutton, and G.C. Ainsworth, 1983, *Ainsworth & Bisby's Dictionary of the Fungi*, 7th edn, Commonwealth Agricultural Bureaux, Slough.

Henssen, A., B. Büdel, and A. Titze, 1987, *Euopsis* and *Harpidium*, genera of the *Lichinaceae* (*Lichenes*) with rostrate asci, *Botanica Acta* 101: 83-89.

Henssen, A., and H.M. Jahns, 1973 ["1974"], *Lichenes. Eine Einführung in die Flechtenkunde*, G. Thieme, Stuttgart.

Hertel, H., and G. Rambold, 1985, *Lecidea* sect. *Armeniacae*: lecideoide Arten der Flechtengattungen *Lecanora* und *Tephromela* (*Lecanorales*), *Botanisches Jahrbücher für Systematik* 107: 469-501.

Honegger, R., 1978, The ascus apex in lichenized fungi I. The *Lecanora*-, *Peltigera*- and *Teloschistes*-types, *Lichenologist* 10: 47-67.

Honegger, R., 1980, The ascus apex in lichenized fungi II. The *Rhizocarpon*-type, *Lichenologist* 12: 157-172.
Honegger, R., 1982a, Ascus structure and function, ascospore delimitation, and phycobiont cell wall types associated with the *Lecanorales* (lichenized ascomycetes), *Journal of the Hattori Botanical Laboratory* 52: 417-429.
Honegger, R., 1982b, The ascus apex in lichenized fungi III. The *Pertusaria*-type, *Lichenologist* 14: 205-217.
Kärnefelt, I., J.-E. Mattsson, and A. Thell, 1992, Evolution and phylogeny of cetrarioid lichens, *Plant Systematics and Evolution* 183: 113-160.
Letrouit-Galinou, M.-A., 1973, Les asques des lichens et le type archaeascé, *Bryologist* 76: 30-47.
Magne, F., 1946, Anatomie et morphologie comparées des asques de quelques lichens, *Revue bryologique et lichénologique* 15: 203-209.
Poelt, J., 1974 ["1973"], Classification, *In*: *The Lichens* (V. Ahmadjian and M.E. Hale, eds): 599-632. Academic Press, New York and London.
Poelt, J., 1987, On reductions of morphological structures in lichens, *Bibliotheca Lichenologica* 25: 35-45.
Poelt, J., and J. Hafellner, 1975, Schlauchpforten bei der Flechtengattung *Thelocarpon*, *Phyton, Horn* 17: 67-77.
Rambold, G., and D. Triebel, 1992, The inter-lecanoralean associations, *Bibliotheca Lichenologica* 48: 1-201.
Sérusiaux, E., 1986, The nature and origin of campylidia in lichenized fungi, *Lichenologist* 18: 1-35.
Vezda, A., 1986, Neue Gattungen der Familie *Lecideaceae* s.lat. (*Lichenes*), *Folia Geobotanica Phytotaxonomia, Praha* 21: 199-219.

PROBLEM GENERA AND FAMILY INTERFACES IN THE *EUPYRENOMYCETES*

J.D. Rogers

Department of Plant Pathology
Washington State University
Pullman, WA 99164-6430, USA

SUMMARY

Perplexities related to the systematics of *Eupyrenomycetes* are due to the relatively few taxa that have been studied, and the many fewer taxa that have been investigated in depth. Moreover, the apparent great age of the ascomycetes, the general scarcity of fossil records, and the supposition that early genera have not survived to the present, accentuate problems associated with developing a natural (phylogenetic) classification. Two specific problem areas of eupyrenomycete systematics discussed are: the interfaces among *Eupyrenomycetes* and *Loculoascomycetes*; and selected examples of problem interfaces among eupyrenomycete families and genera. In the latter problem area the following topics are considered: generic relationships in *Xylariaceae*; relationships between *Xylariaceae* and *Diatrypaceae*; relationships between *Calosphaeriaceae* and some other families; the status of *Boliniaceae*; the status of *Coniochaetaceae*; *Ceratocystis* and *Ophiostoma*; interfaces between *Amphisphaeriaceae* and *Xylariaceae*; problems in taxa of *Diaporthales*; the status of *Cainiaceae*; and parallel versus convergent evolution among *Eupyrenomycetes*.

INTRODUCTION

I was asked to consider problems in ascomycete systematics, emphasizing the *Eupyrenomycetes*. There are, of course, many scientific and technical problems at the ordinal, familial, generic, and specific levels. There are, however, some problems that transcend specific groups and influence ascomycete systematics and, indeed, the systematics of fungi in general. Dick and Hawksworth (1985) and Hawksworth (1985) have written particularly stimulating papers dealing with ascomycetes.

The first problem is that comparatively few eupyrenomycetes have been carefully examined and many fewer have been studied in any detail. Luttrell (1989) truly stated that mycological research is obsessed with novelty. Too many fungi are examined only once and often superficially. As Luttrell has written: "The vastness of our ignorance of even the simplest morphological facts in the fungi is not generally appreciated." If there are many more *Eupyrenomycetes* to be discovered and described among the total undescribed fungi estimated by Hawksworth (1991), then our ignorance will indeed be compounded.

With current emphasis on biodiversity, great numbers of undescribed or poorly described fungi are arriving from the tropics. It is often difficult to describe them and place them, even tentatively, into families. Needless to say, such fungi are seldom investigated in more than a rudimentary way and add to our overall "ignorance load". We are surely building an inventory of the Earth's mycobiota, but our understanding of it is largely nil.

A second problem is the apparent great age of the ascomycetes and the fact that early genera have probably not survived to the present (Cain, 1972). In Cain's opinion, mycologists have been overly influenced by dichotomous keys, leading workers to think of relationships based upon single features. As he has truly noted, it is relatively easy to find characters that appear to be connecting links between taxa of any level, but, on analysis, a particular taxon is seen to be linked in all directions, rather than in an orderly ascending phylogenetic tree.

A third problem, closely tied to the previous ones, is the tendency to formalize taxonomic systems on the basis of too few data. For example, the centrum types put forward by Luttrell (1951) on comparatively few studies have become blurred as many more taxa have been investigated and, indeed, within the pyrenomycetes there seems to be a continuum rather than sharp demarcation among centrum types (Uecker, 1976). Another example is a system of hyphomycetes based primarily on conidiogenesis and conidiogenous cell proliferation. In the beginning it appeared that, for example, phialidic and annellidic conidiogenesis were easily separable. As more taxa were investigated, the distinctions among the variants and permutations of these types of conidiogenesis became less clear. It became obvious that virtually every taxon producing conidia from, for example, phialides did so at least a bit differently than other taxa, i.e. there seems also to be a continuum in conidiogenesis. It has become apparent that centrum characteristics as well as conidial production phenomena must be weighed and used in conjunction with other characters. There are lessons here for the molecular biologist based upon the experiences of structural biologists. First, there is not likely to be any single character, including the molecular or biochemical, that is so basic that it negates the value of other potential characters. Second, restraint should be used in erecting formal taxonomic systems based upon data from too few taxa. Molecular, morphological, and other data must be used together with great care in an attempt to get at the truth.

Progress in understanding *Eupyrenomycetes* will come only as individual taxa are studied in great detail using a variety of methodologies and techniques. Certain particularly well-studied taxa, such as certain *Neurospora* species, can serve as guides or as the basis of hypotheses. It should not be assumed, however, that the structure, behaviour, or genetics of *Neurospora* or any other genus is necessarily typical of *Eupyrenomycetes*.

The broad synopsis of problems in eupyrenomycete systematics, presented above, is perhaps of less real interest and importance to the working systematist than those more discrete problems of taxa at various levels. The remainder of this contribution will discuss some of them.

INTERFACES AMONG *EUPYRENOMYCETES* AND *LOCULOASCOMYCETES*

The subject of this contribution is taxonomic problems in *Eupyrenomycetes*, those fungi with true perithecial ascomata and *functionally* unitunicate asci. These fungi are contrasted with the *Loculoascomycetes*, especially those fungi with ascostromatic ascomata that resemble true perithecia and have *functionally* bitunicate asci. The unitunicate ascus is sometimes described as non-fissitunicate as opposed to the bitunicate ascus which is often described as fissitunicate (Eriksson, 1981; Reynolds, 1989). Chadefaud, a great student of ascus structure, and his students and colleagues have designated the kinds of unitunicate asci found in most *Eupyrenomycetes* as annellasceous on the basis of the apical ring, and the types of bitunicate asci found in most *Loculoascomycetes* as nassasceous on the basis of the apical nasse (Bellemère and Letrouit-Galinou, 1981; Parguey-Leduc and Janex-Favre, 1981).

Luttrell (1951, 1973, 1981) was the most influential of modern workers who emphasized the correlation between ascus types and ascomatal structure in taxonomy; Reynolds (1981) designated Luttrell's system as the "Luttrellian concept". As useful as the Luttrellian concept has proven to be, it is sometimes difficult to apply. It is notably difficult in some fungi to determine the status of asci and the nature of hamathecial elements (von Arx, 1979).

The order *Coryneliales*, family *Coryneliaceae* has long been a paradoxical group, apparently possessing unitunicate asci and loculoascomycetous ascomata (Johnston and Minter, 1989). In an elegant study, Johnston and Minter have shown that representatives of the *Coryneliaceae* have two functional ascus wall layers, the outer layer breaking during ascus elongation long before ascospore maturity and release. Thus, it appears that the

Coryneliales can now be assigned to the *Loculoascomycetes* with a high degree of confidence.

Huhndorf (1992) described *Hypsostroma* as a new genus of *Loculoascomycetes*. The ascomatal wall structure and the pseudoparaphysate hamathecium supported the loculoascomycetous affinities; asci, however, were described as thin-walled, but probably two-walled, with dehiscence unknown. Asci of both described species were shown to have apical rings that fluoresced in the fluorescence brightener Calcofluor (Huhndorf, 1992). As Huhndorf noted, the presence of an apical ring does not necessarily indicate that the *Hypsostroma* ascus is unitunicate. Ascus rings have been noted sporadically among undoubted loculoascomycetous fungi (Eriksson, 1981; Luttrell, 1973).

Huhndorf (1992) speculated that *Hypsostroma* might be related to *Xylobotryum*. The *X. andinum* ascus has been considered unitunicate by Rossman (1976) and bitunicate by Barr (1990b). G.J. Samuels (pers. comm.) has reported fissitunicate dehiscence in *X. andinum*, and Huhndorf (1992) has observed a fluorescing apical ascus ring. Y.-M. Ju and I have been studying *X. andinum* from pure culture. Using the fluorescence procedure of Rogers (1975), we can report that ascus dehiscence is fissitunicate, probably semifissitunicate in the sense of Eriksson (1981). Moreover, the ascomatal wall is loculoascomycetous, i.e. is continuous with the stroma. The hamathecium, however, is composed of essentially non-branching hyphae that definitely originate at the base of the centrum; the apices appear to be free. It is probable that these are paraphyses.

I have examined the asci and hamathecial elements of a number of taxa and can report the following findings. *Leptosphaerulina argentinensis*, several species of *Leptosphaeria*, *Sporormiella intermedia*, *Lophiostoma* sp., *Cucurbitaria berberidis*, *Pseudovalsaria foedans*, and *Trichodelitschia munkii* have fluorescing rings in the endotunica. I suspect that the presence of apical rings among undoubted loculoascomycetes will be the norm. The presence of a fluorescing apical ring alone will thus be of little utility in determining whether asci are unitunicate or bitunicate. Interestingly, Chadefaud recognized an ascus type possessing both a nasse and an apical ring which he called the archaeascus type (Parguey-Leduc and Janex-Favre, 1981; Parguey-Leduc et al., this volume). According to Parguey-Leduc and Janex-Favre (1981), the archaeascus type is an archaic type not known in pyrenomycetes, i.e. extant bitunicate asci. It is, however, present in certain pyrenolichens (Letrouit-Galinou, 1973). It seems probable that, in fact, the archaeascus type is very common among loculoascomycetes! This might add credence to the hypothesis of Parguey-Leduc and Janex-Favre (1981) that the unitunicates are derived from the bitunicates because in some asci the bitunicate structure becomes unitunicate in the mature form. In the course of evolution the two walls would be fused into one (Parguey-Leduc and Janex-Favre, 1981), at least functionally, I presume. On the other hand, it might be hypothesized that unitunicate asci, known to be structurally composed of more than one wall layer, became functionally bitunicate by the loss (or gain) of elasticity of one or more wall layer(s).

PROBLEM INTERFACES AMONG EUPYRENOMYCETE FAMILIES AND GENERA

Generic Relationships in the *Xylariaceae*

The assignment of genera to eupyrenomycetous families can, in some cases, be done with high confidence that related taxa are grouped together. For example, genera of *Xylariaceae* usually show perithecia embedded in a stroma, cylindrical asci with an amyloid apical ring, paraphysate hamathecium, ascospores with a germ slit, and hyphomycetous anamorph with holoblastic conidiogenesis and sympodial conidiogenous cell proliferation (Rogers, 1979, 1985). In some cases, interfaces among genera are somewhat unclear: e.g. *Hypoxylon* and *Rosellinia*; *Poronia* and *Podosordaria*; *Hypoxylon* and *Daldinia*. In other cases, a genus seems to be isolated from other genera on a combination of correlated characters. For example, *Camillea* is highly distinctive on the basis of the pale-colored, intricately ornamented ascospores that lack germ slits, the diamond-shaped to rounded ascus ring, and the *Xylocladium* anamorph (Læssøe et al., 1989). *Leprieuria* has a cylindrical stroma reminiscent of *Camillea*, but the ascus lacks an apical ring and the anamorph is not of the *Xylocladium* type (Læssøe et al., 1989). *Thamnomyces* stromata are unlike those of any other xylariaceous genus and the asci lack an apical ring. A

Nodulisporium anamorph of *T. chordalis* has been produced in culture (Samuels and Müller, 1980). *Thuemenella* lacks most of the characters of undoubted xylariaceous fungi, but was placed in the family by Samuels and Rossman (1992), primarily on the morphology of the anamorph. Two genera that have been placed in family *Xylariaceae* on, in my opinion, very tenuous criteria are *Phylacia* and *Pulveria*. The former genus lacks almost every character associated with xylariaceous fungi, save the teleomorphic stroma and the *Nodulisporium* anamorph (Rodrigues and Samuels, 1989). The latter genus has a teleomorphic stroma much like the xylariaceous type that lacks a natural opening; the ascospores have germ slits (Malloch and Rogerson, 1977).

The examples given here should be sufficient to indicate that several evolutionary courses have been followed in this group. A number of other examples could be cited. Large and complex groups such as the red *Hypoxylon* species have apparently arisen from one or a few ancestral taxa and radiated widely in terms of taxa and distribution. On the other hand, *Xylaria* is clearly polyphyletic. *Thamnomyces* and *Leprieuria* occupy isolated positions with uncertain relationships with other taxa. *Pulveria*, if it is truly xylariaceous, apparently has become isolated from the core of the *Xylariaceae* by development of non-ostiolate ascomata and globose non-explosive asci.

Relationships Between the *Xylariaceae* and *Diatrypaceae*

Mycologists have long been suspicious of an interface between *Xylariaceae* and *Diatrypaceae* (Munk, 1957). Cardinal characters of the latter family include perithecia embedded in a more or less well-developed stroma, coloured allantoid ascospores, and hyphomycetous or coelomycetous anamorphs usually featuring elongated wet conidia produced holoblastically from sympodially or percurrently proliferating conidiogenous cells (Glawe and Rogers, 1986; Munk, 1957; Rappaz, 1987). Y.-M. Ju, F. San Martín, and I have recently cultured *Lopadostoma turgidum* and discovered that it produces conidiomata that bear slimy elongated conidia similar to the type expected in the *Diatrypaceae*, often placed in *Libertella* (Ju et al., 1993). In other respects this species is typically xylariaceous. Moreover, wet elongated conidia have been reported in *Hypoxylon sassafras* (Petrini and Müller, 1986), leading us to transfer this species to *Creosphaeria* (Ju et al., 1993). Still another xylariaceous species, probably close to *Nummularia viridis*, produces elongated wet conidia. This taxon probably represents a new genus (Ju et al., 1993). *Hypoxylon microplacum* has previously been reported to produce elongated dry conidia in culture and its possible relationship to the *Diatrypaceae* noted (Jong and Rogers, 1972; Glawe and Rogers, 1986).

Several of these taxa certainly represent heretofore unrecognized evolutionary lines in the *Xylariaceae*. It will be interesting to culture other taxa to discover the extent of these lines and to evaluate the extent to which the *Xylariaceae* and *Diatrypaceae* show convergent evolutionary patterns or, indeed, interfamilial relationships.

Relationships Between the *Calosphaeriaceae* and Some Other Families

Munk (1957) erected the family *Calosphaeriaceae* primarily on the peculiar production of asci in fascicles or whorls from an ascogenous hypha without the intervention of typical croziers. Barr accepted this family and erected order *Calosphaeriales* to accommodate it (Barr, 1990a). Among the genera placed in the *Calosphaeriaceae* by Barr, primarily on the morphology of the ascogenous system, is *Graphostroma* with the single species *G. platystoma*. This species greatly resembles *Diatrype stigma* in stromatal characters, but produces permanently hyaline ascospores that are not truly allantoid and a *Nodulisporium* anamorph (Pirozynski, 1974; Glawe and Rogers, 1986). Pirozynski (1974) assigned the genus to the *Xylariaceae*, a disposition favoured by Rogers (1979).

The question here is this: to what extent can the morphology of the ascogenous system be used to characterize a family. There are not enough studies to answer this question definitively, but there are some interesting data. Romero and Samuels (1991) have shown that *Jattaea stachybotryoides* produces the *Calosphaeria*-type ascogenous system; the anamorph in unknown. *Endothia metrosideri* also produces asci from a calosphaeriaceous ascogenous system, but produces a *Cytospora*-like anamorph in culture (misinterpreted as a *Phomopsis* by Roane et al., 1986). It thus appears that if one recognizes the calosphaeriaceous centrum to be of cardinal importance, family *Calosphaeri-*

aceae as now constructed includes taxa with vastly different anamorphs, e.g. *Cytospora* and *Nodulisporium*.

Recently, Barr et al. (1993) have moved *Endothia metrosideri* to another genus, *Pachytrype*, to reflect its calosphaeriaceous affinities. *Graphostroma* was moved into a new family of the *Calosphaeriales*, *Graphostromataceae*, to reflect its non-phialidic anamorph.

Are there data that bear on this problem? Perhaps, indirectly. In the "core" of the *Xylariaceae*, within genera and species that are undoubtedly related, two general and intergrading ascogenous system types are found. One is an extensive system of long ascogenous hyphae sending off asci at intervals *via* croziers. This kind of system was well-described by Ingold (1954) in *Daldinia concentrica* and is common in the coloured *Hypoxylon* species, *Thamnomyces*, etc. (Rogers, unpubl.). The other type involves separate limited ascogenous systems arising directly from ascogenous cells and extending by crozier proliferation. This kind of system has been described for *Hypoxylon microplacum* and *H. cohaerens* by Rogers (1971, 1972) and for *H. serpens* by Jensen (1981). *Pulveria porrecta* produces an ascogenous system without the intervention of typical croziers (Malloch and Rogerson, 1977). *Phylacia bomba* produces asci from long geniculate ascogenous hyphae without the intervention of typical croziers, whereas *Phylacia sagraeana* produces asci from long hyphae *via* croziers (Rogers, unpubl.), as does *P. poculiformis* (Rodrigues and Samuels, 1989).

On the basis of these limited data it appears that the extent of the ascogenous system, and whether or not croziers are produced, depends upon the need, or lack of need, for extensive ascogenous hyphal distribution in the centrum of individual taxa and might cut across generic or family lines. Certainly, additional studies are required to clarify this issue.

The Status of the *Boliniaceae*

The family *Boliniaceae* was considered by Nannfeldt (1972) to be monotypic, containing *Camarops*; Samuels and Rogers (1987) added *Apiocamarops*. Barr (1990a) accepted *Rhynchostoma* as well as *Camarops*, but did not formally accept *Apiocamarops*. Interestingly, *Rhynchostoma minutum* has recently been shown to have a coelomycetous anamorph (Constantinescu and Tibell, 1992). More recently, Barr (1993) accepted *Endoxyla* and *Pseudovalsaria*. In any case, I presently accept *Camarops* and *Apiocamarops* in the family. The cardinal characters of the family thus delimited include one-celled brownish ascospores that are usually somewhat flattened and bear a germ pore at one end, or both ends in the case of *C. biporosa* (Rogers and Samuels, 1987), or apiosporous ascospores with a germ pore in the much larger brown cell; small asci with definite non-amyloid apical rings; small to large stromata that are generally well-developed in *Camarops* and poorly developed in *Apiocamarops*; polystichous perithecia in a number of taxa; and lack of anamorphs. Until Nannfeldt's (1972) treatment members of the *Boliniaceae* were usually placed in the *Xylariaceae* (Müller and von Arx, 1973; Munk, 1957). Munk (1957) included *Camarops* in tribe *Camaropeae* of the *Xylariaceae* and considered these fungi to be a connecting link with the diatrypaceous fungi (Munk's tribe *Diatrypeae* of the *Xylariaceae*).

The relationships of the *Boliniaceae* with other families remains obscure, in my opinion. The discovery of an anamorph associated with any taxon would be of great value. Interestingly, a number of *Camarops* species and one *Apiocamarops* species have been cultured. Several species of *Camarops* produce the teleomorph in culture, but no anamorph has been reported. Obviously, new characters should be sought that could illumine the relationships of this most curious assemblage.

The Status of the *Coniochaetaceae*

The *Coniochaetaceae* was erected to accommodate *Coniochaeta* and *Coniochaetidium* (Malloch and Cain, 1971). As pointed out by those authors, it differs from the *Sordariaceae* in the ascospore germ slit and from the *Xylariaceae* in the non-stromatic ascomata. It is likewise noteworthy that it differs from the *Xylariaceae* also in the phialidic anamorphs and the (usually) flattened ascospores. Ascus rings are well-developed and inamyloid. Doguet's account of ascocarp development indicates that its structure might well differ from either the sordariaceous or xylariaceous type (Doguet, 1959).

Coniochaeta is obviously a successful genus in terms of numbers of species and distribution. Mahoney and LaFavre (1981) and Checa et al. (1988) have produced useful papers on the genus. For many years, *Coniochaeta* species were considered to be small-spored *Rosellinia* and are filed with *Rosellinia* in many herbaria. Most of these are lignicolous, coprophilous, or are from herbaceous plants. During the last 30 years *Coniochaeta* species have commonly been isolated from soils.

As far as I am concerned, the *Coniochaetaceae* is an isolated family whose relationship with other ascomycetes is obscure. Taxa need to be studied more comprehensively and additional characters sought in order to clarify the affinities of this group.

Ceratocystis and *Ophiostoma*

The taxonomic history of *Ceratocystis* and *Ophiostoma*, is turbulent. Until relatively recently, there was a tendency to place relevant taxa in the *Ophiostomataceae*, recognizing either two genera or placing *Ophiostoma* into synonymy with *Ceratocystis*. Whether or not two genera were recognized, the involved taxa seemed closely related in producing long-necked ascomata and broad evanescent asci, in being intimately associated with insects, and in causing insect-vectored wilt diseases of angiospermous trees, e.g. oak wilt caused by *Ceratocystis fagacearum* and Dutch elm disease caused by *Ophiostoma novo-ulmi*.

Currently, most mycologists accept *Ceratocystis* as a genus having a phialidic anamorph (*Chalara*), lacking rhamnose and cellulose in the walls, and showing marked sensitivity to cycloheximide (Harrington, 1981; de Hoog and Scheffer, 1984). *Ophiostoma*, and its cleistocarpous counterpart *Europhium*, are accepted as having anamorphs other than *Chalara* (anamorphs that are usually considered to be non-phialidic), having rhamnose in the cell walls, and lacking sensitivity to cycloheximide (Harrington, 1981; de Hoog and Scheffer, 1984). Moreover, Hausner et al. (1992) analyzed small subunit ribosomal DNA (SSrDNA) from species of *Ophiostoma* and *Ceratocystis* with galeate ascospores. They concluded that these genera are not closely related, i.e. did not cluster together. Barr (1990a) accommodates *Ophiostoma* in the *Ophiostomataceae* in the *Microascales*, and *Ceratocystis* in the *Lasiosphaeriaceae* of the *Sordariales*. It seems noteworthy that the family *Lasiosphaeriaceae* contains genera that are largely considered to be saprophytic; the activities of *Ceratocystis* species as vascular parasites seem incongruous.

In my opinion, the relationship (or lack of relationship) between *Ceratocystis* and *Ophiostoma* remains unclear. I predict that future investigations will reveal additional facets of great interest to the evolution-oriented mycologist.

Interfaces Between the *Amphisphaeriaceae* and the *Xylariaceae*

The family *Amphisphaeriaceae*, as interpreted by Müller and von Arx (1973), includes taxa with (mostly) amyloid ascus rings and one- to several-septate ascospores that either do not have germination sites or have germination pores. Barr's (1990a) concept of the family is more restrictive and includes only taxa with coelomycetous anamorphs. In any case, there are some interesting interfaces with the *Xylariaceae*.

Samuels et al. (1987) discussed *Collodiscula japonica* and described its *Acanthodochium* anamorph. *Collodiscula* has asci with a well-developed amyloid apical ring and two-celled brown ascospores that lack a germination site. The anamorph shows one-celled conidia produced holoblastically from sympodially proliferating conidiogenous cells. Ju and Rogers (1990) investigated two species of *Astrocystis* and concluded that they could be accommodated in *Rosellinia*. Interestingly, the anamorphs of these species are of the genus *Acanthodochium* and, on natural substrates and in culture, greatly resemble that of *Collodiscula*. Moreover, the hosts of all of these fungi are bamboos. In spite of the obvious close relationship between *Collodiscula* and the two *Rosellinia* species, it is difficult to accept *Collodiscula* as truly xylariaceous or, on the other hand, to consider the *Rosellinia* species as amphisphaeriaceous. It is, however, somewhat unsatisfying to keep these clearly related taxa separated in different families.

It has recently been discovered that some taxa buried in *Hypoxylon* as *Hypodiscus* Lloyd (non Nees), which Miller (1961) considered a synonym of *Hypoxylon*, in fact represent a discrete genus that could be accommodated in the family *Amphisphaeriaceae*. These taxa have well-developed stromata of the kind associated with *Hypoxylon* or *Dal-*

dinia. Ascospores, however, have hyaline polar caps (probably equivalent to secondary appendages) and lack a germ slit. In at least three species ascospores have a polar germ pore. Asci have minute, iodine-negative apical rings. A new genus will be erected to accommodate these taxa.

Some Problems in Families of *Diaporthales*

The order *Diaporthales*, as usually delimited embraces taxa with a pseudoparenchymatous hamathecium that is usually lysed by developing asci and few, if any, true paraphyses; asci that detach from the ascogenous system and float free in the centrum; asci with apical refractive rings that are non-amyloid, but often chitinoid; stromata that often include host tissue as well as hyphal elements; and coelomycetous anamorphs. My studies of taxa usually assigned to this order indicate that it is far less homogenous than sometimes depicted. In particular, the nature of the ascus apex and the nature of hamathecial elements requires the kind of careful study that Jensen (1983) and Uecker (1989) have applied to *Diaporthe* species. Several examples follow.

Barr (1990a) placed *Sydowiella fenestrans* and *Pseudovalsa lanciformis* in the *Melanconidaceae*. The former species has an apical ascus apparatus that, by fluorescence microscopy, resembles point-to-point triangles; very broad paraphyses are intermingled with the asci (Rogers, unpubl.; von Arx, 1979). The latter species has a dome-shaped apical ascus apparatus apparently composed of several stacked rings. The ascus wall is notably thick for a eupyrenomycete and occasionally is seen to be composed of a number of layers. Conspicuous narrow paraphyses are intermingled with the asci (Rogers, unpubl.; von Arx, 1979).

Cryphonectria parasitica and *Anisogramma anomala* are assigned to the *Gnomoniaceae* by Barr (1990a). Asci of the former species show a tiny fluorescing apical ascus ring and conspicuous, rather narrow, paraphyses. Asci of the latter species show only a trace of a fluorescing ring and no paraphyses (Rogers, unpubl.). Other examples could be cited (Bellemère et al., 1987), but it is probably sufficient to say that the *Diaporthales* offers an excellent opportunity for detailed studies along a number of lines.

The Status of the *Cainiaceae*

The family *Cainiaceae* was erected to accommodate *Cainia* (Krug, 1977). This genus has asci with an amyloid apical apparatus and two-celled ascospores with *c*. 6-8 longitudinal germ slits. *Cainia* is often placed in the *Amphisphaeriaceae* (Müller and von Arx, 1973). It has obvious affinities with the *Amphisphaeriaceae* s.lat. and with the *Xylariaceae*. Krug (1977) has noted the similarity of the *Cainia* ascospores with those of *Entosordaria perfidiosa* in that the latter have short germ slits near the apex of the larger and darker of the two cells of the ascospore (Eriksson, 1966). *Amphisphaeria pardalina* has somewhat similar ascospores, i.e. in having short germ slits in each of the two more or less equal cells of the ascospore (Barr, 1989). She noted this ascospore resemblance, but rejected a close relationship with *E. perfidiosa*.

In my opinion the *Cainiaceae* is worthy of recognition, although its limits are currently unclear. Cultural and anamorphic data would be useful in clarifying the taxa that should be included in it. Müller and Corbaz (1956) reported a coelomycetous state (*Rhabdospora*) for *Cainia demazieresii*, but Krug (1977) was unable to induce an anamorph in his cultures. As far as I know, an anamorph has not been reported for *Entosordaria perfidiosa*. A coelomycetous and a hyphomycetous state have been reported, respectively, for two putative species of *Amphisphaeria* (Kendrick and DiCosmo, 1979). I am unaware of an anamorph being reported for *A. pardalina* (syn. *Didymosphaeria pardalina*).

Parallel Versus Convergent Evolution

One of the greatest problems in assessing relationships among taxa and ultimately developing a phylogenetic classification is deciding whether characters shared by taxa at various ranks are the result of parallel or of convergent evolution. Parallel evolution is accepted here as changes occurring in groups with common ancestry *because* they have common ancestry (Davis and Heywood, 1963). Convergent evolution is the similarity of changes occurring in groups that do not have a common ancestry (Davis and Heywood,

1963). It is often difficult to decide whether similar characters are the result of parallel or of convergent evolution because the ancestries of taxa in question are unknown. If all ascomycetes originated from a common ancestor (or ancestors), how ancient or recent are the points of divergence of the groups in question?

An example of common characters is the bipartite stroma, the outermost layer of which is shed to expose perithecial ostioles. This occurs in the xylariaceous genera *Biscogniauxia* and *Camillea* and some species of the diatrypaceous genus *Diatrype*, e.g. *D. stigma*. It is uncertain if the families *Xylariaceae* and *Diatrypaceae* are evolving in parallel or if the evolution of certain features is convergent owing to the exploitation of common substrates, e.g. angiosperm bark (and see earlier discussions of these families earlier herein).

Examples of characters that seem to be convergent are associated with dung-inhabiting *Eupyrenomycetes*. Apparently unrelated genera such as *Neurospora* and *Poronia* show ascospores with more than one nucleus and dormancy that can be broken by heat. Ascospores of many coprophilous taxa have sticky sheaths, and some additional secondary appendages, that are of value in adhering them to herbage.

Hausner et al. (1992) present molecular evidence that the galeate ascospore found among various taxa of yeasts and eupyrenomycetes is a product of convergent evolution. Taxa with galeate ascospores usually are associated with insects and often occur in similar habitats.

Sherwood (1981) has written a thought-provoking discussion on convergent evolution in discomycetes from wood and bark. In an earlier paper (Sherwood, 1977) she discussed the great similarities of the long cylindrical asci with distinctive apical caps and filiform ascospores of the *Clavicipitales* and *Ostropales*, respectively. As she pointed out, filiform ascospores are found in various families of ascomycetes and they tend to be associated with cylindrical asci with thickened apices. On the other hand, she believed that the relationship, or lack of relationship, between these orders could be assessed on the basis of meaningful studies of ascomatal development (Sherwood, 1977), i.e. if evolution is in some respects parallel or convergent.

There has likewise been the suggestion of a relationship between the *Clavicipitales* and *Hypocreales*. In particular, the soft, light-coloured stromata, tendencies to parasitize other fungi and insects, and phialidic anamorphs of these respective groups have intrigued taxonomists. Rogerson (1970) reviewed the taxonomic placement of these groups. He accepted the orders *Hypocreales* and *Clavicipitales* as the "hypocrealean fungi" (Rogerson, 1970). A primary reason for maintaining the hypocreaceous and clavicipitaceous fungi as separate at the ordinal, or at least the familial level, is on the basis of hamathecial morphology. Undoubted hypocreaceous fungi have downward-growing apical paraphyses, whereas clavicipitaceous fungi are said to have upward-growing paraphyses with free ends (Rogerson, 1970), based on few studies. At the present time, it is uncertain if the hypocreaceous fungi and the clavicipitaceous fungi have, in some characters, undergone parallel evolution or, if in the process of exploiting some similar ecological niches, they show the results of remarkable convergent evolution.

Other interesting examples of apparent convergent evolution could be explored, including: the tendency of ascomata of various fungi growing on hard monocot substrates such as bamboo to have poorly developed flattened bases; the evolution of the valsoid habit among apparently unrelated fungi; and the development of ascospore germ slits among unrelated fungi. The previous discussion, however, should at least indicate some fascinating and critical facets of evolution that need to be understood in the context of a natural classification system.

ACKNOWLEDGEMENTS AND DEDICATION

PPNS No. 0153, Department of Plant Pathology, Washington State University, Agricultural Research Center, College of Agriculture and Home Economics. Project 1767. This work was supported in part by National Science Foundation grant BSR-9017920. I am grateful to Jane Lawford for typing the manuscript and Lori M. Carris for reading it.

I dedicate this paper to my friend, Françoise Candoussau of Pau, France, a most enthusiastic and knowledgeable student of ascomycetes!

REFERENCES

Arx, J.A. von, 1979, Ascomycetes as Fungi Imperfecti, *in*: *The Whole Fungus* ([W.] B. Kendrick, ed.) 1: 201-213. National Museums of Canada, Ottawa.

Barr, M.E., 1989, Some unitunicate taxa excluded from *Didymosphaeria*, *Studies in Mycology, Baarn* 31: 23-27.

Barr, M.E., J.D. Rogers, and Y.-M. Ju, 1993, Revisionary studies in the *Calosphaeriales*, *Mycotaxon* 98: 529-535.

Barr, M.E., 1990a, Prodromus to nonlichenized, pyrenomycetous members of class *Hymenoascomycetes*, *Mycotaxon* 39: 43-184.

Barr, M.E., 1990b, *Melanommatales* (*Loculoascomycetes*), *North American Flora* 2(13): 1-129.

Barr, M.E., 1993, Redisposition of some taxa described by J.B. Ellis, *Mycotaxon* 46: 45-76.

Bellemère, A., M.-C. Janex-Favre, and A. Parguey-Leduc, 1987, Marius Chadefaud et les asques: données in édites études ultrastructurales complémentaires, *Bulletin de la Société Botanique du France* 134 (*Lettres botanique*) : 217-246.

Bellemère, A., and M.A. Letrouit-Galinou, 1981, The lecanoralean ascus: an ultrastructural preliminary study, *In*: *Ascomycete Systematics: The Luttrellian Concept* (D.R. Reynolds, ed.): 54-70. Springer-Verlag, New York.

Cain, R.F., 1972, Evolution of the fungi, *Mycologia* 64:1-14.

Checa, J., J.M. Barrasa, G. Moreno, F. Fort, and J. Guarro, 1988, The genus *Coniochaeta* (Sacc.) Cooke (*Coniochaetaceae*, *Ascomycotina*) in Spain, *Cryptogamie, Mycologie* 9: 1-34.

Constantinescu, O., and L. Tibell, 1992, Teleomorph-anamorph connections in ascomycetes 4. *Arthropycnis praetermissa* gen. and sp. nov., the anamorph of *Rhynchostoma minutum*, *Nova Hedwigia* 55: 169-177.

Davis, P.H., and V.H. Heywood, 1963, *Principles of Angiosperm Taxonomy*, Oliver & Boyd, Edinburgh & London.

de Hoog, G.S., and R.J. Scheffer, 1984, *Ceratocystis* versus *Ophiostoma*: a reappraisal, *Mycologia* 76: 292-299.

Dick, M.W., and D.L. Hawksworth, 1985, A synopsis of the biology of the *Ascomycotina*, *Botanical Journal of the Linnean Society* 91: 175-179.

Doguet, G., 1959, Organogénie du périthèce du *Coniochaeta ligniaria*. Comparison avec l'organogénie des *Xylaria* et des discomycétes angiocarps, *Revue de Mycologie* 24: 18-38.

Eriksson, O., 1966, On *Anthostomella* Sacc., *Entosordaria* (Sacc.) Höhn. and some related genera (*Pyrenomycetes*), *Svensk Botanisk Tidskrift* 60: 315-324.

Eriksson, O., 1981, The families of bitunicate ascomycetes, *Opera Botanica* 60: 8-220.

Glawe, D.A., and J.D. Rogers, 1986, Conidial states of some species of *Diatrypaceae* and *Xylariaceae*, *Canadian Journal of Botany* 64: 1493-1498.

Harrington, T.C., 1981, Cycloheximide sensitivity as a taxonomic character in *Ceratocystis*, *Mycologia* 73: 1123-1129.

Hausner, G., J. Reid, and G.R. Klassen, 1992, Do galeate-ascospore members of the *Cephaloascaceae*, *Endomycetaceae* and *Ophiostomataceae* share a common phylogeny?, *Mycologia* 84: 870-881.

Hawksworth, D.L., 1985, Problems and prospects in the systematics of the *Ascomycotina*, *Proceedings of the Indian Academy of Sciences, Plant Sciences* 94: 319-339.

Hawksworth, D.L., 1991, The fungal dimension of biodiversity: magnitude, significance, and conservation, *Mycological Research* 95: 641-655.

Huhndorf, S.M., 1992, Neotropical ascomycetes 2. *Hypsostroma*, a new genus from the Dominican Republic and Venezuela, *Mycologia* 84: 750-758.

Ingold, C.T., 1954, The ascogenous hyphae in *Daldinia*, *Transactions of the British Mycological Society* 37: 108-110.

Jensen, J.D., 1981, The developmental morphology of *Hypoxylon serpens* in culture, *Canadian Journal of Botany* 59: 40-49.

Jensen, J.D., 1983, The development of *Diaporthe phaseolorum* variety *sojae* in culture, *Mycologia* 75: 1074-1091.

Johnston, P.R., and D.W. Minter, 1989, Structure and taxonomic significance of the ascus in the *Coryneliaceae*, *Mycological Research* 92: 422-430.

Jong, S.C., and J.D. Rogers, 1972, Conidial states of some *Hypoxylon* species, *Washington State University, College of Agriculture Technical Bulletin* 71: 1-51.

Ju, Y.-M., and J.D. Rogers, 1990, *Astrocystis* reconsidered, *Mycologia* 82: 342-349.

Ju, Y.-M., F. San Martín Gonzalez, and J.D. Rogers, 1993, Three xylariaceous fungi with scolecosporous conidia, *Mycotaxon* 47: 219-228.

Kendrick, B., and F. DiCosmo, 1979, Teleomorph-anamorph connections in ascomycetes, *In*: *The Whole Fungus* ([W.] B. Kendrick, ed.) 1: 283-359. National Museums of Canada, Ottawa.

Krug, J.C., 1977, The genus *Cainia* and a new family, *Cainiaceae*, *Sydowia* 30: 122-133.

Læssöe, T., J.D. Rogers, and A.J.S. Whalley, 1989, *Camillea*, *Jongiella*, and light-spored species of *Hypoxylon*, *Mycological Research* 93: 121-155.

Letrouit-Galinou, M.-A., 1973, Les asques des lichens et le type archaeascé, *Bryologist* 76: 30-47.

Luttrell, E.S., 1951, Taxonomy of the pyrenomycetes, *University of Missouri Studies* 24 (3): 1-120.

Luttrell, E.S., 1973, *Loculoascomycetes*, *In*: *The Fungi: An advanced treatise* (G.C. Ainsworth, F.K. Sparrow and A.S. Sussman, eds) 4A: 185-219. Academic Press, New York.

Luttrell, E.S., 1981, The pyrenomycete centrum - *Loculoascomycetes*, *In*: *Ascomycete Systematics: The Luttrellian Concept* (D.R. Reynolds, ed.): 124-137. Springer-Verlag, New York.

Luttrell, E.S., 1989, The package approach to growing peanuts, *Annual Review of Phytopathology* 27: 1-10.

Mahoney, D.P., and J.S. LaFavre, 1981, *Coniochaeta extramundana* with a synopsis of other *Coniochaeta* species, *Mycologia* 73: 931-952.

Malloch, D., and R.F. Cain, 1971, New cleistothecial *Sordariaceae* and a new family, *Coniochaetaceae*, *Canadian Journal of Botany* 49: 869-880.

Malloch, D., and C.T. Rogerson, 1977, *Pulveria*, a new genus of *Xylariaceae* (ascomycetes), *Canadian Journal of Botany* 55: 1505-1509.

Miller, J.H., 1961, *A Monograph of the World Species of Hypoxylon*, University of Georgia Press, Athens.

Müller, E., and J.A. von Arx, 1973, *Pyrenomycetes*: *Meliolales*, *Coronophorales*, *Sphaeriales*, *In*: *The Fungi: An advanced treatise* (G.C. Ainsworth, F.K. Sparrow, and A.S. Sussman, eds) 4A: 87-132. Academic Press, New York.

Müller, E., and R. Corbaz, 1956, Kulturversuche mit Ascomyceten III, *Sydowia* 10: 181-188.

Munk, A., 1957, Danish pyrenomycetes, *Dansk Botanisk Arkiv* 17: 1-491.

Nannfeldt, J.A., 1972, *Camarops* Karst. (*Sphaeriales*, *Boliniaceae*) with special regard to European species, *Svensk Botanisk Tidskrift* 66: 335-376.

Parguey-Leduc, A., and M.-C. Janex-Favre, 1981, The ascocarps of ascohymenial pyrenomycetes, *In*: *Ascomycete Systematics: The Luttrellian Concept* (D.R. Reynolds, ed.): 102-123. Springer-Verlag, New York.

Petrini, L.E., and E. Müller, 1986, Haupt-und Nebenfruchtformen Europäischer *Hypoxylon* - arten (*Xylariaceae*, *Sphaeriales*) und verwandter pilze, *Mycologia Helvetica* 1: 501-627.

Pirozynski, K.A., 1974, *Xenotypa* Petrak and *Graphostroma* gen. nov., segregates from *Diatrypaceae*, *Canadian Journal of Botany* 52: 2129-2135.

Rappaz, F., 1987, Taxonomic et nomenclature des Diatrypacées à asques octospores, *Mycologia Helvetica* 2: 285-648.

Reynolds, D.R., 1981, *Ascomycete Systematics: The Luttrellian Concept*, Springer-Verlag, New York.

Reynolds, D.R., 1989, The bitunicate ascus paradigm, *Botanical Review* 55: 1-52.

Roane, M.K., G.J. Griffin, and J.R. Elkins, 1986, *Chestnut Blight, other Endothia diseases, and the genus Endothia*, American Phytopathological Society Press, St Paul, Minn.

Rodrigues, K.F., and G.J. Samuels, 1989, Studies in the genus *Phylacia* (*Xylariaceae*), *Memoirs of the New York Botanical Garden* 49: 290-297.

Rogers, J.D., 1971, Observations on the ascogenous system of *Hypoxylon microplacum*, *Canadian Journal of Botany* 49: 1075-1077.

Rogers, J.D., 1972, *Hypoxylon cohaerens*: cytology of the ascus, *Mycopathologia et Mycologia Applicata* 48: 161-165.

Rogers, J.D., 1975, *Hypoxylon serpens*: cytology and taxonomic considerations, *Canadian Journal of Botany* 53: 52-55.

Rogers, J.D., 1979, The *Xylariaceae*: systematic, biological and evolutionary aspects, *Mycologia* 71: 1-42.

Rogers, J.D., 1985, Anamorphs of *Xylaria*: taxonomic considerations, *Sydowia* 38: 255-262.

Rogers, J.D., and G.J. Samuels, 1987, *Camarops biporosa* sp. nov. from French Guiana, *Mycotaxon* 28: 415-417.

Rogerson, C.T., 1970, The hypocrealean fungi (ascomycetes, *Hypocreales*), *Mycologia* 62: 865-910.

Romero, A.I., and G.J. Samuels, 1991, Studies on xylophilous fungi from Argentina. VI. *Ascomycotina* on *Eucalyptus viminalis* (*Myrtaceae*), *Sydowia* 43: 228-248.

Rossman, A.Y., 1976, *Xylobotryum andinum*, a tropical pyrenomycete from norther California, *Mycotaxon* 4: 179-183.

Samuels, G.J., and E. Müller, 1980, Life histories of Brazilian ascomycetes 8. *Thamnomyces chordalis* (anam.: *Nodulisporium*) and *Camillea bacillum* (anam.: *Geniculosporium*) with notes on taxonomy of the *Xylariaceae*, *Sydowia* 33: 274-281.

Samuels, G.J., and J.D. Rogers, 1987, *Camarops flava* sp. nov., *Apiocamarops alba* gen. et sp. nov., and notes on *Camarops scleroderma* and *C. ustulinoides*, *Mycotaxon*, 28:45- 59.
Samuels, G.J., J.D. Rogers, and E. Nagasawa, 1987, Studies in the *Amphisphaeriaceae* (sensu lato). 1. *Collodiscula japonica* and its anamorph, *Acanthodochium collodisculae*, *Mycotaxon* 28: 453-459.
Samuels, G.J., and A.Y. Rossman, 1992, *Thuemenella* and *Sarawakus*, *Mycologia* 84: 26-40.
Sherwood, M.A., 1977, The ostropalean fungi, *Mycotaxon* 5: 1-277.
Sherwood, M.A., 1981, Convergent evolution in discomycetes from wood and bark, *Botanical Journal of the Linnean Society* 82: 15-34.
Uecker, F.A., 1976, Development and cytology of *Sordaria humana*, *Mycologia* 68: 30-46.
Uecker, F.A., 1989, A timed sequence of development of *Diaporthe phaseolorum* (*Diaporthaceae*) from *Stokesia laevis*, *Memoirs of the New York Botanical Garden* 49: 38-50.

THE DEVELOPMENT OF HOLOMORPHIC CONCEPTS IN OPHIOSTOMATALEAN ASCOMYCETES

M.J. Wingfield[1], B.D. Wingfield[1] and W.B. Kendrick[2]

[1]Department of Microbiology and Biochemistry
University of the Orange Free State
P.O. Box 339
Bloemfontein 9300, South Africa

[2]Department of Biology
University of Waterloo
Waterloo, Ontario N2L 3G1
Canada

SUMMARY

The ophiostomatoid fungi, including, the genera *Ceratocystis* s.str., *Ophiostoma* and *Ceratocystiopsis*, have strong anamorph/teleomorph connections and provide an outstanding model for evaluating taxonomic schemes based on morphology. Recent evidence from extended collections as well as ultrastructural and molecular studies suggests that these fungi have evolved several, if not many times. Primary taxonomic characters for the group such as ascomatal structure and ascospore shape now appear to have been misleading. A great deal of convergence has evidently occurred and re-evaluation of morphological characters based on additional molecular studies is needed before a new and more meaningful classification can be established for this group. Ultimately, we expect that numerous genera, belonging to at least two and perhaps more distantly related groups of fungi will emerge.

INTRODUCTION

The ophiostomatoid fungi, including the well known genera *Ceratocystis* and *Ophiostoma*, have been known since early this Century. They include many important plant pathogens particularly of trees, symbionts of insects, species of medical importance as well as species with potential industrial and biotechnological significance (Wingfield et al., 1993). The taxonomy of the ophiostomatoid fungi has been the subject of considerable controversy, virtually since their initial discovery. This controversy continues today and there are many conflicting opinions concerning the subdivision of the group, not only at the generic and species levels but also at higher taxonomic rankings (Upadyhyay, 1993).

Ceratocystis s.lat. encompasses approximately 116 species of superficially similar ascomycetes which may be only distantly related. This taxonomic aggregate is commonly subdivided into three genera: *Ceratocystis* s.str., *Ophiostoma* and *Ceratocystiopsis* (de Hoog and Scheffer, 1984; Wingfield et al., 1993). *Ceratocystis* s.str. can be distinguished from the other two genera by four features: (1) the presence of *Chalara* anamorphs with ring wall building conidial development (Minter et al., 1982); (2) the absence of rhamnose and cellulose in cell walls (Rosinski and Campana, 1964; Jewell, 1974; Weijman and de Hoog, 1975); (3) sensitivity to even low concentrations of the antibiotic cycloheximide (Harrington, 1981); (4) an association with non-specific insect vectors such as flies and nitidulid beetles (Dowding, 1973; Juzwik and French, 1983). In contrast, *Ophiostoma* and *Ceratocystiopsis* species have: (1) anamorphs in a number of genera such as *Leptographium*, *Graphium*, *Sporothrix* and *Hyalorhinocladiella* with apical wall building conidial development; (2) rhamnose and cellulose in their cell walls; (3) tolerance to high levels of cycloheximide; (4) in many cases, an association with specific insect vectors such as scolytid (*Coleoptera: Scolytidae*) bark beetles (Upadhyay, 1981). *Ceratocystiopsis* has been separated from *Ophiostoma* on the basis of its falcate ascospores although this subdivision is beset with a number of incongruities (Wingfield, 1993).

The major characteristic unifying the ophiostomatoid fungi is an ascoma usually with a neck of variable length containing evanescent asci and ascospores that are extruded from the ascomata in gloeoid masses. They are thus elegantly adapted to insect dispersal and often co-exist with their specific arthropod vectors. Adaptation to insect dispersal has apparently resulted in convergence and the assumption of superficial morphological similarity of ascomata in even distantly related fungi. This problem is further complicated by the presence of anamorph genera typical of ophiostomatoid fungi but also found with connections to other unrelated teleomorphs or of unknown affinity. Dependance on superficial morphological characteristics in taxonomic schemes has therefore led to a great deal of confusion regarding the phylogeny of these fungi.

The ophiostomatoid fungi include outstanding examples of contemporary problems in the taxonomy of the ascomycetes. The aim of this contribution is to illustrate some of these examples, provide a summary of recent developments and to sketch a perspective of future prospects.

ANAMORPHS OF *CERATOCYSTIS* S.LAT.

Anamorphic states have had a considerable influence on the taxonomy of ophiostomatoid fungi. Münch (1907) recognized the presence of "endoconidia" in some species of *Ceratocystis* s.lat. and established *Endoconidiophora* for *E. coerulescens* because it had a *Chalara* anamorph (i.e. *Chalara ungeri*). Similarly, Melin and Nannfeldt (1934) recognized two groups within *Ophiostoma* based on the presence of *Chalara* anamorphs in one and anamorphs that produce conidia exogenously in the other.

Anamorph species currently accommodated in *Ophiostoma* and *Ceratocystiopsis* have had a tremendously complex taxonomic history. Goidánich (1936) recognized that species of *Ophiostoma* with *Leptographium* anamorphs in the *Graphium/Sporothrix* complex were distinct from those with *Chalara* anamorphs and established *Grosmannia* for the former. The introduction of mode of conidium development as a primary taxonomic characteristic by Hughes (1953) led to a proliferation of generic names associated with ophiostomatoid fungi. Hughes (1953) initiated this trend by distinguishing the genus *Verticicladiella* from *Leptographium* based on sympodial *vs* percurrent proliferation of conidiogenous cells. Kendrick (1961) followed this lead and established *Phialocephala* for species in the *Leptographium* complex with phialidic conidium development. This trend continued during the 1960s and 1970s and ultimately no fewer than 18 generic names were recognized for anamorphs of ophiostomatoid fungi (Wingfield et al., 1993; Mouton and Wingfield, 1993). Many of these names were based on analogous anamorphs: colourless *vs* pigmented; mononematous *vs* synnematous; phialidic *vs* sympodial *vs* percurrent conidiogenesis.

Wingfield (1985) undertook a comprehensive study of conidium development in species of *Leptographium* and *Verticicladiella* using scanning (SEM) and transmission (TEM) electron microscopy. Results of this study showed that percurrent proliferation of conidiogenous cells was the norm in these fungi and that delayed secession of conidia leads to an illusion of sympodial development when they are viewed using light microscopy (Wingfield, 1985; van Wyk et al., 1988; Mouton et al., 1993a). Subsequent studies on other anamorphs of ophiostomatoid fungi (Wingfield et al., 1987; Wingfield et al., 1991; Mouton et al., 1992; Mouton and Wingfield, 1993; Seifert and Okada, 1993) have also shown that conidium development as interpreted by light microscopy has led to an unnecessary proliferation of names for anamorphs of ophiostomatoid fungi. Electron microscopic studies of conidium development in the remaining anamorphs of ophiostomatoid fungi are continuing in our laboratory. We ultimately expect to reduce current names for anamorphs of *Ceratocystis* s.lat. to approximately five reasonably well defined genera including *Chalara*, *Knoxdaviesia*, *Leptographium*, *Graphium*, *Sporothrix* and *Hyalorhinocladiella* (Mouton et al., 1993b).

While anamorph characters have been particularly useful in the taxonomy of ophiostomatoid fungi, they have, in some cases also resulted in confusion. For example, ophiostomatoid fungi with *Chalara* anamorphs have easily been segregated in *Ceratocystis* s.str. One exception has been *Ceratocystiopsis falcata* which has a *Chalara* state but, in terms of its teleomorph, is very different from other species of *Ceratocystis* s.str. A similarly enigmatic situation is encountered in *Ceratocystis autographa*. This fungus has a well developed *Chalara* state suggesting that it would best be accommodated in *Ceratocystis* s.str. It, however, is also reported to have a *Sporothrix* state characteristic of typical *Ophiostoma* species. In a recent SEM and TEM study of the anamorphs of *C. autographa* (Benade, 1993), it has been found that the purported *Chalara* state is only superficially similar to that genus. Although it would have been difficult to make these assessments using light microscopy, electron microscopic observations have clearly shown that conidia possess single attachment points and are produced in false chains atypical of *Chalara*. The so-called *Chalara* anamorph of *C. autographa* clearly deserves a new generic disposition and we suspect that the same might be true of other anamorphs of *Ceratocystis* s.str.

Convergent evolution and adaptation to transmission by insects has apparently led to the adoption of superficially similar morphologies in distantly related anamorph states. For example, we believe that *Ceratocystis autographa* is more typical of *Ophiostoma* than *Ceratocystis* s.str. Its *Chalara*-like anamorph is in all probability no more than a manifestation of convergence.

A second fascinating example of convergence in ophiostomatoid fungi is encountered in the *Knoxdaviesia* anamorph of *Ceratocystiopsis proteae* which occurs in *Protea* infructescences found only at the southern tip of Africa (Wingfield et al., 1988). In terms of anamorphs of ophiostomatoid fungi, *Knoxdaviesia* superficially resembles *Leptographium* with dark mononematous conidiophores terminating in gloeoid masses of spores. However, this fungus produces conidia from typical phialidic conidiogenous cells which are unique amongst ophiostomatoid anamorphs (Mouton and Wingfield, 1993). We suspect that *Ceratocystiopsis proteae* and the more recently described *Ophiostoma capense* (Wingfield and van Wyk, 1993) with *Knoxdaviesia* anamorphs from *Protea* infructescences are unrelated to other ophiostomatoid fungi, and their resemblance to these fungi has evolved as an adaptation to insect transmission. Similar examples of convergence might well be found amongst other species of *Ophiostoma* and *Ceratocystiopsis* when less subjective characters than morphology based solely on light microscopy are used to examine these fungi.

Anamorph genera associated with ophiostomatoid fungi are also commonly encountered in distantly related groups of fungi. The genera *Chalara* and *Sporothrix* perhaps provide the best examples of this complex situation. *Chalara* species are, for example found in at least seven ascomycete genera representing five different orders (Nag Raj and Kendrick, 1993) and *Sporothrix* species occur in the *Endomycetes*, *Ascomycetes* and *Basidiomycetes* (de Hoog, 1993). Similarly, many species of these genera that either have not been connected to teleomorphs, or simply exist as apparent anamorphic holomorphs, might or might not be related to ophiostomatoid fungi It is possible that a more detailed examination of these fungi might lead to the discovery of new morphological characters which will enable us to separate them into phylogenetically meaningful groups. Certainly, there is sufficient evidence to believe that similar morphological forms have evolved many times and that any attempt to establish natural groups based on these morphs is likely to be beset by many problems.

TELEOMORPH CHARACTERISTICS

Ascomata of ophiostomatoid fungi are found in niches frequented by insects and many species are formed in the galleries of scolytid bark beetles with which they are associated (Upadhyay, 1981; Wingfield et al., 1993). They are typically beaked and have necks that vary in length from being virtually non-existent to extremely long. The apex of these necks are often surrounded by ostiolar hyphae that may be either straight, divergent, or convergent. Ascospores are produced in gloeoid masses and accumulate at the apices of the ascomatal necks facilitating attachment to their vectors.

Ascomatal Morphology

The morphology of the ascomata is the most important characteristic on which the taxonomy of the ophiostomatoid fungi is based. This characteristic has also led to considerable confusion in the classification of the group. Indeed, the first ophiostomatoid fungus to be described was misidentified as a coelomycete, and it was not until the study of Elliott (1925) that this error was corrected. The incorrect placement of this group of fungi resulted from the fact that asci are evanescent and ascospores are extruded from the ascomata without any evidence of the presence of asci.

"Perithecium" is perhaps not a fully appropriate term for the ascomata of ophiostomatoid fungi. Strictly speaking, perithecia are found in pyrenomycetous fungi where asci are formed in unitunicate asci which, in many cases, have apical structures to facilitate the forceful discharge of ascospores. Indeed, considerable controversy has surrounded the placement of these fungi, which have been either considered to be either pyrenomycetes because of their long necked ascomata, or plectomycetes because of their evanescent and irregularly shaped asci.

The ascomatal form in the ophiostomatoid fungi obviously has evolved in association with insects. Therefore, the ecological pressure has been on these fungi to assume a morphology that will ensure their dispersal. From the taxonomic standpoint this has led to considerable confusion as all fungi with typical ophiostomatoid ascomata have tended to be treated as being phylogenetically related. A common trend has thus been to treat all species in the single genus *Ceratocystis*. Other authors, have accepted one or a number of segregate genera such as *Ophiostoma*, *Ceratocystis* s.str., *Ceratocystiopsis*, and *Europhium* (Wingfield et al., 1993). These have, however, usually been considered as being related and treated in the family *Ophiostomataceae*, order *Microascales* of the *Plectomycetes* (Upadhyay, 1981).

Contemporary studies have provided substantial evidence to suggest that members of the genus *Ceratocystis* s.str. are phylogenetically distinct from species of *Ophiostoma* and *Ceratocystiopsis* as discussed above. More recently, application of non-subjective molecular techniques have confirmed that species of *Ophiostoma* and *Ceratocystiopsis* are phylogenetically distinct from species of *Ceratocystis* s.str. For example, Hausner et al. (1992), Spatafora and Blackwell (1993; this volume) have compared sequence of the small subunit ribosomal RNA and have shown that *Ceratocystis* is phylogenetically distant from *Ophiostoma*. There also appears to be reasonable congruence between the results of these studies and the results of recent observations on the ultrastructure of centrum development in *Ceratocystis* s.str. and *Ophiostoma* (van Wyk and Wingfield, 1991a, 1991b, 1991c; van Wyk et al., 1991; van Wyk and Wingfield, 1993; van Wyk et al., 1993).

The fascinating group of ophiostomatoid fungi recently discovered in the infructescences of *Protea* species in the Cape Fynbos Biome of South Africa also provide an interesting example of convergent evolution in the ophiostomatoid fungi. The first of these fungi to be described was disposed in the genus *Ceratocystiopsis* as *C. proteae* because it has falcate ascospores (Wingfield et al., 1988). The second species of ophiostomatoid fungus from *Protea* infructescences is very similar to *C. proteae* and also has a *Knoxdaviesia* anamorph (Wingfield and van Wyk, 1993). This fungus is unusual in having ascospores that are not falcate and it has thus been placed in *Ophiostoma* as *O. capense*.

Morphologically, *C. proteae* and *O. capense* are very similar to each other and distinct from *Ceratocystis* s.str. They are, however, sensitive to cycloheximide in the growth media which is strong evidence to suggest that they do not reside in *Ceratocystis* s.str. Evidence from molecular studies (Hausner et al., 1993a, 1993b; authors unpubl.) also support the view that these fungi should reside in a genus other than *Ophiostoma*, *Ceratocystis* or *Ceratocystiopsis*. The taxonomic ramifications of these studies will be published later. Of interest in this particular discussion is the fact that the ophiostomatoid fungi with *Knoxdaviesia* anamorphs provide another example of convergent evolution in this group of fungi. All indications are, therefore, that ecological pressure to ensure dispersal by insects has led to similar morphological manifestations in numerous unrelated fungi.

Ascospore Morphology

Ascospore morphology has been one of the most important characteristics upon which taxonomy of ophiostomatoid fungi has been based. Diverse ascospore forms are encountered in this group, among which galeate or hat-shaped ascospores are perhaps best known. Other than galeate ascospores, species in this group of fungi can have ascospores without sheaths (extended outer wall layers) or with a wide range (inequilateral, orange section, pillow, falcate) of sheath forms (Upadhyay, 1981; van Wyk et al., 1993). These ascospore forms have thus been used as a basis for subdividing *Ceratocystis* s.lat. at the sectional level (Griffin, 1968; Olchowecki and Reid, 1974; Upadhyay, 1981) and even at the generic level in the case of *Ceratocystiopsis* (Upadhyay and Kendrick, 1975; Upadhyay, 1981).

Galeate ascospores are encountered both in *Ceratocystis* s.lat. and in a number of yeast genera. This has been the source of considerable debate as to whether relatedness between these groups of fungi might be implied. For example, Redhead and Malloch (1977) included *Ophiostoma* and *Ceratocystis* with yeasts in the *Endomycetales* primarily on the basis of this character. This grouping was contested by Benny and Kimbrough (1980) because it was based solely on morphology as observed using light microscopy, and did not take into consideration available developmental, ultrastructural and chemical evidence. Von Arx and van der Walt (1987) recognized that the ophiostomatoid fungi were a polyphyletic group yet still contended that the galeate ascospores were indicative of some relatedness. Kendrick et al. (1993) added further commentary on these possible relationships.

Galeate ascospores in *Ceratocystis* and *Ophiostoma* as determined by light microscopy have been the subject of recent ultrastructural study (van Wyk and Wingfield 1991a, 1991b, 1991c; van Wyk et al., 1991; van Wyk et al., 1993). It was determined that the apparently similar galeate ascospores in species of *Ceratocystis* and *Ophiostoma* are fundamentally different in morphology. Ascospores of *Ceratocystis fimbriata* and *C. moniliformis* are typically bowler hat-shaped and develop in pairs (van Wyk et al., 1991). In contrast, those of *Ophiostoma cucullatum* and *O. davidsonii* develop individually, and their brims are triangular (van Wyk and Wingfield, 1991a, 1991b). These observations provide additional evidence that species of *Ceratocystis* and *Ophiostoma* are only superficially similar.

Hausner et al. (1992) have recently considered the evolution of the apparent galeate ascospore using analysis of small subunit ribosomal genes sequences. This study reaffirms the fact that species of *Ceratocystis* and *Ophiostoma* with apparently similar ascospores are distantly related. Furthermore, these authors provide evidence that the ophiostomatoid fungi are also distantly related to yeasts with galeate ascospores.

In an analysis of sequence data derived from ribosomal RNA, B. Wingfield et al. (1994) have considered the relatedness of various species of *Ceratocystis* s.str. Results of this study have shown that ascospore shape, even within this more clearly defined subgroup of *Ceratocystis* s.lat. is a poor taxonomic character. Indeed, all indications are that *Ceratocystis* s.str. itself represents a phylogenetically diverse group of fungi.

The occurrence of two different ascospore forms in *C. proteae* and *O. capense* which are very similar fungi has led us to examine the ascospores of *C. proteae* more carefully. In an ultrastructural examination of ascospores in this fungus (van Wyk and Wingfield, 1993), it has been shown that the ascospores are reniform without sheaths. The falcate outline detected by light microscopy is apparently an illusion resulting from the adherence of ascomatal remains to the apices of the ascospores when they are released. This study has provided additional evidence that *O. capense* and *C. proteae* are very closely related and should reside in the same genus.

CONCLUSIONS

The ophiostomatoid fungi are one of our best known groups of fungi with strong anamorph teleomorph connections. Their apparently diagnostic morphology has clearly evolved several if not many times, and thus the taxonomic schemes which we have followed for this group have been wholly misleading. The most important and supposedly conservative taxonomic characters for the ophiostomatoid fungi, ie ascomatal structure and ascospore shape, appear to have been relatively useless. We are now faced with the challenge of finding meaningful groupings and characteristics on which to base a future classification for these fungi.

Although molecular studies have given us very clear indications that the taxonomic schemes that we have followed for the ophiostomatoid fungi in the past have been misleading, we require substantially more data before an ideal classification can be developed. We now know that there has been a great deal of convergence in the evolution of the ophiostomatoid fungi although the exact extent of this must still be determined. Many more examples of these fungi now need to be examined in order to determine the limits of the various natural groups. Once this has been accomplished, we will be in a position to re-evaluate morphological characters and thus to provide a meaningful classification of members of the various groups. At this level, we suspect that numerous genera, belonging to at least two, and perhaps more, distantly related groups of fungi will ultimately result.

ACKNOWLEDGEMENTS

We thank the Foundation for Research Development, South Africa and the National Science and Engineering Research Council, Canada for financial support which also enabled the senior and junior authors to participate in the First International Workshop on Ascomycete Systematics. The last author also acknowledges support from NATO which facilitated his attendance at the workshop.

REFERENCES

von Arx, J.A., and J.P. van der Walt, 1987, *Ophiostomatales* and *Endomycetales*, *In: The Expanding Realm of Yeast-Like Fungi* (G.S. de Hoog and A.C.M. Weijman, eds): 167-176. Elsevier Science Publishers, Amsterdam.

Benade, E., 1993, *A Study of Conidium Development in Mycelial Anamorphs of Ophiostoma*. MSc thesis, University of the Orange Free State.

Benny, G.L., and J.W. Kimbrough, 1980, A synopsis of the orders and families of *Plectomycetes* with keys to genera, *Mycotaxon* 12: 1-91.

de Hoog, G.S., and R.J. Scheffer, 1984, *Ceratocystis* versus *Ophiostoma*: a reappraisal, *Mycologia* 76: 292-299.

de Hoog, G.S. 1993, *Sporothrix*-like anamorphs of *Ophiostoma* species and other fungi, *In*: *Ceratocystis and Ophiostoma: Taxonomy, Ecology and Pathogenicity* (M.J. Wingfield, K.A. Seifert and J.F. Webber, eds): 49-56. American Phytopathological Society Press, St Paul.

Dowding, P., 1973, The evolution of insect-fungus relationships in the primary invasion of forest timber, *In: Invertebrate-microbial Interactions* (J.M. Anderson, A.D.M. Rayner and D.W.H. Walton, eds): 133-153. Cambridge University Press, New York.

Elliott, J.A., 1925, A cytological study of *Ceratostomella fimbriata* (E. & H.) Elliott, *Phytopathology* 16: 417-422.

Goidánich, G., 1936, Il genere di ascomiceti *Grosmannia,Bollettino della Stazione di Patologia Vegetale, Roma, n.s.* 16: 26-60.

Griffin, H.D., 1968, The genus *Ceratocystis* in Ontario, *Canadian Journal of Botany* 46: 689-718.

Harrington, T.C., 1981, Cycloheximide sensitivity as a taxonomic character in *Ceratocystis Mycologia* 73:1123-1129.

Hausner, G., J. Reid, and G.R. Klassen, 1992, Do galeate-ascospore members of the *Cephaloascaceae*, *Endomycetaceae* and *Ophiostomataceae* share a common phylogeny? *Mycologia* 84: 870-881.

Hausner, G., J. Reid, and G.R. Klassen, 1993a, *Ceratocystiopsis*: a reappraisal based on molecular criteria, *Mycological Research* 97:625-633.

Hausner, G., J. Reid, and G.R. Klassen, 1993b, On the subdivision of *Ceratocystis* s.l., based on partial ribosomal DNA sequences, *Canadian Journal of Botany* 71:52-63

Hughes, S.J., 1953, Conidiophores, conidia and classification, *Canadian Journal of Botany* 31: 577-659.

Jewell, T.R., 1974, A qualitative study of cellulose distribution in *Ceratocystis* and *Europhium*, *Mycologia* 66: 139-146.

Juzwik, J., and D.W. French, 1983, *Ceratocystis fagacearum* and *C. piceae* on the surface of free-flying and fungus mat inhabiting nitidulids, *Phytopathology* 73: 1164-1168.

Kendrick, W.B., 1961, The *Leptographium* complex. *Phialocephala* gen. nov., *Canadian Journal of Botany* 39: 1079-1085.

Kendrick, W.B., J.P. van der Walt, and M.J. Wingfield, 1993, Relationships between the yeasts with hat-shaped ascospores and the ophiostomatoid fungi, *In: Ceratocystis and Ophiostoma: Taxonomy, Ecology and Pathogenicity* (M.J. Wingfield, K.A. Seifert and J.F. Webber, eds): 67-70. American Phytopathological Society Press, St Paul.

Melin, E., and J.A. Nannfeldt, 1934, Researches into the blueing of ground woodpulp, *Sveriges Skogsvårdsförenings Tidskrift* 32: 397-616.

Minter, D.W., P.M. Kirk, and B.C. Sutton, 1982, Holoblastic phialides, *Transactions of the British Mycological Society* 79: 75-93.

Mouton, M., and M.J. Wingfield, 1993, Conidium development in synnematous anamorphs of *Ophiostoma*, *Mycotaxon* 46: 429-436.

Mouton, M., M.J. Wingfield, and P.S. van Wyk, 1992, The anamorph of *Ophiostoma francke-grosmanniae* is a *Leptographium*, *Mycologia* 84: 857-862.

Mouton, M., M.J. Wingfield, and P.S. van Wyk, 1993a, Conidium development in the *Knoxdaviesia* anamorph of *Ceratocystiopsis proteae*, *Mycotaxon* 46: 363-370.

Mouton, M., M.J. Wingfield, and P.S. van Wyk. 1993b, Conidium development in anamorphs of *Ceratocystis sensu lato*: a review, *South African Journal of Science*: in press.

Münch, E., 1907, Die Blaufaule des nadelholzes, *Naturwissenschaftliche Zeitschrift für Forst- und Landwirtschaft* 5: 531-573.

Nag Raj, T.R., and W. B. Kendrick, 1993, The anamorph as generic determinant in the holomorph: the *Chalara* connection in the ascomycetes, with special reference to ophiostomatoid fungi, *In: Ceratocystis and Ophiostoma: Taxonomy, Ecology and Pathogenicity* (M.J. Wingfield, K.A. Seifert and J.F. Webber, eds): 57-70. American Phytopathological Society Press, St Paul.

Olchowecki, A., and J. Reid, 1974, Taxonomy of the genus *Ceratocystis* in Manitoba, *Canadian Journal of Botany* 52:1675-1711.

Redhead, S.A., and D.W. Malloch, 1977, The *Endomycetaceae*: new concepts, new taxa, *Canadian Journal of Botany* 55: 701-1711.

Rosinski, M.A., and R.J. Campana, 1964, Chemical analysis of the cell wall of *Ceratocystis ulmi*, *Mycologia* 56: 738-744.

Seifert, K.A., and G. Okada, 1993, *Graphium* anamorphs of *Ophiostoma* species and similar anamorphs of other ascomycetes, *In: Ceratocystis and Ophiostoma: Taxonomy, Ecology and Pathogenicity.* (M.J. Wingfield, K.A. Seifert and J.F. Webber, eds): 25-39. American Phytopathological Society Press, St Paul.

Spatafora, J.W., and M. Blackwell, 1993, The polyphyletic origins of ophiostomatoid fungi, *Mycological Research* 98: 1-9.

Upadhyay, H.P., 1981, A monograph of *Ceratocystis* and *Ceratocystiopsis*. University of Georgia Press, Athens, Georgia.

Upadhyay, H.P. 1993, Classification of the ophiostomatoid fungi, *In: Ceratocystis and Ophiostoma: Taxonomy, Ecology and Pathogenicity* (M.J. Wingfield, K.A. Seifert and J.F. Webber, eds): 7-13. American Phytopathological Society Press, St Paul.

Upadhyay, H.P., and W.B. Kendrick, 1975, Prodromus for a revision of *Ceratocystis* (*Microascales*: *Ascomycetes*) and its conidial states, *Mycologia* 67: 798-805.

van Wyk, P.S., M.J. Wingfield, and W.F.O. Marasas, 1988, Differences in synchronization of stages of conidial development in *Leptographium* species, *Transactions of the British Mycological Society* 90: 451-456.

van Wyk, P.W.J., and M.J. Wingfield, 1991a, Ultrastructure of ascosporogenesis in *Ophiostoma davidsonii*, *Mycological Research* 95: 725-730.

van Wyk, P.W.J., and M.J. Wingfield, 1991b, Ascospore ultrastructure and development in *Ophiostoma cucullatum*, *Mycologia* 83: 698-707.

van Wyk, P.W.J., and M.J. Wingfield, 1991c, Ultrastructural study of ascospore development in *Ophiostoma distortum* and *O. minus*, *Canadian Journal of Botany* 69: 2529-2538.

van Wyk, P.W.J., and M.J. Wingfield, 1993, Fine structure of ascosporogenesis in *Ceratocystiopsis proteae*, *Canadian Journal of Botany*: in press.

van Wyk, P.W.J., M.J. Wingfield, and P.S. van Wyk, 1991, Ascospore development in *Ceratocystis moniliformis*, *Mycological Research* 95: 96-103.

van Wyk, P.W.J., M.J. Wingfield, and P.S. van Wyk, 1993, Ultrastructure of centrum and ascospore development in selected *Ceratocystis* and *Ophiostoma* species, *In: Ceratocystis and Ophiostoma: Taxonomy, Ecology and Pathogenicity* (M.J. Wingfield, K.A. Seifert and J.F. Webber, eds): 126-131. American Phytopathological Society Press, St Paul.

Weijman, A.C.M., and G.S. de Hoog, 1975, On the subdivision of the genus *Ceratocystis*, *Antonie van Leeuwenhoek* 41: 353-360.

Wingfield, B.D., W.S. Grant, J.F. Wolfaardt, and M.J. Wingfield, 1994, Ribosomal RNA sequence phylogeny is not congruent with ascospore morphology among species in *Ceratocystis sensu stricto*, *Molecular Biology and Evolution*: in press.

Wingfield, M.J., 1985, Reclassification of *Verticicladiella* based on conidial development, *Transactions of the British Mycological Society* 85: 81-93.

Wingfield, M.J., 1993, Problems in delineating the genus *Ceratocystiopsis*, *In: Ceratocystis and Ophiostoma: Taxonomy, Ecology and Pathogenicity* (M.J. Wingfield, K.A. Seifert and J.F. Webber, eds): 20-24. American Phytopathological Society Press, St Paul.

Wingfield, M.J., W.B. Kendrick, and P.S. van Wyk, 1991, Analysis of conidium ontogeny in anamorphs of *Ophiostoma*. *Pesotum* and *Phialographium* are synonyms of *Graphium*, *Mycological Research* 95: 1328-1333.

Wingfield, M.J., K.A. Seifert, and J.F. Webber (eds.), 1993, *Ceratocystis and Ophiostoma: Taxonomy, Ecology and Pathogenicity*. American Phytopathological Society Press, St Paul.

Wingfield, M.J., and P.S. van Wyk, 1993, A new species of *Ophiostoma* from *Protea* infructescences in South Africa, *Mycological Research*: in press.

Wingfield, M.J., P.S. van Wyk, and W.F.O. Marasas, 1988, *Ceratocystiopsis proteae* sp. nov. with a new anamorph genus, *Mycologia* 80: 23-30.

Wingfield, M.J., P.S. van Wyk, and B.D. Wingfield, 1987, Reclassification of *Phialocephala* based on conidial development, *Transactions of the British Mycological Society* 89: 509-520.

PROBLEMS IN THE CLASSIFICATION OF FISSITUNICATE ASCOMYCETES

O.E. Eriksson

Institute of Ecological Botany
University of Umeå, S-90187 Umeå, Sweden

SUMMARY

The classification of orders and subclasses in *Ascoloculares* by Barr (1987) is compared with the classification adopted in *Systema Ascomycetum*, and problems in classification of orders and higher taxa in fissitunicate and related fungi are discussed.

INTRODUCTION

The term "fissitunicate" was proposed by Dughi (1956) for the "Jack-in-the-box" type of ascus, with a distinctly layered wall, with an outer rigid and an inner extensible layer. However, this paper will include also some other types of asci which have been referred to as "bitunicate", notably the rostrate or extenditunicate type.

Pringsheim (1858) published the first illustration of an ascus that we know today has a fissitunicate dehiscence. Many nineteenth century mycologists saw such asci, described their structure and even based generic or specific names on them (e.g. *Diplotheca*), but these observations did not result in any major changes in ascomycete classification.

Nannfeldt (1932) described the subclass *Ascoloculares*, later called *Loculoascomycetes*. His concepts were partly based on scattered information on this group of fungi in many papers by von Höhnel, H. Sydow, Theissen, and Petrak. The most important criteria were the ascomatal morphology and ontogeny, but Nannfeldt was fully aware of the connection between the morphology of the ascomata and the asci (*op. cit.*: 27). He gave a detailed description of the function of a fissitunicate ascus.

Luttrell (1951) coined the term "bitunicate" and erected the group *Bitunicatae*, a name that is usually considered synonymous with *Loculoascomycetes*. He also defined some ascomatal developmental types and extended the studies by earlier workers, but in principle his concepts were the same as Nannfeldt's.

Many mycologists have contributed to the development of the classification of ascolocular ascomycetes since the early 1950s. Müller and von Arx have published several important papers, the most comprehensive on ascolocular ascomycetes being "A re-evaluation of the bitunicate ascomycetes with keys to families and genera" (von Arx and Müller, 1975). This work, probably more than any other, provided an impetus to several more recent studies of this group of ascomycetes, and indirectly it was important for the genesis of the *Systema Ascomycetum* project.

Many other mycologists deserve to be mentioned here, but I restrict myself to one, Margaret E. Barr. She has produced a large number of important papers, and independently proposed an outline of the ascomycetes, and classifications of almost all orders characterized by perithecioid ascomata. In many instances her classifications differ from those adopted in the *Outlines*. This is discussed and explained in more detail below.

PROBLEMS IN THE CLASSIFICATIONS

In the 1990 *Outline* (Eriksson and Hawksworth, 1991), the ascomycetes discussed in this paper comprised about 7200 species in 818 genera, 72 families, and six orders. There are innumerable problems in the classification of all these taxa, but I will largely restrict my discussion to the problems of classifying orders and supraordinal taxa.

Barr (1987) recognized 11 orders in the class *Loculoascomycetes*. She divided the class into four subclasses, characterized by either different types or the absence of hamathecium. This classification has been accepted by a large number of mycologists. The classification in *Systema Ascomycetum* differs in two respects: it does not use taxonomic units above the ordinal levels, and it provisionally uses a very wide concept of the order *Dothideales*. The following orders, accepted by Barr, were included in *Dothideales* s.lat. by Eriksson and Hawksworth (1991): *Asterinales*, *Capnodiales*, *Chaetothyriales*, *Dothideales* s.str., *Melanommatales* (including *Pyrenulales*, *Microthyriales*), *Myriangiales*, and *Pleosporales* (including *Hysteriales*). The main reason for this more cautious classification has been the lack of molecular data for inference of the phylogeny of this and other groups of ascomycetes.

As the classification in *Systema Ascomycetum* does not suggest any hypothesis on supraordinal taxa, the following discussion will follow the subdivision of the ascolocular fungi proposed by Barr.

Asci Irregularly Scattered in Uniascal Locules (Loculoplectoascomycetidae)

Myriangiales. *Myriangium duriaei* is the basic type of *Myriangiales*. It is a characteristic fungus associated with scale insects and has stromata with rounded outgrowths. Each such outgrowth is a compact stroma with firmly embedded asci at different levels, with the youngest asci in the basal parts and with gradually older asci the closer they are to the upper surface. The outgrowths disintegrate in the surface region and by that the mature asci become free. Asci are rounded. The ectotunica bursts and the endotunica extends a little.

There are a number of similar fungi on various substrates, and several families have been recognized in *Myriangiales*. Barr (1987) accepted five families in that order. The only features they have in common are the usually globose to ovoid shape of the asci and the pseudoparenchymatous tissue, firmly embedding the asci.

Some mycologists, including Barr, regard the type of ascoma we see in *Myriangium* as "primitive", or more correctly as with plesiomorphic features. Another alternative is that they are the result of an arrested development, paedomorphosis. The only possibility to solve this problem is by using molecular data.

Arthoniales. *Arthonia radiata* is the basic type of *Arthoniales*. *Arthonia* is a large genus of mainly lichenized fungi on bark. The type species has densely crowded, radially branched or lobed, flat, black apothecia, with a poorly developed lateral wall. Most of the apothecium is a hymenium on a thin subhymenium. Barr considered *Arthoniales* to be related to *Myriangiales*, as both orders were characterized by an indeterminate habit of growth. However, the asci are separated by paraphysoids as in the *Opegraphales*, which Barr placed in another subclass. Most lichenologists seem to prefer to include the latter order in *Arthoniales*.

The *Arthoniales* differ from the *Opegraphales* in having usually smaller ascomata with a very thin excipule, and comparatively broader asci. Janex-Favre (1964) interpreted the ascomata of *Arthonia* as neotenic. They may have evolved from a fungus with *Opegrapha*-like ascomata. This hypothesis and the question whether two orders should be united or not will probably be solved by using molecular data.

Asci Separated by Thin-Walled Interthecial Tissue, Cells of the Locule, or These Cells Disintegrating Early (Loculoparenchymatomycetidae)

Dothideales. *Dothidea*, the type genus of *Dothideales*, is a rather small genus of species with distinct stromata on branches of various trees and shrubs. Each stroma contains several locules, each opening by a pore, and containing numerous asci, but no hamathecium.

A majority of the genera in *Dothideales* s.lat. have a hamathecium and are not closely related to *Dothidea*. When this order is split up, *Dothideales* s.str. will be a rather small order or include some ascomycetes with a hamathecium. The lack of a hamathecium is not necessarily a very important criterion. The function of the hamathecium is to support the asci during the dehiscence of the asci. This apparatus could have been remodelled many times during the evolution. Barr (1987) accepted six families in the order.

Asterinales. *Asterina* is a large genus of mainly tropical foliicolous fungi with dimidiate ascomata, seated on a superficial mycelium, and which is usually provided with hyphopodia forming haustoria that penetrate the underlying host cell. The asci are few, thick-walled, and contain eight, 1-septate, dark-brown ascospores. The hamathecium consists of some physes that often dissolve early. A very thorough thesis of the Australian *Asterinaceae* was presented in 1992 by Dr Rahayu (unpubl.). She saw a hamathecium in many species, but in some species not, probably because the physes had dissolved.

The lack of a hamathecium is probably a secondary condition in the *Asterinales*. The order will certainly be recognized as a separate order when their separation from other ascomycetes and their position has been determined by the use of DNA data. They differ from most other *Dothideales* s.lat. by asci having the rostrate type of dehiscence (extenditunicate sensu Reynolds, 1989). They are probably not closely related to *Dothideales* s. str.

Capnodiales. *Capnodium* is one of the sooty mould genera. The type species, *C. salicinum*, is widespread and occurs on thin twigs of *Salix*, etc. The branches become covered by the sooty black subiculum. The ascomata are subglobose or pear-shaped and deeply buried in the subiculum. The asci are slightly ventricose and almost sessile. Both the endo- and ectotunica are thin, the endotunica even invisible in the lower half of the ascus. The ascospores have both trans- and longisepta. Reynolds (1989) found periphysoids in *Capnodium* ascomata; I did not find any in the material available to me, so they probably dissolve.

There is a number of sooty moulds with *Capnodium*-like ascomata in subicula, some have a hamathecium of periphysoids, others not. This was important in Barr's (1987) classification, and she kept those lacking a hamathecium in *Capnodiales*, whereas those having periphysoids were placed in *Chaetothyriales* in another subclass. However, as mentioned above, *Capnodium* may also have periphysoids, so the situation is uncertain. Again, we need molecular data.

Hamathecium of Periphysoids (Loculoanteromycetidae)

Verrucariales. *Verrucaria* is a large genus of mainly crustose lichens on rocks. All have perithecioid ascomata, a hamathecium of periphyses and periphysoids, and fissitunicate asci. Extruded parts of the endotunica soon become more or less dissolved.

Henssen and Jahns (1973) stated that the ascomatal development resembles that in *Gyalectales* and *Ostropales*. Parguey-Leduc and Janex-Favre (1981) found that *Verrucariaceae* s.str. is close to the ascolocular fungi, but that *Dermatocarpon miniatum* is ascohynenial.

Chaetothyriales. *Chaetothyrium* species are small foliicolous fungi, often with setose, perithecioid ascomata covered by a thin hyphal membrane, a pellicle. The hamathecium consists of periphysoids. The small asci are thick-walled in the upper half. The ascospores are several-septate and hyaline.

Barr (1987) used this order for six families. One of them, *Herpotrichiellaceae*, contains small saprobes on old wood, etc. They have more or less superficial, globose, thin-walled, commonly setose ascomata. The asci are very thick-walled and narrow in the upper parts, and produce greyish phrago- or dictyospores. The hamathecium consists of periphysoids. This family has been discussed as possibly related to the *Verrucariaceae*, but other features do not support this classification. Molecular data are needed.

Hamathecium of "Cellular" or "Trabeculate" Pseudoparaphyses (Loculoedaphomycetidae)

Pleosporales (incl. Hysteriales). *Pleospora herbarum* is a common saprobe on various herbaceous stems, leaves, etc. The subglobose, perithecioid ascomata are about 0.5 mm across and very thick-walled. They open by a lysigenous pore, contain rather numerous asci, each producing eight dictyospores. The physes between the asci have been described as pseudoparaphyses, originating in the upper part of the locule, growing down to the bottom, anchoring, stretching and finally bursting in the upper part. However, others have described the physes as originating from the stretching of an interascal tissue, with no free ends in a young stage. In other words, the physes should be paraphysoids. Whatever their origins are, they are comparatively wide and have been termed cellular pseudoparaphyses, one of the features characterizing the order *Pleosporales* sensu Barr (1987). Other characteristics were "asci in a basal layer, peridium usually pseudoparenchymatous, ascospores that exhibit bipolar asymmetry" (Barr, 1987).

Melanommatales (incl. Pyrenulales, Microthyriales). *Melanomma pulvis-pyrius* is the basal type of the order *Melanommatales*. It is one of the most common pyrenomycetes on old wood in the North Temperate zone. The often densely crowded pseudothecia are almost superficial. They have a rather thick, and dark-coloured wall, contain numerous, thin-walled, fissitunicate asci producing pale brown, 3-septate ascospores. The asci are separated by very narrow trabeculae ("trabecular pseudoparaphyses", "trabeculate pseudoparaphyses"), originating through the stretching of a central pseudoparenchymatous tissue.

There is, no doubt, a great evolutionary distance between *Pleospora* and *Melanomma*. When I published my thesis on graminicolous pyrenomycetes from Fennoscandia (Eriksson, 1967), I arranged the ascolocular taxa into three groups, one corresponding to *Dothideales*, one to *Pleosporales*, and one partly to *Melanommatales*, the two latter mostly on the basis of the structure of the ascomatal wall, influenced by ideas first proposed by Munk (1957). Barr has then found other features supporting a classification that was discussed in detail in her "Prodromus" (Barr, 1987), and in her more recent works on the *Melanommatales* and loculoascomycetes with muriform spores. Some mycologists have accepted this as the final system, but Barr has pointed out herself that, for example, the *Pleosporales* "includes the most complex array of organisms in the *Loculoascomycetes*, consequently the arrangement of genera and families is not yet satisfactory".

Pyrenulales has been treated as a separate order in the *Outlines*. It could just as well have been included in the *Dothideales* s.lat. It was recognized as a disperate order as it seemed to constitute a separate branch of mainly lichenized fungi, with characteristic pseudostromata, hamathecia, asci, and ascospores. Some non-lichenized members were accepted, for example, the genus *Massaria*, but as some members of the *Dothideales* s.lat. with a similar hamathecium are probably closely related, either they should have been transferred to the *Pyrenulales*, or that order should have been united with the *Dothideales*. In Barr's classification it falls within *Melanommatales*.

Opegraphales. Most species in the *Opegraphales* are lichenized. They have apothecioid ascomata, either circular in outline or lirelliform, opening by a longitudinal slit. The hamathecium consists of paraphysoids. The asci have an IKI+ pink or KOH/IKI+ blue zone. As mentioned above, this order is probably closely related to the *Arthoniales* (see also Tehler, this volume).

Patellariales. The *Patellariales* are non-lichenized, have apothecioid ascomata with a rounded outline, a hamathecium of paraphysoids, and IKI- asci. The relationships of this order are unknown.

CONCLUSIONS

Some important problems in the classification of the fissitunicate ascomycetes will probably soon have answers through the use of molecular methods. This is so especially for problems on the relationships between the orders and supraordinal taxa. This will give information about the direction of morphological evolution, and so will be of great help in the classification of lower taxa.

REFERENCES

von Arx, J.A. and E. Müller, 1975, A re-evaluation of the bitunicate ascomycetes with keys to families and genera, *Studies in Mycology, Baarn* 9: 1-159.

Barr, M.E., 1987, *Prodromus to Class Loculoascomycetes*, M.E. Barr, Amherst.

Dughi, R., 1956, La signification des appareils apicaux des asques de *Gymnophysma* et de *Chlamydophysma*, *Compte Rendu Hebdomadaire des Séances de l'Academie des Sciences, Paris* 243: 750-752.

Eriksson, O. [E.], 1967, *Studies on Graminicolous Pyrenomycetes from Fennoscandia*. Inaugural Dissertation. Almquist & Wiksells, Uppsala.

Eriksson, O.E. and D.L. Hawksworth, 1991, Outline of the ascomycetes - 1990, *Systema Ascomycetum* 9: 39-271.

Henssen, A., and H.M. Jahns, 1973 ["1974"], *Lichenes. Eine Einführung in die Flechtenkunde*, Thieme, Stuttgart.

Janex-Favre, M., 1964. Sur les ascocarpes, asques et la position systématique des lichens du genre *Graphis*, *Revue Bryologique et Lichénologique* 33: 244-288.

Luttrell, E.S., 1951, Taxonomy of the pyrenomycetes, *University of Missouri Studies* 24: 1-120.

Munk, A., 1957, Danish pyrenomycetes, *Dansk Botanisk Arkiv* 17: 1-491.

Nannfeldt, J.A., 1932, Studien über die Morphologie und Systematik der nicht-lichenisierten, inoperculaten Discomyceten, *Nova Acta Regiae Societatis Scientiarum upsalienses, ser.* IV 8(2): 1-368.

Parguey-Leduc, A. and M.-C. Janex-Favre, 1981, The ascocarps of ascohymenial pyrenomycetes, *In*: *Ascomycete Systematics. The Luttrellian Concept* (D.R. Reynolds, ed.): 102-123. Springer Verlag, New York.

Pringsheim, N., 1858, Über das Austreten der Sporen *Sphaeria scirpi* aus ihren Schläuchen, *Jahrbuch für Wissenschaftliche Botanik* 1: 189-192.

Reynolds, D.R., 1989, An extenditunicate ascus in the ascostromatic genus *Meliolina*, *Cryptogamie, Mycologie* 10: 305-320.

Outline of the Ascomycetes

SYSTEMA ASCOMYCETUM: THE CONCEPT

D.L. Hawksworth[1] and O.E. Eriksson[2]

[1]International Mycological Institute
Bakeham Lane, Egham
Surrey TW20 9TY, UK

[2]Institute of Ecological Botany
University of Umeå
S-901 87 Umeå, Sweden

SUMMARY

The problems in developing an overall system for ascomycete classification are addressed through "Outlines of the ascomycetes" published at intervals in *Systema Ascomycetum*, a journal devoted to the construction of an improving system for the group. The background and concept are discussed, the controversy occasioned is reviewed, and future plans for the "Outline" and the journal are noted.

THE PROBLEM

By 1980 there were several fundamentally different systems of ascomycete classification in use. For example, Ainsworth et al. (1973) accepted 21 orders in six classes, Barr (1976) 38 orders in three classes and nine subclasses, and von Arx (1979) eight orders and no classes. Further examples are tabulated in Hawksworth (1985).

At the same time, for more than half a century, the only work in which all fungal orders, families and genera were included in a single system was that of Clements and Shear (1931). According to the authors, their classification was "an endeavour to approximate the natural system in several respects". It was, however, a highly unnatural system, and a new classification of the ascomycetes (and most of the other fungal groups), including all new genera and higher taxa, and considering all new concepts, was desperately needed (Eriksson, 1992: 8). A modest gesture towards this end was made in the sixth edition of the *Dictionary of the Fungi* (Ainsworth et al., 1971) by the incorporation of lichenized and non-lichenized fungal genera in a single volume, but at that time the systems were not integrated - and indeed many "mycologists" greeted this inclusion with surprise and even shock. It is gratifying that such integration has become the norm.

The problem of the need of biologists for a framework suitable for general use, which at the same time would both accurately reflect the current level of knowledge and be capable of progressive refinement without seemingly dramatic changes, just had to be addressed.

In this context, it is vital to distinguish between a reasonably robust "generalist system" suited to widespread use, and a "specialist system" proposed as a hypothesis for testing and to be subjected to healthy debates (Hawksworth, 1985). "Specialist systems" are an integral part of a healthy scientific process, while "generalist systems" are what most consumers of the results of systematic research require. In practice there have always been generalist systems - those used in widely distributed textbooks and encyclopedia's to which non-specialists refer.

Systematists are unusual amongst biologists in that the results of their science also provide the communication and reference points for discussions of all aspects of life on Earth. There is a social dimension to their work. The failure of systematists to respond to user needs has been identified as a major reason for the diminished support that they often receive (Hawksworth and Bisby, 1988; Bisby and Hawksworth, 1991).

BACKGROUND

We independently reached the same point in 1982. A synopsis of the course that led to this situation may contribute to a clarification of the *Systema Ascomycetum* today.

Cooke and Hawksworth (1970) compiled a preliminary list of all family names proposed for fungi (including lichens). This was of importance when "The families of bitunicate ascomycetes" (Eriksson, 1981) was in preparation, and our first correspondence was on family name nomenclature in 1979. We first met when Eriksson visited IMI in that year to use its collections - convenient as both lichenized and non-lichenized ascomycetes had been placed in a single alphabetical sequence by genera in 1975, but we did not discuss collaboration at that time.

Eriksson (1981) classified all ascomycete families in 109 putatively monophyletic groups, termed "clades". In the autumn of 1981, Eriksson presented plans for the development of a new *Outline* of the ascomycetes to an international panel of scientists evaluating projects for the Swedish Natural Science Research Council (NFR). The project was recommended to be accepted, and has been supported by them since 1982.

The work on the project was already started by Eriksson in 1981. A classification that would evolve stepwise in annually revised *Outlines* was conceived. At first this would have only orders and families, but then accepted genera, and finally also all synonymous generic names. During 1981-2, all ascomycete entries in *Index Nominum Genericorum* (Farr et al., 1979) were compared with those in the *Dictionary of the Fungi* (Ainsworth et al., 1971), and several hundreds of papers were screened for information on the position of the genera.

Dissatisfied with the ascomycete treatment in the 1971 edition of the *Dictionary*, Hawksworth was determined to have a fully integrated system for the ascomycetes, including lichenized genera, in the 1983 edition. But which of the competing systems should be used for the higher categories?; and how many and what orders should be recognized ? Separate slips for each entry had been typed in 1979-80, checks had been made against *Index Nominum Genericorum,* and in 1981 computerization of the slips had started. The arrival of Eriksson (1981) in the fall of 1981 seemed like a breath of fresh air to Hawksworth - not only for the system adopted but as some of the ideas of the antiquity of some lichenized groups had been pointed out by him quite independently in a lecture at the 1981 International Botanical Congress in Sydney that summer (Hawksworth, 1982).

Aware of Hawksworth's interests, Eriksson sent a draft of the first *Outline*, which included only families and orders, to him for comment in early 1982. This was published later that year, and contained 38 orders and 228 families (plus 288 synonyms) in a "transitional classification, which may be used provisionally until more is known" (Eriksson, 1982a: 204).

We met in Kew in June 1982, and agreed that it was scientifically most honest at the current stage of ascomycete systematics to accept only ranks of order and below, and to place them in an alphabetical system. As a consequence, 37 orders and no higher ranks were accepted in the seventh edition of the *Dictionary* (Hawksworth et al., 1983). At that meeting we also resolved to minimize the duplication of effort between us wherever possible, bearing in mind that we had common needs for our different projects.

As the *Outline* started to move towards the publication of genera and their synonyms, the magnitude of the project was increasing. Following correspondence in December 1984-February 1985, we agreed to work as co-authors on the *Outline* for 1985. The *Systema Ascomycetum*, which included the *Outline*, was at the same time increasing in both size and demand - and thus becoming more expensive to print and distribute. Following a visit by Hawksworth to Umeå in July 1986, it was agreed that IMI should help with the costs of production and distribution from the 1986 issue.

THE CONCEPT

A sound and non-speculative route to refining the classification of ascomycetes is required. A downward reconstruction of phylogenies seemed the pragmatic way to proceed, clarifying the concepts of orders, families, and then genera, concentrating on the recognition of monophyletic groups.

The annually revised *Outlines* in *Systema Ascomycetum*, a journal "which aims at a more natural and generally acceptable classification of the genera and higher taxa of lichenized and non-lichenized ascomycetes" (Eriksson, 1983: 1), addressed this fundamental problem. The key concepts and principles in preparing the annual *Outlines* were explained by Eriksson (1983: 2-5):

(1) Only taxa that very probably are monophyletic are accepted.
(2) The classification is not influenced by hypothetical assumptions on the origin of ascomycetes or on reconstructions of the main evolutionary pathways within the group.
(3) The *Outline* includes all validly published legitimate names of ascomycete families and some other names used *ad interim*.
(4) The *Outline* will be revised annually with an ambition to gradually approach as natural and generally acceptable classification as possible.
(5) In several cases well-known family names were preferred instead of older little-known synonyms.

But when should a change be made ? And how could input from as many mycologists as possible be secured ? In 1985, we redefined the objectives and methods of operation of *Systema Ascomycetum* to satisfy these questions (Eriksson and Hawksworth, 1985: 2):

> "The objective . . . is to produce a consensus of current views as a system that reflects the current state of knowledge that can be commended for general use". . . "All new taxonomical concepts and changes which have to be considered for future *Outlines* will be presented in a separate paper before they are accepted, unless it is clear that the changes are unavoidable" . . . "This *Outline*, and its future editions, are most appropriately regarded as a collaborative effort to make the most recent accepted decisions generally available to all who work with these fungi." . . . "The aim is also to provoke a debate on various problems in the classification . . . and then, if possible, to arrive at a consensus in each matter."

The *Notes*, which we started to publish in the following year (Eriksson and Hawksworth, 1986a), were to be the vehicle for open interactive debates and drew on both published and unpublished sources, including new observations and research. All mycologists were "urged to inform us of any changes they could not support, giving the reasons for their views", and "proposals for further modifications, together with notes explaining their necessity" were invited. Changes were not to be adopted in the *Outlines* until at least a year had been allowed for comments to be submitted. The interactive process had begun.

Progress was greatly facilitated by the development of increasingly compatible PCs. We progressively used mail, electronic links, floppy disks, and diskettes to compile our *Notes*. Scanning different journals, editing each others *Notes*, and combining the files. With the advent of readily compatible personal computers, fax and express mail, the process has become increasingly efficient and rapid.

THE PRODUCT

Outlines were published annually for the eight years 1982 to 1990 (Fig. 1). In 1982 and 1983 only names at the ranks of order and family were considered; in 1984 and 1985 the names of accepted genera were added, with indications of the numbers of synonyms but not listing the names themselves. The synonymous names were held on computer, and for the 1986 to 1990 *Outlines* synonyms of generic names were presented in full. The most recent *Outline*, that for 1990, (Eriksson and Hawksworth, 1991) comprised five elements: (1) the outline of orders and families; (2) the outline of families

Publication	Vol.	Year	Ord.	Fam.	Gen.	Syn.	Notes	Ind.
Clements & Shear	-	1931	x	x	x	x	-	-
Opera Botanica	60	1981	-	x	-	-	-	-
Mycotaxon	15	1982	x	x	-	-	-	-
Systema Ascomycetum	1	1982	-	-	-	-	-	-
	2	1983	x	x	-	-	-	-
	3	1984	x	x	x	-	-	-
	4	1985	x	x	x	-	-	-
	5	1986	x	x	x	x	1-224	-
	6	1987	x	x	x	x	225-551	-
	7	1988	x	x	x	x	552-803	-
	8	1990	x	x	x	x	804-968	-
	9	1991	x	x	x	x	969-1127	-
	10	1991	-	-	-	-	1128-1293	-
	11	1993	-	-	-	-	1294-1529	-
	12	1993	x	x	x	x	1530-xxxx	x

Figure 1. The development of the contents of *Systema Ascomycetum.*

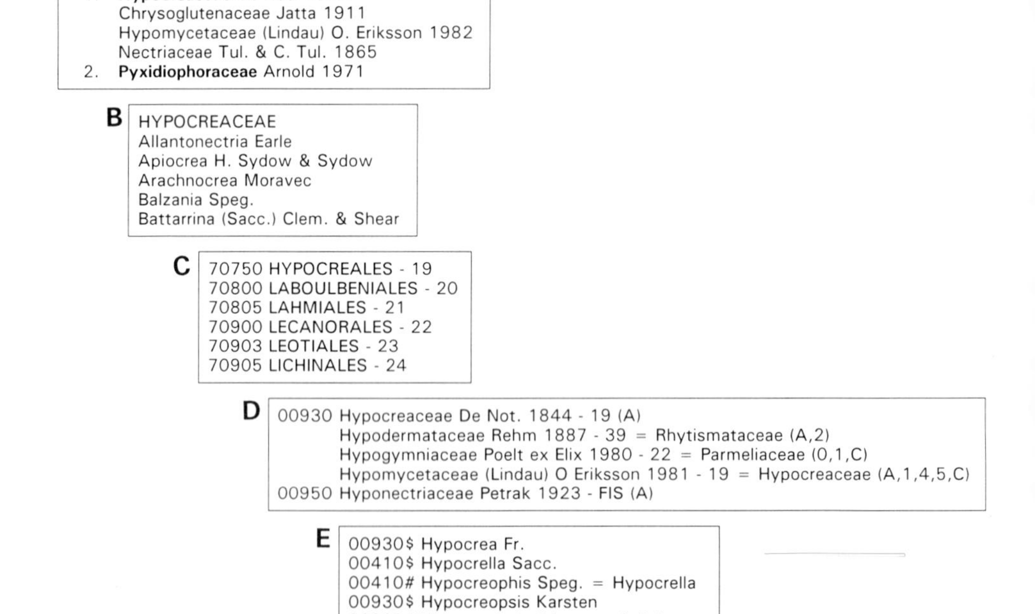

Figure 2. Examples of the elements included in the *Outline*: A, orders and families; B, families and genera; C, index to orders; D, index to families; E, index to genera. Extracted from Eriksson and Hawksworth (1991).

and genera; (3) an index to orders; (4) an index to families; and (5) an index to genera. Examples of the text from each of these sections is presented as Fig. 2.

Original papers in line with the objective of the *Systema Ascomycetum*, dealing with genera and higher ranks, have been actively encouraged since 1986. Papers are accepted on the basis of peer review by at least two persons.

The *Notes* have also appeared annually or twice-yearly from 1986, and at the end of volume 12 in 1993, 1610 had been published. Original *Notes* submitted, as they reflect an individual's views, are not peer reviewed, although the co-editors may propose changes for consideration.

As a consequence of these three elements, original papers, *Notes*, and the *Outlines*, *Systema Ascomycetum* has evolved so that it is not only a producer of a "generalist" system for the largest and most diverse group in the fungal kingdom, but also fulfils a current awareness function for researchers on these fungi, serves as a regularly updated checklist of all family and generic names proposed, and provides an outlet for the rapid publication of papers that will be sure to reach most ascomycete specialists in a timely manner.

CONTROVERSY AND CONSENSUS

Radical approaches to the solution of a long-established problem not unnaturally provoke controversy. Objective debate is an integral component of the scientific process and should always be encouraged. It can promote clearer thinking and be a spur to progress. Reynolds (1989) questioned the word "consensus", asking: "Why consensus? Who are the consensus formers? [and] What purpose does [*Systema Ascomycetum*] serve?" He offered four suggestions "for removing doubt concerning the means of carrying out what is intended to be noble in spirit" (Reynolds, 1989: 63), to which Eriksson (1989) replied in detail. These suggestions are cited below, together with abbreviated versions of the published responses:

(1) The editorial management of the journal should reflect the international status of the publication's purpose. *Response*: This had been discussed, and we concluded that it would make our work much more complicated and that the results would then perhaps appear too final; original papers are in any case peer-reviewed.
(2) The input from specialists should be clearly attributed to the original authors. *Response*: This practice has been adopted from the start of the *Notes*.
(3) Abandon the consensus crusade. *Response*: We had already stressed that we aimed "to approach a consensus" and that 100% could never be achieved (Eriksson and Hawksworth, 1986b: 186).
(4) Cease the declaration of a taxon as "accepted" or "unaccepted". *Response*: We regard the provision of information as to whether a name is currently considered the correct name of a taxon or just a synonym as one of the most important tasks of the project. This information is needed desperately by users of ascomycete names.

It is the word "consensus" which causes most difficulty, and for precision we consider another to be more appropriate: "eclectic", the collecting of what appears to be the best, or composed of elements drawn from different sources (Eriksson, 1989: 64). Indeed, the "eclectic" approach had already been adopted and used by Eriksson in the early 1980s (1981: 183, 1984: 2).

In order to widen the debate on this issue, a special open session on the *Systema Ascomycetum* was convened during the Fourth International Mycological Congress (IMC4) in Regensburg on 28 August 1990. We were gratified at the extent of the support received, and proposed that the *Outline*, instead of being revised annually, should in future be produced at intervals of about four years. At that time, we suggested a meeting be held of interested specialists to enable them to present their views on the changes proposed since the last *Outline*. The First International Workshop on Ascomycete Systematics in Paris on 11-14 May 1993 provided a unique opportunity for such an open debate.

Our peers will be able to judge the results of this experiment in the *Outline* published in December 1993. This will also form the basis of the treatment in the eighth edition of the *Dictionary of the Fungi*, due for publication shortly. The opportunity of such a wide counsel is especially welcome in the latter case as each edition of the *Dictionary* is planned to have a 10 year life-span.

As Professor A. Burgess, Chairman of the *Flora Europaea* editorial committee remarked of their product in March 1993, the *Outline* and *Dictionary* are surely also "marvellously imperfect products needed by all".

THE FUTURE

In future, in addition to publishing roughly four-yearly *Outlines*, we have plans to produce comprehensive keys to orders, families, and eventually genera. In addition, we aim to eventually aspire to the production of an *Index Ascomycetum*.

Draft keys, unpublished, to the orders and most families of ascomycetes accepted in the seventh edition of the *Dictionary* were prepared by Minter, Cannon and Hawksworth at IMI in 1986. These are being updated for inclusion in the eighth edition.

Eriksson plans to work towards an *Index Ascomycetum* which would include information on the type species of each generic name, a discussion of important concepts, anamorph information and references.

As to supraordinal taxa, we feel that it is wisest, bearing in mind the speed at which molecular data is accumulating, to wait until towards the end of the decade before proposing acceptance of such taxa in any "generalist system" such as the *Outline* (and *Dictionary*).

At the same time, we will have to consider how best to integrate mitotic fungi as their position in the overall fungal systems become clearer - especially through the application of molecular methods (Reynolds and Taylor, 1993).

The *Outline* has already had another purpose, quite separately from its original intention, to provide the basis for the ascomycete entries in the List of Generic Names in Current Use (Greuter et al., 1993) for which protection against changes for nomenclatural reasons is being sought at the XIV International Botanical Congress in Japan in August 1993. When this mechanism to reduce the numbers of name changes for non-scientific reasons is in place, we can perhaps look forward to an even more smoothly evolving system for the ascomycetes.

But we have also come to the conclusion that the success of this project has a much wider significance throughout systematic biology. We know of no similar interactive approach to the development of a system for such a large number of taxa. We have produced a model which we believe has the potential to be applied in all groups of organisms. Further, we have shown that long-term collaboration over large distances is possible, fruitful, and indeed synergistic.

ACKNOWLEDGEMENTS

Eriksson acknowledges the continuing support received from the Swedish Natural Science Research Council (NFR) for his work on ascomycete systematics. We both remain indebted to the numerous mycologists throughout the world who have shown their support for the *Systema Ascomycetum* initiative by sending copies of their papers, observations and comments for the *Notes*, original papers for publication, and more particularly be subscribing to the journal. Our future visions will not be realized without such continuing support.

REFERENCES

Ainsworth, G.C., D.L. Hawksworth, and P.W. James, 1971, *Ainsworth & Bisby's Dictionary of the Fungi*, 6th edition. Commonwealth Mycological Institute, Kew.

Ainsworth, G.C., F.K. Sparrow, and A.S. Sussman, eds, 1973, *The Fungi: An Advanced Treatise*, Vol. 4A, Academic Press, New York.

von Arx, J.A., 1979, Ascomycetes as fungi imperfecti, *In: The Whole Fungus*, [W.] B. Kendrick, ed., 201-213, National Museums of Canada, Ottawa.

Barr, M.E., 1976, Perspectives in the *Ascomycotina*, *Memoirs of the New York Botanic Garden* 28: 1-8.

Bisby, F.A., and D.L. Hawksworth, 1991. What must be done to save systematics?, *In: Improving the Stability of Names: Needs and Options* (D.L. Hawksworth, ed.): 323-336. [*Regnum Vegetabile* No. 123.] Koeltz Scientific Books, Königstein.
Clements, F.E., and C.L. Shear, 1931, *The Genera of Fungi*, H.W. Wilson, New York.
Cooke, W.B., and D.L. Hawksworth, 1970, A preliminary list of the families proposed for fungi (including the lichens), *Mycological Papers* 121: 1-86.
Eriksson, O. [E.], 1981, The families of bitunicate ascomycetes, *Opera Botanica* 60: 1-219.
Eriksson, O. [E.], 1982, Outline of the ascomycetes - 1982, *Mycotaxon* 15: 203-248.
Eriksson, O. [E.], 1983, Outline of the ascomycetes - 1983, *Systema Ascomycetum* 2: 1-37.
Eriksson, O. [E.], 1984, Outline of the ascomycetes - 1984, *Systema Ascomycetum* 3: 1-72.
Eriksson, O.E., 1989, Eclectic mycology: a response to Reynolds, *Taxon* 38: 64-67.
Eriksson, O. E., 1992, *The Non-Lichenized Pyrenomycetes of Sweden*. SBT-förlaget, Lund
Eriksson, O. [E.], and D.L. Hawksworth, 1985, Outline of the ascomycetes - 1985, *Systema Ascomycetum* 4: 1-79.
Eriksson, O. [E.], and D.L. Hawksworth, 1986a, Notes on ascomycete systematics. Nos 1-224, *Systema Ascomycetum* 5: 113-174.
Eriksson, O. [E.], and D.L. Hawksworth, 1986b, Outline of the ascomycetes - 1986, *Systema Ascomycetum* 5: 185-324.
Eriksson, O.E., and D.L. Hawksworth, 1991, Outline of the ascomycetes - 1990, *Systema Ascomycetum* 9: 39-271.
Eriksson, O.E., and D.L. Hawksworth, 1993, Outline of the ascomycetes - 1993, *Systema Ascomycetum* 12: 51-257.
Farr, E., J.A. Leussink, and F.A. Stafleu, 1979, *Index Nominum Genericorum (Plantarum)*, 3 vols [*Regnum Vegetabile* Nos 100-102.] Bohn, Scheltema and Holkema, Utrecht.
Greuter, W. *et al.* eds, 1993, *Generic Names in Current Use*. [*Regnum Vegetabile* No. 127.] Koeltz Scientific Books, Königstein.
Hawksworth, D.L., 1982, Co-evolution and the detection of ancestry in lichens, *Journal of the Hattori Botanical Laboratory*, 52: 323-329.
Hawksworth, D.L., 1985, Problems and prospects in the systematics of the *Ascomycotina*, *Proceedings of the Indian Academy of Science, Plant Science*, 94: 319-339.
Hawksworth, D.L., and F.A. Bisby, 1988, Systematics: the keystone of biology, *In*: *Prospects in Systematics* (D.L. Hawksworth, ed.): 3-30. Clarendon Press, Oxford.
Hawksworth, D.L., B.C. Sutton, and G.C. Ainsworth, 1983, *Ainsworth & Bisby's Dictionary of the Fungi*, 7th edition, Commonwealth Agricultural Bureaux, Slough.
Reynolds, D.R., 1989, Consensus mycology, *Taxon* 38: 62-63.
Reynolds, D.R., and J.W. Taylor (eds), 1993, *The Fungal Holomorph: Mitotic, Meiotic and Pleomorphic Speciation in Fungal Systematics*, CAB INTERNATIONAL, Wallingford.

DISCUSSION

Introduction

Whalley: I believe I would probably be speaking for everyone here to say that we are in admiration of the energy and stamina that have been put into the *Outline* for the past ten years. An unparalleled wealth of talent has been gathered at this Workshop and it is up to those of you that have ideas, proposals or disagreements with the *Outline* to come forward at this time. As participants we also appreciate the "red book" compilation of the numerous proposals made since to modify the ascomycete system since 1990.

Hawksworth: At this point we would welcome feedback on a variety of issues that relate to the *Systema Ascomycetum*, for example the way in which the different listings in the *Outline* are presented, including the coding system.

The Product

Taylor, J.: I like the way the data are organized, but it would be helpful to make fingerholes in the margins to facilitate access to the main components.

Hawksworth: *Systema Ascomycetum* has a rising circulation, indicating that a real need is being met. With so many of its subscribers at this Workshop, views on the publication and its distribution would be appreciated. Perhaps a change from a single- to a differential pricing policy would be beneficial?

Kendrick: Libraries in North America now have to undergo an annual pruning exercise. Personal subscriptions would be welcome, but a raised institutional price could be harmful.

Untereiner: A reduced rate for registered graduate students should be considered.

Lodha: The problems of costs are even more acute in developing countries, and the supply of information inhibits good research. The costs should be kept as low as possible.

Hawksworth: I am very concerned about this issue, and it is one reason I am enthusiastic as to the potential of CD-ROM delivery systems (see below). Further, our experience is that aid agencies are more favourably disposed to buying new technology than printed books and journals.

Kendrick: It would be helpful if this and other CAB INTERNATIONAL/ IMI products were available on disk. For example, the *Index of Fungi* which is one of the most useful compilations ever made in mycology would be much less difficult to use in such a format. If files could be prepared in ASKSAM format it could be sent out with INFOSIFT (the fast search engine of ASKSAM) quite cheaply. If the price were reduced, many more copies would be sold.

Hawksworth: I believe that this would be technically feasible, but as IMI is required to operate on a fee-for-service basis disks would have to be priced at a level that would ensure we could continue to prepare the product. The Institute has about 65 000 species names of fungi, with places of publication, in a single relational database and we are conscious of how best to validate data and make it more widely available. Validation is a particular problem as I am concerned that the Institute does not put out unverified data that others then rely on. As a first step, the *Bibliography of Systematic Mycology* is planned to be incorporated into the CAB ABSTRACTS database in 1994; the hard copy is indexed down to genera. CAB INTERNATIONAL now has considerable experience of issuing CD-ROMs; these can be operated from portable computers even in a Land Rover, and have immense potential for applications in less developed countries.

Pfister: It would be best to have the information on-line rather than as a disc or hard copy.

Hawksworth: I agree, if we can find a mechanism for this. I have been impressed by the power of INTERNET, and understand "toll-gates" can be incorporated into the system.

Rossman: Systematic databases need to be available to all free of charge.

Pfister: Korf and I consider that electronic mail, and electronic mail bulletin boards, are one of the best ways to facilitate inputs (i.e. comments on draft Notes) as well as outputs.

Blackwell: The Louisiana State University has not subscribed to a new journal for ten years, and I have to subscribe to this and others myself. If it could be placed on INTERNET mycologists could then easily access it.

Eriksson: I would like so see lists such as those produced by Læssøe on *Xylariaceae* included in *Systema Ascomycetum*. The nomenclature in some of the sequence databanks also needs attention.

Biosystematic Databases

Cannon: We already have a sophisticated set of relational bases at IMI, incorporating the fungal generic names from the NCU-3 listing and the *Dictionary of the Fungi*, and which could be extended to include the further data types Eriksson suggests. The IMI databases also assimilate information on bibliography, nomenclature, associated organisms, distributions, descriptions, differing taxonomic opinions, and other notes. The *Bibliography of Systematic Mycology* and *Index of Fungi* are incorporated into this database suite. Data can be output in a wide variety of formats. Exchange of data other than in a well-ordered database format is very inefficient. The data held at IMI is potentially available over networks.

Whalley: The compilation prepared by Læssøe is very useful, not least because it provides other specialists with something to "attack". It will increase debate and help us approach a consensus as to a more natural grouping, whether it is in a hard copy or database format.

In the Natural Environment Research Council (NERC) *Ascomycetes of Great Britain and Ireland* project initiated in 1992, a relational database is being constructed which includes descriptions, ecology, distribution, etc. Keys produced will be tested on field meetings and courses as the project progresses to ensure they can be employed by non-specialists. The possibility of making the data on the 5100 species eventually to be treated more generally available will be looked at later by the project's Advisory Committee.

Gams: I wish to stress the tremendous value of such databases, especially ones that cover all published names. Computers are the only way in which all information can be stored and updated. To ensure success, databases must be accessible free of charge to all potential contributors. Only in this way can optimal input and exchange of data be attained. Further agreements about data formats and information exchange need to be reached.

Periodicity

Hawksworth: Are there any views on the period of one year we have allowed before adopting changes in the *Outline*? There is a possibly apocryphal story that IMI's first mycologist, E.W. Mason, would not change any name in our collections for ten years; that is surely excessively conservative, but perhaps the minimum of one year we have adopted is too short a time for others to react.

Hafellner: The one year period for the adoption of new changes seems reasonable, but as the *Outlines* will appear less frequently fewer names will be affected.

Hawksworth: Following discussions on the *Systema Ascomycetum* during the Fourth International Mycological Congress (IMC4) at Regensburg in 1990, we proposed a four-yearly interval, to coincide with future Congresses, instead of annually as was the case from 1982-90. That was supported by the meeting, and this procedure has been adopted. We plan to issue a new *Outline* at the end of 1993 now we have had this Workshop, but perhaps that could be viewed as a "draft" for honing after the Fifth International Mycological Congress (IMC5) in Vancouver in 1994.

Supraordinal Groupings

Hawksworth: We have preferred to be conservative with regard to the recognition of supraordinal taxa until more data, especially molecular data, is available, but perhaps some more informal system could be adopted that might better show hypothesized relationships. Possibly the inclusion of figures on the lines of Venn diagrams used by Hafellner (*in* M. Galun (ed.) *CRC handbook of Lichenology* 3: 41-52, 1988) would be of value.

Taylor, J.: While there are uncertainties in many areas, there are some where there are at least strong likelihoods and where supraordinal taxa would be appropriate. If in the future they were proved wrong, then they could be changed.

Hawksworth: I fear this could lead to a somewhat confusing situation operationally as question marks would have to be used that might not be understood by many of its users.

Berbee: We should not expect too much from molecular methods, but they can challenge our hypotheses. New methods usually answer some, but not all, questions. We have a good idea what type of questions 18S RNA can resolve. It has proved particularly strong in relation to the recognition of two major monophyletic groups, for which I have used the names *Plectomycetes* and *Pyrenomycetes*. These two groups can reasonably be accepted now.

Eriksson: I do not think we should accept supraordinal taxa in the *Outline* yet, but in each issue we could perhaps try to provide, for example, a consensus cladogram of those available.

Schumacher: In trying to approach a classification system emphasizing the importance of defining monophyletic groups, cladistic methods other than those using parsimony analysis rooted by an outgroup should also be considered. While that method is probably the best way of handling molecular information, it is not superior to some other cladistic methods. Character compatibility analyses, based on various morphological character sets, can be particularly instructive.

Taylor, J.: I work on supraordinal taxa because I do not know enough about any single group. When I work with lower taxa, I work with morphologists. The time has come for systematic specialists to do the molecular work. I think that the distinction between molecular and morphological systematists should disappear. The need for cooperation between molecular and morphological systematics was a strong recommendation of *The Fungal Holomorph* conference in 1992 (Reynolds, D.R. and J.W. Taylor (eds), 1993): we need a computer databank including both phenotypic data and names; this "PhenBank" would be a counterpart to the existing "GenBank" sequence database.

Hawksworth: Those depositing sequences should always deposit cultures in service culture collections so that the identities of the fungi used can be verified and further work can be done on the same strains.

Lodha: I do not feel that morphological means of classification have yet been fully exhausted in the ascomycetes, and I am concerned at the amount of energy now flowing into molecular systematics, and to some extent ultrastructural investigations. The sophisticated methods may not solve all the problems, and yet it makes the cost of systematic research in less-developed countries prohibitive.

Blackwell: I am concerned at the tendency to differentiate between molecular and morphological approaches. We are all systematists, and, whatever techniques we use, we all need to understand at least enough of the others so that we can evaluate them properly.

Gams: I wish to draw attention to the Expert Taxonomic Information (ETI) system being developed at the University of Amsterdam. It uses computerized expert systems, including coloured illustrations. It started with birds and some aquatic organisms, but CBS is initiating pilot projects in collaboration with them on *Aspergillus, Penicillium*, and yeasts.

The balance between molecular and morphological taxonomists is of concern, especially because of the shortage of the latter. Because of the good reproducibility, a database of systematic studies in progress where molecular techniques are used would help avoid duplication.

Taylor, J.: Duplication is important within and between projects. The independent verification of results is an integral aspect of science.

Editorial Arrangements

Hawksworth: Eriksson and myself have discussed the possibility of enlarging the editorship of the journal periodically, or formalizing links with some over-seeing body, perhaps by a new subcommission of the IUBS/IUMS International Commission on the Taxonomy of Fungi (ICTF). I can see both advantages and disadvantages; your opinions are invited.

Reynolds: In 1988 questions were raised by me (*Taxon* 38: 62-63) concerning the manner in which *Systema Ascomycetum* is being "foisted on the mycological community". The questions were: Why consensus?, Who are the consensus formers?, What purpose does the journal really serve? The remedy I suggested was the formation of a better editorial practice including improved involvement of the international mycological community and by an abandonment of the consensus crusade, and of the declaration of a taxon as "accepted" or "unaccepted".

Eriksson's response (*Taxon* 38: 64-67, 1989) was written from the viewpoint that there were a number of misunderstandings and incorrect conclusions. Some of those questions have been addressed, some are unanswered, and some of my original criticisms remain.

The major conclusion of the discussions on the project during the Fourth International Mycological Congress (IMC4) in 1990 was that the published Notes were a valuable contribution. Yet questions remained as to how the *Outline* was prepared.

The present Workshop has aimed to further involve the international community in the assessment of the literature. The review of proposals by taxonomic groups with specialists as discussion leaders is indeed an improvement. Nonetheless, it seems clear that the editors will make the final decision as to whether a taxon is "accepted" or "not accepted"; this is not simply a question of nomenclatorial correctness as the data supporting an author's conclusions are interpreted.

As the stated aim is to provide a guide to eclectic mycology, the conclusions of leading specialists and their approval by peers is enough. Work such as that of Aptroot and Barr should be accommodated in the *Outline* and their results used to the fullest extent.

The stingy incorporation of some categories of recent data can only reduce the journal's potential contribution to mycology. To continue to ignore contributions because of editorial bias will diminish its standing. An attractive alternative to overcome this charge of censorship is an independently published criticism and correction of the most recent *Outline*.

The *Systema Ascomycetum* has demonstrated value in that a large body of literature is summarized in a useful format. It now needs to explore mechanisms by which seeming biases of self-appointed arbitrators can be obviated so that insights are not hindered. A diversity of opinion is the essence of ascomycete systematics in the 1990s.

Hilber: I recommend that a draft of the next *Outline* is sent to all Workshop participants, and that critical points are discussed at IMC5 in Vancouver before it is published.

Malloch: I see a danger that the *Outline* may increasingly act as an "establishment statement" and suppress the expression of new or unpopular views. I wish to urge the commentators on Notes to pay particular care as to their phraseology so they do not inhibit others.

Eriksson: I agree we have to be very careful, and this is of great concern to us. We strive to present concepts without favouring our own, and certainly have no intention of being unfair. There are many examples in science of major new theories being vigorously attacked. In mycology, for example, Bonorden fiercely attacked the Tulasne brothers because he did not believe their work connecting anamorphs and teleomorphs.

Kendrick: Recognizing that many mycologists are not familiar with many orders of ascomycetes, either because they are young and inexperienced or because they specialize on a particular group, it would be helpful to have more illustrations of the fungi involved. Were these available, it would make our debates and the documentation even more instructive and enjoyable.

Hawksworth: If such a compilation existed it would be of tremendous value. In 1986, Cannon. Minter and myself prepared a document for use on IMI courses which has illustrations up to and including representatives of accepted families (see above).

Kendrick: It would be good to have illustrations on screen when taxa are being discussed.

General Considerations

Rogers: We must stress that we need more data of all types, not just molecular information. We must not give the impression that we only need molecular data as that may give the wrong impression to funding bodies.

Hawksworth: We also wish to encourage the publication of hypotheses in *Systema Ascomycetum*, based on molecular or other methods, so that they are available for testing. This is an important part of the development of our science.

Rossman: Given the interest in biodiversity, and Hawksworth's (*Mycological Research* 95: 641-655, 1991) estimate of 1.5 million species of fungi, I wondered how many ascomycetes there might be on Earth. Hawksworth and myself attended an All-Taxa Biodiversity Inventory (ATBI) workshop in Philadelphia in March 1993 when we attempted to work out how a large area might be inventoried. During that meeting I went through the fungi group by group estimating how many species there might be, and that came to 1.2 million.

Kohlmeyer: Marine fungi are being described in increasing numbers. We know about 500, and I suspect that the total would not be more than 5000 because there are limited numbers of habitats (algae, sea grasses, marsh plants, mangroves, and submerged wood). However, marsh plants are little investigated; within the first few months collecting on *Juncus roemerianus*, a very common plant on the east coast of North America, we found 40 species all of which appear to be new genera, many of which cannot be placed in existing families.

Hawksworth: My figure of 1.5 million is conservative, for reasons explained in my paper. It is gratifying that this figure has not been seriously challenged, and especially that my calculations have been accepted as not unreasonable by entomologists. Observations such as Kohlmeyer's substantiate our case.

Rossman: I guesstimated 300 000 species of living ascomycetes.

Hawksworth: I did not prepare my calculations on a group-by-group basis but suspect a figure of 300 000 could be too low. In the UK about half all the fungi known are ascomycetes. However, there is immense scope for refining estimates of species numbers by critical studies in the tropics, especially in relation to host specificity on both insects and plants. Interestingly, while the lichenized ascomycetes comprise just under half the known ascomycetes, Galloway (*Biodiversity and Conservation* 1: 312-323, 1992) has estimated about 17-20 000 species in total; I have no reason to doubt his opinion.

Lodha: While much information is available from temperate Europe and North America, knowledge of ascomycetes in most other parts of the world remains poor. This situation will continue until conditions for mycological investigations are strengthened in those regions. Increasing sophistication in research methods may even be counterproductive for those in areas where taxonomy has hardly developed. As new technologies are developed, we need to develop simultaneously mechanisms to deliver that knowledge where it is needed and in a form in which it can be employed.

Wingfield: The biodiversity issue is not confined to scientific questions as to species numbers or conservation. In southern Africa we are urged to discover novel microorganisms for screening purposes. Cultures and not only specimens are required, and many isolates are anamorphs. These states need to be considered much more by ascomycetologists. There are other issues also to be discussed in future, for example whether the International Code of Botanical Nomenclature is serving us well and if we need a Code independent from that covering plants.

Hawksworth: I am personally in favour of increased harmonization between the different nomenclatural Codes, rather than a proliferation of more specialized ones. All Codes have more common issues to confront than at any previous time in their evolution, and they are starting to work towards similar solutions. This matter is of particular concern to IUBS who are to sponsor an exploratory inter-Code discussion in March 1994.

Conclusion

Hawksworth: The success of the *Systema Ascomycetum* concept, which represents a new approach to the development of a classification for a major group, depends on all our inputs. My Professor when I was a graduate student, T.G. Tutin, once remarked that you could only really know about 1000 plant species at one time; as you thoroughly learnt an unfamiliar species, your grasp on another lessened. That is perhaps an exaggeration, but it reinforces the view that one problem we have to confront is the enormous variation the ascomycetes encompass. We are striving to produce a service, and thank you for both your encouragements and your criticisms. Your scientific inputs and your suggestions as to how the project can continue its evolution and be improved are most welcome, and we will endeavour to address your legitimate concerns.

DISCUSSION 1

ENDOMYCETALES, *PROTOMYCETALES*, AND *TAPHRINALES*

Leader: C.P. Kurtzman

National Center for Agricultural Utilization Research
Agricultural Research Service
US Department of Agriculture
1815 N. University Street
Peoria, Illinois 61604, USA

INTRODUCTION

Kurtzman: The current state of yeast systematics is somewhat different from that in most filamentous fungi. We have had the benefit of having undertaken DNA reassociation studies for over 20 years and correlating these data with genetic definitions of species. The species concept in yeasts is thus now approaching that of a biological one. Further, if we have a certain amount of DNA relatedness we can now be confident we can define a species. The species concept is not a major problem in yeasts, but definition of genera can cause difficulties.

Sequencing work has attempted to correlate extent of divergence with phenotypic characteristics. In many cases this can be done, clades are isolated and can be recognized as genera, but for other taxa the result is unclear.

At higher taxonomic levels, the rRNA sequencing work consistently shows budding yeasts in a clade distinct from that of the filamentous fungi, and also separate from *Taphrina* and *Protomyces*.

Ascomycete yeasts have traditionally been regarded as part of the *Hemiascomycetes* and perceived as budding- and fission-yeasts. Molecular studies show that these are two diverse groups which do not now seem to belong to a single order. It has been suggested that these be termed *Endomycetales* and *Schizosaccharomycetales*. This division is consistent with other biochemical comparisons.

However, the genus *Endomyces*, and in particular *E. decipiens*, is a parasite of mushrooms. Some cultures are available under this name but appear to be wrongly identified. No authenticated culture is known and no sequence data are available for the species. If the sequence were congruent with other members of the order there is a long precedent for the use of the ordinal name *Endomycetales* for the budding yeasts. The name *Saccharomycetales* was introduced in 1960 by Kudriavzev as it was felt the budding yeasts must be fundamentally different from a genus such as *Endomyces*. However, the molecular data now suggest that the difference is not as great as then imagined. The Code allows flexibility as to which ordinal name to use. My preference would be *Saccharomycetales*, but as *Endomycetales* has been in use for a century we should perhaps retain that name.

Within the *Saccharomycetales*, families are presently uncertain, but the following may be accepted: *Ascoideaceae, Cephaloascaceae, Dipodascaceae, Endomycetaceae, Lipomycetaceae, Metschnikowiaceae*, and *Saccharomycetaceae*.

Turning to *Taphrinales* and *Protomycetales*, sequence data are showing a close linking of the two orders. However, generally only two species have been sequenced in each so major changes may be premature. However, the potential exists that *Protomycetales* might be a synonym of *Taphrinales*.

DISCUSSION

Eriksson: I and two of my students (Eriksson, Svedskog and Landvik, *Systema Ascomycetum* 11, 119-162, 1993) have retrieved all homologous DNA sequences of *Saccharomyces cerevisiae* and *Schizosaccharomyces pombe* from the EMBL Data Library and compared them, and found that two orders should be recognized. We described the new order *Schizosaccharomycetales* for the fission yeasts. If we continue to use the name *Endomycetales,* for the budding yeasts, it is not clear whether that is in the old sense including the *Schizosaccharomycetales* or the newer more restricted sense. The use of *Saccharomycetales* would immediately imply the budding yeasts. Moreover, as the basic type of the order is *Saccharomyces cerevisiae,* about 2490 sequences in that are known - more than in any other fungus.

Kurtzman: I think many yeast specialists would accept *Schizosaccharomycetales* as a separate order because of its genetic distance, but the historical precedent of *Saccharomyces* being included with *Endomycetales* remains.

Kendrick: We should stop using the terms "budding" and "fission" yeasts. They are not only outdated but not very informative as there at least five different kinds of conidiogenesis amongst the yeast fungi as demonstrated by von Arx and coworkers. Further, fission yeasts do not undergo fission as seen in prokaryotes but exhibit percurrent extension growth and bear annellations after the daughter cell secedes. I also support the use of *Saccharomycetales.*

Malloch: As *Saccharomycetales* is well-defined and understood, it may be best to reserve the use of the name *Endomycetales* until field mycologists can elucidate the true nature of *Endomyces.*

Blackwell: I wondered how you could have a biological species concept for asexual yeasts.

Kurtzman: The biological species concept is difficult in terms of proving it, but the phylogenetic species concept also concerns me as ever finer cuts can be made depending on the molecule being studied. Reproductive isolation can be demonstrated without too much difficulty in heterothallic yeasts and a good correlation in terms of DNA-relatedness emerges. In homothallics, most are to some extent outbreeding and similar levels of DNA relatedness occur. For asexual yeasts that appear to be well-defined, similar levels of DNA relatedness are found. We have felt it reasonable to apply these criteria to asexual yeasts generally at least to provide an approximation of the biological species. Siblings may be missed on occasion, but I doubt that any taxa with significant distances will be combined by this approach.

Blackwell: At present in the US and Europe there is a strong emphasis on the phylogenetic and not the biological species concept, to the extent that using the latter term might inhibit successful grant applications. Mycologists need to consider these issues further.

Hawksworth: The need to have a critical reassessment of species concepts in all microbial groups, including fungi, has been recognized in the IUBS/IUMS MICROBIAL DIVERSITY 21 programme. It is anticipated that a major international meeting on this crucial and fundamental topic will be convened in 1995 or 1996. The problem of asexual species is not unique to mycology and is seen also in many bacteria and microalgae. Exchanges of experiences and practices between systematists in these different groups promise a stimulating occasion.

Rogers: How widespread is the *a*-α mating system of *Saccharomyces cerevisiae* amongst allied genera ? This is likely to be important in view of the complicated and conservative nature of mating systems.

Kurtzman: An analagous system is quite widespread in heterothallic yeasts. Most have not been studied in detail but there are indications that they will be similar. I would not expect to see any vastly different systems.

With respect to generic limits in the *Saccharomycetales*. Even *Zygosaccharomyces* drifts a little into *Saccharomyces*. A little shifting of certain species may be needed when more results are available. RNA work has shown that four genera form a very tight clade: *Ashbya, Eremothecium, Hollyea* and *Nematospora.* All are plant pathogens and two produce riboflavin. The extent of divergence amongst these four taxa is no greater than we find in other tightly-knit genera, and to me they appear to be members of the same genus. All are characterized by elongated ascospores with small whip- or stub-like tails. As two are of particular economic importance, could these be combined without considerable controversy and confusion.

Rossman: I think you would do a service to those who work with these economically important fungi by putting them in one genus. Those that work with them need to know that they are closely related. Despite an initial reluctance to change the names, the names should reflect their biology. The changes need to be well-documented and made only at the end-point of definitive data-gathering.

Hawksworth: The key issue is that if as taxonomists we make a change, it should be only when the evidence in support of that change is overwhelming and likely to stand the test of time. What annoys pathologists is when such taxa are united one year, split the next, and then reunited.

Kendrick: As we have seen that the ascomycete-like yeasts evolved rather separately from the main line of ascomycetes, should we really be calling the meiosporangia of yeasts asci ? To me an ascus is a kind of meiosporangium which developed in order to shoot spores, while those of yeasts have not done that.

Kimbrough: If you compare ascosporogenesis in *Saccharomyces* and *Nannizia* you will find it to be identical.

Taylor, J.: The ancestral state does not appear to be asci that shoot their spores. Excepting *Taphrina*, one does not see ejaculating asci until the hyphal ascomycetes.

Rogers: Are the needle-shaped spores of the *Spermophthoraceae* always ascospores? In at least one there appears to be conjugation and then the production of an 8-spored ascus on a stalk.

Kurtzman: They look like asci and are free in culture (rather than attached to hyphae), but I do not know if meiosis has ever been verified.

Cannon: Considering structure and function, you cannot have an elongated ascus actively releasing spores which is not supported by a well-developed hyphal structure to hold that ascus in order to maintain directionality at spore release.

Malloch: The ascus is always the end-point of a dikaryon. In that case, perhaps this could be considered in relation to the yeasts. If no fungi ancestral to the yeasts have a dikaryon than Kendrick may have a structure ancestral to both the ascus and basidium.

Kendrick: I do not think the "ascus" of a yeast is an ascus any more than is an oogonium; it is a another kind of meiosporangium. As Malloch has pointed out, since the "asci" of yeasts are not the sequel to a dikaryophase, the logic for the term "ascus" in the group is faulty.

Kurtzman: There are several further items to be noted in the next *Outline*:

(1) *Zygozyma* is a member of the *Lipomycetaceae.*

(2) The *Spermophthoraceae* is a confused name as the type was based on a mixed culture.

(3) The close 18S rRNA relationship between *Dekkera, Issatchenkia* and *Pichia membranaefaciens* needs further investigation that includes all species of the genera before any changes are made. *Cyniclomyces* may belong to the same clade.

(4) *Schwanniomyces* should remain in *Debaryomyces.*

(5) *Octosporomyces* was shown to be a synonym of *Schizosaccharomyces* by Kurtzman and Robnett in 1991.

(6) *Clavispora* clusters with *Metschnikowia* as a member of the *Metschnikowiaceae.*

(7) *Cephaloascus* belongs in the *Saccharomycetales.*

(8) At present there are insufficient data to separate *Hyphopichia* from *Pichia.*

(9) *Waltozyma* was shown to be a synonym of *Lipomyces* by Kurtzman and Liu in 1990.

(10) *Yarrowia, Zygoascus*, and possibly other genera, may require a new family.

DISCUSSION 2

ASCOSPHAERALES, *ELAPHOMYCETALES*, *ERYSIPHALES*, *EUROTIALES*, *GLAZIELLALES*, *LABOULBENIALES*, *MICROASCALES*, *ONYGENALES*, AND *SPATHULOSPORALES*

Leaders: D.W. Malloch[1] and J. Mouchacca[2]

[1]Department of Botany
University of Toronto
25 Wilcocks Street
Toronto, Ontario M5S 3B2, Canada

[2]J. Mouchacca
Laboratoire de Cryptogamie
Museum National d'Histoire Naturelle
12, rue Buffon, 75005 Paris, France

INTRODUCTION

Malloch: In this session we have a formidable array of taxa to consider, many of which few of us will be personally familiar with. We propose to discuss these in turn.

DISCUSSION

Ascosphaerales

Berbee: I wondered if *Ascosphaera*, *Eremascus* and *Monascus*, together with the *Eurotiales*, would be placed under *Eurotiales* or *Onygenales*.

Malloch: *Eurotiales* in the current *Outline* are very close to the *Plectomycetes* of Berbee and Taylor (this volume). Also pertinent is the suggestion of Currah (this volume) to transfer the *Gymnoascaceae* into the *Trichocomaceae*; nomenclatural issues then must be considered.

Currah: My suggested rearrangement aimed to provide a more realistic basis for the application of molecular data to test the phylogenetic hypothesis I proposed. I.e. that *Gymnoascus* is more closely related to the *Trichocomoideae* than to other *Onygenales*. I also suggested dismantling and re-arranging species in *Gymnascella*, and moving *Gymnascoideus* to *Arthrodermataceae*. I would like to await more molecular data before making formal changes.

Blackwell: Currah has a phylogenetic hypothesis and with a different data set one can look for congruence with his hypothesis.

Kimbrough: I would discourage the merger of *Ascosphaerales* into *Eurotiales* as *Ascosphaera* has thick, uniperforate septa without Woronin bodies, and lacks a true cellular ascomatal wall; an enlarged ascogonial cell holds the spore-balls together.

Berbee: *Ascosphaera apis* is a weird fungus, but while its unique characters do not link it with other ascomycete groups, the molecular data shows it to be closely related with the plectomycete group.

Kimbrough: Has this been compared with *Conidiobolus* or *Basidiobolus* ? I would predict that *Ascosphaera* may be more closely related to that group of fungi than to the ascomycetes.

Berbee: The molecular data shows very little divergence between *Ascosphaera* and other *Plectomycetes*. Interestingly, Olive and Spiltoir placed the genus with the *Plectomycetes* on the basis of a shared derived character of gametangial formation.

Eriksson: We have been discussing some of the most difficult fungi to classify and cannot expect to solve this problem now. The *Outlines* are a dynamic system so I feel we can include *Ascosphaeraceae* and *Eremascaceae* as families within *Eurotiales*, but with a "?".

Cannon: As Kimbrough has stated, the number of characters unique to *Ascosphaera* is enormous. In considering those, remember we are looking at the expression of the whole genome and not only a small sequence within it. I suspect the genus is very ancient and that it merits separation from other members of the *Ascomycota* at a higher level than order.

Taylor, J.: The argument is not about molecular data, but rather unique and shared characters. It is shared derived characteristics that are the useful ones for phylogenetic reconstruction. In this case, shared derived molecular characters connect *Ascosphaera apis* to the "plectomycetes".

Elaphomycetales

Malloch: Landvik and Eriksson (this volume) have pointed out that the *Elaphomycetales* have operculate discomycetous relationships. Should the order continue to be maintained ?

Kimbrough: Investigations by Benny and myself on *Elaphomyces* show that it has a unique type of ascosporogenesis different from that in other "*Tuberales*" or *Pezizales*. However, because of the robust and large size of the ascomata we have to eliminate it from the *Plectomycetes*. I feel it is a relic as amongst the ectomycorrhizal fungi no others appear to be associated with ferns, and the genus also has species associated with different conifers and woody dicotyledons plants. It might be viewed as an evolutionary projection from "*Tuberales*".

Malloch: I am pleased to see it no longer considered as a cleistothecial plectomycete.

Eriksson: Our results with *Elaphomyces* should only be regarded as preliminary as we intend to sequence some other molecule to test whether the *Elaphomycetales* should be included in *Pezizales* or not.

Landvik: One sequence cannot be expected to solve a problem that has existed for decades. Our data provides a hint in one direction, but other sequences may hint in others. We do not therefore propose any change in taxonomy at this time.

Malloch: I agree, but the reality is that no-one else may analyze any other sequences for a considerable time.

Erysiphales

Malloch: This group has been the subject of considerable speculation. Cain had suggested they might be aligned with the inoperculate discomycetes, mainly because of the ascospore type, but others have suggested placements with loculoascomycetes.

Poelt: If the orders including parasites of vascular plants are considered, some are restricted to evergreens in the tropics and subtropics (e.g. *Coryneliales, Meliolales*), while orders including parasites of deciduous trees and short-lived plants have many taxonomic problems (e.g. *Erysiphales, Protomycetales, Taphrinales*). The adaptation to a life-style with a short active and a long passive phase promotes changes in propagation, especially towards conidial stages or in *Erysiphales* special ascoma types. As in the case of Mediterranean adaptations in flowering plants, perhaps part of the fungal genome is lost during the adaptive process from evergreen to summergreen plants.

Kurtzman: As some fungi have very wide host ranges, at least in some cases the amount of genetic change involved must be minor. RNA sequences do, however, have the potential to track a history of host adaptation. At the infraspecific level, history has been inferred through molecular substitutions in some plant pathogens.

Taylor, J. and Berbee: G. Saenz has sequenced the 18SrRNA gene from *Blumeria graminis* (syn. *Erysiphe graminis*). It forms part of the early major radiation of filamentous ascomycetes, along with apothecial and ascolocular forms. We are concerned that not enough note is being taken of existing molecular data that could be used to group orders into higher taxa. We believe that there are a few well-supported hierarchical branches that define monophyletic supraordinal taxa; not to recognize these relationships is a disservice to users of the *Outline*. At a minimum, we ought to recognize plectomycetes, pyrenomycetes, hemiascomycetes, and some name embracing *Protomycetales, Schizosaccharomycetales, Taphrinales* and *Pneumocystis* (the "early ascomycetes" of Berbee and Taylor, *BioSystems* 28, 117-125, 1992). We should go further and provide names for yet larger clades: euascomycetes for the filamentous ascomycetes, ascomycotina for hemiascomycetes plus euascomycetes, and "*Ascomycota*" for "ascomycotina" plus "early ascomycetes". The tree shown in Fig. 1 summarizes these ideas.

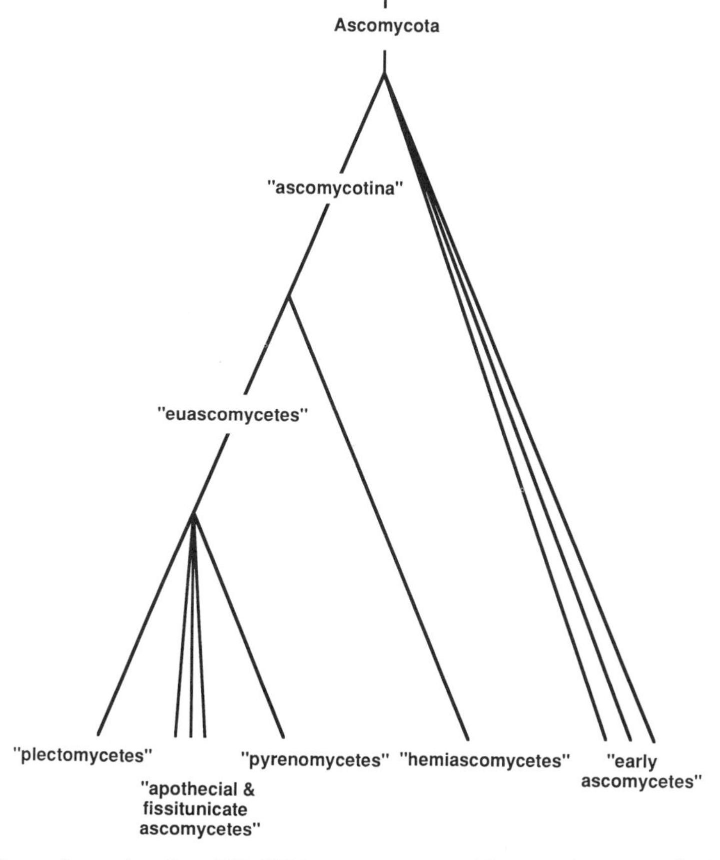

Figure 1. Phylogenetic tree based on 18S rRNA gene sequences with names for supraordinal taxa. Each name embraces the taxa below it. With the exception of *Ascomycota*, all the names allocated are to be treated as informal.

Eriksson: We need a higher category that we are sure is monophyletic, and that is how order is used in the *Outline*. The supraordinal situation is too unstable to formalize now and I feel we should wait ten years until many more taxa have been studied and the situation is clear. For practical purposes, we can use the term "plectomycetes" just as we use the word "lichens", but not *Plectomycetes* as a formal class.

Kimbrough: The septal structure in *Erysiphales* recalls much more that of the loculoascomycetes that I have reviewed than of the pyrenomycetes. *Sclerotinia* also falls into the loculoascomycete group on this feature. When we refer to apothecial fungi, we may need to make a distinction. In 1977 I suggested that the inoperculate discomycetes are probably more closely related to certain groups of pyrenomycetes than they are to operculate discomycetes, despite having apothecia.

Schumacher: Taylor has raised the issue of principles of classification. Molecular biologists produce hypotheses by cladistic methods using parsimony but based on limited DNA sequences. However, 95 % of those who work at the genus and species level in mycology have no overriding wish to classify only monophyletically. The contributions of van Brummelen and Kimbrough identified important characters that need to be weighted in producing a classification. Many of us feel that a phylogenetic classification is not just one using molecular and cladistic methods. There is also a school within cladistics that wishes to classify by principles of character compatibility; this is not appropriate for those using PAUP and parsimony analysis working with the single character of DNA. Those who want to use molecular characters along with other data sets and wish to place an evolutionary rationale behind what we are doing by character weighting need a quite different methodology. Another option would be to focus on such a combined approach at the generic level and below, and to rely only on molecular data for the recognition of families and higher taxa.

Kendrick: The *Erysiphales* are very different from any other group of fungi, whether one considers the assimilative phase, the anamorphs or the teleomorphs, and are certainly monophyletic. I wish all groups were that easy to keep together.

Eurotiales

Mouchacca: The genera of *Tricochomaceae* mostly have anamorphs in *Aspergillus* and *Penicillium*. I wondered if the anamorphic genera had characters that might merit their subdivision so that they correlated more cleanly with particular teleomorph genera of the family.

Taylor, J.: LoBuglio and Berbee have molecular data supporting two distinct groups of fungi with *Penicillium* anamorphs, those similar to subgenus *Penicillium* (e.g. *Eupenicillium*) and those similar to subgenus *Biverticillium* (e.g. *Talaromyces*). *Talaromyces* species are well-separated from *Eupenicillium* species at the molecular level, but we do not think this is the right time to split *Penicillium*. The situation in *Aspergillus* is worse due to its larger number of teleomorphs.

Mouchacca: Having the same anamorph names can cause confusion in identifications because of the connection to different teleomorphs.

Reynolds: In the case of mitotic species that do not show such connections but could be placed by molecular data, would you be able to place them, with an inferred ascus, on the basis of the morphology of the anamorph ?

Mouchacca: I could not be confident of this, but because of the economic importance of these genera more predictive possibilities based on anamorphs alone would be desirable.

Gams: The consequences of splitting large anamorph genera and integrating anamorphic species into a single system were considered at the meeting on *The Fungal Holomorph* in Newport, Oregon, in 1992. Names for anamorphs have been coined for practical purposes and to facilitate identification. While *Penicillium* subgen. *Biverticillium* coincides nicely with *Talaromyces*, the situation with *Eupenicillium* is more complex and cannot be resolved by resurrecting *Citromyces*. In *Aspergillus* the associated teleomorph genera correlate with subgenera and sections, but subdividing *Aspergillus* seems too great a sacrifice of workability in relation to the amount of information gained. The issue needs to be considered in relation to nomenclatural stability, especially when proposals are underway to conserve *Aspergillus nidulans* as an anamorph name by those who are not even prepared to change to *Emericella*. The mycological community is not ready to accept the consequences of major changes based on limited data. In these cases, and also *Acremonium* and *Fusarium*, it is better to use subgenera and sections.

Hennebert: It is too early to contemplate dividing the genera *Aspergillus* and *Penicillium* to match their teleomorphs in a comparable manner attained in the *Sclerotiniaceae*. The infrageneric classifications that have been proposed already are a step in this direction, but more morphological as well as molecular data are needed to proceed further.

The species characterizations need to be refined by defining the terms used in strain descriptions more accurately. A programme to harmonize and redefine descriptive terminology has been initiated in Louvain. Terms will then be able to be used more precisely in future monographs and papers.

The diversity of physiological responses of *Penicillium* strains within the same laboratory and the variable composition of "standardized" substrates, means that the physiological characters used at a primary level used by Pitt are of relative value. Strict redefinitions of media are needed.

Electrophoretic studies of total proteins, now in progress in Louvain, may also assist in refining species characterizations in *Penicillium*. These profiles appear to be more significant than either secondary metabolites or extracellular enzymes.

Udagawa: From the teleomorph/anamorph connections, *Eupenicillium* (anam. *Penicillium*) is homogenous, but *Talaromyces* (anam. *Geosmithia, Paecilomyces, Penicillium,* etc.) is complex. A solution needs to be sought in the future when more molecular and chemotaxonomic information is available.

Taylor, J.: In this case we have good information that mitosporic fungi are very close relatives of those with teleomorphs. If we do not include such mitosporic species in the ascomycetes, we are ignoring that information. Such fungi should be included in the *Ascomycota*.

Korf: The reality of anamorph-teleomorph connections differs from group to group in terms of the usability of separate names for the anamorph. In *Bipolaris/ Cochliobolus*, *Drechslera/ Pyrenophora,* and *Exserohilum/ Setosphaeria*, it is possible to guarantee that from a *Bipolaris* one will obtain a *Cochliobolus* teleomorph and not a *Pyrenophora* nor a *Setosphaeria*. In other groups we are unable to achieve such a level predictability because the anamorphs are not sufficiently distinct, as in *Aspergillus* and *Penicillium*. Perhaps a more careful search for previously overlooked characters in the anamorph would enable such connections to be made.

Taylor, J.: We should not restrict ourselves to microscopic features. Thin-layer chromatography of secondary metabolites and growth on defined media can facilitate correlations, and molecular work will show exactly where they fit.

Korf: I have no problem with accepting *Bipolaris* species as ascomycetes even though their teleomorph may not be known. Similarly, species of *Aspergillus* and *Penicillium* are ascomycetes even if they no longer produce a teleomorph. For example, *A. niger* is an ascomycete.

Rogers: There can be no objection to making information retrieval easier and more relevant, but what is of concern is the formality of placing holomorphic anamorphs into teleomorphic genera because of the formidable nomenclatural problems. An informal, but informative, system might be workable.

Hawksworth: I see complete integration as the logical long-term aim, but, as recognized at the *Fungal Holomorph* workshop (J.W. Taylor and D.R. Reynolds, eds., *The Fungal Holomorph*, 1993), I feel this to be premature at this time. The full nomenclatural ramifications would need to be carefully scrutinized, including the possibility of abandoning Art. 59 of the Code. Interestingly, mycologists working with some groups of pleomorphic fungi (e.g. *Mucorales, Uredinales*), do not retain separate Latinized anamorph names once a teleomorph is known; that practice is to be encouraged as also practiced by M.B. Ellis (*Dematiaceous Hyphomycetes*, 1971).

Reynolds: The terms anamorph and teleomorph are now used in diverse and often confusing ways, and examples of groups that have no need of this include lichens and many basidiomycetes. The possibility of removing Art. 59 is worthy of serious consideration. Without it there would be lessened concern about the creation of a dual system of classification where mitotic fungi are pulled into a natural system as real or inferred ascal connections are made. I favour the formal withdrawal of Art. 59 from the Code.

Laboulbeniales

Malloch: These fungi have been the subject of interesting phylogenetic speculations over the last century. Work by Blackwell and myself strongly suggests that *Pyxidiophora* is very close to the order both morphologically and biologically. It might be viewed as a heteroecious version of the monoecious *Laboulbeniales*.

Kimbrough: I feel there is a great deal of convergent evolution in entomogenous fungi, for example the development of a melanized cell at the point of attachment to the exoskeleton. Further, in *Pyxidiophora* the teleomorph is a mycoparasite on dung while the anamorph is attached to the insect.

Blackwell: I do not believe that the similarity between *Pyxidiophora* and other minute arthropod ectoparasites and *Laboulbeniales* results from convergence as many of the genera Spegazzini and Thaxter described (e.g. *Endosporella*) seem to be either *Pyxidiophora* ascospores, as already demonstrated for *Acariniola* and *Thaxteriola*. The zygomycetes *Amphoromorpha* and *Basidiobolus* may, however, be convergent in this feature.

Weir: The proposal to include *Pyxidiophora* in the *Laboulbeniales* seems to be well-supported both on morphological and ecological grounds. The interrelationships between fungi, dung, insects and mites are highly complex and one could certainly conceive a *Pyxidiophora*-type ancestor to the order in this situation.

While little work at the family level is currently in progress, at the generic and species level there is much instability. In recent years there has been a proliferation of new generic names based on limited material, and some will undoubtedly be reduced to synonymy (e.g. *Fanniomyces* to *Stigmatomyces*). The situation is even worse at the species level, especially in *Peyritschiella* where unreliable characters such as the number of perithecia produced are used in species delimitation.

Microascales

Malloch: This order seems to be holding together as a group in the molecular data presented. I have doubts about the inclusion of the family *Chadefaudiellaceae* in the order, which I believe was based on an early report of a dextrinoid ascospore reaction. Some genera within the *Microascaceae* may, however, require transfer elsewhere.

Onygenales

Malloch: Both Currah (this volume) and Tehler have independently suggested that *Myxotrichaceae* should be excluded from *Onygenales* and perhaps aligned with the inoperculate discomycetes. That family should certainly be excluded, and the evidence for the linking of *Gymnoascaceae* with *Trichocomaceae* is growing.

Currah: Preliminary molecular data by Guého further confirms that the *Myxotrichaceae* do not belong here. The family needs to be raised to the rank of order.

I would discourage the inclusion of *Nannizzia* in *Arthroderma*. There is a clear dichotomy between the species currently in each genus, and lumping will cause a credibility problem with medical mycologists. A parallel proposal was to include *Keratinophyton* in *Aphanoascus*; the genera are similar, but again there is still a clear separation on the basis of ascospore and anamorph characters.

It is premature to transfer *Amauroascus* into the *Ascodesmidaceae* in *Pezizales*; a position in *Onygenaceae* is supported by the punctate ascospores, keratinolytic abilities, and the arthroconidial anamorphs.

Spathulosporales

Jones: Ultrastructural studies on *Spathulospora* point to a position close to the *Halosphaeriales*. I do not believe it to be as primitive as has been suggested, but the order is good and it is unique in both habitat and distribution.

Kohlmeyer: Nakagiri (*Mycologia* 85, in press, 1993) is describing a new genus *Hispidicarpomyces* (on *Galaxaura falcata*, *Rhodophyta*) for which the new family *Hispidicarpomycetaceae* in *Spathulosporales* is being introduced. The family is distinguished from the *Spathulosporaceae* by forming spermodochia with spermatiophores, the ascomata developing from collapsed sporodochia, and in having true hyphae.

DISCUSSION 3

CALOSPHAERIALES, CLAVICIPITALES, CORYNELIALES, DIAPORTHALES, DIATRYPALES, HALOSPHAERIALES, HYPOCREALES, MELIOLALES, OPHIOSTOMATALES, PHYLLACHORALES, SORDARIALES, TRICHOSPHAERIALES, AND *XYLARIALES*

Leaders: M.E. Barr[1] and P.F. Cannon[2]

[1]9475 Inverness Avenue
Sidney, British Columbia V8L 5G8, Canada

[2]International Mycological Institute
Bakeham Lane, Egham, Surrey TW20 9TY, UK

INTRODUCTION

Cannon: We propose to consider each of the above orders in turn.

DISCUSSION

Calosphaeriales

Barr: We now believe that there are two families in this order, *Calosphaeriaceae* and *Graphostromataceae*. The order appears to be extremely isolated.

Clavicipitales

Spatafora: Our work on the small rRNA subunit, involving a fairly broad sampling of this order and *Hypocreales*, shows the *Clavicipitales* to be nested within that order. The similarities in anamorphs I believe to be homologies. Differences in centrum development have been claimed to distinguish the two orders, but lateral paraphyses in the *Clavicipitales* are not homologous to those of *Chaetomium* (Luttrell, *University of Missouri Studies* 24, 1-120, 1951). It falls in a clade dominated by mycoparasitic fungi (e.g. *Hypocrea. Hypomyces*), and the obligately biotrophic nutritional mode of the *Clavicipitales* is consistent with this placement.

Cannon: The *Hypomycetaceae* has sometimes also been linked with the *Hypocreaceae*.

Rogers: I have long suspected that the *Clavicipitaceae* and *Hypocreales* in the broad sense are closely related. Rogerson almost implied this in 1970 when referring to the two together as "hypocrealean fungi". Too much emphasis was placed on the hamathecial elements in making the original separation.

Cannon: The differences in ascus morphology concern me but I am attracted by the narrow niche they occupy as opposed to that of most *Hypocreales* which tend to be less closely allied to their plant or fungal partners.

Spatafora: While it is true that the *Clavicipitales* are much more restricted in host ranges, they are biotrophic. Our results show them to be a sister group to the *Hypomyces/ Hypocrea* clade represented by a large group of fungal parasites.

Rossman: I favour recognizing the *Clavicipitaceae* as a family of *Hypocreales* rather than an order. The character of bright-coloured pigmentation unites this group, but not exclusively. For example, an immersed ascoma will tend to become colourless and non-melanized and thus be considered light- to bright-coloured but may not be at all related to *Hypocreales*. The ascal apex could be viewed as an adaptation related to the elongated ascospores. In *Myriogenospora*, where the ascospores are not elongate, the thickened ascus apex thins at maturity.

Malloch: There is a report of a *Cordyceps* that also, or alternatively, parasitises nematodes. I wondered if that might facilitate their dispersal.

Kimbrough: I suspect that *Cordyceps* is entirely entomogenous. If you examine the *Elaphomyces* host of *Cordyceps* you will probably find they are infected with insect larvae and pupae.

Minter: *Hirsutella* has species known to live on nematodes.

Kendrick: One adaptation to the very small insect targets of *Clavicipitales* is the part-spores, 800 rather than 8 per ascus and numerous perithecial cavities on each stroma. The net result is millions of part-spores from each ascoma, increasing the probability of hitting small targets. These are derived rather than basal characters and do not impede the recognition of *Clavicipitaceae* as a part of *Hypocreales*.

Rogers: Do the *Clavicipitales* and *Ostropales* have a connection ?

Spatafora: *Stictis radiata,* the only species of *Ostropales* we have sequenced, did not come close to any of the clavicipitaceous fungi we examined.

Bellemère: The ascus structure in *Clavicipitales* and *Ostropales* is quite different. The former has a well-developed plug with no axial canal, while the latter have a large axial canal and probably also a unitunicate ascus. The *Clavicipitales* ascus may also have a *d*-layer.

Gams: I cannot support Kimbrough's suggestion that *Cordyceps canadensis* and allied species are entomogenous rather than mycogenous. They are regularly associated with *Elaphomyces*. This is evidence that these fungi are really adapted to a chitinous substrate rather than to animal dispersal. Entomogenous, mycogenous and nematogenous capacities are closely associated within this order.

Cannon: But are the grass-inhabiting *Clavicipitales,* including *Claviceps* and *Epichloë*, as closely related to *Cordyceps* as has been supposed ?

Spatafora: We have sequenced species from *Aciculosporium, Balansia, Claviceps, Cordyceps, Epichloe, Hirsutella,* and *Hypocrella*. They are monophyletic and appear as a sister group to *Hypocreales*.

Coryneliales

Barr: This appears to be a well-defined and isolated order.

Kimbrough: Benny and I concluded that the ascomata were lysigenous in development, and had *Rhytidhysterion*-like spermogonia. These, and other features led us to suggest that this was a highly modified group of loculoascomycetes that had developed deliquescing asci.

Cannon: Johnston and Minter (*Mycological Research* 92, 422-430, 1989) have shown that the asci in this order are multilayered. The outer layer breaks at a very early stage and it is the inner layer which deliquesces.

Diaporthales

Barr: There are different opinions on how this order should be divided, and I am currently contemplating revising my own and separating out a group with cylindric asci and uniseriate ascospores which remain more firmly attached. The majority of fungi in this order have short and readily detached asci floating free within the centrum.

Rogers: Fluorescence microscopy reveals that there is a great diversity in the ascus apices of members of this order which has not been generally appreciated. Further, paraphyses are rather common and vary from narrow to wide, and from evanescent to persistent. Reports of a basically pseudoparenchymatous hamathecium are at least partly fiction, based on broad band-like paraphyses which have been pushed together. I believe the order is very heterogeneous.

Scheuer: The name *Botanamphora*, of which I am a co-author, was a mistake. Only a drawing by Berlese of the type species, *Leptosphaeria pachycarpa*, was studied and now I have seen material I find that this is correctly *Trematosphaeria pachycarpa*. The new name *Herbampulla* has been introduced for our concept (Scheuer and Nograsek, *Mycotaxon* 47, 415-424, 1993). In spore morphology it recalls species of *Melanconidaceae* but that family is stromatic while *Herbampulla* has no indication of a stroma; I support its provisional inclusion in that family. It occurs on *Carex firma* in the Alps.

Diatrypales

Barr: From my own observations on field collections I would agree with the suggestion of Rogers that there is quite a close relationship between *Diatrypales* and *Xylariaceae*.

Rogers: I am not certain how much of the similarity is real and how much the result of convergent evolution. I would like to see some molecular data on this. The two families also seem connected at points by the anamorph types.

Blackwell: Spatafora has sequenced the small subunit of rDNA in *Diatrype disciformis* and found it came close to *Daldinia concentrica* and *Hypoxylon* and *Xylaria.*

Whalley: It would be interesting to have a result for one of the applanate *Hypoxylon* species now referred to *Biscogniauxia.*

Kendrick: Should the two orders then be combined ?

Rogers: Merging would be unwise as it would make the two most unwieldy. The core of both seem to be independent natural groups and it is mainly taxa on the "fringes" that cause the debate.

Cannon: The *Xylariales* as currently circumscribed is probably somewhat heterogeneous.

Hilber: I do not consider *Biscogniauxia* to be related to *Diatrypales*.

Rossman: In separating the two orders, weight should be given to the anamorphs. These are relatively homogeneous in the *Xylariaceae*. For example, that of *Thuemenella* is clearly xylariaceous. Where anamorphs suggest a grouping the teleomorph characters should be re-evaluated.

Rogers: I agree. We now have an anamorph for *Lopadostoma turgidum* which supports the prediction of Nitschke that it would have a coelomycete state; this proves to be *Libertella*-like. A similar type of elongated anamorph occurs in a species close to *Nummularia viridis*, an *Anthostomella* species, and also in *Hypoxylon sassafras* (which we recently returned to *Creosphaeria*).

Eriksson: The reason for the recognition of the *Diatrypales* was that the French mycologists had described a separate ontogenetic type, the *Diatrype*-type, separate from the *Xylaria*-type. A close relationship was suspected and I have been waiting for molecular data, which we now have for a few species.

Sutton: The developmental processes involved in mitospore production in the two orders are fundamentally no different from the perspective of basic ontogenetic events. This does not deny that there are minor differences in other aspects, such as pigmentation and delimitation. If the *Diatrypales* were placed with the *Xylariales*, there would be no serious objection from anamorph specialists.

Bellemère: Along with Parguey-Leduc, Janex-Favre and Meléndez-Howell (*Cryptogamie, Mycologie* 13, 215-246, 1992), we examined the ascospore walls in these pyrenomycetes. The ascospore wall in *Xylariales* is well-differentiated, while that in *Diaporthales* and *Diatrypales* is not.

Halosphaeriales

Kohlmeyer: This order includes the largest group of marine fungi and is rather well-circumscribed. It contains a single family, but may need to be divided into further families in future. New genera are being described each year so it would be premature to describe new families at this time.

Jones: I don't think all the fungi assigned to the *Halosphaeriales* really belong there, and that they are an example of convergent evolution in response to the marine environment. The fungi included have been greatly modified by the environment and it will take some time to disentangle them. I agree that we will probably eventually have to divide the order into more families. We have initiated some molecular work and expect this to provide a clearer idea of how closely related the genera really are.

Cannon: I wonder where the affinities of the order are. I also suspect that several of the characters that unify the group are a response to their unusual environment, for example hyaline ascospores and complex appendages.

Kohlmeyer: The ascospore colour and septation is not so important. For example, the well-circumscribed genus *Corollospora* has species with 1- to multiseptate and even muriform ascospores, and also others that are hyaline or brown.

Blackwell: The *Halosphaeriopsis mediosetigera* small subunit rDNA sequence Spatafora and myself studied came out in a clade with *Petriella setifera* and *Microascus trigonosporus,* but the taxon sampling was incomplete.

Spatafora: All taxa within that clade possess a pseudoparenchymatous centrum, have evanescent asci, and other gross similarities in ascomatal morphology.

Kendrick: These fungi are obviously descendants of terrestrial forms. The asci must have lost the ability to function as spore-guns and I suspect that they may have been derived from several different terrestrial orders, converging morphologically and functionally as a result of their secondary invasion of a new and very different habitat.

Malloch: *Pseudallescheria boydii* is commonly isolated from brackish and estuarine habitats, and *Petriella* species have been obtained from wood in deep ocean waters by Schatz. The *Microascales* could therefore have some marine affinities.

Kohlmeyer: With regard to the possible terrestrial origin of the *Halosphaeriales,* we have developed the hypothesis that there are two groups of marine ascomycetes: one that primarily originated in marine habitats (parasites of red algae), and one that invaded the marine environment secondarily. The latter group includes many loculoascomycetes.

Jones: I agree with Kendrick that the *Halosphaeriales* are basically terrestrial fungi that have migrated into the marine habitat. *Aniptodera* and *Halosphaeria* have a range of ascus types from those with a well-developed apical apparatus, through those which are persistent but have none, to ones which are deliquescing. Further, physiological studies by Jennings indicate that they have many physiological characteristics in common with terrestrial fungi. Interestingly, members of both these genera, and also *Nais aquatica*, are also to be found in freshwater habitats.

Hyde: The degradation in ascus apical structures from freshwater to marine species of *Aniptodera* and *Halosarpheia* may indicate a common terrestrial ancestry in the *Halosphaeriales* (e.g. from *Sordariales*). However, it is equally likely that the genera referred to the *Halosphaeriales* really belong within several unrelated families.

Hypocreales

Barr: I would favour *Niessliaceae* being transferred here. The centrum and ascospores agree well with the order but the ascomata are dark.

Malloch: *Emericellopsis* is a typical cleistothecial ascomycete with a very simple centrum containing uniformly disposed sphaerical asci. The combination of such ascomata, simple brown ascospores, and an *Acremonium* anamorph, suggests affinities with *Pseudeurotiaceae.* I do not believe it belongs in *Hypocreales.*

Gams: I would not support the classification of *Emericellopsis* in the *Eurotiales*. I have used the criterion of connected conidial chains to segregate *Sagenomella* (teleom. *Sagenoma*) from *Acremonium* which has disconnected chains or slimy conidial heads. *Acremonium* is still heterogeneous and related to several orders of ascomycetes. *Emericellopsis* is a challenge for molecular workers.

Eriksson: What kind of hamathecium should be accepted in *Hypocreales* ? Would you accept paraphysoids, for example as in *Thyridiaceae.*

Rossman: The hypocrealean fungi have a unique centrum with apical paraphyses, which may be slightly modified as in *Hypomyces* and *Clavicipitaceae*, although there are no detailed recent studies on the latter. Young ascomata are needed to see this. The apically free "paraphyses" in *Thyridiaceae* are quite different.

Samuels considers that *Sarawakus* is a *Hypocrea* with non-septate ascospores; it has *Gliocladium*- or *Trichoderma*-like anamorphs. I would not have recognized the genus as I put less emphasis on ascospore septation.

The colourless perithecia of *Payosphaeria* do not necessarily indicate that it belongs in *Hypocreales*. As the genus was described as having true paraphyses it would be better placed in the *Thyridiaceae* or *Trichosphaeriales* based on the ascospore characters. The fungus should be cultured for potential clues from an independent data set, the anamorph.

I support the removal of *Pyxidiophoraceae* from the order. *Sphaeronaemella* and *Viennotidia* should also be removed. The former is mycoparasitic as Malloch and Blackwell suggested *Pyxidiophora* can be.

Hypocreopsis appears to be unique in a biological sense and has a characteristic lobed thallus-like stroma, and fungicolous habit.

Cannon: We have grown ascospores from *Hypocreopsis* at IMI and found a white, slow-growing, *Trichoderma*-like anamorph which is different from those found in *Hypocrea.*

Dissing: *Sphaeronaemella fimicola* has been claimed to have a germ-slit when studied by SEM, but this seems to be an artifact. If so, there is no reason to recognize *Viennotidia.*

Blackwell: Molecular data suggests that *Sphaernonaemella* is a sister taxon to *Ceratocystis* s. str. and is not hypocrealean. The *Pyxidiophoraceae*, *Kathistes* and *Subbaromyces* appear to be basal to other perithecial ascomycetes in their rDNA.

Hafellner: Can *Pronectria* and *Nectriella* be recognized by their anamorphs ? *Hobsonia* anamorphs are often found together with *Pronectria* teleomorphs, although slightly in advance of the perithecia. If *Hobsonia* is not known in *Nectriella* that could help justify the recognition of *Pronectria* for lichenicolous species.

Rossman: These genera are separated simply on the lichenicolous habit of *Pronectria,* and the woody and herbaceous substrates of *Nectriella.* No *Hobsonia* is known associated with *Nectriella* s. str. The taxonomy of immersed nectriaceous fungi is complicated by morphological modifications based on the habitat.

Hawksworth: *Hobsonia* anamorphs have not been proved by single ascospore isolations to be produced by any *Pronectria* species. All *Pronectria*'s so far cultured have *Acremonium* anamorphs. It is not uncommon for different lichenicolous fungi to be closely associated on a single host, especially if one is pathogenic and has reduced the resistance of the host lichen to invasion by other species.

Meliolales

Kendrick: These fungi are a melanic counterpart to the *Erysiphales* in having superficial assimilative hyphae and superficial ascomata. This is a fine example of convergent evolution and both groups are clearly different from any other living fungi.

Ophiostomatales

Wingfield: There is now overwhelming evidence that the two central genera, *Ceratocystis* and *Ophiostoma* are not closely related on morphological grounds as well as in their cell-wall constituents, and this has been supported by molecular data. I would maintain the order for *Ophiostoma,* which I interpret as an aggregate of several genera, including *Ceratocystiopsis* (excl. *C. falcata*, *C. proteae*). It might be related to *Diaporthales*. *Ceratocystis* s. str. could be moved to *Microascales*, along with *Sphaeronaemella,* etc.

Spatafora: Of the taxa we have sampled from *Ceratocystis* str., they all group within *Microascales* in our rDNA analyses. The *Ophiostoma* species we studied appear to be a sister group to those we sampled from *Diaporthales,* but without a high bootstrap value; this conforms to the method of ascus development, in both cases asci are produced from the base and released into the central cavity. *Ophiostoma* has been described as having swollen paraphyses-like structures similar to those mentioned by Rogers (above) in *Diaporthales.*

Rogers: That is true in some cases, but the *Diaporthales* is so large that while broad paraphyses are widespread a variety of hamathecial types is to be expected.

Blackwell: Hausner has sequenced about 100 species in this group and always finds that all the taxa with a *Chalara* anamorph belong to *Ceratocystis* s. str. The *Knoxdaviesia* anamorph is also close to *Ceratocystis* s. str. He also showed this to have a connection with the *Microascales*. The genera *Ceratocystiopsis, Europhium,* and *Ophiostoma* fell in a separate clade. *Ceratocystiopsis proteae* is an exception. Berbee and Taylor have shown that *Leucostoma* is in the same clade as *Ophiostoma*. Spatafora and myself have confirmed these results, and in addition find *Sphaeronaemella* to be a sister taxon to two species of *Ceratocystis* on a very long branch supported by jack-knifing.

Wingfield: *Ceratocystiopsis proteae*, with a *Knoxdaviesia* anamorph, needs to be segregated and is nearer to *Ceratocystis* than to *Ophiostoma*. Our studies on centrum development support a *Diaporthales* link, but rather few have been studied. While *Ceratocystiopsis* s.str. has *Chalara* anamorphs, *Ophiostoma* has ones in *Graphium, Leptographium, Sporothrix* and *Hyalorhinocladiella*. However, cycloheximide tolerance and cell wall components distinguish the *Ophiostoma* anamorphs.

Phyllachorales

Cannon: I have speculated that the order is linked to *Diaporthales*.

Scheuer: E. Müller mentioned to me in about 1983 that some of the genera are somewhat intermediate in centrum structure between *Phyllachora* and *Gnomoniaceae*, for example *Phomatospora* - a genus he placed near *Phyllachora* and not in *Amphisphaeriaceae*, noting also the similarities in ascus apices. This might lend support to your hypothesis.

Barr: *Phomatospora* does seem to belong with *Amphisphaeriaceae* in its other morphological features.

Eriksson: It seems premature to merge the *Phyllachorales* and *Diaporthales*. Berbee and coworkers now have molecular data that *Glomerella* (anamorph *Colletotrichum*) cluster with *Neurospora* (*Sordariales*) on a separate branch from *Diaporthe* suggesting they are not closely related.

Cannon: I am not certain that *Glomerella* is closely related to *Phyllachora*.

Hyde: *Marinosphaera* has many characteristics typical of the *Phyllachorales,* and was tentatively placed in *Phyllachoraceae*. If it is accepted that marine fungi have terrestrial ancestors, it is conceivable that *Marinosphaera* evolved from a terrestrial phyllachoraceous taxon. Ultrastructural studies now in progress may provide more information on its placement.

Sordariales

Cannon: I broadly agree with the five families currently accepted in the *Outline,* but the *Lasiosphaeriaceae* and *Sordariaceae* are so close that subfamilial rank might be more appropriate. The differences mainly relate to ascospore shape and septation, and ascus shape. The *Melanosporaceae* do not seem close to *Chaetomiaceae* and that is now supported by some molecular data. The *Nitschkeaceae* is probably close to *Lasiosphaeriaceae*. There are identifiable links between *Lasiosphaeria, Cercophora, Podospora* and other lasiosphaeriaceous genera.

Barr: My 1990 interpretation of *Lasiosphaeriaceae* was extremely wide as the characters seemed to flow from one to another. I am unclear how else to separate out some of the genera.

Hilber: I recommend that *Cercophora* be treated as a subgenus of *Lasiosphaeria* as they are only separated by the swollen upper part of the ascospore.

Lundqvist: I think neoteny occurs in the centre of the order around *Lasiopshaeriaceae*. *Lasiosphaeria* could be regarded a paedomorphic genus where the species have become physiologically mature in the hyaline stage. *Cercophora* and its allies have a pigmented cell in the ascospores that could be related to their ecological niche. The question is the weight to be given to the character of pigmentation in relation to others. Perhaps the generic limits between *Cercophora* and *Lasiosphaeria* should be drawn in a different way using, for example, ascoma wall characters. Cladistic and molecular studies have yet to be carried out in this group.

The order has expanded over the years incorporating some families of which I am sceptical, particularly the *Melanosporaceae*. The *Coniochaetaceae*, whose ascospores have germ-slits, also seems isolated.

The *Sordariaceae* itself seems to be a good clade, centred on *Sordaria, Neurospora, Gelasinospora* and *Copromyces*, and united also by good ascomatal characters.

Malloch: The *Tripterosporaceae* are doubtfully distinct from *Lasiosphaeriaceae*. Both *Tripterospora* and *Zopfiella* seem to be very close to, or members of, *Podospora* and *Triangularia*. They may be polyphyletic, representing simplifications that have occurred several times. Retention of the family seems unjustified.

Lodha: The *Tripterosporaceae* was established when the cleistothecial character was thought to be particularly important. I agree that it is no longer needed and that its members should be merged with other sordariaceous fungi.

The *Chaetomiaceae* should either be kept here or perhaps placed in a separate order with the *Melanosporaceae*. Within the *Chaetomiaceae*, genera should not be segregated only on anamorph differences, but germ pore(s) seem to be a good character.

Cercophora and *Lasiosphaeria* are different, but overlap. Perhaps the *Lasiosphaeria* species with a dark cell in the ascospores should be studied more critically to see if they can be accommodated in *Cercophora*.

Lundqvist: I am unhappy about the *Chaetomiaceae* being close to the core groups here but have no other suggestion.

Dissing: I agree that the *Melanosporaceae* should perhaps be removed from *Sordariales*. I also wish to point out that the name *Melanospora fimicola* has been misapplied; many fresh collections from musk ox dung from Greenland are in C and it is distinct from *Sphaerodes ornata*. I made some SEM micrographs of *Sphaeronaemella fimicola* and *S. helvellae*; there is definitely no germ slit in the latter.

Cannon: In terms of nomenclature, the family name *Melanosporaceae* is not validly published, and the *Ceratostomataceae* has now been found to be based on a fungus that has nothing to do with the current family concept.

Blackwell: Molecular data shows the type species of *Melanospora (M. zamiae)*, and also *M. fallax*, to be part of a radiation including *Clavicipitales* and *Hypocreales*. The family needs to be there. However, *Sphaerodes ornata* is far outside this region of the tree from the rDNA sequences we have obtained.

Rogers: I think *Coniochaeta* is isolated and merits an independent family.

Hawksworth: The presence of *Geniculosporium-Nodulisporium*-like anamorphs in certain species of *Coniochaeta* (Hawksworth, *Norw. J. Bot.* 25, 15-18, 1978; Udagawa & Horie, *Reports on the Cryptogamic Study in Nepal*, 97-104, 1982) either suggests a link with *Xylariaceae* or that the genus is polyphyletic (i.e. that a new genus should be introduced for those fungi).

Rogers: *Coniochaeta* should be restricted to taxa with *Phialophora*-like anamorphs. Those with *Nodulisporium*-anamorphs probably do belong in *Xylariaceae*, but I agree that a new genus is possibly required.

Malloch: What of *Ascotricha,* a *Coniochaeta*-like genus with germ-slits, ascomatal hairs, and anamorphs not unlike *Nodulisporium* ? Considering that *Neurospora* includes the world's best-known filamentous fungus, does the *Neurosporaceae* merit recognition separately from the *Sordariaceae* ?

Hawksworth: *Ascotricha* has one species with an I+ blue apical ring reported in the ascus (Khan & Cain, *Mycotaxon* 5, 409-414, 1977), which, taken together with the germ-slits and *Dicyma* anamorph, clearly indicates a placement in *Xylariaceae*.

Whalley: That may be so, but it is on the boundaries of the family.

Cannon: I am concerned at the placement of *Apiospora* in *Sordariales*, let alone in *Lasiosphaeriaceae* - a family which I feel is too broadly conceived by Barr. The anamorph of *Apiospora* is so unusual that the genus is best referred to *incertae sedis*. *Acrospermoides* should probably also be excluded from *Lasiosphaeriaceae*.

Taylor, J.: Comparisons of the mitochondrial small subunit rRNA gene and ITS sequences show that *Gelasinospora, Neurospora* and *Sordaria* have almost no mutations relative to each other. However, *Podospora* showed ten times as many substitutions when compard to the others. Surprisingly, *Cercophora* had the same higher number of mutations as *Podospora,* but was equally distant from it and the

Neurospora group. Thus, molecular data offer no support for creating a separate family for *Neurospora*; the *Sordariaceae* accommodates *Neurospora* very nicely.

Hafellner: The family position of the lichenicolous genera of *Sordariales* remains unclear, although they clearly belong there. No appropriate families appear to be described.

Trichosphaeriales

Barr: If *Niessliaceae* is removed to *Hypocreales*, the group contains only *Trichosphaeriaceae* which comprises many fungi placed in the former broad concept of *Sphaeriales* and is currently incoherent. I would add *Acanthostigma*; remove *Chaetosphaeria* to *Sordariales* (perhaps eventually to its own family); remove *Litschaueria* and *Phaeotrichosphaeria* to the *Lasiosphaeriaceae* s.lat.; and place *Trichosphaerella* and *Valetoniella* in *Niessliaceae*. With the removal also of *Xylobotyrum*, that reduces the family to a more coherent group in its own order. *Eriosphaeria* and *Trichosphaeria* are scarcely separable.

Eriksson: We accepted your order, and agree it is heterogeneous. Many of the genera could be left in *incertae sedis* until we have more information. *Rhynchostoma* probably merits a separate family - it has a most fantastic cigarette-glow colour near the ostiole and a *Calicium*-like ecology; Tibell and Constantinescu have recently demonstrated it has an anamorph.

Gams: I cannot favour the inclusion of *Niessliaceae* in *Hypocreales* as in culture and ascoma structure they are so different, but I do not have an alternative suggestion.

Xylariales

Cannon: The family *Xylariaceae* is mostly well-circumscribed, but some other taxa in the order are more suspect.

Rogers: The *Boliniaceae* appears to be an isolated family with obscure relationships. Nannfeldt restricted it to *Camarops,* but Samuels and myself added *Apiocamarops* although that may not be correct.

Whalley: The secondary metabolites found in *Boliniaceae* are quite different from those in the *Xylariaceae* s. str.

Eriksson: Nannfeldt was convinced that the family was not related to *Xylariaceae*, mainly because of the shape of the ascospores, which are flattened in different directions.

Rogers: Ju and I interpreted *Astrocystis* as a highly modified *Rosellinia* on bamboo.

Læssøe and Spooner: We support the retention of *Astrocystis* in *Xylariaceae*, as does Petrini, based on differences in stromatal and ascus apical characters; all species occur on monocotyledons, including palms as well as bamboos. The genus is close to *Rosellinia*, but there seem not to be intermediates. There are many new species to be described in it. Rogers has a broad concept of *Rosellinia* and also suggested *Helicogermslita* might belong there, but again we retain it. We also support the inclusion of *Collodiscula* in the *Xylariaceae*, which differs from *Astrocystis* in having 1-septate ascospores lacking a germ-slit.

Rogers: *Collodiscula* also has the same type of *Acanthostigma* anamorph as *Astrocystis*. It is of interest with respect to possible interfaces between the *Amphisphaeriaceae* and *Xylariaceae*. *Astrocystis* and *Collodiscula* may be better referred to *Amphisphaeriaceae*, currently a highly heterogeneous family.

DISCUSSION 4

LECANORALES

Leaders: J. Hafellner[1], H. Hertel[2], G. Rambold[2] and E. Timdal[3]

[1]Institut für Botanik
Karl-Franzens Universität Graz
Holteigasse 6, A-8010 Graz, Austria

[2]Botanische Staatssammlung
Menzinger Strasse 67
D-8000 Munchen 19, Germany

[3]Botanical Garden and Museum
University of Oslo
Trondheim Veien 23 B, N-0562 Oslo, Norway

INTRODUCTION

Hertel: A general introduction to the problems of systematics in this group has been provided by Hafellner (this volume). The *Lecanorales* are a huge order with around 6000 species, including the majority of lichenized species. We consequently have many problems to confront. In order to facilitate discussion, the four of us have tabled a document synthesizing our collective knowledge. We could not agree on all details, so this again is something of a compromise. The system we propose is mainly based on morphological and anatomical characters that can be seen by light microscopy, ascus characters playing a central role. We regard ascomatal ontogeny as fundamental but could not use it extensively because of a lack of published data. The numbers of species we have studied is small in comparison to those that exist, and we concentrated on the type species of the genera. In some cases data remained incomplete, for example that on the conidial apparatus. Complete character sets are currently available for only about 1 % of *Lecanorales*. We plan to publish an annotated version of the tabled document, including key characters for the suborders after receiving comments at this Workshop.
A major change we propose is the use of *Lecanorales* in a broad sense, but including eight suborders. Several of those suborders were treated as separate orders in the latest *Outline*, i.e. the *Peltigerales* and *Teloschistales*. Those are to be debated in Discussions 5 and 6.

Poelt: The system for the lichenized ascomycetes was for a long time completely artificial, but during the last 20 years much more basic understanding has been achieved. A small group of workers have not been able to study more than single species of many of the genera, but the system proposed is a clear compromise. It is an important step towards a better understanding and therefore should be intensely discussed. The approaches in lichenized and non-lichenized groups have in many ways been different but we should try to put these together.

The leaders of this discussion started with very different opinions on many issues. They met in Munich in preparation for this Workshop and debated them, often fiercely, but the resulting compromise is a major contribution towards a better understanding of these lichenized fungi.

Hafellner: Here I summarize some of the features of those suborders we propose which were not treated as orders in the last *Outline.* More detailed descriptions will be published later elsewhere.

Acarosporineae: This includes families with biatorine, aspicilioid or lecanorine apothecia, and asci with a well-developed tholus only faintly reacting with iodine or Lugol's reagent. The method of ascus dehiscence is still not understood. Ascospores are 1-celled and hyaline, and polyspory often occurs. The photobionts are coceoid green algae. Most are saxicolous, some occur on inorganic soils, and others may be lichenicolous. They favour drier areas of the world.

Agyriineae: This suborder is more problematic. The *Agyriaceae* and the former *Trapeliaceae* may form the core group. Some further entities do not fit as well. The asci normally have small tholi and small cap-like structures, and often react pale bluish uthiodine or Lugol's reagent in the apical tholus; tube structures may also occur in *Rimulariaceae.* Polyspory is unknown and the ascospores are unicellular and usually not pigmented. The species grow on various substrates but are unknown from calcareous rocks; they are often pioneers of fresh substrates.

Cladoniineae: This includes species with crustose, squamulose and stipitate thalli, while the foliose type is very rare and probably only occurs in *Heterodea.* The ascomata are often biatorine, but aspicilioid or rarely lecanorine apothecia also occur. The ascus apex normally has a tholus with an amyloid tube or a cap-like structure, or these may be combined. All members so far studied have a rostrate ascus dehiscence. In placing *Rhizocarpon* here we regard the special kind of partly fissitunicate ascus as a variant of the rostrate dehiscence type. The ascospores are hyaline, rarely brownish, and the asci never polyspored. They occur on various substrates and lichenicolous fungi are rather common in the suborder. The core of the suborder is the *Micareaceae* group of families. There is also a group centred around *Lecideaceae* and *Porpidiaceae*, and another around *Rhizocarpaceae* and *Squamarinaceae*.

Lecanorineae: A diverse suborder, very rich in species, with crustose, squamulose, foliose and fruticose and rarely stipitate growth forms. The key hymenial character is the pale axial body in the reactive tholus; all taxa so far investigated have rostrate dehiscence. The ascospores are mostly hyaline, except in the *Physciaceae*.

Umbilicariineae: This suborder is based on two genera in one family. The hymenial characters are difficult to study, and spore formation is often sparse. *Umbilicaria* has an isolated position and crustose relatives are unknown.

Uncertain position: A large number of families cannot be accommodated into the above suborders. Some have ascal features in common, while others are isolated. Further suborders may need to be recognized later.

DISCUSSION

Acarosporineae

Poelt: This group was traditionally characterized by polyspored asci and there is always a question as to if this is a basal character. It now also includes 8-spored taxa.

Tehler: Is polyspory also present in *Agyriineae* and *Lecanorineae* ? Further, you mentioned that *Acarosporineae* had coccoid green algae but did not mention this character for the other suborders.

Hafellner: In *Lecanorineae* polyspory is quite common, as in *Maronea* and *Pleopsidium.* In *Agyriineae* cyanobacteria are also present in *Placopsis,* and in some *Cladoniineae* cyanobacteria are also quite common. In the *Lecanorineae* all but the *Harpidiaceae* live just with green algae.

Jφrgensen: *Acarospora* is not a uniform genus. I am concerned because its typification by Fink in 1911 on *A. schleicheri* cannot stand because of the "first-species" rule, and Clements and Shear selected a species not in the protologue in 1981.

Hafellner: I have cited *A. schleicheri* myself as type so that could be taken as a lectotypification instead. *Pleopsidium* (the *A. chlorophana* group) is a polyspored group with some resemblance to the true *Acarospora*'s with tetronic acids. However, true *Acarospora* species also differ in other features. *Pleopsidium* is linked to the *Lecanora polytropa* group in *Lecanoraceae*.

Timdal: Most brown *Acarospora*'s belong in the genus, but a few belong elsewhere.

Jϕrgensen: *Bouvetiella* is a very strange lichen which appears not to belong to *Hymeneliaceae*. It is probably a good genus but where it belongs is hard to say. It should be listed as of uncertain position.

Agyriineae

Lumbsch: Why was *Anamylopsora* included in *Agyriaceae* ? The ascus is superficially similar but the development of the ascoma is entirely different as well as the structure of the pycnidia and chemistry.

Timdal: Mainly because of the ascus type, which is similar to that of *Trapelia* and *Trapeliopsis*. It is a problem to place it anywhere.

Rambold: Bellemère and Letrouit-Galinou have reported that they observed the eversion type of dehiscence known in *Leotiales* in *Placynthiella*. This is not assumed for all taxa in the suborder, but ultrastructural studies may lead to transfers elsewhere. The proposed synonymization of *Trapeliaceae* with *Agyriaceae* is not yet solved. The outward appearance of ascomata should not be used as a major family character.

Tehler: Are the cap-like structures in the asci of *Agyriineae* and *Cladonineae* homologous ?

Hafellner: The main feature of the *Agyriineae* ascus is the pale amyloid reaction very far down its flanks, which may be combined with a small internal cap. It is almost a hemi-amyloid reaction and may not necessarily be the same.

Henssen: I placed *Trapelia* close to *Pertusariaceae* and cannot understand why *Trapelia* and *Trapeliopsis* appear so close together as they have different ontogenies.

Cladoniineae

Ahti: *Heterodea* may well be correctly placed in *Cladoniaceae*, although it has no stipes, rather than as a separate family. *Ramalea cochleata* in Australia undoubtedly belongs to the *Cladoniaceae,* but the type specimen of the type species of the genus (*R. tribulosa*) is fragmentary and hardly belongs there; a "?" should be used until this is clarified. *Sphaerophoropsis* does not seem to belong to *Cladoniaceae* but I can suggest no alternative placement; again a "?" should be added. *Thysanothecium* was treated as belonging to a separate family by Duvigneaud, and close to *Baeomycetaceae*; that may well be correct but the family name was published invalidly. The status of *Heteromyces* is also uncertain. Some molecular work supports the separation of *Cladina*, and also some morphology, but these are insufficient for a decision.

Hafellner: *Ramalea cochleata* is certainly *Cladoniaceae*; we left a "?" as we had not studied the type species. I am not so convinced that *Heterodea* belongs to the *Cladoniaceae* as it is the only member with a foliose thallus. *Sphaerophoropsis* has a perispore which is strange in the family.

Ahti: A foliose thallus is also found in some *Gymnoderma* species.

Lumbsch: I wondered why *Neophyllis* was omitted ?

Ahti: Yoshimura included *Neophyllis* in *Gymnoderma,* but is now being recognized as distinct by Australian lichenologists because of differences in ascoma ontogeny.

Henssen: Döring has studied the ontogeny of *Neophyllis* and found more of a relationship to *Stereocaulaceae*.

Hertel: Except for *Cryptodictyon* and *Pseudopannaria*, of which we have not seen material, there seems to be no reason for further division of *Lecideaceae*. It is now well-circumscribed.

Rambold: The treatment of *Rhizocarpaceae* is controversial because of similarities to the *Lecideaceae* and the presence of a perispore similar to that of *Porpidiaceae*. Honegger's data on the dehiscence method of the asci caused us to keep them separate.

Bellemère: Why are the *Lecideaceae* and *Porpidiaceae* placed in the same group ? In my opinion they each have a different apical apparatus in the asci.

Rambold: Transitional ascus types can be seen in, for example, the lichenicolous *Cecidonia umbonella*. Amyloid cap-like structures may be interpreted as rudimentary amyloid tubes.

Matzer: *Badimia* has already been placed in the *Cladoniineae,* so why are not other genera of the *Ectolechiaceae* also included ?

Hafellner: In *Badimia* and *Ectolechiaceae* (incl. *Lasiolomataceae*) muriform ascospores and 1-spored asci are common. There may be correlations between this and secondary reductions of ascal structure. The occurrence of campylidia could provide important information. I think that the *Ectolechiaceae* may indeed belong to the *Cladoniineae* in our sense.

Lumbsch: I would be cautious in using campylidia as a family character. The chemistry of *Badimia* is more like that of the *Pilocarpaceae* than of the *Ectolechiaceae.*

Aptroot: Sipman and myself are even describing a pyrenocarpous lichen which has campylidia. They are perhaps an ecological adaptation to rainforest habitats.

Lecanorineae

Blackwell: Does Lugol's solution have the same staining properties as Melzer's iodine solution?

Hafellner: Melzer's reagent often gives no reaction where Lugol's does. This is discussed in Baral's paper on hemi-amyloidy, where he points out than many differences in observed characters depend on the solution used. Lugol's solution is now generally used by lichenologists. After pre-treatment of hemi-amyloid structures with potassium hydroxide prior to the application of Lugol's iodine, blue colours are also formed immediately.

Aptroot: We all agree that in the order there are various lineages proceeding from crustose to foliose and fruticose thalli. At present the *Lecanoraceae* is paraphyletic and the *Parmeliaceae* is polyphyletic; if these are combined, monophyletic groups can be recognized.

Rambold: We do not intend to combine them into one family as the *Parmeliaceae* is well-characterized by a cupulate excipular structure.

Lumbsch: The excipular structure in the *Lecanora subfusca* group, the type of the *Lecanoraceae*, is different from that of *Parmeliaceae*. Whether some genera in *Lecanoraceae* (e.g. *Protoparmelia*) should be transferred to the *Parmeliaceae* is a separate question.

Ahti: In general what has been presented corresponds to the majority opinion. I understand that suborders and groups are not easily accommodated in the format of the *Outline* so some confusion will be inevitable. I wonder about the treatment of *Parmelia* s. ampl. as many genera have been recognized, especially in North America, and all are not equal in significance. I understand that further data is awaited although Swedish workers are finding data that might support some of Hale's genera. All are not just "vegetative" genera as some have assumed.

Rambold: We remain uncertain about the arrangement within some suborders. For example, within *Cladoniineae*, there could be three, four, or just one group. These are proposals to encourage the search for possible linkages between the subgroups. Many of the genera in *Parmeliaceae* are listed to show that they belong in that family, and not to suggest that they should be accepted as genera or not. In general, we kept to the last version of the *Outline*.

Lumbsch: Which characters correlate with the different structural elements of the ascus in *Cladoniineae* and *Lecanorineae* ? I.e. which characters do the very different genera *Amygdalaria, Cladia* and *Psora* have in common which do not occur in the *Lecanorineae* apart from ascus structure.

Hafellner: The *Parmelia*- or *Lecanora*-type ascus is a highly significant group of characters. The paler part within the tube structure is believed to be directly derived from an axial body. We excluded from *Lecanorineae* genera in which such a structure was not detected.

Rambold: There are, with a few exceptions, some correlating characters. For example the different spectrum of thallus types, and the absence of cephalodia and cyanobacterial photobionts in *Lecanorineae* (with one exception). Physiological and ecological characters should also be considered when seeking correlations.

Aptroot: I am concerned that many of the now commonly used parmeliaceous genera are not accepted in the last *Outline*. At present there seem to be many of the cetrarioid but few of the parmelioid names recognized.

Hawksworth: The practice for the acceptance of genera in these groups in the *Outline* has been to do so where evidence from thalline or secondary chemical characters was supported by ones related to the fungal fructifications, for example the globose ascospores of *Tuckermannopsis* s. str. In other cases we preferred to wait until further studies tested their robustness. Approaches on the lines of those by Kärnefelt and his co-workers (*Pl. Syst. Evol.* 180, 181-204, 1992; 183, 113-160, 1992) on the alectorioid and cetrarioid lichens would be especially valuable in this regard.

Hale was an intensive collector who amassed a huge amount of important material, especially from the tropics, and made an immense contribution to our knowledge by his discovery and description of so many new species. He explained to me in 1982 that his initial segregation of *Parmelia* s. ampl. was motivated by a need to rapidly organize his rich collections into manageable units in the herbarium. This was primarily based on characters such as thallus form, rhizine type, and colour, and later secondary chemistry and cortical structure were increasingly used. These characters are ones which do not necessarily reflect phylogeny. However, since the early 1980s we have known that characters such as exciple structure, carbohydrate chemistry, and pycnidia were a source of further features that could augment those in establishing natural groupings. In the absence of adequate data sets it is prudent to remain cautious until that is available.

Classifications are scientific hypotheses for testing, and as further tests are passed our confidence in them increases. When our confidence in them has been established is the time to encourage the widespread adoption of the resultant name changes. To have to explain to students on successive annual field courses that some of the commonest European lichens are in yet another genus does not inspire confidence in our subject.

There are certainly additional monophyletic groups within *Parmelia* s. ampl., and these will be recognized gradually in the *Outline* as corroborating evidence is obtained.

Pittam: I believe that not including many of these genera now to be more confusing than to eventually have to synonymize and eliminate them in future. Some have been in use for several years, and if it is found necessary to reject them this should be on the basis of scientific evidence published in an appropriate journal. As a minimum, lengthier notes should be included in *Systema Ascomycetum* to explain the decisions more fully.

Poelt: I am sure some of the parmeliaceous genera are good while others are not and this has to be studied much further.

Letrouit-Galinou: I thought that the *Ramalinaceae* had links with the *Physciaceae* rather than the *Lecanoraceae*.

Rambold: I examined the hyphal septa of *Ramalina* and some foliose lichens and found that those of the *Parmeliaceae* were bone-shaped as in the *Physciaceae* and *Teloschistaceae,* but I could not find this type in *Ramalina.* I do not know how to interpret this. The *Ramalinaceae* have a *Biatora-* or *Buellia*-type ascus, while the foliose genera of the *Physciaceae* have a *Lecanora*-type ascus; *Diriniaria* has a *Buellia-* or *Biatora*-type.

Lumbsch: I did not understand why the suborder *Cladoniineae* was necessary if there was only one character. What does *Byssoloma* have in common with *Cladonia* apart from the structure of the ascus? You must have correlations with other characters to circumscribe suborders.

Hafellner: The suborder was already proposed by Henssen and Jahns in 1973.

Rambold: In the order *Lecanorales*, there exists a central group of families including the *Lecanoraceae, Parmeliaceae*, and *Physciaceae*. They are closely related and regarded as the core of the order, representing the suborder *Lecanorineae*. On the other hand, it is not yet clear whether the proposed suborder *Cladoniineae* is monophyletic or not. As mentioned above, the two suborders are not only separated by ascus characters, but also by morphological and physiological ones, such as the occurrence of stipitate ascomata and the presence of cyanobacteria in cephalodia. In taxa around *Biatora* and *Mycobilimbia*, there are problems which suggest these suborders may not be so distant.

Tibell: You have mentioned that "tendencies" such as those towards polyspory or a certain growth form which are expressed in some species of a taxon are characteristic of the groups. If that is the case, it could lead us astray from the task of finding monophyletic groups. Characters can be used to find monophyletic groups only where they are manifest.

Hafellner: Some very good characters are negative ones, so the absence of some growth forms in a group could be a distinguishing character. If groups which have hymenial characters in common in addition share some ecological ones presenting in only more evolved genera, that provides an indication that the group is real. The arrangement of suborders attempts to provide structure to many groups of genera called families.

Poelt: In phanerogams there are many groups bound together by a few morphological traits and tendencies of dispersal, pollination and growth forms. Tendencies are important additions to morphological and other characters.

Matzer: You have recognized two groups in the *Physciaceae,* the *Buellia* group and the *Rinodina* group, based on different ascus types. I do not think it is possible to divide the *Physciaceae* into two families at present.

Rambold: I will follow your ideas if you are sure that *Bacidia-* and *Lecanora*-type asci occur side by side in closely related species. A similar difficulty exists with *Bacidiaceae* and *Lecanoraceae*; there are slight correlations with ascospore septation and conidial characters, but it is not yet clear whether these tendencies are sufficient to maintain the separation of these two families.

Sipman: The two main characters of the *Parmeliaceae* are the structures of the exciple and ascus apex. The *Alectoriaceae* and *Anziaceae* seem to have the same basic structure and I would suggest that they are united with the *Parmeliaceae.*

Letrouit-Galinou: I would support that view for *Anziaceae* based on the ascus tip.

Hafellner: While that is so, the *Anziaceae* has unique ascospore and thallus characters and lives in environments where few other members of the family live. The *Alectoriaceae* also has some characters in common, but as shown by Kärnefelt and coworkers (see above) there are also significant differences in the hamathecium.

Sipman: In relation to the delimitation of genera there seem to be two schools. One studies the asci of type species and mainly on that basis decides if the genus is good. The other looks at one species and its variation in the herbarium, proceeds with related species to identify marked discontinuities, and that is where generic boundaries are drawn; the second was the approach of Hale and seems a logical approach to come to natural genera.

Hafellner: We did not wish to say there were no natural groups, but as Hawksworth explained there are mycological characters to observe in the hymenia. It is premature to accept these genera until that is done thoroughly, otherwise many will at first be included and then synonymized in the *Outline.*

Hawksworth: The lichenicolous fungi can also be of assistance here. Several will support certain of the *Parmeliaceae* segregates but not others. They are also useful in the *Cladoniaceae* with respect to the *Cladina*/ *Cladonia* question where both "genera" have many such fungi in common. We have to work gradually, taking on all possible data sets, before recommending changes.

Lumbsch: What are the diagnostic characters of the *Candelariaceae*?

Hafellner: In *Candelaria, Candelariella* and *Placomaronea,* we can recognize a line of increasingly complex thallus organization. While these seem a completely natural group, from the mycological standpoint they are not so separate from *Lecanoraceae.*

Poelt: The *Candelariaceae* also have different hyphal structures and the cortex is quite distinct from many *Lecanoraceae* types; I am sure it will remain a good family.

Hafellner: The *Lecanorales* share some characters with other lichenized orders, for example they live with algae as symbionts or lichen parasites, they form apothecial ascomata with ascohymenial development, they are long-living. The distinguishing features from orders such as *Graphidales* or *Leotiales* relate to the lecanoralean ascus construction: mainly the reactive "*a*" layer, almost inactive and rigid "*c*" layer, and the "*d*" layer thickened at the apex with many structural elements.

Ahti: I wondered why the genus *Ahtia* was listed. It is a nomenclatural synonym of *Cetrariopsis* and should be removed. Also, *Saccomorpha* should be called *Placynthiella* for nomenclatural reasons.

Henssen: I suggest that *Protoparmelia* is transferred to *Parmeliaceae* because of the meristematic exciple. The recently described genera *Omphalodiella* and *Placoparmelia* also need to be included in *Parmeliaceae. Austropeltum* has recently been described in *Stereocaulaceae,* but we do not think *Muhria* should be included in that family.

Hafellner: We hesitated to change the placement of *Protoparmelia* as both *P. badia* and *Lecanora rupicola* share the same lichenicolous fungus, hinting that *Protoparmelia* may not be so different from the *Lecanoraceae.*

Poelt: A cupular exciple is known in some species of *Caloplaca* and *Lecanora* subgen. *Placodium* and it seems dangerous to make a completely different family on this one character.

Rambold: The cupular exciple of *Protoparmelia* is seen from the earliest ontogenetic stages onwards, and resembles that of *Parmeliaceae.* However, *Protoparmelia* is currently heterogeneous so a transfer now would place some true *Lecanoraceae* species in the *Parmeliaceae.*

Galloway: I was surprised to see *Austropelta* placed in *Stereocaulaceae* as to me there seems to be no evidence for this.

Jϕrgensen: In *Muhria* the ontogeny of the apothecium is not correlated with the chemistry, hymenial characters, ascospores and pycnidia, all of which recall the *Stereocaulaceae.* Further data is needed to decide the case, but I prefer to keep it in *Stereocaulaceae* at the moment.

Umbilicariineae

Kärnefelt: Is the suborder defined on morphological or ascus characters ? If it is defined on morphology, there is a dilemma of describing suborders only on such features.

Hafellner: The ascus characters may not be very significant in this case as the asci often do not function in the normal manner. Undifferentiated tholi can be found in the "*c*" layer but it is uncertain if that is juvenile or all that develops.

Uncertain Subordinal Placements

Aptroot: I wonder how you feel about *Normandina,* which I believe is a member of the *Verrucariaceae.*

Hafellner: We have not seen fruiting material so membership in *Lecanorales* could still be a possibility.

Hawksworth: To resolve this question an objective and independent test with a different character is needed. Aptroot has attempted a cultural approach, unfortunately unsuccessfully, but this is a correct way to proceed.

Aptroot: Molecular approaches may resolve whether the perithecia really do belong to the lichen.

Timdal: Indications that *Normandina* does not belong to *Verrucariales* are the presence of soredia and zeorin.

Sipman: I wonder why two genera have been removed from the *Megalosporaceae* and another included in it.

Hafellner: I checked the type species of *Austroblastenia* and found it to have ascal characters of the *Teloschistineae,* where it may represent a separate family. *Megaloblastenia marginiflexa* shares characters with *Austroblastenia* but may indeed be a good genus. They belong in the same family. *Lopezaria* (the *Catillaria versicolor* group) was included in *Megalosporaceae* as it shares hymenial characters with *Megalospora,* for example the asci do not have additional structural elements, but I am not sure of that.

Jϕrgensen: *Ropalospora* was placed as a synonym of *Fuscidea* in Purvis *et al.* (*The Lichen Flora of Great Britain and Ireland,* 1992), but here the family *Ropalosporaceae* is accepted and not even placed close to the *Fuscideaceae.*

Rambold: I do not regard these genera as confamilial or even congeneric. We hesitated to place the *Ropalosporaceae* in the *Teloschistineae* as the concept of the

Teloschistes-type ascus then has to be extended slightly. There are some hints that reductions are occurring and that several other families might be included in *Teloschistineae,* for example *Ophioparmaceae*.

Henssen: I wondered how *Pachyascus* and *Vezdaea* differed.

Poelt: *Pachyascus* is known from little material. The ascus structure is extremely complicated, and it has two ascomata bound by jelly. *Vezdaea* has no true ascomata but only bundles of asci, each ascus is a functional entity; jelly is absent and there are some paraphysoidal hyphae which include the asci in a net-like manner. If water is added to the ascomata of *Pachyascus* they swell and are almost invisible and perhaps function more like *Vezdaea*. The goniocysts also differ. The two genera are not closely related.

Aptroot: I saw that *Schaereria* had been returned from the *Pezizales* and wondered what the reason was for that.

Lumbsch: My colleague T. Lüuke is studying the *Schaereriaceae*, and his ontogenetic investigations indicate that it belongs to the *Agyriaceae.*

Hertel: *Orceolina* has a rather similar ascus type, which supports the idea that there are tendencies of reduction in this family.

Hafellner: The extremely narrow asci without any hymenial gel and hardly a tholus recall members of the *Pezizales* that lack a functional operculum.

Eriksson: A recent Note in *Systema Ascomycetum* (12, 42, 1993) suggests the transfer of *Schaereria* to *incertae sedis* as the asci differ from the operculate type when treated with sodium hypochlorite.

Henssen: I do not think that the *Collemataceae* should be kept with the *Peltigerineae,* and wondered why *Homothecium* and *Leciophysma* are excluded; both genera have the thallus structure and ontogeny of other *Collemataceae* and asci with an apical tube.

Rambold: The cupulate excipular structure you described as characteristic of *Collemataceae* does not appear to be present in *Homothecium,* and the asci in that genus and in *Leciophysma* have a different type of amyloid tube. The *Collemataceae* and *Placynthiaceae* characteristically both have a special type of tube combined with an apical cap in the ascus.

The Subordinal System

Hawksworth: I am not sure that to adopt a system of suborders would be helpful in the *Outline* at this time. We are anxious to clearly distinguish monophyletic groups, and from the discussion today it is clear there are still many uncertainties about the natural groupings within *Lecanorales* s. lat. I feel it could also be rather confusing to users who prefer a neatly presented system.

Aptroot: A disadvantage of the suborder system as proposed is that there is now a large list of families and genera of uncertain position, and it is no longer possible to recognize quickly, for example, that the *Coccotremataceae* has always been considered associated with the *Pertusariales*.

Hertel: A consequence of not using the suborders would be to have to describe *Cladoniales* and I feel very uneasy about that as there are so many uncertainties. Also, it would be necessary to keep moving some genera between orders which would give an even greater impression of instability. If we keep to a concept of *Lecanorales* defined by the lecanoralean ascus that is really distinct from other ascomycete orders perhaps such transfers will not appear so dramatic.

Hawksworth: You may be correct that some of these groups are more closely allied than they appear in the current *Outline*. You have presented an important hypothesis in your revised classification which now merits examination by molecular and other data sets.

Timdal: I should add that we have some problems in separating the *Lecanorineae* and *Cladoniineae*, that is not clear cut to all of us.

Sipman: When we started studying asci it became evident that there were very distinct types. This seemed a good reason to place the corresponding groups well apart. Now it is becoming clear that there are strong connections between many of these types. The proposal to use *Lecanorales* in a wide sense is a logical consequence of this new information, therefore, I support it. Otherwise, the impression is given that, for example, *Teloschistales* are no more closely related to *Lecanorales* than to *Dothideales* or *Graphidales*.

Tehler: I was interested in Timdal's remark about problems in separating the *Lecanorineae* and *Cladoniineae*. On the basis of the data set provided by Hafellner *et al.* (see above), that is about the only thing that is clear, namely that the *Agyrineae* and *Cladoniineae* are more closely related to each other than to the *Lecanorineae* based on the characters of polyspory and a cap-like structure in the ascus tip (Fig. 1).

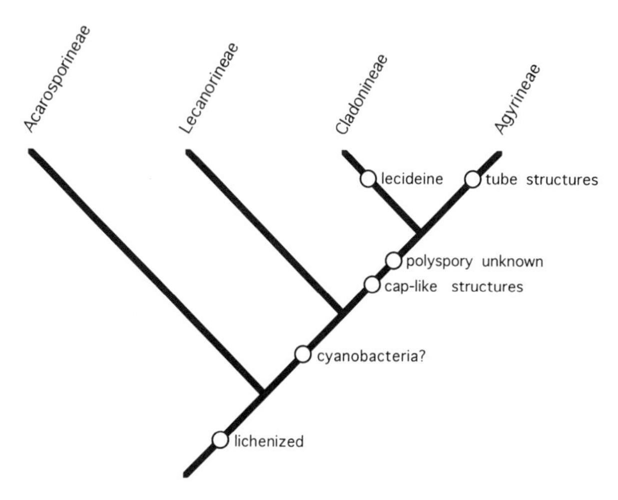

Figure 1. Tentative phlogeny for the *Lecanorales* constructed by Tehler on the basis of the taxa and characters communicated to the Workshop by Hafellner *et al.*

Tibell: I am still left with an uneasy feeling I have had for a long time. The proposed classification is one-dimensional, based on one data set allegedly without homoplasy, *viz.* ascus structure. If you have managed to identify monophyletic groups or not I do not know, but other data sets should be consulted such as ontogeny, chemistry, and molecular data. I wish to plead restraint in describing new taxa. I would like to see two types of nomenclature around, one for use in our discussions and another which is more conservative for public consumption, perhaps with a moratorium of ten years. If we continue to proceed as at present we will put our science further into disrepute.

Eriksson: The treatment of the lecanoralean groups in the first *Outline* in 1981 mainly followed the works of Henssen and Jahns of 1973 and Poelt of 1974 and recognized a large number of clades. In the 1982 *Outline*, clades characterized by a special type of ascus (e.g. *Peltigera*, *Pertusaria*) were treated as separate orders. It was not practical to use suborders only for lichenized groups as I had to be consistent for the whole ascomycete system in which I used only the ordinal level. In principle I have nothing against suborders nor supraordinal taxa. However, I am not at all sure that the *Peltigerales* are not a distinct group and so for the moment wish to be conservative.

DISCUSSION 5

GYALECTALES, LICHINALES, PELTIGERALES, AND *PERTUSARIALES*

Leaders: A. Henssen[1] and P.M. Jϕrgensen[2]

[1]Herbarium Botanik, Fachbereich Biologie
Philipps-Universitat
Karl-von-Frisch-Strasse
D-3550 Marburg/Lahn, Germany

[2]Botanisk Institut
Universitetet i Bergen
Allegt. 41, N-5007 Bergen, Norway

INTRODUCTION

Jϕrgensen: The situation in *Peltigerales* is rather different to that in *Lecanorales* as the order seems to include several distinct lines that are not closely related. The lines are very old and it is not easy to determine connections. It is doubtful if a satisfactory system in that order can be achieved now, and I would prefer to see more informal groupings rather than a rush to introduce formal names. There are also issues to discuss in *Lichinales*, but there has been little recent research on *Gyalectales* and *Pertusariales*.

DISCUSSION

Gyalectales

Henssen: I wish to confirm that *Ramonia* belongs to *Gyalectales*; Vezda is also convinced of this.

Jϕrgensen: *Belonia* is listed with a "?", but Vezda and myself wondered whether it should be included *Ostropales* because of the peculiar periphyses. I have studied more material and am now convinced the genus belongs in *Gyalectales* and the "?" can be removed. We previously placed too much weight on the periphyses.

Lichinales

Henssen: I have worked with this family since 1960 and feel I have a clear view of it. I was surprised to see the removal of the *Harpidicaeae* and *Pyrenopsidaceae* proposed. The ascus structure can vary within a family. Most members have a prototunicate ascus, but some, especially in *Pyrenopsis*, differ as to whether an amyloid structure is present or not. The ascomatal ontogeny and thallus structure correlate well with the order.

Ascomycete Systematics: Problems and Perspectives in the Nineties
Edited by D.L. Hawksworth, Plenum Press, New York, 1994

Lumbsch: The difference in ascus structure between *Pyrenopsis* and *Lichinaceae* could be explained by the prototunicate type having been derived from the *Pyrenopsis* type by the loss of the tholus. That would explain the occurrence of both in *Lichinales*.

Hafellner: The *Harpidiaceae* clearly have features of a lecanoralean fungus which separate it from *Lichinaceae*. If this amount of variation in ascus structure is permitted, there is no reason not to include the *Lichinaceae* in the *Lecanorales*.

Peltigerales

Jørgensen: In the *Peltigeraceae* there is no doubt that *Hydrothyria* and *Peltigera* are closely related; they are held together by the very special semi-fissitunicate *Peltigera*-type ascus and the ontogeny. Many, but not all, of the other genera placed in the order/ suborder have hemiangiocarpic development. In the *Outline Massalongia* has a "?"; I agree with that and feel it should be transferred out as the ascus structure is quite different and nothing else correlates it with *Peltigerales* in my opinion. Negative characters are a particular problem in this group, also including a lack of secondary chemistry in *Massalongia*.

The suggestion to include *Siphulina* seems very doubtful according to its description and I feel that should be placed with the uncertain genera.

We should consider whether to accept *Solorinaceae*. There is a slight difference in the ascus tip of *Solorina* in that the amyloid region is a more elongated tube than that seen in *Peltigera*, the apothecia are sessile but I understand have the same ontogeny, and there are some differences in the ascospores.

Rambold: The suggestion that *Solorinaceae* might be resurrected was based on slight differences in the ascus structures.

Timdal: I did not observe any major differences between the asci of *Solorina* and *Peltigera*.

Letrouit-Galinou: There is quite a range of similarities and only small differences between *Solorina* and *Peltigera* and the two genera are better kept together. The ascogonia in both are large-celled.

Hawksworth: *Peltigera* and *Solorina* share numerous lichenicolous fungi not known on other lichens. Further, it is not an uncommon phenomenon for a characteristic "*Peltigera*" fungus to be found occasionally on *Solorina* (e.g. *Corticifraga peltigerae*); this suggests a very close relationship.

Galloway: Some years ago I suggested that *Sticta* might be separated out into a separate family, basically because it tends not to have any secondary chemistry. However, I have subsequently found that some *Sticta* species, for example *S. laciniata*, do have a varied chemistry. The species is rich in pigments turning red in potassium hydroxide. Cyphellae and pseudocyphellae, although different in structure, have similar physiological functions. There is no difference in the pycnidia and the ascospores are also similar. On reflection all are better kept together.

Hafellner: There is no reason to maintain *Dendriscocaulon* as separate from *Lobaria*.

Hawksworth: *Dendriscocaulon* was retained because James and Henssen argued (in D.H. Brown *et al.*, *Lichenology: Progress and Problems*: 27-77, 1976) that it was useful to retain this where connections were not known.

Henssen: I recommend that *Dendriscocaulon* is retained in the family, perhaps indicated to be an "imperfect" stage because there are species for which connections are unknown and these must be placed somewhere.

Hafellner: To keep an imperfect lichen genus in the *Outline* makes no sense in comparison with the treatment of anamorphs of non-lichenized fungi.

Hawksworth: *Dendriscocaulon* would not be an anamorph in the strict sense as it does not produce mitospores. Also, remember that Art. 59 does not apply to lichenized fungi so sterile and anamorphic lichens compete nomenclaturally with those where ascomata are known. Other sterile lichens are included in the *Outline*, either in a family or order where there are good reasons to accept such a hypothesis, or in the list of those of uncertain position (e.g. as for *Thamnolia*).

Galloway: The occurrence of *Plectocarpon* on *Dendriscocaulon* also supports its retention here.

Jørgensen: We know that the type species of *Dendriscocaulon* is linked to *Lobaria* so the names are synonyms whatever you wish to call the unlinked species.

Letrouit-Galinou: I agree that the *Lobariaceae* is quite different from the *Peltigeraceae* in the ascus tips, ascoma ontogeny, conidiomata structure and ontogeny, and further in the structure and development of the ascogonial apparatus which is a mass of filaments.

Jϕrgensen: The *Nephromataceae* seems to be quite distinct and should be maintained.

The position of the *Placynthiaceae* is more problematic, partly because the family itself is heterogeneous. Its members are all hemi-angiocarpic, and there is some relationship between the genera included; its nature is obscure. The type species, *Placynthium nigrum*, has a rather special ascus structure which Rambold believes to be close to *Collema*; Keuck has previously showed that all *Placynthium* species do not have this same structure. *P. aspratile* has no internal amyloid structure and a small cap outside. However, I know there are other genera in which variations in amyloid reactions are accepted.

Vestergrenopsis has very strange polyspored asci with no cap or other structures. *Leptochidium* seems to be rather distantly related. *Hertella, Koerberia*, and *Polychidium* s. str. also need to be considered. These are a very difficult group of genera.

Henssen: We used to place the *Placynthiaceae* in the *Peltigerineae* on account of a similar structure, the very thick hyphae. In addition, they have rather large ascogonia, a hemiangiocarpic ontogeny, and a distinctive type of true paraphyses. We therefore placed the genera together in the *Placynthiaceae*, but I was never content with its delimitation. In the meantime, some genera have been omittd and others added. The genera are best left in the *Peltigerineae* or *Peltigerales*, but without any formal groupings until we have more information.

Galloway: We are dealing with a group which has cyanobacteria as a primary or secondary symbiont, which while it is better able to exploit a wide range of ecologies, at the same time is mandatorily tied to liquid water. Most also include green algae, although they are commonly overlooked, which enables them to exploit a wider range of light regimes. The group has had a particularly long evolution so a variety of "blind alleys" is perhaps not unexpected.

Verkley: With regard to ascus amyloidy, when I compare reactions to ultrastructure in genera of *Leotiales*, some that lack blueing as a general feature have very different structures at the EM level, whereas others are very similar in structure even though no reaction was obtained. In the discussion of *Lecanorales*, discussions focussed on amyloidity rather than ultrastructural details; I think that is very difficult character to use without real structural information.

Jϕrgensen: I agree that we have to be very careful here, but it seems best to retain *Placynthiaceae* in the order with a "?" for the moment. The *Collemataceae* have *Nostoc* in common with many *Peltigerales*, but there are differences in ascomatal ontogeny and the asci which suggests it is best kept outside. *Collema* itself was once divided on the basis of differences in the apical apparatus of the asci, leading to the recognition of *Synechoblastus*; that needs to be re-examined.

Henssen: The thallus structure in *Collemataceae* is different in being very uniform. *Homothecium* and *Leciophysma* also belong to the family; these genera were kept in an artificial group in the arrangement of the *Lecanorales* proposed by Hertel *et al.* (see Discussion 4).

Jϕrgensen: *Ramalodium* and *Staurolemma* lack any apical ascal characters, again highlighting the problem of negative features. However, on the basis of other characters they seem to belong in *Collemataceae*.

Hafellner: Do you think that *Collemataceae* and *Placynthiaceae* might be linked ? At least the type species of the type genera of the families share quite significant characters.

Jϕrgensen: That is a possibility, but there is a clear difference in the ontogeny and structure. The basis of the amyloidity also needs to be determined. Unfortunately there is no chemistry to help resolve this issue.

Henssen: The ontogeny in *Collemataceae* is so peculiar, and different from *Placynthiaceae* and *Peltigerales* s.lat. that I would never put *Collemataceae* here.

Letrouit-Galinou: I do not object to the exclusion of *Collemataceae* from *Peltigerales,* but I am sure that the differences between *Lobariaceae, Peltigeraceae* and *Nephromataceae* are of an equal weight. The only characters the three families really have in common are the non-gelantinous structure of the thallus and frequency of cyanobacterial photobionts. We do not always separate families with similar arguments.

Jϕrgensen: The *Coccocarpiaceae* and *Pannariaceae* are rather difficult. They are probably associated, but whether they are inside or outside the *Peltigerales* is doubtful. *Pannaria* s. str. has no apical apparatus and its chemistry and ontogeny also separate it from *Peltigerales.* However, some members of the *Pannariaceae* have an apical tube and also fatty acids and triterpenes and are not so dissimilar in their ontogeny. Parts of that family belong here but others do not.

I do not believe that *Leproloma* belongs here; only a single apothecium is known and when that was sectioned its apical apparatus was not described; only a particular chemical difference was the reason for the placement and I suggest it is placed in *incertae sedis*.

The *Coccocarpiaceae* is not evidently *Peltigerales;* the apical apparatus of their asci includes a sheet-like structure. A similar sheet is seen in *Erioderma* and this is not incongruent with the *Peltigerales.*

Hafellner: If the *Lobariaceae, Placynthiaceae* and *Stictaceae* are kept in *Peltigerales*, there is hardly any evidence to support a separation from *Lecanorales.*

Henssen: There is no similar thallus structure and ontogeny in *Lecanorales.*

Hafellner: You already explained that the ascomatal ontogeny of the type species of *Placynthiaceae* differed from that of the type family of the order. It follows that you accept two ontogenetic types within *Peltigerales.*

Henssen: They are hemiangiocarpic and have true paraphyses of a very similar structure; in addition, the hyphal structure is different. Some variations in ontogeny and thallus structure should be allowed, just as they are in ascus types.

Letrouit-Galinou: I do not agree as there is a great range of ascomatal ontogeny models in the *Lecanorales,* so this appears to be a variable character. I believe there is a problem with the *Peltigera* ascus, as in some way it is synthetic as it is associated with ascohymenial development and has an elongate extension at discharge. The ascus can be interpreted as between the rostrate and the true fissitunicate type. It is the "*d*" layer which extends, but this does not extend to the base of the ascus. The ascus type is very isolated.

Henssen: In earlier papers you referred to these asci as "semifissitunicate", and applied this also to *Nephroma* which has a cap and not a ring structure. Asci of *Lobaria* and *Sticta* have not yet been studied in such detail.

Pertusariales

Aptroot: There was a suggestion that *Monoblastiaceae* be accepted in the next *Outline*, but not in *Pertusariales.* It should be included in *Pyrenulales* rather than *Dothideales* if the *Melanommatales* is not to be accepted.

Lumbsch: I recommend that the *Pertusariales* continues to be accepted as an order rather than a suborder of *Lecanorales.* They are more distinct from other suborders of *Lecanorales* in their ascus structure, as documented by Honegger, and also in their ascoma ontogeny. Also, I can confirm that *Ochrolechia* is closely related to *Pertusaria* and should not be excluded as *Ochrolechiaceae* as proposed by Harris. *Thamnochrolechia* is somewhat intermediate between *Ochrolechia* and *Pertusaria* having some characters of each.

Hafellner: If *Ochrolechia,* the *Pertusaria velata* group, and *Varicellaria* are kept in one group with other *Pertusaria*'s, there is no reason to exclude them from *Lecanorales* as they have ascal characters as defined in my paper (this volume) for lecanoralean fungi.

DISCUSSION 6

CALICIALES, GRAPHIDALES, AND *TELOSCHISTALES*

Leaders: A. Aptroot[1], I. Kärnefelt[2] and L. Tibell[3]

[1]Centraalbureau voor Schimmelcultures
P O Box 273
3740 AG Baarn, The Netherlands

[2]Department of Systematic Botany
Lund University
Ö. Vallgatan 18-20, S-223 61 Lund, Sweden

[3]Department of Systematic Botany
University of Uppsala
Villavagen 6, S-752 36 Uppsala, Sweden

DISCUSSION

Caliciales

Tibell: The *Caliciales* were thought to be a prime example of a monophyletic group, the main character uniting the group being the presence of a mazaedium; the loose ascospore mass, passive dispersal, and "prototunicate" asci. In comparison to the turmoil in *Lecanorales*, the order seems to be scarcely controversial. However, that this appears as a discrete order in the *Outline* has to be seen as a capitulation to our ignorance with respect to the phylogeny of the group.

Based on studies of morphology, anatomy, chemistry, ecology, and particularly ascospore ontogeny, I concluded nearly a decade ago that the order was highly polyphyletic, including some 25 genera of lichenized and non-lichenized fungi. My problem was to find the relatives of the families recognized, and my task to deconstruct the *Caliciales*. I have been pleased to see that a few mainly tropical genera that I could not assign to families in 1984 have been weeded out: *Pyrgillus* and *Pyrgillocarpon* were placed in *Pyrenulaceae,* and *Nadvornikia* has recently been suggested to belong to *Thelotremataceae.*

Seven families remain in the *Outline*. Three may be monophyletic, the *Caliciaceae, Mycocaliciaceae,* and *Sphinctrinaceae.* These families share the synapomorphies of stalked ascomata, and a dark pigmented ascospore wall, but not passive ascospore dispersal nor the "prototunicate" ascus. However, even this may be in doubt to judge from the molecular studies of Gargas, who studied the 18s rDNA of *Calicium, Mycocalicium* and *Sphaerophorus*. In her analysis these genera were all rather far apart. This is welcome, but the embarrassment as to where they really belong remains.

In the *Caliciaceae* the genera need revision. There is a monophyletic group within *Cyphelium*, sometimes called *Acolium*, which I now think should be recognized.

Ascomycete Systematics: Problems and Perspectives in the Nineties
Edited by D.L. Hawksworth, Plenum Press, New York, 1994

nized. *Acrocyphus, Texosporium*, and *Thelomma* also need redelimiting but form a monophyletic group.

In *Coniocybaceae*, *Chaenotheca* and *Sclerophora* are probably not closely related and I prefer to keep the latter separate as *Sclerophoraceae*. There are differences in ascospore wall pigmentation, ornamentation, ascospore ontogeny, and secondary chemistry. It has been suggested that *Cybebe* should be included in *Chaenotheca*; a preliminary analysis I made of these genera supports that view.

The *Calycidiaceae* is probably closely related to *Sphaerophoraceae*. This is supported by the presence of sphaerophorin in some populations of *Calycidium*.

The *Microcaliciaceae* is a small, enigmatic, monotypic family with no close relatives.

The *Mycocaliciaceae* seems to be reasonably well delimited. Its members have active ascospore dispersal and a well-developed apical apparatus in the asci. However, the genera within the family are not clearly delimited. It is particularly difficult to segregate *Chaenothecopsis* from *Mycocalicium*. In the last six years I have cultivated the mycobionts of this family and obtained a diversity of anamorphs, most previously totally unknown and capable of reproducing the species asexually; I suspect most exist in a non-lichenized state in nature. The anamorph data supports the family delimitation, and am confident that it will help resolve some anomalies in the family at the generic level.

Wedin has investigated the *Sphaerophoraceae* in detail using comprehensive new ontogentic and chemical data. He suggests that *Sphaerophorus* is divided into three genera differing in characters such as thallus symmetry, apical or non-apical apothecia, chemistry, conidial shape, etc.: *Bunodorophus*, *Leifidium* (for *S. tener*), and *Sphaerophorus* s.str.

The *Sphinctrinaceae,* with the *Caliciaceae*, may form the monophyletic core of the order. Nevertheless, in some ways it appears to be more closely related to the *Mycocaliciaceae.*

The questions remaining to be answered are: What is the monophyletic core of *Caliciales* ? Do the other families belong in any of the described orders ? Where do *Allophoron, Heterocyphelium, Roeslerina, Tylophorella,* and *Tylophoron* belong ?

I suggest we keep *Caliciaceae* and possibly *Sphinctrinaceae* in *Caliciales* and move the remaining families and genera into *incertae sedis.*

Blackwell: Do the anamorphs in *Mycocaliciaceae* belong to any familiar genera ?

Tibell: These include the hyphomycetes *Phialophora,* *Catenomycopis,* and the coelomycete *Asterophoma.* The coelomycetous anamorph of *Microcalicium* appears not to have an anamorph name.

Tehler: On what evidence does the *Mycocaliciaceae* seem to recall the *Caliciacae.*

Tibell: There are only similarities in ascospore ontogeny between *Mycocalicium* and the *Sphinctrinaceae.*

Tehler: Fungi with "prototunicate" asci are often considered secondary reductions and not just plesiomorphic, although that would initially be a more simple hypothesis.

Tibell: Both situations may be present in *Caliciales*. Congruent but disparate characters assist in showing groupings. These include morphology, anatomy, ascospore ontogeny, and ecology. Comparisons with sister groups, when these have been identified, will assist in the resolution of the significance of the "prototunicate" ascus in this group.

van Brummelen: You showed a cladogram showing *Calicium* very close to *Ascobolus,* and I wondered on what this was based.

Tibell: This cladogram was from rDNA data by J. Taylor's group and there was not much support for its detailed resolution. It included few taxa and very few lichens so should not be over-emphasized.

Graphidales

Aptroot: This is one of two groups in the *Outline* which still have an entirely schematic Saccardoan generic system, the other being the *Pyrenulales*. The two families *Graphidaceae* and *Thelotremataceae* are mainly tropical and each includes 800-1000 species. The core genera in *Graphidaceae* are based on ascospore colour and septation, yet there are single specimens that show more than one of these generic combinations. A revision of the whole unsatisfactory system is required.

There has been uncertainty as to whether the order is distinct from *Ostropales*, and Harris and Lumbsch have both proposed that the two are combined.

Lumbsch: Gilenstam, in his 1969 study on *Conotrema*, was able to show that the two orders were closely related on the basis of ascomatal development and ascus structure, and I cannot see how they can be distinguished.

Aptroot: *Fissurina* seems to be a natural group and to merit reintroduction, but there are perhaps many such groups so it may be premature to do this now. In due course many further genera will need to be distinguished, but that should be after most species have been studied. *Helminthocarpon* has been suggested to belong to the *Roccellaceae*; it clearly has nothing to do with *Ostropales*. In the *Thelotremataceae,* the foliicolous *Chroodiscus* is being accepted. *Nadvornikia* seems to be a good member of *Thelotremataceae*. *Asterothyrium,* type of the family *Asterothyriaceae*, also belongs to *Thelotremataceae*; the family *Solorinellaceae* has been introduced to include most of the former *Asterothyriaceae*. *Solorinella* is close to *Gyalidea*; more species are being found, one 16-spored.

Sipman: I think it is premature to accept *Fissurina*, which I interpret as having a reduced exciple and a distinctive ascospore type. *Acanthographis* has very conspicuous paraphyses apices otherwise unknown in the order. The ascospores of *Stictis* are very different and I question the proposed union of the order with *Ostropales*.

Johnston: Cultural studies have shown *Stictis* to have an unusual anamorph, which might be compared with those of graphidalean genera.

Aptroot: As far as I know no cultural studies have been carried out with *Graphidales*.

Eriksson: *Graphidales* and *Ostropales* were kept separate in the *Outline* following the views of Sherwood who discussed this question in detail in her thesis of 1977. I discussed the question of generic concepts in *Graphidales* with Santesson many years ago, who recommended that we continued the present system so as to avoid hundreds of new combinations.

Hafellner: The *Gomphillaceae* do not belong here as *Gomphillus calycioides* has bitunicate asci; Vezda has shown this to be accompanied by a second unique character, the hyphophores. I suggested, but did not validly publish, the name *Gomphillales* for this family in 1988. Most of the former members of the *Asterothyriaceae* belong in that family, leaving *Asterothyriaceae* s. str. as a small group with unitunicate asci best left in *Graphidales* but with a "?". *Psorotheciopsis* seems also to belong in the *Asterothyriaceae*, as possibly may *Linhartia*.

Aptroot: *Gyalidea* and *Solorinella* in *Solorinellaceae* (the old concept of *Asterothyriaceae* without *Asterothyrium*) are probably synonymous since an intermediate species has been described (*G. multispora*).

Tehler: You indicated that the main differences between *Graphidales* and *Ostropales* were lichenization and macro- *vs.* microcephalic ascospores, yet you united them.

Aptroot: Other characters are more-or-less in common, including the structure of the apothecia, the paraphyses, and similar range of ascospore variation.

Poelt: In this group and *Teloschistales,* enormous numbers of species have never been studied. We should not draw conclusions when we really know nothing about the groups.

Teloschistales

Kärnefelt: This order appears to be a good monophyletic group based on a number of correlated characters in the asci and also of the secondary metabolic products, particularly anthraquinones. The asci have an apically thickened and broad internal beak, and both the inner part of the ascus apex and an external diffuse cap turn blue in iodine. The ascospores are ellipsoid and hyaline.

The *Letrouitiaceae* are mainly tropical and characterized by apothecia with anthraquinones and a diffuse outer amyloid cap and a well-developed internal apparatus. The ascospores are multiloculate to muriform and ascus discharge is by an elongation of all apical layers. It includes a single genus with 11 species.

The *Teloschistaceae* comprises 11 genera and 5-600 species world-wide, and has crustose, placodioid, foliose, fruticose and umbilicate members. The apothecia are usually orange with anthraquinones and the asci are characterized by a well-developed amyloid cap and a rudimentary internal apical apparatus. The ascospores are polarilocular, simple or 1-septate, and discharged by a longitudinal crack.

Cyanodactylon, *Follmannia* and *Leproplaca* should be deleted.

The *Fuscideaceae* have been suggested to belong to the order on the basis of the ascal characters which are similar those of *Teloschistaceae*. However, these lichens lack anthraquinones. The *Umbilicariaceae* also has a similar ascus structure but has an umbilicate thallus and again lacks anthraquinones.

We should try to define monophyletic groups based on a variety of characters, not only the asci but also the secondary chemistry. In this sense the order *Teloschistales* includes only the *Letrouitiaceae* and *Teloschistaceae,* and not the *Fuscideaceae* nor the *Umbilicariaceae.*

Lumbsch: I was surprised to see the *Umbilicariaceae* placed close to the *Teloschistaceae*, but there are anthraquinone pigments present in that family, notably skyrin and valsarin. That might support your ideas, but skyrin is also present in *Lecanorales.*

Kärnefelt: While to a large extent anthraquinones occur in *Teloschistales* they also occur in other groups. The dominance of anthraquinones and polarilocular ascospores in the family, combined with the ascal characters, seem to provide a set of synapomorphic character states than can be used to define this monophyletic group.

Hafellner: Anthraquinones and polarilocular ascospores are no longer key characters for the group. Anthraquinones are not present in all *Teloschistaceae,* including *Austroblastenia* and *Megaloblastenia* recently transferred there. The absence of anthraquinones is no reason to exclude *Fuscideaceae.* The suggested inclusion of *Umbilicariaceae* is premature; it would not then be possible to circumscribe the group as in that family the asci lack the external apical cap which otherwise holds the other families of the *Teloschistales/ Teloschistineae* together.

Hawksworth: No mention has been made of the conidiomata of the order. Those of *Xanthoria* are of a distinct multilocular type only otherwise known amongst lichenized ascomycetes in *Dermatocarpon.* I have not studied those of *Caloplaca*, but suspect that this character may assist in the delimitation of the order.

Kärnefelt: The conidiomata are actually very similar in the genera of *Teloschistaceae.*

Hertel: You excluded the *Fuscideaceae* because of the absence of anthraquinones and the simple ascospores, yet you accept *Apatoplaca* and *Cephalophysis* both of which also lack anthraquinones and have simple ascospores.

Kärnefelt: I am aware of those problems in the system, but they are exceptions. Certain *Caloplaca* species also lack anthraquinones or have simple ascospores.

Jφrgensen: According to studies by Ekmann, the *Ropalosporaceae* should be placed close to *Fuscideaceae,* at least with a "?". Its members have anthraquinones and the ascus tip is a slightly modified *Teloschistes*-type, perhaps because of the elongated ascospores.

DISCUSSION 7

PEZIZALES

Leaders: H. Dissing[1] and T. Schumacher[2]

[1]University of Copenhagen
Farimagsgade 2D, Copenhagen, Denmark

[2]Botany Division, Department of Biology,
University of Oslo
P O Box 1045, Blindern, N-0316 Oslo 3, Norway

INTRODUCTION

Korf: Coming back to Paris was an emotional experience for me, because it was 45 years ago as a graduate student I came on a pilgrimage to meet one of the most influential of people who have ever worked with discomycetes, Mme Marcelle LeGal. She greeted me and treated me with the kind of courtesy and interest that I hope all of us can express towards those who come green and wanting to learn. It was an experience I will never forget, and one which led me in future years to propose a genus *Galiella* in her honour, and to copropose *Marcellina* also in her honour, and on whose back many of us stand today. The other great figure of that period on whose contributions much of our classification lies is Jan A. Nannfeldt, another mycologist I met that same year in Uppsala where I went to meet and learn from him.

I chose not to present a paper here as I have already published one on what I think are some of the problems.

Dissing: The system for the *Pezizales* in the *Outline* for 1990 was mainly based on that proposed by Korf in 1973, with important additions by Trappe who suggested that most families of the *Tuberales* should be incorporated into *Pezizales*. Van Brummelen (this volume) has provided an historical account of the systematics of *Pezizales*, and also surveyed the ascus ultrastructure in the order, while Kimbrough (this volume) has reported on investigations into septal ultrastructure.

A few molecular and cladistic studies have been carried out, but many more are wanted.

Schumacher: As pointed out by van Brummelen, the suprageneric system, between genus and order, has been little considered. There are scattered investigations focussed on specific points, particularly the ascus apical apparatus and septal pore plugs. Based on these results, we need to realign the genera using cladistic methods, particularly character compatibility, to better express their relationships at the suprageneric level. Currah (this volume) recommended that systems be proposed so that they could be tested by the very powerful tools of molecular biology. However, most workers in *Pezizales* have been little concerned with the rearrangement and recognition of families. The 18S rDNA molecular studies by Landvik and Eriksson (this volume), and also the recent unpublished thesis of Gargas, are especially important recent contributions as they included a wide variety of discomycete fungi.

DISCUSSION

Ascobolaceae

van Brummelen: I would restrict this family to *Ascobolus, Saccobolus*, and *Thecotheus*. Other genera referred here, including *Ascophanus* (a synonym of *Thelebolus*), *Ascozonus, Pseudascozonus,* and *Leptokalpion* could be transferred to *Thelebolaceae* on the basis of ascus structure.

Bellemère: I agree with your proposal to transfer those genera to *Thelebolaceae*.

Dissing: If *Thecotheus* is placed here, why not also *Iodophanus* ?

Kimbrough: The placement of *Thecotheus* here is supported by the pore-plugging mechanism which is very consistent in the family. The ascospore ontogeny and ascus apex are different from that of *Iodophanus* in *Pezizaceae. Thecotheus* lacks the accumulation of epiplasmal pigments but the deposition of secondary wall material is like that of certain *Ascobolus* species. The ascospore perispore ornamentation of *Iodophanus* is also characteristic of the *Pezizaceae.*

van Brummelen: The anamorph also supports such a placement of *Iodophanus* in *Pezizaceae. Iodophanus* and *Peziza* have *Oedocephalum*-type blastoconidia unknown in any other family.

Ascodesmidaceae

van Brummelen: The proposal to include *Amauroascus* should not be accepted as studies by Currah indicate a relationship with *Onygenales.*

Currah: I agree that *Amauroascus* should remain in *Onygenales*, but would welcome information on the circumscription of the family.

Kimbrough: We were probably premature in suggesting the transfer of *Amauroascus* as that was based on the literature which indicated similarities with *Ascodesmis*. I agree that the family should be restricted to *Ascodesmis* and *Eleutherascus*; both share particularly characteristic radiating tubular septal pore plugs. *Ascodesmis* starts as pairs of ascogonia which remain gymnohymenial, but post-staining studies indicate that the outer wall of the ascus changes chemically as it matures to form the operculum, whereas it is the inner wall of the ascus that changes chemically in other families of *Pezizales*.

Balsamiaceae

van Brummelen: The number of nuclei in the ascospores, and the structure of the plugs in both the pore at the base of the ascus and vegetative hyphae are particularly important characters for family delimitation within *Pezizales*. On this basis, *Balsamiaceae*, with four nuclei in each ascospore, can be placed as a synonym of *Helvellaceae*.

Eoterfeziaceae

van Brummelen: For the time being, *Cleistothelebolus, Lasiobolidium, Orbicula,* and *Warcupia* were also included here by Benny and Kimbrough.

Malloch: The genus *Eoterfezia* is very enigmatic, being based on a single glycerine-jelly microscope slide. It has a gelatinous ascoma with basally produced asci and is doubtfully a member of *Pezizales* It is unfortunate that the prototunicate genera *Lasibolidium* and *Orbicula* with more clearly pezizalean characteristics are often aligned here. Those two genera might be better included in *Pyronemataceae* or included in *Pezizales incertae sedis.*

Schumacher: This illustrates that we have many doubtful families and genera, and also monotypic genera, in the order.

Geneaceae

van Brummelen: The family can be included in *Pyronemataceae* according to the independent studies of both Pfister and Zhuang.

Helvellaceae

van Brummelen: *Balsamia* (see above) and *Gymnohydnotria* can be included; both have 4-nucleate ascospores.

Karstenellaceae

van Brummelen: No operculum has been observed in *Karstenella* and the family is best excluded until it can be studied further.

Monascaceae

Schumacher: The 18S rDNA sequence for *Monascus* obtained by Berbee and Taylor was compared with many pezizalean fungi by Landvik and Eriksson (this volume). That analysis shows that *Monascus* is not really related to the discomycetous groups.
Cannon: While I support the removal of the family from *Pezizales*, I do not favour its inclusion in *Eurotiales* as Landvik and Eriksson proposed. The ascoma ontogeny, ascomatal wall structure, asci, ascospores and anamorph all have clear differences from those seen in *Eurotiales*. I suggest that the family is transferred to *incertae sedis*.

Morchellaceae

Schumacher: Molecular studies are currently in progress in Kurtzman's laboratory and we await those results.
van Brummelen: The uncertain genus *Cidaris* would be better placed with a "?" in *Helvellaceae.*

Neolectaceae

Schumacher: On the basis of molecular and critical morphological studies, the new order *Neolectales* was introduced for this monogeneric family. This seems to be acceptable on the evidence we now have.

Pezizaceae

van Brummelen: Studies by Zhuang, Kimbrough and Wu have revealed that *Hydnobolites, Pachyphloeus* and *Terfezia* can be included here, while *Ruhlandiella* has been shown by Dissing and Korf to belong to *Pyronemataceae.*

Pyronemataceae

Korf: *Pyronema* is not central to the family as now circumscribed, which was reduced as a result of Kimbrough's suggestion that this represented a suborder of *Pezizales.* The family had become something of a waste-basket in which there are obviously a number of different clades, around 15 of which had previously been considered as possible families. Some will need to be pulled-out. In a preliminary treatment of Macaronesian discomycetes I accepted the *Pyronemataceae* in a very limited sense, including also genera like *Coprotus*. I was then left without a name for the remainder; *Humariaceae* of Velenovsky had been suggested but was invalid, and so I took up the next valid name, *Otideaceae*, but I am not sure that we should pick up *Otideaceae* as another waste basket, and perhaps we should retain *Pyronemataceae* in the broad sense.
van Brummelen: If we keep only *Pyronema* in the family that would be acceptable. *Coprotus* should be excluded as it has a different ascus structure. Arpin suggested

some rearrangements, but his family schemes were not generally accepted because most of them were based on quantitative differences in α- and ß-carotene pigments. Due to geographical differences within single species, this work needs to be extended.

Schumacher: I propose that we accept Korf's suggestion to retain a heterogeneous and polyphyletic *Pyronemataceae* in the next *Outline*. Further results are starting to be obtained, for example the cladistic analysis included by Wang and Kimbrough in a study of *Octospora*.

Kimbrough: The work of Landvik and Eriksson (this volume) substantiates the view that there is a separation within *Pezizales* and I was fascinated that Momol's data using 5.8S and 18S rDNA showed a gap predicted earlier on the basis of septal ultrastructure, and dome- *vs*. torus-building. Interestingly, in the basidiomycetes the septal structures line up better with molecular results than many other characters.

The proposal to merge *Lamprospora* and *Octospora,* was based on advice on cladistic interpretations. I have some concerns over this. I do not know if the type of *Octospora* will prove to be congeneric with the species with highly ornamented ascospores currently placed within it; *Lamprospora,* all species of which have ornamented ascospores, may show a relationship with those. However, molecular data could show alignments with smooth-spored genera such as *Pulvinula*. I do not place much reliance on ornamentation characters in view of the extent of variation seen within genera such as *Peziza*.

Dissing: I would be sorry to see only *Octospora* in the next *Outline*.

Pfister: I am taking up the name *Byssonectria* for three species previously referred to *Inermisia* and the monotypic *Pseudocollema*. *P. cartilagineum* is being combined into the genus.

van Brummelen: *Genea* should be included (see above) and the following excluded: *Cleistothelebolus* and *Lasiobolidium* to *Eoterfeziaceae*; *Coprotus* (syn. *Leporina*, of which I have studied the type), *Coprotiella, Dennisiopsis, Lasiobolus, Mycoarctium, Ochotrichobolus* and *Trichobolus* to *Thelebolaceae* based on ascus structure; and *Hydnocystis* to *Helvellaceae*.

Geopyxis resembles *Helvella* in the ascus tip, septal type, ascospore structure, and ploidy level, but has only a single nucleus in the ascospores.

Sarcoscyphaceae

Schumacher: We have no problem with the proposal to separate this family out from the *Sarcosomataceae*.

Bellemère: Ultrastructural studies on the asci, carried out independently from Cabello's cladistic work, supports this distinction. These two families seem to be in the same evolutionary line with regard to the large and coloured apothecia, but there are differences in the ascus wall, ascus apex, and dehiscence (Bellemère *et al.*, *Cryptogamie, Mycologie* 11, 203-238, 1990). Molecular work will be of interest here. The recently introduced genus *Donadinia* also needs to be added.

Korf: As I proposed the recognition of these two families, I am pleased to see they are now being taken up as cladistics, wall structure and ascal studies reinforce this separation.

Schaereriaceae

Aptroot: This family can be accepted in *Lecanorales s. ampl.* (see Discussion 4), where its best place is probably at the end of the *Teloschistineae* together with *Ophioparmaceae, Ropalosporaceae*, and *Sarrameanaceae*.

Pfister: I have also examined material of *Schaereria* and do not find it to be operculate.

Thelebolaceae

Lundqvist: At least the polyspored species of *Thelebolus* appear to have asci which are bitunicate in some way, although this is less distinct in the 8-spored ones. I wonder how the family should be circumscribed and if there is any connection with the true loculoascomycetes? Some other genera placed in the family, however, undoubtedly have unitunicate asci.

van Brummelen: I regard *Thelebolaceae* as a family in which the ascus structure is rather variable and that it should be placed at the base of the system of *Pezizales*. In *Thelebolus* the inner wall layer is gelatinized, multilayered, and, particularly in the polyspored species, the asci rupture at the top, but the splitting is prohibited from extending downwards by a strengthened ring in the ascal wall. The "bitunicate" structure is not functional, and similar phenomena can be found in *Lasiobolus*, *Scutellinia* and *Trichobolus* where the external ascus wall is very rigid while the inner wall may be elastic and swell. In species with lower ascospore numbers the type of rupture approaches that of an operculum.

Kimbrough: Our extensive studies on *Thelebolus*, *Trichobolus* and *Lasiobolus* showed each genus to have a different ascus structure. The opercular indentation in the ascus wall of *Lasiobolus* is persistent. When mounted in water the asci of all these genera function as a bitunicate ascus. Sometimes the asci can burst and look like an operculum, but even in 8-spored *Thelebolus* species the wall is two-functional, the inner layer protruding before the ascospores are liberated. The ontogeny of the apothecia is also distinctive. The family should be linked more closely with the loculoascomycetes. What we are seeing in many coprophilous discomycetes, including *Caccobius*, is evolution towards a large projectile.

van Brummelen: *Caccobius* is distinguished by a special pore in the ascus tip stained by Waterman fountain-pen ink, but ultrastructure shows this to be based on an incidental cavity in the great amount of inner wall material at the tip. It is very close to *Thelebolus*.

Montemartini: I have studied numerous strains of *T. microsporus* from Antarctica which vary in many characters, including ascus wall thickness. I think it is possible to interpret these fungi as very primitive and at the base of a divergent evolution towards both the *Pezizales* and loculoascomycetes. I have a paper on this in press in *Mycotaxon*.

Kimbrough: I agree.

van Brummelen: *Chalazion* is better placed in *Pezizales incertae sedis*. However, in addition to genera to be excluded from *Pyronemataceae* (see above) *Ascozonus*, *Leptokalpion* and *Pseudascozonus* can also be included in *Thelebolaceae* on the basis of the ascus structure.

Tuberaceae

Bellemère: I think the family should be placed in the reinstated order *Tuberales* as Parguey-Leduc and Janex-Favre have shown that ascospore formation differed from that in *Pezizales* (Janex-Favre and Parguey-Leduc, *Cryptogamie*, *Mycologie* 6, 87-99, 1985; Parguey-Leduc *et al.*, *ibid.* 11, 47-68, 1990). It is as important a difference as that of the type of septal pore in the mycelium.

Kimbrough: The presence of lamellate structures in the septal pores of vegetative cells in *Tuber* leads me to believe that the family is well-placed in *Pezizales* as that appears to be a solely pezizalean character. I follow Trappe and merge these fungi.

van Brummelen: There is a lot of evidence from ascus organization, ascospore initiation and ornamentation, and septal pores of the *Sordaria*-type in the vegetative hyphae, as demonstrated by Parguey-Leduc and coworkers (see above), to maintain the probably monotypic order *Tuberales*.

DISCUSSION 8

CYTTARIALES, *LAHMIALES*, *LEOTIALES*, *MEDEOLARIALES*, *OSTROPALES*, *PATELLARIALES*, *RHYTISMATALES*, AND *TRIBLIDIALES*

Leaders: A. Raitviir[1] and B.M. Spooner[2]

[1]Mycological Herbarium
Institute of Zoology and Botany
Academy of Sciences of Estonia
Box 93, SU-202400 Tartu, Estonia

[2]The Herbarium
Royal Botanic Gardens
Kew, Richmond, Surrey TW9 3AE, UK

INTRODUCTION

Spooner: Having considered the impact molecular data is now making in some groups, and also the state of the inoperculate groups, there is a major problem in making substantial progress. The overall situation can only be described as chaotic. Molecular data is minimal and our understanding is based largely on traditional morphological approaches. Additional data is at last starting to be produced on ascus ultrastructure, cytology, chemistry, ontogeny, anamorphs, etc. In addition we have to reconcile what we know now with the examination of fresh material, and comparing that in culture from what is seen in nature. Further, at most perhaps 20-30 % of the fungi in these groups have yet been described.

There is a huge system to be developed, but the information on which to do that very largely does not exist. It is therefore inevitable that revisions will need to be made as data becomes available and that it will be a considerable time before a stable classification can be realized.

The orders range in size from two with only single species to the vast array of the *Leotiales*.

Raitviir: I have followed work on the phylogenetic reconstruction using molecular data, and also that on species and populations. The *Systema Ascomycetum* project endeavours to produce a phylogenetic system, but it also aims to provide a framework for working taxonomists - especially in relation to the discovering of biodiversity. We have to use different sets of techniques for these purposes. The inputs into the *Outline* will consequently also differ, and the result will need to be a compromise.

While new data are being collected we will have to continue in effect to classify life-forms. For example, the *Hyaloscyphaceae* probably do not reflect a compact evolutionary lineage and it will be of interest to see how our understanding develops in the future.

Hennebert: I agree that molecular and traditional methods may lead to different taxonomies, and that phylogenetic classifications cannot be practical for those using

Ascomycete Systematics: Problems and Perspectives in the Nineties
Edited by D.L. Hawksworth, Plenum Press, New York, 1994

only morphological characteristics - for example in less-developed tropical countries that have most of the biodiversity. I wish to emphasize the need to refine the morphological characters so that we can make our results accessible to all who need them. We must avoid building two different taxonomies: a molecularly based phylogenetic system only accessible in the developed world, and a classical system available to all.

DISCUSSION

Cyttariales

Spooner: This order includes fungi that have coevolved with *Nothofagus* in the Southern Hemisphere, some forming galls. Its position in the system has been somewhat uncertain as the structure of the ascus is unclear.

Bellemère: The apical apparatus in *Cyttaria* is like that of *Bulgaria*, and I consider that it merits a family but not an order on the same evolutionary line as *Bulgaria*. I would prefer to include it in *Leotiales*.

Verkley: I have studied *Bulgaria* but not *Cyttaria*. There seem to be some similarities but I would not support a transfer to *Leotiales* based only on the apical apparatus.

Kimbrough: The ascal tip in *Cyttaria* is comparable not only to that in *Leotiales* but also in some pyrenomycetes. The excipular structure comprises extremely thick-walled and "glassy" cells. One tribe of *Helotiaceae* recognized by Dennis had species erumpent through the bark and with similar excipular cells; the possibility of a linkage here merits investigation.

Eriksson: The 18S rDNA data clusters *Cyttaria* with *Leotiales*, but we would like to compare it with more genera before making any formal transfer.

Schumacher: According to a cladistic study by Criski and coworkers even this apparently very uniform genus was paraphyletic based on the character set employed.

Minter: I wondered if any other members of *Leotiales* have pycnidial states arising on the developing stroma but becoming separated as the stroma grows. The conidial state might provide some useful extra characters.

Raitviir: In building up any classification we must distinguish between grouping and ranking. Mayr has discussed this in detail with respect to zoological taxonomy. In the case of *Cyttariales*, the grouping seems satisfactory but the ranking appears to be too high. During the coevolution with *Nothofagus*, there have not been opportunities to evolve ordinal characters. We can expect other studies to support the rank of family.

Lahmiales

Spooner: This order is based on a single species with closed apothecioid ascomata and bitunicate asci.

Eriksson: The fruit body is closed from the beginning, the wall is rubber-like and continuous with the hamathecium and opens in a star-like manner, the asci have a long base and appear to be bitunicate although I have not seen the dehiscence, and the ascospores are boomerang-shaped and arranged helically. It occurs on *Populus* bark.

Hafellner: I know some *Patellariales* with rather similar ascomata, for example *Buelliella*, splitting at first in a star-like manner but with the upper part of the peridium then crumbling; all are bitunicate and I wonder if they might belong there.

Eriksson: *Lahmia* has two types of hamathecium. Before the ascomata burst this is a periphysoidal network ending in paraphysis-like hyphae. That is quite different from *Patellariales*.

Bellemère: Does *Lahmia* have any relationship with *Odontotremataceae* ? I suggest this because the apical structure of the ascus appears to be similar.

Rossman: I also wonder about a link with the *Acrospermataceae* as *Acrospermum* has a similar rubbery texture to the ascomata and a questionably bitunicate ascus.

Eriksson: The asci are quite different from those of *Acrospermum*, a genus that at various times has been placed in six or seven different orders. However, in that genus the asci are cylindrical, the ascospores are thread-like, and if broken the "endotunica" is a jelly-like extrusion and not sharply delimited. *Lahmia* has been cultured in Canada and described there under the name *Parkerella*.

Leotiales

Spooner: Eleven families are definitely included in the last *Outline*, but the order has problems that require study at all levels. Most of the recognized families are unnatural, but how to restructure these is uncertain.

Scheuer: Huhtinen (this volume) made many of us feel guilt, but I wish to express caution with regard to cultures. In particular, anamorphs should not be reported only on the basis of growth obtained from ascospore deposits. Also, it would be wise to always make more than two cultures.

Huhtinen: I was discussing a situation where a monographer knows the species in one area well, but has the chance to collect in another with few facilities apart from a camp-fire. In monographing *Hyaloscypha* I obtained over 250 strains by micromanipulation methods, especially using 2% malt agar. The additional field isolates are a particularly valuable approach when making geographic comparisons between strains. Of course the results obtained need to be verified by comparison with other single-ascospore isolates in the laboratory prior to publication. I also wish to stress that it takes only a few minutes to learn micromanipulation techniques.

Hawksworth: If cultures are not attempted when material is still fresh and in a functioning condition, experience with lichenized ascomycetes in particular suggests that there is a greater probability you will be unsuccessful. Pharmaceutical companies are also well aware of this.

Korf: For many of the fungi I have worked with, the best field results have been obtained using water agar to which has been added chloramphenicol. Richer media can encourage the growth of contaminants. I make transfers to a nutrient medium after germination occurs.

Cannon: When cultures have been examined, representative isolates should not be discarded but deposited in service culture or genetic resource collections. This is as important as depositing the dried material in herbaria.

Raitviir: The blue reaction in Melzer's solution is widely used in some groups. While this has some taxonomic value, reactions can be found in unexpected contexts. For example, we have found *Lachnum* species with strongly amyloid paraphyses, a two-layered exciple of a *Cistella* species from Baja California in which only the inner was strongly amyloid, and a strongly amyloid *textura oblita* occurs in an undescribed *Crocicreas* species yet to be studied in detail.

Gelatinized tissues also have taxonomic value in some groups, but in others they appear to be an adaptation to dry ecological conditions as they can occur in genera where species from different habitats lack this feature.

Caution is therefore needed even in the application of traditionally used characters.

Rambold: The *Baeomycetaceae* belongs to *Leotiales* and contains only the genus *Baeomyces*; the asci have an eversion type of dehiscence. *Icmadophila* has a different ascus type and belongs to a separate family, *Icmadophilaceae* (to be described formally shortly and also referrable to *Leotiales*).

Triebel: We include several lichenized genera in *Icmadophilaceae*, i.e. *Dibaeis*, *Icmadophila*, *Knightiella*, *Pseudobaeomyces*, and *Siphulella*. These genera have asci with a thin cup-like amyloid apical structure which is sometimes prolonged into a short ring-like structure. Furthermore, the *Icmadophilaceae* are distinguished from the *Baeomycetaceae* by ascospore characters, the secondary chemistry, and the lichenicolous fungi.

Spooner: The small dark-coloured *Dermateaceae* are difficult, perhaps mainly due to adaptation to grasses. The family is clearly heterogenous as it stands.

Baral: The *Dermateaceae* comprises two groups; the *Dermea-Pezicula* complex should remain in the family, while the *Mollisia-Pyrenopeziza* complex may be closer to the *Hyaloscyphaceae*. The *Dermateaceae* is based mainly on the ectal exciple being of *textura angularis*, other features differing markedly amongst the groups.

Scheuer: A third group in the family which also may be quite isolated is the *Hysteropezizella-Hysteronaevia* complex, as outlined by Nannfeldt and Hein. Amongst its special features are characteristic exudates on the paraphyses, and the plump and broad somewhat *Phacidium*-like asci. It probably merits an independent family.

Bellemère: I agree that the *Dermateaceae* is heterogeneous and that it should be limited to the *Dermea-Pezicula* group. The *Mollisiaceae* might be taken up for numerous genera which have a *Mollisia*-like apothecial structure with a well-developed parathecium recalling that of the *Hyaloscyphaceae*, but instead with rather dark ascomata.

The *Geoglossaceae* contains genera with different ascus structures recalling, for example, *Dermateaceae*, *Leotiaceae*, and *Phacidiaceae*, and needs to be dispersed. Ascomatal shape is not a sufficient criterion for the definition of a family. The asci of *Geoglossum* itself are most similar to those of *Dermateaceae*.

Verkley: The ultrastructure of the apical apparatus in *Geoglossum* and *Trichoglossum* is quite different from that in other genera and families now recognized. There are also similarities between the asci of *Microglossum* and *Leotia*. The situation is difficult to resolve and molecular characters are likely to be very important.

Cannon: It is not only the ascus apex of *Geoglossum* but also the ascospores which are unusual for the *Leotiales*; in addition the paraphyses are extraordinary. There may be grounds for separating the family out from *Leotiales*. However, many of the other genera in the family as currently circumscribed, including *Mitrula*, could simply be regarded as stalked members of the *Leotiaceae*.

Verkley: The ascus apex of *Mitrula* has some similarities with that of the *Sclerotiniaceae* which I have investigated recently. In this respect it is of interest that some *Mitrula* species have sclerotia.

Raitviir: *Cudonia* and *Spathularia* have been referred to *Geoglossaceae* but do not belong there. I was interested to see that the molecular data presented by Landvik and Eriksson (this volume) showed these two genera to be closely linked as this supports my previously published suggestion that they should be placed in a separate family.

Korf: I have noted the great amount of work on ascus apices and the weight now being accorded to this feature in the delimitation of genera, families, and orders. I wish to draw your attention to *Rhabdocline*, a remarkable genus of the *Hemiphacidiaceae*. The genus includes two species, both causing a snow-mould disease of *Pseudotsuga* (Parker and Reid, *Canadian Journal of Botany* 47, 1533-1545, 1969). The very reduced apothecia arise on the needles of this conifer, throwing back a scale to expose the asci and paraphyses but scarcely any exciple. The ascospores are didymoid, hyaline and 1-celled, later becoming 1-septate, and much later one cell becomes darkly melanized - that cell will still much later germinate.

The asci are most exciting. In the literature they are variously reported as having an apex bluing in Melzer's solution, and one which is not reactive. In one of the two species the thickened ascus apex is hemispherical, and when rehydrated in water and then mounted in Melzer's solution there is no bluing reaction. But, if rehydrated in potassium or sodium hydroxide and then mounted in Melzer's solution, a very complex apical apparatus that blues intensely is seen (hemi-amyloid). This might explain the competing reports, but the discovery of a second species, biologically very similar to the first, sheds new light on the situation. The second species has apically unthickened ascus apices, and gives no blue reaction in Melzer's solution, with or without pretreatment with potassium hydroxide, and no pore mechanism is discernible. The apex ruptures irregularly, sometimes almost forming an operculum.

Those who weight the ascus apex as the clue to relationships would be well-advised to study these sister taxa that share so many strange characters. To suggest that this was a case of convergent evolution would be preposterous to me. It seems to me an error to complain that molecular systematists often examine only a tiny piece of the DNA molecule, while traditional morphologists over-stress a single morphological feature in their delimitation of phylogenetic hypotheses.

Bellemère: In my studies of *Rhabdocline* I found that the "*d*" layer of the lateral wall becomes thinner at the apex; perhaps in a closely allied species this same layer could be better developed. A flat apex may be present in genera lacking a specialized operculum, as is probaly the case in *Orbilia*.

Verkley: In the case of these *Rhabdocline* species, the apical apparatus should be studied at the ultrastructural level to investigate if there really is a specialized operculum-like structure. It could be that this species has lost the ability to form an apical apparatus by a mutation.

Minter: I suggest that *Rhabdocline* may belong to the *Rhytismatales* as that order, by definition has asci with variable apices and contains endophytes. Korf omitted to mention that one species of *Rhabdocline* has ascospores with a mucous sheath when still colourless, a further feature agreeing with the *Rhytismatales*.

Korf: I have an open mind and would be happy to accept that placement. The *Hemiphacidiaceae* as I conceived it is primarily an ecological rather than a phylogenetic group.

Baral: Within *Lachnellula occidentalis* (*Hyaloscyphaceae*), a single apothecium can contain mature asci with a thick hemi-amyloid annulus, and also mature asci devoid of any apical thickening and iodine reaction. The ascospores inside the two types of asci are the same.

Huhtinen: The *Hyaloscyphaceae* is heterogeneous but practical. So many genera may be unknown that energy devoted to developing a more "accurate" classification is almost certainly wasted at this stage of our knowledge. This can be compared to attempting to construct a jig-saw of 500 pieces when we have only five or ten of them.

Phialina, a genus with yellow pigments I previously treated as close to the *Hyaloscyphaceae*, does not seem to be really related to it.

Raitviir: The genera currently included in *Hyaloscyphaceae* could be grouped into about five more or less natural clusters based on correlations between several characters, including hair structure, exudates, paraphysis type, ascus structure, etc. I hope to propose a new classification of the genera currently placed in *Hyaloscyphaceae* at a future meeting.

Kimbrough: In our ultrastructural studies on *Orbiliaceae* we were unable to find any ascal pores. Other organelles in the ascospores and excipular cells had previously only been found in some lichenized fungi. The discovery of algal nests in rooting hyphae arising from the ectal excipulum confirmed to us that *Orbilia* species were lichens in disguise. We did not study *Hyalinia*.

Baral: I agree that many species of *Orbilia* have no apical thickening and no ascal pore. However, some species, for example *O. occulta*, have distinct inamyloid apical domes with an apical chamber. Some species may be lichenized, but the placement in *Leotiales* seems the best solution for the time being.

Spooner: The *Sclerotiniaceae* have been the subject of some molecular studies. They have been considered homogeneous, but some information on ascus structure displayed at this Workshop appears to conflict with that view. This highlights our ignorance over the significance of the development of sclerotia and stromata.

Schumacher: The *Sclerotiniaceae* was thought to be well-founded until recent work at the ultrastructural and molecular level. In many groups where stromata are well-developed, the fungi are rarely cultured, while in this family sclerotia form after a few days on an agar plate. In some other families we have genera which we believe do not form stromata, but some can do this if left for many weeks in culture.

Minter: I have spent three years trying to get *Lophodermium* species to produce ascomata in culture without any success and many forest pathologists have had the same experience. However, Johnston has cultured some New Zealand species and obtained ascomata.

Medeolariales

Spooner: This order contains a single species, apparently unique in many ways, and parasitic on the liliaceous genus *Medeola*. The asci develop between paraphyses which arise on the second-year growth stems of the host. The ascospores are large and coloured, but little else is known about this fungus.

Pfister: I have collected this species and examined fresh asci but could never determine the method of dehiscence satisfactorily. There is no evidence of active spore dispersal and I was not able to culture it.

Ostropales

Spooner: Aptroot (see Discussion 6) has suggested that the *Graphidales* be united with this order. It contains two families as currently circumscribed, the *Odontotremataceae* and *Stictidaceae*. The *Odontotremataceae* is fairly well delimited, but the asci of *Rhymbocarpus* are quite unlike those to be expected in the family.

Hawksworth: It was suggested in a Note on *Rhymbocarpus* s.str. (*Systema Ascomycetum* 10, 51, 1991) to place it in *incertae sedis* pending the collection of fresh material.

Korf: I wish to call attention to the failure of the recently published *Family Names in Current Use for Vascular Plants, Bryophytes and Fungi* (NCU-1; *Regnum Vegetabile* 126, 1993) to list the family name *Stictidaceae*. This is precisely the problem I foresaw with such lists, which have a chilling effect of setting in concrete what is one opinion unless alternatives are clearly spelled out. I suspect this is just the tip of an iceberg of names now to be forgotten because of a prevailing taxonomic arrangement appearing in NCUs. I made some other comments on a draft of this list which did result in additions of names that are in current use but which were not included at that stage. Where are the cross-reference safeguards necessary to see alternative classifications?

Hawksworth: The omission of the name *Sticidaceae* was an oversight by the compiler, and this particular name is being included in an addendum to the list being presented in Tokyo in August 1993. While such lists do not prohibit any unlisted name from being used, they aim to allow for alternative taxonomies. Their success in doing that will depend on the inputs each of us makes. In the event of a list being accepted for protection, procedures for amending and extending that list will be established by the International Association for Plant Taxonomy (IAPT).

Patellariales

Hafellner: The *Patellariaceae* are highly heterogeneous, and probably all but the type genus will eventually be classified outside as most genera have very few characters in common. Some might belong to *Dothideales* s.ampl. in which the upper part of the ascomata crumbles away to produce apothecia-like structures (e.g. *Buelliella*, *Karschia*, *Poetschia*, *Heterosphaeriopsis*). However, true apothecia may also occur in this bitunicate family as true paraphyses are rather common as they keep on growing after the hymenial layer has been opened (e.g. *Schrakia*), further filaments growing upwards from the subhymenial layer.

Rossman: There seem to be quite a number of discomycetes that have distinctly bitunicate asci which although very diverse biologically have been "dumped" into the *Patellariaceae*. There is a real need for this group to be thoroughly revised.

Rhytismatales

Spooner: This is a very large order, and in collecting in the tropics I have found large numbers of new species. However, the generic concepts in some cases remain unclear. Ontogenetic studies are beginning to pull apart some traditionally recognized genera such as *Lophodermium*, hopefully leading to more natural groupings.

Johnston: In the study of this and many other groups of fungi, there is a bias towards the Northern Hemisphere taxa. A great diversity of fungi is now being gathered in the Southern Hemisphere and the tropics, not least in this order which until about ten years ago was regarded as primarily Northern Hemisphere. This new material is likely to be at least as important as molecular studies in developing a natural system.

Bellemère: There is diversity of ascus tip structure in the *Rhytismataceae*, but it appears, however, to be a natural evolutionary group. There is a central core of genera with *Rhytisma*-type asci, but also others not so directly tied to *Rhytisma*. It would be of interest to study the asci and ascogenous structures of those genera by ultrastructural methods.

Minter: I have been working the order for over ten years and wish to make the following points:

(1) I agree that the *Rhytismataceae* is a natural family in which the ascus apex happens to be rather variable.

(2) In the 1983 edition of the *Dictionary of the Fungi*, a separation was made between the *Hypodermataceae* and the *Rhytismataceae* on the basis of the anamorphs being phialidic or not. I do not accept a separation on that basis.

(3) I have been fortunate to have been able to study fresh material of *Cryptomyces*, a large fungus obligately parasitic on *Salix* which fruits only occasionally in the UK. The ascus apex is no more variable than other *Rhytismataceae*, and like that family the ascospores also have a gelatinous sheath. The separation of the *Cryptomycetaceae* on the basis of the lack of a gelatinous sheath cannot be sustained. The anamorphs are also similar so it is clear that the two families are very close.

(4) *Ascodichaena* clearly belongs in a family other than the *Rhytismataceae* and might eventually be found to belong to a different order.

(5) The *Ascodichaenaceae* and *Cryptomycetaceae* have been little-studied, but I have the impression that they are a hotch-potch of fungi that require transfer elsewhere (e.g. *Cryptomyces peltigerae* transferred to *Corticifraga* in *incertae sedis*). New genera and even orders may emerge.

(6) The biology of the order is of interest as in the late 1960s it became clear that many species on conifers were endophytes and there may be some link between these and forest decline in Europe.

(7) The relationship of the *Rhytismatales* to other orders is unclear. It was formerly separated from *Phacidium* in *Leotiales* because of the lack of an iodine positive ascus apex, but Romero has found a new genus in Argentina which definitely belongs to the *Rhytismataceae* but has an ascus apex turning blue in iodine. Proposed links with *Phacidiaceae* and *Hysteriaceae* are untenable, and I believe similarities with *Phyllachoraceae* result from convergent evolution. I currently have an open mind as to where its relationships may be.

(8) A tremendous amount remains to be discovered in this group. Baral has cautioned on the care needed in microscopic studies, and indeed when I first mounted a species in water I thought I had found a new one it looked so different. Rayner (this volume) has also demonstrated the wealth of information that might be obtained from mycelia. Molecular data will also surely afford many new insights. Coppins made a collection of *Lophodermium conigenum*, which normally occurs on *Pinus* needles, on bare *Pinus* wood; had I not written my PhD thesis on this fungus it is doubtful if I would have recognized it as it appeared so totally different. Further, Descals and Webster discovered a star-spored aquatic hyphomycete as an anamorph of a *Coccomyces*. Such re-evaluations make me wonder how significant much of the morphology in the group is. Johnston has made a tremendous contribution to our knowledge of the order in the Southern Hemisphere, but we still know little of its members in Australia, South America, China, Mongolia, southern Africa, etc.

(9) I believe that the *Rhytismatales* is one of the very first groups for which a real attempt has been made to gather all data properly and systematically in a structured suite of databases. At IMI a fully relational set of databases on the order has been developed, containing over 20 000 items of information on bibliography, nomenclature, descriptions, and individual collections. Programs can mechanically produce monographs and maps on a world or regional basis on demand. My focus is currently on data-gathering rather than describing new families and genera, and a monographic statement of what is known at the moment will be published through IMI shortly.

(10) I wish to congratulate Huhtinen (this volume) on his common-sense appeal for the better handling of information. We need more discussion as to how we can store in a standard way all the data we are now generating. This question needs to be considered by us all.

Bellemère: When examined by TEM, the ascus apex in members of the *Rhytismataceae* is found to have an annular structure recalling that of the *Mollisiaceae*. It is therefore possible that the group is to some extent on the same evolutionary line as some leotialian fungi.

Triblidiales

Eriksson: The family *Triblidiaceae* contains *Triblidium* and *Pseudographis* and was formerly tentatively placed in *Graphidales*. I also discovered a third genus *Huangshania* from China. All occur on the bark of conifers, starting as a small elevation that increasingly becomes semiglobose. The hamathecium consists of paraphysoids and the asci are extremely thin-walled, folding rather like a *Pertusaria* ascus after discharge; in Congo-red a transverse striation of the ascus wall is evident which resembes that in *Sarcoscyphaceae*. *Huangshania* differs from the other two genera in the extremely long and verrucose ascospores, which exceed 100 μm.

Rossman: Does your new genus have any iodine reactions such as those found in *Pseudographis* ?

Eriksson: *Pseudographis* has a fantastic blue colouration of the ascospores in iodine, like that seen in *Graphidaceae*, but the whole ascospore reacts and not just the septa as in *Graphis*. *Huangshania* and *Triblidium* both lack this reaction.

Bellemère: I think it is useful to have this order in which we can place families and genera with similar ascus tips (i.e. with apical walls getting thinner) and which are currently dispersed through different groups.

DISCUSSION 9

ARTHONIALES, DOTHIDEALES, OPEGRAPHALES, PYRENULALES, AND *VERRUCARIALES*

Leaders: O.E. Eriksson[1] and D.R. Reynolds[2]

[1]Department of Ecological Botany
University of Umea
S-901 87 Umea, Sweden

[2]Natural History Museum of Los Angeles County
900 Exposition Boulevard
Los Angeles, California 90007, USA

INTRODUCTION

Eriksson: The total number of species included in the orders under discussion is about 7200, distributed through 1818 genera. Some of the problems we have to confront can be exemplified by a particular example. In 1957 Holm defended a thesis in which he divided *Leptosphaeria* into segregate genera; that represented a revolution in the classification of this group of fungi. That genus is characterized by perithecioid ascomata, with a comparatively thick wall of scleroplectenchymatous cells, bitunicate asci, and brownish several-septate ascospores.

Until that time, almost any pyrenomycete with a similar morphology was placed in *Leptosphaeria*. One of the genera recognized by Holm was *Phaeosphaeria*, which has a much thinner ascomatal wall of reddish brown cells, and ascospores which are pale yellowish brown. No original material of the type species, described by Miyake from Japan in 1909, was found, and Holm accepted the genus based on the published account even though that did include obviously incorrect information on the hamathecium.

When I started to study this genus in 1963, I found what proved to be part of the original material in Stockholm, with locality details in Japanese, and this was selected as lectotype. Major studies of the genus have been performed by Hedjaroude, Leuchtman, and Shoemaker and Babcock. The two latter authors provided important information on the type material.

How can this type of information, and also all current alternative concepts of the genus, be retrieved? My suggestion is that we try to compile an index of all ascomycete genera, families and orders. An example of how this data might be presented is tabled.

A manuscript preliminary synopsis of the generic names in *Xylariaceae* has been compiled by Læssøe, and this is an example of the type of data that could be included in the new *Index Ascomycetum*.

DISCUSSION

Arthoniales

Tehler: Where to draw the boundaries of orders and families is necessarily subjective, but the current *Outline* recognizes both *Arthoniales* and *Opegraphales*. Cladistic studies conducted in recent years show that these orders are monophyletic, and to attempt to divide the assemblage in other ways would necessarily make one paraphyletic. I believe it is best to unite these under the name *Arthoniales*, and also to combine *Opegraphaceae* and *Roccellaceae* into one family,

Grube: What is the position of *Cryptothecia* and *Stirtonia* ?

Tehler: *Cryptothecia* and *Stirtonia* have not yet been included in my phylogenetic analysis.

Grube: It is difficult to speak about *Arthonia* if only a few species are studied as it is extremely heterogeneous. There are also cross-links with *Arthothelium*, so if the latter is excluded it will be impossible to form a satisfactory group.

Tehler: I used *Arthothelium spectabile* as an outgroup, and at least that species must be excluded from *Arthoniales* on the data we have. It would be surprising if *Arthonia*, with 5-600 species, was a monophyletic group.

Egea: The order *Arthoniales* is heterogeneous and it is necessary to restudy those genera not yet monographed according to modern systematic criteria before proposing new families. *Lecanactis* as used by Zahlbruckner is also heterogeneous, but *Lecanactis* s. str. and some allied genera clearly have the characteristics of *Arthoniales*.

Tehler: I agree that we have to be very careful in proposing new formal groups, and analyses should be re-run as additional genera are included in the data set.

Aptroot: I wondered where you placed *Helminthocarpon* ? It is of particular interest as it is lirellate and I suspect it could be near to *Roccellaceae*.

Tehler: I have not studied that genus so it has not been included in the data set used for cladogram construction.

Eriksson: The cladistic analyses made by Tehler are very impressive, using some 92 characters in the case of the *Arthoniales* analysis. Where such large data sets are used we can be more confident about the conclusions.

Dothideales

Reynolds: The treatment of the *Dothideales* in the current *Outline* can be viewed as one of two competing systems. Some five or six orders in Barr's system are collapsed into one. As a result, we have lost some rather clear "core" groups. When Luttrell defined the *Loculoascomycetes* in the 1950s, he stressed the ascomatal development, a character not seriously challenged since that time; the bituncate ascus, in contrast, has been shown to have a considerable amount of variation. Barr, through the publication of evolving classifications based on Luttrell's system, has been defining natural groups - yet this is apparently being ignored.

Barr: In 1987 Eriksson proposed an informal "subordinal" grouping for *Dothideales* s. ampl. That grouping was very similar to that which I had presented. My arrangement comes from the observation of variations among the many fungi I study; variations that lead to family and then to higher groupings. The separation of families and orders is based on more than a single set of characters, i.e. on a set of states; sometimes one or two characteristics may be missing in a particular taxon, but the preponderance leads to the disposition. The hamathecium is important, but so is the peridial structure.

Schumacher: A character compatibility analysis should be performed to try to justify your classification, based on the importance of your data in a phylogenetic sense. In this method only characters considered to be of phylogenetic importance are included. This method analyzes ways in which phylogenetic information derived from different characters conflicts or agrees. A phylogenetic tree, a cladogram, can be constructed from the largest number of apomorphic characters that are mutually compatible, i.e. the largest cliques. The most parsimonious cladogram is constructed so as to minimize the number of homoplasic characters.

Reynolds: When I attempted a cladistic analysis of this group using Barr's characters, the reaction of non-mycological cladists was that my approach was naive. Yet, it showed that many of the characters were not predictable and open to alternative interpretations.

Rossman: I strongly disagree with the maintenance of *Dothideales* s. ampl. If Eriksson is prepared to recognize the orders *Lahmiales* and *Neolectales* when we have so little molecular data on these groups, surely at least groups such as *Melanommatales* should also be accepted. Some such segregates will surely be heterogeneous, but that is the situation in many other orders at this time. I am concerned that in the preparation of the *Outline*, the evidence does not seem to be evaluated evenly in making final recommendations.

Kimbrough: Having taught graduate courses in ascomycetes for almost 30 years, I also see some very distinct groupings within the loculoascomycete-bitunicate fungi. However, those looking at other people's forests from our hillside can always see more distinct groups in other people's forests. Does a key exist to this large *Dothideales*; it is very necessary for our students ?

Eriksson: I have no doubt that *Dothideales* as we currently have it must be divided into several orders just as separate as *Pezizales* and *Leotiales*. However, I do not see that we need hurry and I am awaiting molecular data so that they can be divided with more confidence. We will have such data very soon.

Taylor, J.: It seems that we are victims of the philosophy that only orders are allowed and nothing higher. Here, with the *Dothideales*, there are clearly several closely related orders, but their recognition would imply a supraordinal taxon. I think that the time is right to do just that and to agree on some supraordinal taxa.

Blackwell: Has anyone attempted a cladistic analysis of even part of this complex in order to try to define monophyletic groups ?

Eriksson: In which orders would you place *Microthyrium* and *Trichothyrina* ?

Barr: In *Microthyriales* and *Chaetothyriales* respectively, based on the centrum structure.

Eriksson: In the past *Microthyrium* and *Trichothyrina* have been placed in monotypic families, and the latter was separated out as *Trichothyriales* by Nannfeldt. This separation was made because it was thought the asci arose from different directions. The asci are the same and the ascospores are most unusual with cilia-like appendages arising at or near the septum. The only difference is an extremely thin subhymenial layer in *Microthyrium,* whereas the ascomata of *Trichothyrina* have a definite lower wall. *Microthyrium* may have evolved from *Trichothyrina* by the loss of this layer. If these very closely related genera are placed in different orders, this illustrates the difficulties we have to address.

Barr: You did not mention the hamathecial structures. Trabeculate pseudoparaphyses occur in *Microthyrium* (and other *Microthyriaceae*), while *Trichothyrina* (and other *Trichothyriaceae*) has short down-growing periphysoids.

Eriksson: A main difference is that you feel the hamathecium is the most important criterion in making divisions.

Barr: Certainly "considerable", but not "most". In looking at fresh collections, differences can be seen in this group just as a hand section is being cut. The centrum is clear in *Trichothyrium*, and solidly white in *Microthyrium.* This is also one of my reasons for separating *Platystomum* from *Lophiostoma.*

Berbee: Whether the trabeculae are a central character needs to be studied, but it is much more interesting to attack a clear hypothesis than it is to begin with none. We prefer to have somebody's "best guess" that can then be challenged by new methods. Barr's system is much more useful for this than the system of narrowly delimited orders in the current *Outline*.

Taxonomists are often accused of working as stamp collectors. When we present our work as a succession of descriptions of species, genera and families we lose the ear of other biologists. However, when our work is presented in terms of hypotheses challenging current beliefs, we then have a chance of communicating the excitement of systematics.

Hawksworth: This relates to a more general issue that has arisen in earlier discussions. We have to distinguish between scientific hypotheses for testing and refining, and ones which have been tested, found to be robust, and can be recommended for general use. Barr's ideas are an excellent example of the first and in the best traditions of science; her concepts develop as she studies more material, and molecular and other data sets can be applied as tests. I am especially concerned that numerous characters which we suspect may be relevant remain unstudied in so many groups of ascomycetes, for example septal structure and numbers of ascospore nuclei; as these data become available they can be used for such testing.

I have referred to these two types of classification as "specialist" and "generalist" systems (*Proceedings of the Indian Academy of Science, Plant Science* 94, 319-339, 1985). The *Outline* aims to be a generalist system that represents what we have a high level of confidence in. A scenario of constantly changing specialist systems is certainly exciting for the scientist and a vital part of the scientific method, but it is not what our user constituency demands.

Eriksson: This is certainly the most difficult of all the ascomycete orders. Note for example the large number of genera *incertae sedis* placed here and not yet referred to a family. There is a tremendous amount to be gained by studying these. When I discovered the complexity of the ascus tip in *Pyrenophora* I was amazed; this was achieved using a variety of stains on fresh material. When asked if I believe in all the structures Chadefaud and his colleagues described (see Parguey-Leduc *et al.*, this volume), I say that I am sure they have seen these structures even if others have not. They examined a wide range of stages of maturity for hours using many different stains. In comparison, often nothing can be seen in lactophenol. It is an immense task to consider how to transfer such ascus details into a cladistic analysis.

Reynolds: What groups do you think will eventually prove to be good monophyletic units in *Dothideales* s. ampl.

Eriksson: *Asterinales, Dothideales* s. str., *Melanommatales, Microthyriales,* and *Pleosporales*; I am not sure about any others. When I published on graminicolous pyrenomycetes in 1967, before Barr had published her classification, I divided the bitunicates into three groups, corresponding to *Dothideales, Melanommatales,* and *Pleosporales.* This was based particularly on differences in the ascomatal wall, first discovered by Munk. Barr has found further characters and made a more detailed classification. I am convinced this is the kind of classification we need but I do not feel prepared for this yet.

The proposal to conserve the family name *Dimeriaceae* over *Pseudoperisporiaceae* has been rejected by the Committee for Fungi, yet the latter has not been used since its introduction. I am uncertain what should be done in this case.

Hawksworth: The situation over *Dimeriaceae* is more complex than originally supposed as the generic name *Dimerium* has proved to be based on a mixture of species, none of which conform to the current concept of the family. The Committee correctly took the view that the application of the generic name needed to be clarified before the family name was protected ! The possibility of proposing the conservation of *Dimerium* with a conserved type species has to be investigated.

Eriksson: The ascus in *Trichodelitschia* illustrates that it is not always easy to see if an ascus is bitunicate or not. I have demonstrated it to be fissitunicate, but Barr regards it as unitunicate and she may have seen that as it very difficult to observe the discharge. In *Sporormiella* also the ascus wall can be so thin that it is hard to imagine that there are two functional wall layers; however, in fresh material I have clearly observed "jack-in-the-box" discharge in water.

Barr: I interpret the asci of *Trichodelitschia* as unitunicate, based on a study of both dried and fresh material of both species of the genus, whereas those other members of the *Phaeotrichaceae* are bitunicate. I cannot convince myself that they are other than unitunicate,

Lundqvist: The asci in *Trichodelitschia* are a good example of "jack-in-the-box" fissitunicate discharge. I have seen hundreds of collections in the living state. That the tip of the inner ascus contains a thickened ring does not change my opinion. I suspect that the genus is not unique in this respect. The placement of the genus in *Phaeotrichaceae* seems to be well-founded, even though *Phaeotrichum* itself is cleistocarpic and has a somewhat different ascus morphology. However, if a fissitunicate ostiolate taxon becomes cleistocarpic, the asci can loose this mechanism and even evolve into deliquescing asci.

Rogers: *T. munkii* certainly has a double wall.

Minter: Is it possible that both these apparently conflicting observations are correct and that asci can sometimes be functionally bitunicate and at other times not ? In this character we have something of a sacred cow in mycology in that asci cannot vary in a single taxon. However, Baral (see below) pointed out that he has seen a tremendous variation of ascus morphology within a single ascoma.

Eriksson: I am prepared to believe that experienced mycologists have really seen the structures they report. However, differences could be due to the material, mounting medium, or other factors. I do not exclude the possibility that different herbarium material of one species could be interpreted as having different types of asci. By combining different stains you can see much more that just by using, for example, water and Congo red. In fresh material there should be no problem, but in many cases we have no alternative but to work with old dried herbarium material and then we must be cautious.

Baral: I do not know *Trichodelitschia*, but the easiest explanation for me would be that a bitunicate ascus is always thin-walled in the living state, but very thick-walled in the dead state irrespective of its stage of maturation. The controversy might arise from this single fact. This effect, which seems not to be part of our common knowledge, was described by de Bary (1887, *Comparative Morphology and Biology of the Fungi, Mycetozoa and Bacteria,* Clarendon Press, Oxford; Baral, *Mycotaxon* 44, 333-390, 1992).

Kendrick: Fresh collections need to be shared so that the specialists can agree on the interpretation. If we always correspond through journals we are never going to achieve finality.

Barr: *Daruvedia* may be removed from *Dothideales*; *D. bacillata* is closely related to species of *Melomastia* and *Saccardoella.* It has unitunicate asci and is close to the *Amphisphaeriaceae* or *Clypeosphaeriaceae*.

Aptroot: I consider that the *Aspidotheliaceae* belong to *Melanommatales*, just outside *Pyrenulales*. This is discussed in a paper now in press. The families *Phyllobatheliaceae* and *Thelenellaceae* also belong there.

Pyrenulales

Aptroot: I strongly recommend that the families *Monoblastiaceae, Pyrenulaceae, Strigulaceae*, and *Trypetheliaceae* are recognized within this order in the next *Outline*. The delimitation of these mainly tropical lichenized groups, which had been hardly studied for 100 years, has been clarified to a great extent by Harris. All four have a different hamathecium structure, and also a different basic pattern of ascospore distoseptation (complicated when muriform). I moved *Eopyrenula* from the group because of the cellular wall structures; other members of the order have thick and melanized walls in which it is difficult to recognize individual cells.

Barr: *Eopyrenula* fits in *Pleomassariaceae*, as does *Kirschsteiniothelia*.

Eriksson: Some of the ideas Harris has have not been documented in detail. For example, it is unclear if he has studied type material of all the genera he synonymized with *Pyrenula*, and we have assumed that this was going to be presented in a forthcoming monograph of the genus. We have no reason to believe that Harris was wrong in this case, but in developing the *Outline* in recent years more stringent criteria have developed for including changes. It is now necessary to explain why a new concept is proposed before it can be accepted in the *Outline*.

Aptroot: The characters separating these families are very clear now that much more material is available from the tropics. There are some problems in generic separations, but no intermediates between the families are known. This suggests that the families are very old.

Eriksson: Type material of *Monoblastia* existed in two herbaria in North America. The ascomata are perithecioid, but that type is also found in *Pertusariaceae*. The asci in that material fold in a *Pertusaria*-like manner. When I studied this in 1981, only *Pertusaria* was known to have bilabiate asci, but these are now known in other groups (e.g. *Triblidiales*). Harris has studied fresh material and has quite a different concept, but this was published only as a note in a privately printed account of the Florida lichens.

Aptroot: I have not seen the type of *Monoblastia*, but I have examined fresh material of other species of the genus which certainly belong to *Pertusariales*. They are clearly related to *Acrocordia*, and also have ornamented ascospores - a feature not seen in *Pertusariales*. The family can also include the large genus *Anisomeridium*.

Monoblastiaceae is an earlier name than *Acrocordiaceae*; I am not concerned which family name is used, but that it is not lost in the *Pyrenulaceae*.

Eriksson: I am sure Harris and yourself are the world experts on this group, but for others it is difficult.

Aguirre-Hudson: Aptroot places *Celothelium* in *Monoblastiaceae*, a genus I suggested has affinities with *Acrocordiaceae* and *Trypetheliaceae* because I saw a meniscus in the apex of the ascus of only one species of the genus and not in the type. Has he further evidence to support his opinion ?

Aptroot: The *Trypetheliaceae* all have a broad ring in the ascus tip, while the *Pyrenulales* can have two, one or no rings. I have seen only one ring in *Celothelium*. The genus is somewhat atypical in having elongated not clearly distoseptate ascospores; all *Pyrenulales* have distoseptate ascospores to a certain extent.

Reynolds: Hawksworth explained to me in 1989 that mycologists are encouraged to disagree with the *Outline*, as you and Barr have done. It is a favour to have these views pulled together in the *Systema Ascomycetum*, but the *Outline* has no formal standing and the decisions in it are those of the editors. You should continue to disagree.

Eriksson: The hamathecium is a difficult criterion. Aptroot accepted the family *Requienellaceae* in *Melanommatales*, but states the hamathecium consists of cellular pseudoparaphyses - generally regarded as a feature of *Pleosporales*.

Barr: The *Requienellaceae* have paraphysoids, i.e. trabeculate pseudoparaphyses, and not cellular pseudoparaphyses. That is one of the bases of the family.

Aptroot: This is a matter of definition. The hamathecium in that family is not the same as that in *Trypetheliaceae*; it is much less anastomosed.

Eriksson: This discussion illustrates the difficulties in making divisions within *Dothideales* s. ampl.

Hawksworth: I corresponded with Harris on generic concepts in *Pyrenula* when I was preparing the account of that genus for *The Lichen Flora of Great Britain and Ireland* (Purvis *et al.*, 1992). I consider that Harris's enlarged concept of the genus should be accepted as it takes us beyond nineteenth century spore-based schemes towards a more natural grouping, but I was hoping for the views of other lichenologists before implementing this in the *Outline* as some familiar names would disappear.

Letrouit-Galinou: I agree with Aptroot concerning the heterogeneity of both *Pyrenulales* and *Pyrenulaceae*. The *Pyrenulaceae* is interesting in including taxa with true paraphyses and fissitunicate asci, but the *Trypetheliaceae* is very different and lacks true paraphyses.

Eriksson: I have accepted the interpretation of Janex-Favre that *Pyrenula* originally has paraphysoids and secondarily develops true paraphyses.

Letrouit-Galinou: The situation is parallel to that in many fissitunicate discomycetes, such as *Lecanactis* and *Patellaria*, where true paraphyses are also secondary.

Tehler: The presence of both ascohymenial development and fissitunicate asci can be explained in that ascohymenial development is plesiomorphic (and cannot be used for grouping), whereas bitunicate asci is apomorphic (and can be used for grouping). In other words, the ascohymenial fungi are paraphyletic. If *Lecanactis* has true paraphyses, there is a possibility it should be removed from *Arthoniales*.

Egea: I have studied about 120 species referred to *Lecanactis*, but only 18 are accepted in *Lecanactis* s. str., including some that are new. The other species belong to a variety of genera, including the new genus *Creponea* for *L. premnea*; some belong to in *Graphidaceae* and *Lecideaceae* s. lat. *Creponea* and *Lecanactis* s. str. belong to *Arthoniales*; they have fissitunicate asci with an apical amyloid reaction, and paraphysoids. I studied *Patellaria atrata* and found it to be quite different.

Verrucariales

Eriksson: The division of the order into two families, *Dermatocarpaceae* and *Verrucariaceae,* has been proposed by French workers who reported that *Verrucaria* has an ascolocular type of development, while *Dermatocarpon* was found to be ascohymenial.

Hawksworth: This separation might also be supported by the conidiomata which are of the multilocular *Xanthoria*-type in both *Dermatocarpon* and *Endocarpon*.

Hafellner: I cannot support the removal of *Dermatocarpon* from *Verrucariaceae*. However, *Endocarpon* and *Staurothele* form a natural group united by hymenial algae and muriform ascospores; these form a separate line of thallus evolution in the order, and the name *Endocarpaceae* is already available.

Letrouit-Galinou: I consider that *Dermatocarpaceae* should be maintained because of the different ascomatal ontogeny. It is very different from the uniform type seen in *Endocarpon* and *Verrucaria*. However, rather few species have been studied ontogenetically.

Triebel: The lichenicolous genera *Endococcus*, *Merismatium*, *Muellerella*, *Norrlinia*, and *Phaeospora* form a natural group within the *Verrucariales*. *Adelococcus* and *Sagediopsis* are different in forming persistent interascal filaments. The family name *Adelococcaceae* (*Verrucariales*) will be introduced to accommodate these two genera.

Aptroot: I would include also *Macentina*, *Microtheliopsis*, *Normandina*, and *Pocsia* in *Verrucariales*.

Eriksson: *Microtheliopsis* is perhaps closer to *Microthyriaceae*. The asci are of the same type and have similar staining reactions, the endotunica becoming reddish with chlorasol black.

Psoroglaena, described from Cuba, turns out to be common in the tropics and subtropics. It has bright-coloured perithecia only yet known on the type specimen. The genus belongs in this order if the perithecia are really those of the lichen. Additional species are to be described shortly by Henssen.

Aptroot: *Agonimia, Flakea*, and *Psoroglaena* share a unique synapomorphy, the papillose cell walls. As *Agonimia* is definitely a member of the *Verrucariaceae* and similar ascomata are rarely found on *Psoroglaena*, they can all be included in the family - and possibly within a single genus. I also consider that *Normandina* belongs to the family, but there are different opinions on that (see Discussion 4).

DISCUSSION 10

FAMILIES AND GENERA OF UNCERTAIN POSITION

Leader: D.L. Hawksworth

International Mycological Institute
Bakeham Lane, Egham
Surrey TW20 9TY, UK

INTRODUCTION

Hawksworth: In this section we focus on the 28 accepted ascomycete families which could not be assigned to an order in the *Outline* for 1990, and also 203 accepted genera which were not placed in any family. The number of unplaced, but yet accepted, taxa in these categories is a salutary reminder of how much work is required before a fully synthetic system for even the already described ascomycetes on Earth can be produced.

In the case of the obligately lichenicolous ascomycetes, there appears to be an unending succession of genera that cannot be placed satisfactorily. I am sure this pattern will increasingly prove to be the case as little studied substrata, habitats, and geographic regions are studied in depth. Indeed, this situation might have been predicted by my hypothesis that only about 5% of the fungi on Earth have yet been described (*Mycological Research* 95, 641-655, 1991).

Nevertheless, the number in these two categories has been reduced significantly since the early issues of the *Outline*, especially as long-forgotten types have been sought out, re-examined, and then either redisposed or synonymized. This is time-consuming and often unrewarding work. However, during this Workshop I have been pleased to see fresh placements proposed for some of these problematic taxa. Here, we consider those not otherwise debated during this Workshop.

DISCUSSION

Boliniaceae

Whalley: I consider that the *Boliniaceae* has little in common with the *Xylariales*, but do not know where it might be placed.

Rogers: I do not object to the family being placed in *Xylariales*, as proposed by Barr, but not within the family *Xylariaceae*.

Untereiner: Barr included *Endoxyla* in *Boliniaceae*. It had previously been placed in *Clypeosphaeriaceae*, but probably should be placed in an independent family. *Ceratostomella* does not seem to be a member of the *Clypeosphaeriaceae* sensu Barr, and its systematic position is unclear to me. *Apiocamarops* and *Camarops* seem to be a more or less natural assemblage.

Ascomycete Systematics: Problems and Perspectives in the Nineties
Edited by D.L. Hawksworth, Plenum Press, New York, 1994

Cephalothecaceae

Cannon: I suggest that the *Cephalothecaceae* and *Pseudeurotiaceae* are merged. Alternatively, the cephalothecoid members of the latter family (e.g. *Fragosphaeria*) should be transferred to the former. The only really distinctive feature seems to be the hyphae around the ascomata.

Malloch: That has been considered for a long time. The family is based on *Cephalotheca sulfurea*, a reduced cleistothecial form not yet studied in culture. It would be unwise to make a change until this fungus has been characterized more fully. The cephalothecoid peridium occurs in many other fungi so too much emphasis should not be placed on that.

Dactylosporaceae

Hafellner: The family can be placed as *Lecanorales incertae sedis*. The apothecia are long-lived, and the asci have an amyloid "*a*" layer and gelatinous outer cap.

Gypsoplacaceae

Aptroot: This family can be accepted in *Lecanorales*, suborder *Lecanorineae.*

Melaspileaceae

Hafellner: This is a very difficult group, but the type species *Melaspilea arthonioides* is definitely a member of *Arthoniales*, although the genus itself is highly heterogeneous.

Pachyascaceae

Aptroot: *Pachyascaceae* can be accommodated in *Lecanorales*.

Phlyctidaceae

Aptroot: This family can be accepted in *Lecanorales*.

Solorinellaceae

Aptroot: This can be placed in *Graphidales*, where it is in effect a replacement name for *Asterothyriaceae* s. lat.

Thyridiaceae

Eriksson: *Thyridium* resembles the *Hypocreaceae* in having secondary metabolites dissolving in potassium hydroxide, and the asci are also similar.

Rossman: This family is characterized by species with bright-coloured perithecia and true paraphyses. Diverse fungi with those characters are now being placed in the family. Bright colours do not necessarily indicate any link with *Hypocreales*, indeed similarly coloured perithecia occur in the loculoascomycete family *Tubeufiaceae.*

Trichotheliaceae

Aptroot: This family should be moved from *Pyrenulales* to the list of families *incertae sedis*.

Genera not Assigned to Families

Barr: *Melomastia, Saccardoella* and *Daruvedia* form a sequence in *Amphisphaeriaceae,* and a proposal to separate them from that family is in preparation. I have collected all three genera on *Lonicera* and cannot distinguish them until I return to the laboratory.

Rogers: In *Pseudovalsaria* I have seen part of the ascus becoming separated and I strongly suspect it to be bitunicate.

Blackwell: *Kathistes, Pyxidiophora*, and *Subbaromyces* do not fit within the perithecial ascomycetes. Small subunit rDNA sequences support a placement basal to the main clade of derived perithecial forms. We did not intend to imply any phylogenetic relationship between *Subbaromyces* and *Klasterskya*. *Klasterskya* belongs with *Ophiostomatales*.

Kurtzman: The situation in *Saccardoella* illustrates that while specialists can agree on what they see, they cannot determine what that means. New sets of characters are needed, perhaps a combination of both molecular and novel morphological characters.

Wingfield: I am concerned that so few ascomycete specialists consider the anamorphs or make cultures to reveal these. The can reveal a wealth of new characters.

Rossman: The best medium for culturing *Hypocreales* ascospores is a weak one, for example corn meal agar with dextrose and an array of antibiotics.

Læssøe: *Pyrenomyxa* is an earlier synonym of *Pulveria*, an accepted genus of *Xylariaceae*.

Annex 1

APPENDIX

OUTLINE OF THE ASCOMYCETES - 1993

INTRODUCTION

This Appendix reproduces the arrangement of the orders and families of the "Outline of the ascomycetes - 1993" prepared after the Workshop and published by O.E. Eriksson & D.L. Hawksworth (*Systema Ascomycetum* **12**: 51-257, 1993), *but excluding* synonymized family names. The original paper should be consulted for full details, and also placement and treatment of generic names and their synonyms.

Note that changes proposed at the Workshop that had not been the subject of discussion in the *Notes* or in other papers have not been included in this revision, but will be taken up in future editions according to inputs and reactions received from mycologists. This is in accordance with the usual practice adopted for the *Outlines* (Hawksworth & Eriksson, this volume).

OUTLINE OF ORDERS AND FAMILIES

1. **ARTHONIALES** Henssen ex D. Hawksw. & O.E. Erikss. 1986
    1. **Arthoniaceae** Reichenb. ex Reichenb. 1841
    2. **Chrysotrichaceae** Zahlbr. 1905
    3. **Opegraphaceae** Stizenb. 1862
    4. **Roccellaceae** Chevall. 1826

2. **CALICIALES** C. Bessey 1907
    1. **Caliciaceae** Chevall. 1826
    2. **Calycidiaceae** Elenkin 1929
    3. **Coniocybaceae** Reichenb. 1837
    4. **Microcaliciaceae** Tibell 1984
    5. **Mycocaliciaceae** A.F.W. Schmidt 1970
    6. **Sphaerophoraceae** Fr. 1831
    7. **Sphinctrinaceae** M. Choisy 1950

3. **CALOSPHAERIALES** M.E. Barr 1983
    1. **Calosphaeriaceae** Munk 1957
    2. **Graphostromataceae** M.E. Barr, J.D. Rogers & Y.M. Ju 1993

4. **CORYNELIALES** Seaver & Chardón 1926
    1. **Coryneliaceae** Sacc. ex Berl. & Voglino 1886

5. **CYTTARIALES** Luttr. ex Gamundí 1971
    1. **Cyttariaceae** Speg. 1887

6. **DIAPORTHALES** Nannf. 1932
    1. **Melanconidaceae** G. Winter 1886
    2. **Valsaceae** Tul. & C. Tul. 1861

7. **DIATRYPALES** Chadef. ex D.L. Hawksw. & O.E. Erikss. 1986
    1. **Diatrypaceae** Nitschke 1869

8. **DOTHIDEALES** Lindau 1897
    1. **Antennulariellaceae** Woron. 1925
    2. **Argynnaceae** Shearer & J.L. Crane 1980
    3. **Arthopyreniaceae** W. Watson 1929
    4. **Asterinaceae** Hansf. 1946
    5. **Aulographaceae** (G. Arnaud) Luttr. 1973 nom. inval.
    6. **Botryosphaeriaceae** Theiss. & H. Syd. 1918
    7. **Capnodiaceae** (Sacc.) Höhn. ex Theiss. & H. Syd. 1917
    8. **Chaetothyriaceae** Hansf. ex M.E. Barr 1979
    9. **Coccodiniaceae** Höhn. ex O.E. Erikss. 1981
    10. **Coccoideaceae** P. Henn. ex Sacc. & D. Sacc. 1905
    11. **Cookellaceae** Höhn. ex Sacc. & Trotter 1913
    12. **Cucurbitariaceae** G. Winter 1885

DOTHIDEALES Lindau 1897 (cont.)
13. **Dacampiaceae** Körb. 1855
14. **Diademaceae** Shoemaker & C.E. Babc. 1992
15. **Didymosphaeriaceae** Munk 1953
16. **Dimeriaceae** E. Müll. & Arx ex E. Müll. & Arx 1975
17. **Dothideaceae** Chevall. 1826
18. **Dothioraceae** Theiss. & H. Syd. 1917
19. **Elsinoaceae** Höhn. ex Sacc. & Trotter 1913
20. **Englerulaceae** P. Henn. 1904
21. **Eremomycetaceae** Malloch & Cain 1971
22. **Euantennariaceae** Hughes & Corlett 1972
23. **Fenestellaceae** M.E. Barr 1979
24. **Herpotrichiellaceae** Munk 1953
25. **Hysteriaceae** Chevall. 1826
26. **Leptopeltidaceae** Höhn. ex Trotter 1928
27. **Leptosphaeriaceae** M.E. Barr 1987
28. **Lichenotheliaceae** Henssen 1986
29. **Lophiostomataceae** Sacc. 1883
30. **Melanommataceae** G. Winter 1885
31. **Mesnieraceae** Arx & E. Müll. 1975
32. **Metacapnodiaceae** Hughes & Corlett 1972
33. **Micropeltidaceae** Clem. & Shear 1931
34. **Microtheliopsidaceae** O.E. Erikss. 1981
35. **Microthyriaceae** Sacc. 1883
36. **Mycoporaceae** Zahlbr. 1903
37. **Mycosphaerellaceae** Lindau 1897
38. **Myriangiaceae** Nyl. 1854
39. **Mytilinidiaceae** Kirschst. 1924
40. **Parmulariaceae** E. Müll. & Arx ex M.E. Barr 1979
41. **Parodiellaceae** Theiss. & H. Syd. ex M.E. Barr 1987
42. **Parodiopsidaceae** (G. Arnaud) ex Toro 1952
43. **Phaeosphaeriaceae** M.E. Barr 1979
44. **Phaeotrichaceae** Cain 1956
45. **Phragmopelthecaceae** L. Xavier 1976
46. **Piedraiaceae** Viégas ex Cif., Bat. & Campos 1956
47. **Pleomassariaceae** M.E. Barr 1979
48. **Pleosporaceae** Nitschke 1869
49. **Polystomellaceae** Theiss. & H. Syd. 1915
50. **Pyrenotrichaceae** Zahlbr. 1926
51. **Schizothyriaceae** Höhn. ex Trotter et al. 1928
52. **Sporormiaceae** Munk 1957
53. **Tubeufiaceae** M.E. Barr 1979
54. **Venturiaceae** E. Müll. & Arx ex M.E. Barr 1979
55. **Vizellaceae** H.J. Swart 1971
56. **Zopfiaceae** G. Arnaud ex D. Hawksw. 1992

9. **ELAPHOMYCETALES** Trappe 1979
   1. **Elaphomycetaceae** Tul. ex Paol. 1889

10. **ERYSIPHALES** Gwynne-Vaughan 1922
   1. **Erysiphaceae** Tul. & C.Tul. 1861

11. **EUROTIALES** G.W. Martin ex Benny & Kimbr. 1980
   1. **Ascosphaeraceae** L.S. Olive & Spiltoir 1955
   2. **Eremascaceae** Engl. & E. Gilg 1924
   3. **Monascaceae** J. Schröt. 1894
   4. **Trichocomaceae** E. Fisch. 1897

12. **GLAZIELLALES** J.L. Gibson 1986
   1. **Glaziellaceae** J.L. Gibson 1986

13. **GYALECTALES** Henssen ex D. Hawksw. & O.E. Erikss. 1986
   1. **Gyalectaceae** (Massal.) Stizenb. 1862

14. **HALOSPHAERIALES** Kohlm. 1986
   1. **Halosphaeriaceae** E. Müll. & Arx ex Kohlm. 1972

15. **HYPOCREALES** Lindau 1897
   1. **Clavicipitaceae** (Lindau) Earle ex Rogerson 1971
   2. **Hypocreaceae** De Not. 1844
   3. **Niessliaceae** Kirschst. 1939

16. **LABOULBENIALES** Engler 1898
   1. **Ceratomycetaceae** S. Colla 1934
   2. **Euceratomycetaceae** I.I. Tav. 1980
   3. **Herpomycetaceae** I.I. Tav. 1981
   4. **Laboulbeniaceae** G. Winter 1886
   5. **Pyxidiophoraceae** Arnold 1971

17. **LAHMIALES** O.E. Erikss. 1986
   1. **Lahmiaceae** O.E. Erikss. 1986

18. **LECANORALES** Nannf. 1932
   1. **Acarosporaceae** Zahlbr. 1906
   2. **Agyriaceae** Corda 1838
   3. **Alectoriaceae** (Hue) Tomas. 1949
   4. **Arctomiaceae** Th. Fr. 1860
   5. **Bacidiaceae** W. Watson 1929
   6. **Biatorellaceae** M. Choisy ex Hafellner & Casares-Porcel 1992
   7. **Brigantiaeaceae** Hafellner & Bellem. 1982
   8. **Candelariaceae** Hakul. 1954
   9. **Catillariaceae** Hafellner 1984
   10. **Cladoniaceae** Zenker 1827
   11. **Coccocarpiaceae** (Mont. ex Müll. Stuttg.) Henssen 1986
   12. **Collemataceae** Zenker 1827
   13. **Crocyniaceae** M. Choisy ex Hafellner 1984
   14. **Ectolechiaceae** Zahlbr. 1905

LECANORALES Nannf. 1932 (cont.)
15. **Eigleraceae** Hafellner 1984
16. **Gypsoplaceae** Timdal 1990
17. **Haematommataceae** Hafellner 1984
18. **Heppiaceae** Zahlbr. 1906
19. **Heterodeaceae** Filson 1978
20. **Hymeneliaceae** Körb. 1855
21. **Lecanoraceae** Körb. 1855
22. **Lecideaceae** Chevall. 1826
23. **Megalosporaceae** Vezda ex Hafellner & Bellem. 1982
24. **Micareaceae** Vezda ex Hafellner 1984
25. **Miltideaceae** Hafellner 1984
26. **Mycoblastaceae** Hafellner 1984
27. **Ophioparmaceae** R.W. Rogers & Hafellner 1988
28. **Pannariaceae** Tuck. 1872
29. **Parmeliaceae** Zenker 1827
30. **Physciaceae** Zahlbr. 1898
31. **Pilocarpaceae** Zahlbr. 1905
32. **Porpidiaceae** Hertel & Hafellner 1984
33. **Psoraceae** Zahlbr. 1898
34. **Ramalinaceae** C. Agardh 1821
35. ?**Rhizocarpaceae** M. Choisy ex Hafellner 1984
36. **Rimulariaceae** Hafellner 1984
37. **Stereocaulaceae** Chevall. 1826
38. **Trapeliaceae** M. Choisy ex Hertel 1970
39. **Umbilicariaceae** Chevall. 1826
40. **Vezdaeaceae** Poelt & Vezda ex J.C. David & D. Hawksw. 1991

19. **LEOTIALES** Carpenter 1988
1. **Baeomycetaceae** Dumort. 1829
2. **Dermateaceae** Fr. 1849
3. **Geoglossaceae** Corda 1838
4. **Hemiphacidiaceae** Korf 1962
5. **Hyaloscyphaceae** Nannf. 1932
6. **Leotiaceae** Corda 1842
7. **Loramycetaceae** Dennis ex Digby & Goos 1988
8. **Orbiliaceae** Nannf. 1932
9. **Phacidiaceae** Fr. 1821
10. **Sclerotiniaceae** Whetzel ex Whetzel 1945
11. **Vibrisseaceae** Korf 1991
- - - - -
12. ?**Ascocorticiaceae** J. Schröt. 1893

20. **LICHINALES** Henssen & Büdel 1986
1. **Lichinaceae** Nyl. 1854.
2. **Peltulaceae** Büdel 1986

21. **MEDEOLARIALES** Korf 1982
1. **Medeolariaceae** Korf 1982

22. **MELIOLALES** Gäum. ex D. Hawksw. & O.E. Erikss. 1986
1. **Meliolaceae** G.W. Martin ex Hansf. 1946

23. **MICROASCALES** Luttr. ex Benny & Kimbr. 1980
1. **Chadefaudiellaceae** Faurel & Schotter ex Benny & Kimbr. 1980
2. **Microascaceae** Luttr. ex Malloch 1970

24. **NEOLECTALES** Landvik, O.E. Erikss., Gargas & P. Gustafsson 1992
1. **Neolectaceae** Redhead 1977

25. **ONYGENALES** Cif. ex Benny & Kimbr. 1980
1. **Arthrodermataceae** Currah 1985
2. **Gymnoascaceae** Baran. 1872
3. **Myxotrichaceae** Currah 1985
4. **Onygenaceae** Berk. 1857

26. **OPHIOSTOMATALES** Benny & Kimbr. 1980
1. **Kathistaceae** Malloch & M. Blackw. 1990
2. **Ophiostomataceae** Nannf. 1932

27. **OSTROPALES** Nannf. 1932
1. **Gomphillaceae** W. Watson ex Hafellner 1984
2. **Graphidaceae** Dumort. 1822
3. **Odontotremataceae** D. Hawksw. & Sherwood 1982
4. **Stictidaceae** Fr. 1849
5. **Solorinellaceae** Vezda & Poelt 1990
6. **Thelotremataceae** (Nyl.) Stizenb. 1862

28. **PATELLARIALES** D. Hawksw. & O.E. Erikss. 1986
1. **Arthrorhaphidaceae** Poelt & Hafellner 1976
2. **Patellariaceae** Corda 1838

29. **PELTIGERALES** W. Watson 1929
1. **Lobariaceae** Chevall. 1826
2. **Nephromataceae** Wetm. ex J.C. David & D. Hawksw. 1991
3. **Peltigeraceae** Dumort. 1822
- - - - -
4. **Placynthiaceae** Å.E. Dahl 1950

30. **PERTUSARIALES** M. Choisy ex D. Hawksw. & O.E. Erikss. 1986
1. **Coccotremataceae** Henssen ex J.C. David & D. Hawksw. 1991
2. **Pertusariaceae** Körb. ex Körb. 1855

31. **PEZIZALES** C. Bessey 1907
1. **Ascobolaceae** Boud. ex Sacc. 1884
2. **Ascodesmidaceae** J. Schröt. 1893
3. **Balsamiaceae** E. Fisch.
4. ?**Carbomycetaceae** Trappe 1971
5. **Eoterfeziaceae** G.F. Atk. 1902
6. **Helvellaceae** Fr. 1823
7. ?**Karstenellaceae** Harmaja 1974

8. **Morchellaceae** Reichenb. 1834

PEZIZALES C. Bessey 1907 (cont.)

9. **Otideaceae** Eckblad 1968
10. **Pezizaceae** Dumort. 1829
11. **Pyronemataceae** Corda 1842
12. **Sarcoscyphaceae** LeGal ex Eckblad 1968
13. **Sarcosomataceae** Kobayasi 1937
14. **Terfeziaceae** E. Fisch. 1897
15. **Thelebolaceae** (Brumm.) Eckblad 1968
16. **Tuberaceae** Dumort. 1822

32. **PHYLLACHORALES** M.E. Barr 1983
    1. **Phyllachoraceae** Theiss. & H. Syd. 1915

33. **PROTOMYCETALES** Luttr. ex D. Hawksw. & O.E. Erikss. 1986
    1. **Mixiaceae** C.L. Kramer 1987
    2. **Protomycetaceae** Gray 1821

34. **PYRENULALES** Fink ex D. Hawksw. & O.E. Erikss. 1986
    1. **Massariaceae** Nitschke 1869
    2. ?**Monoblastiaceae** W. Watson 1929
    3. **Pyrenulaceae** Rabenh. 1870
    4. **Trichotheliaceae** (Müll. Arg.) Bitter & F. Schill. 1927
    5. **Trypetheliaceae** Zenker 1827

35. **RHYTISMATALES** M.E. Barr ex Minter 1986
    1. **Ascodichaenaceae** D. Hawksw. & Sherwood 1982
    2. **Rhytismataceae** Chevall. 1826

36. **SACCHAROMYCETALES** Kudrjanzev 1960
    1. **Cephaloascaceae** L.R. Batra 1973
    2. **Dipodascaceae** Engl. & E. Gilg 1924
    3. **Endomycetaceae** J. Schröt. 1893
    4. **Lipomycetaceae** E.K. Novák & Zsolt 1961
    5. **Metschnikowiaceae** T. Kamienski 1899
    6. **Saccharomycetaceae** G. Winter 1881
    7. **Saccharomycodaceae** Kudrjanzev 1960
    8. **Saccharomycopsidaceae** Arx & Van der Walt 1987

37. **SCHIZOSACCHAROMYCETALES** O.E. Erikss., Svedskog & Landvik 1993
    1. **Schizosaccharomycetaceae** Beij. ex Klöcker 1905

38. **SORDARIALES** Chad. ex D. Hawksw. & O.E. Erikss. 1986
    1. **Batistiaceae** Samuels & K.F. Rodrigues 1989
    2. **Catabotrydaceae** Petrak ex M.E. Barr
    3. **Ceratostomataceae** G. Winter 1885
    4. **Chaetomiaceae** G. Winter 1885
    5. **Coniochaetaceae** Malloch & Cain 1971
    6. **Lasiosphaeriaceae** Nannf. 1932
    7. **Nitschkiaceae** (Fitzp.) Nannf. 1932
    8. **Sordariaceae** G. Winter 1885

39. **SPATHULOSPORALES** Kohlm. 1973
    1. **Spathulosporaceae** Kohlm. 1973

40. **TAPHRINALES** Gäum. & C.W. Dodge 1928
    1. **Taphrinaceae** Gäum. in Gäum. & C.W. Dodge 1928

41. **TELOSCHISTALES** D. Hawksw. & O.E. Erikss. 1986
    1. **Letrouitiaceae** Bellem. & Hafellner 1982
    2. **Teloschistaceae** Zahlbr. 1898

    \- - - - -

    3. ?**Fuscideaceae** Hafellner 1984

42. **TRIBLIDIALES** O.E. Erikss. 1992
    1. **Triblidiaceae** Rehm 1888

43. **TRICHOSPHAERIALES** M.E. Barr 1983
    1. **Trichosphaeriaceae** G. Winter 1885

44. **VERRUCARIALES** Mattick ex D. Hawksw. & O.E. Erikss. 1986
    1. **Verrucariaceae** Zenker 1827

45. **XYLARIALES** Nannf. 1932
    1. **Amphisphaeriaceae** G. Winter 1885
    2. **Clypeosphaeriaceae** G. Winter 1886
    3. **Xylariaceae** Tul. & C. Tul. 1861

FAMILIAE INCERTAE SEDIS

Acrospermaceae Fuckel 1870 - (10)
(**Amorphothecaceae** Parbery 1969)
Aspidotheliaceae Räsenen ex J.C. David & D. Hawksw. 1991
Boliniaceae Rick 1931
(**Brefeldiellaceae** (Theiss.) E. Müll. & Arx 1962)
Cephalothecaceae Höhn. 1917
Cryptomycetaceae Höhn. 1917 nom. inval.
Dactylosporaceae Bellem. & Hafellner 1982.
Epigloeaceae Zahlbr. 1903
Hyponectriaceae Petr. 1923
Koralionastetaceae Kohlm. & Volkm-Kohlm. 1987
Mastodiaceae Zahlbr. 1907
(**Melaspileaceae** W. Watson 1929)
(**Montagnellaceae** Theiss. & H. Syd. 1915)
Moriolaceae Zahlbr. 1903
(**Pachyascaceae** Poelt 1974 nom. inval.)
Phillipsiellaceae Höhn. 1909
Phlyctidaceae Poelt & Vezda ex J.C. David & D. Hawksw. 1991

FAMILIAE INCERTAE SEDIS (cont.)

Phyllobatheliaceae Bitter & F. Schill. 1927
Protothelenellaceae Vezda, H. Mayrhofer & Poelt 1985
Pseudeurotiaceae Malloch & Cain 1970
Saccardiaceae Höhn. 1909
Schaereriaceae M. Choisy ex Hafellner 1984
Seuratiaceae Vuill. ex M.E. Barr 1987
Strigulaceae Zahlbr. 1898
Testudinaceae Arx 1971
Thelenellaceae H. Mayrhofer 1986
Thrombiaceae Poelt ex J.C. David & D. Hawksw. 1991
Thyridiaceae O.E. Erikss. & J.Z. Yue 1987

LIST OF PARTICIPANTS AND OBSERVERS

Aguirre-Hudson, M.B., Department of Botany, The Natural History Museum, Cromwell Road, London SW7 5BD, UK

Ahti, T., Department of Botany, University of Helsinki, Unioninkatu 44, SF-00170 Helsinki, Finland

Ainsworth, A.M., Xenova Ltd, 545 Ipswich Road, Slough, Berkshire SL1 4EQ, UK

Aptroot, A., Centraalbureau voor Schimmelcultures, P.O. Box 273, 3740 AG Baarn, The Netherlands

Arnold, G.R.W., Pilzkulturensammlung, Friedrich-Schiller-Universitat Jena, Freiherr vom Stein Allee 2, D-0-5300 Weimar, Germany

Asta-Giacometti, J., Laboratoire de Biologie Alpine, Université Joseph Fourier (Grenoble 1), BP 53X, 38041 Grenoble Cedex, France

Atienza, V., Departmento de Biología Vegetal, Faculdad de Ciencias Biologicas, Universidad de Valencia, Valencia, Spain

Baral, H.O., Blaihofstrasse 42, D-7400 Tübingen 9, Germany

Bardin, M., Station de Pathologie Végétale INRA, Domaine Saint Maurice, B.P. 94, 84143 Montfavet Cedex, France

Barr-Bigelow, M.E., 9475 Inverness Avenue, Sidney, British Columbia V8L 5G8, Canada

Barrasa Gonzalez, J.M., Departmento de Biología Vegetal (Botánica), Universidad de Alcalá de Henares, 28271 Alcalá de Henares, Madrid, Spain

Bellemère, A., 53 jardins Boieldicu, 92800 Puteaux, France

Berbee, M.L., Department of Botany, University of British Colombia, 3529-6270 University Boulevard, Vancouver, British Columbia V6T 1Z4, Canada

Blackwell, M., Department of Botany, Louisiana State University, Baton Rouge, Louisiana 70803-1705, USA

Brummelen, J. van, Rijksherbarium/Hortus Botanicus, Mycology, P.O. Box 9514, 2300 RA Leiden, The Netherlands

Brygoo, Y., Laboratoire de Cryptogamie, Université Paris-Sud, Bâtiment 400, 91405 Orsay Cedex, France

Canales, R., Rhone Roulenc, Lyon, France

Cannon, P.F., International Mycological Institute, Bakeham Lane, Egham, Surrey TW20 9TY, UK

Cassini, R., INRA Phytopharmacie Versailles, Route de St. Cyr, 78026 FR, France

Checa Blanco, J., Departamento de Biologia Vegetal (Botánica), Universidad de Alcalá de Henares, 28271 Alcalá de Henares, Madrid, Spain

Crous, P.W., Department of Plant Pathology, University of Stellenbosch, Private Bag X 5018, Stellenbosch 7600, South Africa

Culberson, C.F., Department of Botany, Duke University, Durham, North Carolina 27706, USA

Culberson, W.L., Department of Botany, Duke University, Durham, North Carolina 27706, USA

Currah, R.S., Dept. of Botany, University of Alberta, Edmonton, Alberta T6G 2E9, Canada

Deny, J., 37 Parc d'Ardeney, 91120 Palaiseau, France

Dioh, W., Etudiant, DEA

Dissing, H., Institut for Sporeplanter, Kobenhavns Universitat, Farimasgade 2D, Copenhagen, Denmark

Egea Fernandez, J.M., Departamento de Biologia Vegetal (Botanica) Facultad de Biologia, Universidad de Murcia, E-30071 Murcia, Spain

Eriksson, B., Institute of Ecological Botany, University of Umeå, S-901 87 Umeå, Sweden

Eriksson, O.E., Institute of Ecological Botany, University of Umeå, S-901 87 Umeå, Sweden

Feige, G.B., GHS Botanik/Pflanzenphysiologic, Universität Essen, Universitätstrasse 5, D-0-4300 Essen 1, Germany

Fox, H.F., Coursetown House, Athy, Co. Kildare, Ireland

Galán Marquez, A., Departamento de Biología Vegetal (Botánica), Universidad de Alcalá de Henares, 28271 Alcalá de Henares, Madrid, Spain

Galan Marquez, R., Departamento de Biología Vegetal (Botánica), Universidad de Alcalá de Henares, 28271 Alcalá de Henares, Madrid, Spain
Galloway, D.J., Department of Botany, The Natural History Museum, Cromwell Road, London SW7 5BD, UK
Gams, W., Centraalbureau voor Schimmelcultures, PO Box 273, 3740 AG Baarn, The Netherlands
Grube, M., Institut für Botanik, Karl-Franzens-Universität, Holteigasse 6, A-8010 Graz, Austria
Gueho, E., Unité de Mycologie, Institut Pasteur, 25 rue du Docteur Roux, 75724 Paris Cedex, France
Hafellner, J., Institut für Botanik, Karl-Franzens-Universität Graz, Holteigasse 6, A-8010 Graz, Austria
Hawksworth, D.L., International Mycological Institute, Bakeham Lane, Egham, Surrey TW20 9TY, UK
Hein, B., Botanical Garten und Botanisches Museum,Köningin-Luise-Strasse 608, D-1000 Berlin 33, Germany
Hennebert, G.L., Laboratoire de Mycologie Systématique et Appliquée, Université de Louvain-La-Neuve, 3, place de la Croix du Sud, Louvain-La-Neuve, B-1348, Belgium
Henssen, A., Herbarium Botanik, Fachbereich Biologie, Der Phillips-Universität Marburg, Karl-von-Frisch-Strasse, D-3550 Marburg/Lahn, Germany
Hertel, H., Botanische Staatssammlung, Menzinger Strasse 67 D-8000 Munchen 19, Germany
Hilber, O.R.A., Breitendyk 88, D-4150 Kerfeldt, Germany
Holst-Jensen, A., Division of Botany, Department of Biology, University of Oslo, PO Box 1066, Blindern, N-0316 Oslo, Norway
Horie, Y, Natural History Museum and Institute, Chiba 955-2, Aoba-cho, Chuou-ku, Chiba-shi, Chiba 260, Japan
Huhndorf, S.M., The New York Botanical Gardens, Bronx, New York 10458-5126, USA
Huhtinen, S., Department of Biology, University of Turku, SF-20500 Turku, Finland
Hyde, K.D., Department of Botany, University of Hong Kong, Pokfulum Road, Hong Kong
Jahns, M., Botanisches Institüt, Universität Düsseldorf, Universitätstrasse 1, D-4000 Dusseldorf, Germany
Janex-Favre, M.C., Laboratoire de Cryptogamie, Université Pierre et Marie Curie, 7, quai St-Bernard, 75252 Paris Cedex 05, France
Johnston, P.R., Manaaki Whenua Landcare Research, Mount Albert Research Station, Private Bag 92170, Auckland, New Zealand
Jones, E.B.G., School of Biological Sciences, University of Portsmouth, King Henry 1 Street, Portsmouth, Hampshire PO1 2DY, UK
Jϕrgensen, P., Botanical Institute, University of Bergen, PO Box 12, Bergen, N-5014, Norway
Kärnefelt, I., Department of Plant Taxonomy, Lund University, Ö. Vallgatan 18, S-22361 Lund, Sweden
Kendrick, W.B., Mycology Laboratory, Department of Biology, University of Waterloo, Waterloo, Ontario N2L 3GI, Canada
Kimbrough, J.W., Plant Pathology Department, University of Florida, 1453 Fifield Hall, Gainesville, Florida 32611-0322, USA
Kinsey, G.C., International Mycological Institute, Bakeham Lane, Egham, Surrey TW20 9TY, UK
Kloeckner, P., Institut für Botanik, Universitätstrasse 1, D-4000 Düsseldorf, Germany
Kohlmeyer, J., Institute of Marine Sciences, University of North Carolina, 3431 Arendell Street, Morehead City, North Carolina 28557, USA
Korf, R.P., Plant Pathology Herbarium, Cornell University, Ithaca, New York 14853-6409, USA
Kurtzman, C.P., National Centre for Agricultural Utilization Research, USDA, 1815 N University Street, Peoria, Illinois 61604, USA
Læssϕe, T., The Herbarium, Royal Botanic Gardens, Kew, Surrey TW9 3AE, UK
Landvik, S., Institute of Ecological Botany, University of Umeå, S-901 87 Umeå, Sweden
LeClerc, M.C., Unité de Mycologie, Institut Pasteur, 25 Rue Du Docteur Roux, 75724 Paris Cedex, France
Letrouit-Galinou, M.-A., Laboratoire de Cryptogamie, Université Pierre et Marie Curie, 7 quai St.-Bernard, 75252 Paris Cedex 05, France
Llimona, X., Departamento de Biología Vegetal, Facultad de Biología, Universidad de Barcelona, Diagonal 645, 8028 Barcelona, Spain
Lohmeyer, T.R., In der Point 1, D-8229 Fridolfing, Germany
Lumbsch, H.T., Universität Essen, Fachbereich 91 Botanik, Postfach 103 764, NRW, D-4300 Essen, Germany
Lundqvist, N., Department of Cryptogamic Botany, Swedish Museum of Natural History, Box 50007, Stockholm S-104 05, Sweden
Malloch, D.N., Department of Botany, University of Toronto, 25 Wilcocks Street, Toronto, Ontario M5S 3B2, Canada
Mantle, P.G., Department of Biochemistry, Imperial College, South Kensington, London SW7 2AY, UK

Marson, G., Musée National d'Histoire Naturelle, 45 B rue de Bettembourg, L-5810 Hesperange, Luxembourg
Masselink-Beltman, H.A., Department of Plant Cytology and Morphology, Wageningen Agricultural University, Arboretumlaan 4, 6703 BD Wageningen, The Netherlands
Matzer, M., Institut für Botanik, Karl-Franzens-Universität Graz, Holteigasse 6, A-8010 Graz, Austria
Mayrhofer, H., Institut für Botanik, Karl-Franzens-Universität Graz, Holteigasse 6, A-8010 Graz, Austria
Mies, B., Linnicker Strasse 60, D-5000 Köln 41, Germany
Minter, D.W., International Mycological Institute, Bakeham Lane, Egham, Surrey TW20 9TY, UK
Molina, F.I., American Type Culture Collection, 12301 Parklawn Drive, Rockville, Maryland 20852-1776. USA
Montemartini, C.A., Instituto Botanico "Hanbury" Corso Dagali IC, Genova, Italy
Moreau, C., 13 Rue Van Gogh, 29200 Brest, France
Moss, S.T., School of Biological Sciences, University of Portsmouth, King Henry 1 Street, Portsmouth, Hampshire PO1 2DY, UK
Mouchacca, J., Laboratoire de Cryptogamie, Museum National d'Histoire Naturelle, 12 rue Buffon, 75005 Paris, France
Mouillard, C., University Catholique de Louvain, 3 Place Croix de Sud, B 1348 Louvain-La-Neuve, Belgium
Mugnier, J., Rhone Poulenc, Lyon, France
Ott, S., Botanisches Institut, Universität Düsseldorf, Universitatstrasse 1, Düsseldorf, D-4000, Germany
Parguey-Leduc, A., Laboratoire de Cryptogamie, Université Pierre et Marie Curie, 7, quai St.-Bernard, 75252 Paris Cedex 05, France
Paus, S., Universität, Botanical Institüt und Botanisches Garten, Schlossgarten 3, D-4400 Munster, Germany
Pfister, D.H., Harvard University Herbarium, Harvard University, 20 Divinity Avenue, Cambridge, Massachussets 02138, USA
Pittam, S.K, Department of Botany and Plant Pathology, Oregon State University, Cordley Hall, Corvallis, Oregon 97331-2910, USA
Poelt, J., Institut für Botanik, Karl-Franzens Universität Graz, Holteigasse 6, A-8010 Graz, Austria
Raitviir, A., Mycological Herbarium, Institute of Zoology and Botany, Academy of Sciences of Estonia, Box 93, SU-202400 Tartu, Estonia
Rambold, G., Botanische Staatssammlung München, Menzinger Strasse 67, D-8000 München 19, Germany
Rayner, A.D.M., School of Biology and Biochemistry, University of Bath, Claverton Down, Bath, Avon BA2 7AY, UK
Reynolds, D.R., Natural History Museum of Los Angeles County, 900 Exposition Boulevard, Los Angeles, California 90007, USA
Rogers, J.D., Department of Plant Pathology, Washington State University, Pullman, Washington 99164-6430, USA
Romer-Rassing, B., Novo Nordisk A/S, Novo Allee, 2880 Bagsvaerd, Denmark
Roquebert, M., Museum National d'Histoire Naturelle, Laboratoire de Cryptogamie, 12 rue Buffon, 75005 Paris, France
Rossman, A.Y., Systematic Botany and Mycology Laboratory, USDA-ARS, Beltsville, Maryland 20705-2350, USA
Ruoss, E., Natur-Museum, Kasernenplatz 6, CH-6003 Luzern, Switzerland
Samson, R.A., Centraalbureau voor Schimmelcultures, PO Box 243, 3740 AG Baarn, The Netherlands
Scheuer, C., Institut für Botanik, Karl-Franzens Universität Graz, Holteigasse 6, A-8010 Graz, Austria
Schieleit, P., Institut für Botanik, Universität Düsseldorf, Universitätstrasse 1, D-4000 Düsseldorf, Germany
Scholz, P., Unabhängiges Institut für Umwelffragen, Gr. Klausstrasse 11, D-0-4020 Halle/Saale, Germany
Schumacher, T., Department of Biology, Botany Division, University of Oslo, PO Box 1045, Blindern, N-0316 Oslo 3, Norway
Scott, J., Department of Botany, University of Toronto, 25 Wilcocks Street, Toronto M5S 3B2, Canada
Sensen, M., Institute für Botanik, Universität Düsseldorf, Universitätstrasse 1, D-4000 Düsseldorf, Germany
Serusiaux, E., Départment de Botanique, Université de Liège, Sart-Tilman, B-4000 Liège, Belgium
Sipman, H.J.M., Botanische Garten und Botanisches Museum, Köningin-Luise-Strasse 6-8, 1000 Berlin 33, Germany
Spatafora, J.W., Department of Botany, Duke University, Durham, North Carolina 27708, USA
Spooner, B.M., The Herbarium, Royal Botanic Gardens, Kew, Surrey TW9 3AB, UK
Stocker-Worgotter, E., Institute of Plant Physiology, University of Salzburg, Hellbrunnerstrasse 34, A-5020 Salzburg, Austria

Sutton, B.C., International Mycological Institute, Bakeham Lane, Egham, Surrey TW20 9TY, UK
Svedskog, A., Institute of Ecological Botany, University of Umeå, S-901 87 Umeå, Sweden
Taylor, J.W., Department of Plant Biology, 111 Genetics and Plant Biology, University of California, Berkeley, California 94720, USA
Tehler, A., Botanical Institute, University of Stockholm, S-106 91 Stockholm, Sweden
Thell, A., Lund University, Ö. Vallgatan 18, S-223 61 Lund, Sweden
Tibell, L., Department of Systematic Botany, Uppsala University, PO Box 541, Uppsala, S-751 21, Sweden
Timdal, E., The Herbarium, Botanical Gardens and Museum, University of Oslo, Trondheims Veien 23 B, N-0562 Oslo, Norway
Titze, A., Fachbereich Biologie (Botanik), Der Philipps-Universität Marburg, Karl-von-Frisch-Strasse, D-3550 Marburg/Lahn, Germany
Triebel, D., Botanische Staatssammlung München, Menzinger Strasse 67, D-8000 München 19, Germany
Untereiner, W.A., Department of Botany, University of Toronto, 25 Wilcocks Street, Toronto M5S 3B2, Ontario, Canada
Upadhyay, H.P., Departamento de Micologia, Universidade Federal de Pernambuco 50739, Recife, Brasil
Udagawa, S.-I., Nodai Research Institute, Tokyo University of Agriculture, 1-1-1 Sakuragoaka, Setagaya-ku, Tokyo 156, Japan
van Haluwyn, C., Laboratoire de Botanique et de Cryptogamie, Faculté de Pharmacie, Univérsité de Lille, 3 rue du Pr Laguesse, B.P. 83, 59006 Lille Cedex, France
Vasilyeva, L., Institute of Biology and Pedology, Far East Department of the Russian Academy of Sciences, 690022 Vladivostok, Russia
Verkley, G.J.M., Rijksherbarium/Hortus Botanicus, Rijksuniversiteit Leiden, PO Box 9514, 2300 RA Leiden, The Netherlands
Weden, M., Department of Systematic Botany, University of Uppsala, PO Box 541, S-751 21 Uppsala, Sweden
Weir, A., 32 Hartburn Lane, Stockton-on-Tees, Cleveland TS18 3QH, UK
Whalley, A.J.S., School of Natural Sciences, Liverpool John Moores University, Byrom Street, Liverpool L3 3AF, UK
Wingfield, M.J., Department of Microbiology and Biochemistry, University of the Orange Free State, PO Box 339, Bloemfontein 9300, South Africa
Yao, Y.J., The Herbarium, Royal Botanic Gardens, Kew, Surrey TW9 3AB, UK

INDEX